# Combustion for Power Generation and Transportation

Avinash Kumar Agarwal
Santanu De
Ashok Pandey
Akhilendra Pratap Singh
Editors

# Combustion for Power Generation and Transportation

## Technology, Challenges and Prospects

Springer

*Editors*
Avinash Kumar Agarwal
Department of Mechanical Engineering
Indian Institute of Technology Kanpur
Kanpur, Uttar Pradesh
India

Santanu De
Department of Mechanical Engineering
Indian Institute of Technology Kanpur
Kanpur, Uttar Pradesh
India

Ashok Pandey
Eminent Scientist
Center of Innovative and Applied Bioprocessing (CIAB)
Mohali, Punjab
India

Akhilendra Pratap Singh
Department of Mechanical Engineering
Indian Institute of Technology Kanpur
Kanpur, Uttar Pradesh
India

ISBN 978-981-10-3784-9 ISBN 978-981-10-3785-6 (eBook)
DOI 10.1007/978-981-10-3785-6

Library of Congress Control Number: 2017930133



Printed on acid-free paper

This Springer imprint is published by Springer Nature
The registered company is Springer Nature Singapore Pte Ltd.
The registered company address is: 152 Beach Road, #21-01/04 Gateway East, Singapore 189721, Singapore

# Preface

Energy demand has been rising remarkably due to increasing population and urbanization. Global economy and society are significantly dependent on the energy availability because it touches every facet of human life and activities. Transportation and power generation are major examples of the energy. Without the transportation by millions of personalized and mass transport vehicles and availability of $24 \times 7$ power, human civilization would not have attained contemporary living standards.

An international workshop, 3rd ISEES Workshop on "Sustainable Energy, Environment & Safety with Railway Centric Theme", was held at Research Designs and Standards Organisation (RDSO), Lucknow, India during December 21–23, 2015 under the aegis of International Society for Energy, Environment and Sustainability (ISEES). This workshop provided a platform for discussions between eminent scientists and engineers from various countries including India, USA, South Korea, Thailand, and Austria. In this workshop, eminent speakers presented their views related to different aspects of combustion, and alternative energy resource for sustainable development and cleaner environment. This workshop laid the roadmap for technology development, opportunities and challenges in this technology domain. At this stage of technology development, transportation and power generation systems are dependent on IC engines and gas turbines to a great extent. Fundamentals of combustion and pollutant formation are coupled to each other and need to be explored in order to design energy-efficient and environment-friendly combustion systems for power generation sector, and transport sector. Besides changing the existing design of combustion systems, application of different alternative fuels such as biofuels, other renewable fuels, alcohols and synthetic fuels needs to be explored for sustainable global development. Fundamental and applied studies can be carried out using the state-of-the-art optical diagnostic systems and advanced numerical models for turbulent combustion. Such fundamental investigations would result in optimized design of these combustion systems, which would be more efficient and environment friendly.

The editors would like to express their sincere gratitude to the authors for submitting their high quality work in a timely manner and revising it their contributions appropriately at a short notice. We would like to express our special thanks to Prof. Vaibhav Arghode, Prof. Ashoke De, Prof. Sathesh Mariappan, Prof. Atul Dhar, Prof. Dhananjay Kumar Srivastava, Prof. Rakesh Kumar Maurya, Prof. Dr. Malay Karmakar, Prof. Swarnendu Sen, Prof. Swetaprovo Chaudhuri, Prof. Sreedhara Seshadri, Prof. Achintya Mukhopadhyay, Prof. P.K. Bose, Prof. Amtava Datta, Prof. Thierry Poinsot, Prof. R.P. Gakkhar, Prof. Sudarshan Kumar, Prof. William Roberts and Dr. Akhilendra Pratap Singh, who reviewed various chapters of the monograph and provided their valuable suggestions to improve the submitted manuscripts, which helped us bring the monograph in the present form. We acknowledge the support received from various funding agencies and organizations for successful conduct of the ISEES workshop, where these monographs germinated. These include Department of Science and Technology, Government of India (Special thanks to Dr. Sanjay Bajpai); RITES Ltd., India (Special thanks to Sh. Pradeep Gupta); Office of Naval Research Global, Singapore (Special thanks to Dr. Ramesh Kolar); TSI, India (Special thanks to Dr. Deepak Sharma); Caterpiller India; AVL India; Dynomerk Controls, India (Special thanks to Sh. Kishore Raut); CEI Softwares, India; ESi Group, Pune; BHEL India; and Bosch India.

This monograph is intended for combustion practitioners and we hope that the book would be of great interest to the professionals, postgraduate students involved in advanced combustion techniques, experimentation and numerical simulation of combustion and environmental aspects. The main objective of this monograph is to promote a better and more accurate understanding of combustion in IC engines and gas turbines, besides recent advances and challenges in clean combustion systems and technology.

Kanpur, India — Avinash Kumar Agarwal
Kanpur, India — Santanu De
Mohali, India — Ashok Pandey
Kanpur, India — Akhilendra Pratap Singh

# Contents

# About the Editors

**Prof. Avinash Kumar Agarwal** joined Department of Mechanical Engineering, IIT Kanpur in 2001. Professor Agarwal worked at ERC, University of Wisconsin, Madison, USA as a postdoctoral Fellow (1999–2001). His areas of interest include IC engines, combustion, alternative fuels, hydrogen, conventional fuels, lubricating oil tribology, optical diagnostics, laser ignition, HCCI, emission and particulate control, and large-bore engines. He has published more than 200 peer-reviewed international journal and conference papers. He is associate editor of ASME Journal of Energy Resources Technology, and International Journal of Vehicle Systems Modelling and Testing. He has edited "Handbook of Combustion" (5 Volumes; 3168 pages), published by Wiley VCH, Germany. He is a Fellow of SAE (2012), Fellow of ASME (2013) and a Fellow of INAE (2015). He is the recipient of Prestigious Shanti Swarup Bhatnagar Prize-2016 in Engineering Sciences.

**Dr. Santanu De** is Assistant Professor in the Department of Mechanical Engineering, IIT Kanpur since December 2014. He received a Bachelor of Engineering degree from the North Bengal University in 2002, and an M.Tech. from IIT Kanpur in 2004, both in Mechanical Engineering. He received his Ph.D. in Aerospace Engineering from IISc, Bangalore in 2012. Prior to joining IIT Kanpur, he served for two years at MTU as a postdoctoral research associate, and one year at the University of Stuttgart. His primary areas of research include numerical modeling of turbulent combustion, spray atomization, coal gasification and combustion. He is Associate Editor of the 'Journal of Energy, Environment and Sustainability'.

**Prof. Ashok Pandey** is Eminent Scientist at the Center of Innovative and Applied Bioprocessing (CIAB), Mohali. His major research interests include the areas of microbial, enzyme and bioprocess technology. He has to his credit over 1150 publications/communications, including 16 patents, 50+ books, 140 book chapters, 423 original and review papers, an *h* index of 78 and ~25,000 Goggle Scholar citations. Professor Pandey is the recipient of several fellowships such as the Fellowships of Royal Society of Biology (UK), Academician of European Academy of Sciences and Arts, Germany; ISEES; National Academy of Science (India); BRSI; and awards such as Thomson Scientific India Citation Laureate Award, USA; UNESCO Professor; Raman Research Fellowship Award, CSIR; and GBF, Germany. He is Editor-in-chief of *Bioresource Technology*, Honorary Executive Advisor of *Journal of Water Sustainability* and *Journal of Energy and Environmental Sustainability*, Subject editor of *Proceedings of National Academy of Sciences (India)* and serves on the editorial board of several other journals.

**Dr. Akhilendra Pratap Singh** is a CSIR Pool Scientist at Indian Institute of Technology Kanpur. He received his Masters and Ph.D. degrees in Mechanical Engineering from Indian Institute of Technology Kanpur, India in 2010 and 2016, respectively. His areas of interest include advanced low temperature combustion; optical diagnostics with special reference to engine endoscopy and PIV; combustion diagnostics; engine emissions measurements; particulate characterization and their control; and alternative fuels. Dr. Singh has published 20 peer-reviewed international journal papers and six peer-reviewed international and national conference papers. He is a member of professional societies such as SAE, ASME and ISEES. He is an editorial board member of the 'Journal of Energy, Environment and Sustainability'.

# Part I
# General

# Introduction to Combustion for Power Generation and Transportation

**Avinash Kumar Agarwal, Santanu De, Ashok Pandey and Akhilendra Pratap Singh**

**Abstract** Combustion is dominant source of energy ever since the evolution of mankind. Combustion has widespread applications in many industrial and day-to-day applications including in transportation and power generation sectors, such as in gas turbines and internal combustion (IC) engines. Some by-product of fossil-fuel combustion namely NOx, soot and particulate are extremely harmful for the environment and the human health. Clean combustion technologies with high thermal efficiency are the need of the hour for mitigating the environmental implication of such undesirable emissions. This monograph covers a journey starting with fundamentals of combustion, experimental and numerical investigations of combustion in gas turbines and IC engines, optical diagnostics, and emission measurement and control. Several modern combustion technologies with a vast potential of reduction in emission are discussed.

**Keywords** Combustion · Gas turbines · Internal combustion engines · Experimental techniques · Numerical simulations · Environmental aspects · Optical diagnostics

Ever since the industrial revolution, combustion has been the dominant source of energy for development of modern civilisation. With the exception of nuclear and solar power and a few other exotic sources, most power generation units and

A.K. Agarwal (✉) · S. De · A.P. Singh
Department of Mechanical Engineering, Indian Institute of Technology Kanpur, Kanpur 208016, UP, India
e-mail: akag@iitk.ac.in

S. De
e-mail: sde@iitk.ac.in

A.P. Singh
e-mail: akhips@iitk.ac.in

A. Pandey
Center of Innovative and Applied Bioprocessing (CIAB), Mohali 160071, Punjab, India
e-mail: ashokpandey1956@gmail.com



A.K. Agarwal et al. (eds.), *Combustion for Power Generation and Transportation*, DOI 10.1007/978-981-10-3785-6_1

transportation prime movers are fuelled by chemical reactions of fossil fuels and oxygen via combustion. Fossil fuels mainly comprise of carbon and hydrogen. Most transportation systems except vehicles powered by fuel cells and those having electrical powertrains are predominantly powered by combustion of liquid and gaseous fossil fuels. Apart from transportation and power generation systems, combustion is essential in many process in industries, businesses, homes, and space applications. In these energy conversion systems, heat is supplied to the working fluid by burning fossil fuels in presence of an oxidizer. Subsequently, the thermal energy available in the working fluid is converted into the mechanical energy via some mechanisms.

Design and development of modern combustion systems require extensive investigation of underlying physio-chemical processes involving a wide range of scales associated with various physical processes and their complex interactions, such as fuel injection, spray atomization, droplet evaporation, fuel-air mixing, chemical combustion reactions, formation of pollutants such as NOx, soot and unburnt hydrocarbons. Gaining fundamental insights on these complex processes via experimental investigations is not only difficult but sometimes impossible due to lack of optical access, complex flows and extremely fast combustion reactions. Numerical simulations of these multi-scale and multi-physics phenomena are more easily and conveniently obtained via computer simulations. These simulations now conducted at very large scales, are taking an ever greater role in scientific discovery and understating of one of the most complex scientific mysteries since ancient times i.e. combustion. This is possible using high performance computing (HPC) environment and using novel numerical algorithms designed for massively parallel computations.

During energy conversion in industrial burners/gas turbine combustors, enhancement of thermal efficiency along with ultra-low emissions is the most challenging goal for combustion researchers. Moderate or intense low-oxygen dilution (MILD) combustion is one of the promising concepts towards achieving these goals, which is also known as high temperature air combustion (HiTAC)/flameless combustion (FLOX)/colorless distributed combustion (CDC). In these combustion systems, the injected air entrains burned gases to form hot, diluted oxidizer. Subsequently, the injected fuel mixes with the hot and diluted oxidizer before ignition takes place.

These burners are used in gas turbines, which operate on Brayton cycle. Heat addition takes place at constant pressure in gas turbines. If isobaric process of heat addition in gas turbine combustors is replaced by an isochoric process, the rise in stagnation pressure across a gas turbine combustor has a potential to increase overall efficiency of the gas turbine system significantly. Pressure gain combustion using an acoustically resonant pulse combustor may be used to achieve isochoric heat release, causing a rise in stagnation pressure.

Stringent emission regulations have made lean premixed pre-vaporization (LPP) technology desirable in power generation sector and other industrial applications due to its low NOx and CO emission potential. Fuel is usually sprayed directly into the premixer so that the liquid fuel droplets evaporate and mix with air under lean conditions. The premixers and combustors are designed in such a way that they allow rapid mixing and combustion before spontaneous ignition can take place. However the fuel-lean operation in LPP combustion systems can lead to thermoacoustic instabilities, leading to blowout of flames, commonly referred as lean blowout (LBO). Unless we devise strategy to mitigate LBO, these combustors are unsafe for aircraft and land-based operation of gas turbines and this issue offers significant research challenge.

Over little more than a century of sustained development has gone into internal combustion (IC) engines, which has transformed modern IC engines into very robust and sophisticated machines for both stationary and automotive applications. Advances in material science, combustion and fuel technology along with progress in fundamental insights via sophisticated experiments and numerical investigations have made this transformation possible. Air-fuel mixture formation in an IC engine plays a dominant role in controlling the exhaust emissions and engine efficiency. Several such techniques are presented in a chapter in this monograph.

With quantum improvements in laser based optical systems, development of IC engines has been significantly boosted by the application of the state-of-the-art optical diagnostic techniques namely particle image velocimetry (PIV), phase Doppler interferometry (PDI) and engine endoscopy. Recent advances in instrumentation has made it possible to use PIV and PDI in an optical engine, even under firing conditions. In-cylinder flows have a large impact on combustion characteristics and emission formation in an engine environment. With the help of the real-time measurements using laser based systems, it is now possible to find answers to some of the most intriguing in-cylinder processes, which were only speculated until recently. Fuel spray characteristics, such as droplet size distribution and 3D-velocity distributions of spray droplets could be determined using PDI. In-cylinder air-flow characteristics of motored engine may be obtained using time-resolved tomographic particle imaging velocimetry (TPIV) as well. The effect of in-cylinder parameters on combustion can be directly visualized using engine endoscopy, which provides valuable information for radiative flame temperature distribution and soot distribution. These optical techniques have revolutionised the engine development and research.

Besides experimental investigations, various techniques to resolve different level of complexity, such as thermodynamic, phenomenological and multi-dimensional models, are nowadays routinely used in engine development. In these models, various engine processes from intake to exhaust are considered and solved accurately. They provide valuable information for variations in pressure-volume and pressure-crank angle, and other engine performance parameters using different combustion models. The energy analysis based on first-law of thermodynamics provides an overall engine performance, however it cannot provide information on specific areas of losses in an IC engine. Such information may be obtained using exergy analysis based on second-law of thermodynamics. The exergy analysis allows quantification of individual process inefficiencies therefore such an analysis presents an opportunity for improvement in the engine design.

Low temperature combustion (LTC) concept has emerged as a promising approach to achieve higher fuel conversion efficiencies along with lower NOx and particulate emissions compared to the conventional IC engine combustion. Among several LTC strategies, reactivity controlled compression ignition (RCCI) is found to be suitable for highly efficient, low emissions operation over a wide range of loads and speeds. The impact of the fuel pre-mixing ratio and injection timing of high reactivity fuel governs the combustion and emission characteristics in an RCCI engine.

Gasoline direct injection (GDI) engines offer a promising technology solution as next-generation automotive engines with higher fuel economy, improved performance and reduced emissions. Unlike conventional port fuel injection (PFI) engines, gasoline is directly injected into the combustion chamber at high pressure during the compression stroke in GDI engines. Subsequently, the fuel-air mixture formation takes place inside the combustion chamber, similar to that of a diesel engine. Several difficulties are encountered in direct injection of gasoline such as controlling the stratified charge combustion over a wide range of engine operation, complexity in fuel injection strategy for seamless load changes, injector deposits, increased particulate emissions etc., which offer significant research and development challenges with huge opportunity for large scale implementation of GDI engines.

In pursuit of finding a non-depleting and clean burning fuel for IC engines, hydrogen and syngas have always been an attractive option. The low ignition energy requirement for $H_2$–air mixture causes frequent unscheduled combustion events, and high combustion temperatures in case of nearly stoichiometric mixtures, which leads to high NOx emissions. Recent progress in NOx control strategies, engine modeling and cycle analysis have contributed to development of advanced $H_2$ engines with improved power densities. Wider flammability limits of hydrogen in air allows much wider range of IC engine operations including very lean operation of the engine. However higher autoignition temperature of hydrogen in air somewhat limits its application in CI engines. Hydrogen engines have a bright future if these technological challenges can be overcome. Synthetic fuels derived from biomass, have a potential to augment this. Syngas obtained from biomass gasification may be used as a potential fuel for the SI engines. Besides biomass and agricultural wastes, syngas may be produced from gasification of coal and municipal solid waste. Syngas needs to be thoroughly cleaned before it can be introduced in the engines for vehicular application. Moreover, the volumetric energy content of the syngas is lower than hydrocarbon fuels, hence both the energy output and the volumetric efficiency may be relatively lower.

Carbon capture and sequestration (CCS) offers a potential solution to reduce direct emissions of $CO_2$ from power plants and other industrial sources into the atmosphere. The $CO_2$ is captured and is permanently sequestered in nearby geological formations such as the saline aquifers, oil and gas fields, and unmineable coal seams. These geological formations have a highly permeable reservoir capped with a relatively impermeable layer of rock, which prevents $CO_2$ leakage from these reservoirs. The captured $CO_2$ may be utilized as the working fluid for enhanced oil recovery, enhanced gas recovery, or enhanced geothermal systems to benefit further energy production.

In the end, one chapter focuses on urban air quality. Urban air quality has become a cause of concern, primarily due to rapid industrialization, migration of people into urban centers and vehicular emissions leading to deteriorating air quality. $PM_{2.5}$ are known to have adverse impact on human health such as premature mortality, asthma exacerbation, decreased lung function, and increased risk of myocardial infarction among others. Recent studies reported that smaller airborne particles of submicron size ($PM_1$) are more harmful to human body. Low cost instrument needs to be developed and tested for sub-micron particles. Accurate measurement of contaminants in the samples from the field trial will eventually help us to effectively control the urban air quality. Such an attempt made at IIT Kanpur is covered in the last chapter of this monograph.

This research monograph presents both fundamental science and applied innovations on several key and emerging technologies involving fossil and alternate fuel utilization in power and transport sectors. Specific topics covered in the manuscript include:

- A review on autoignition in laminar and turbulent nonpremixed flames
- Langevin simulation of turbulent combustion
- Numerical modeling of mild combustion at high pressure to predict the optimal operating conditions
- Global and local viewpoints to analyze turbulence-premixed flame interaction investigation of forward and reverse flow CDC combustors
- Lean blowout (LBO) prediction through symbolic time series analysis
- Pulse combustor driven pressure gain combustion for high efficiency gas turbine engines
- Theoretical formulation for the investigation of acoustic and entropy-driven combustion instabilities in gas turbine engines
- Chronology of dynamic stall characteristics through simultaneous PIV and surface pressure measurements
- Lasers and optical diagnostics for next generation IC engine development: ushering new era of engine development
- Exergy destruction study on small DI diesel engine
- Impact of fuel premixing ratio and injection timing on reactivity controlled compression ignition engine
- Exergy analysis of hydrogen-fueled spark ignition engine based on numerical investigations
- Thermodynamic modelling of combustion process in a spark ignition engine and its numerical prediction
- Gasoline direct injection—challenges

- Syngas as SI engine fuel: combustion perspective
- Carbon sequestration and optimization of enhanced oil and gas recovery
- Development of an indigenous sensor for sub-micron aerosol monitoring in India

The topics are organized in five different sections: (i) General, (ii) Combustion: Experiments and Simulations, (iii) Gas Turbines, (iv) Internal Combustion Engines, and (v) Environmental Aspects.

# Part II
# Combustion: Experiments and Simulations

# A Review on Autoignition in Laminar and Turbulent Nonpremixed Flames

**Sanjeev Kumar Ghai and Santanu De**

**Abstract** This chapter presents a condensed review on the autoignition in laminar and turbulent nonpremixed flames. Both experimental and numerical aspects are discussed. Fundamental studies on autoignition in turbulent flows revealed that random ignition spots are initially observed in the lean mixtures where the scalar dissipation rate is low. The mixture fraction corresponding to this lean mixture is usually referred as the "most reactive mixture fraction". The increase in initial turbulent intensity and mixing delays autoignition. For most of the fuels, autoignition is observed as a two-stage process with a negative temperature coefficient. Besides, the physical and chemical properties of the fuels, the complex chemical kinetics also affect auto-ignition as well as combustion characteristics. Autoignition is also a dominant flame stabilization mechanism at the base of the lifted flames. Fundamental experimental investigations on autoignition in turbulent flows are very much limited, and most of the previous work is specifically focused on the Berkley vitiated coflow burner and the Cambridge burner. The combustion models developed so far can capture the trends observed in the experiments and the direct numerical simulation (DNS) studies. However, none of the combustion models developed so far can capture the trends quantitatively.

**Keywords** Autoignition · Turbulent combustion · Vitiated coflow · Lifted flame · PDF approach

**List of abbreviations**

| | |
|---|---|
| DNS | Direct numerical simulation |
| HCCI | Homogeneous charge compression ignition |
| LPP | Lean premixed pre-vaporized |

S.K. Ghai (✉) · S. De
Department of Mechanical Engineering, Indian Institute of Technology Kanpur, Kanpur 208016, UP, India
e-mail: snjv@iitk.ac.in

S. De
e-mail: sde@iitk.ac.in



A.K. Agarwal et al. (eds.), *Combustion for Power Generation and Transportation*, DOI 10.1007/978-981-10-3785-6_2

| | |
|---|---|
| DME | Dimethyl ether |
| CI | Compression ignition |
| PM | Particulate matter |
| RCM | Rapid compression machine |
| NTC | Negative temperature coefficient |
| MILD | Moderate and intense low oxygen dilution |
| RANS | Reynolds averaged Navier Stokes simulation |
| LES | Large eddy simulation |
| CMC | Conditional moment closure |
| PDF | Probability density function |
| IEM | Interaction by exchange with the mean |
| EMST | Euclidean minimum spanning tree |
| HRR | Heat release rate |

**Nomenclature**

| | |
|---|---|
| $Y_{\mathrm{I}}$ | Mass fraction of species I |
| $N$ | Scalar dissipation |
| $Z$ | Mixture fraction |
| $\eta$ | Sample space of mixture fraction |
| $\xi$ | Conditional variable |
| $Q(\eta, \mathbf{x}, t)$ | Conditional expectation |
| $\dot{\omega}$ | Reaction rate |
| $\widetilde{P}(\xi, \chi, \mathbf{x}, t)$ | Joint scalar PDF |

## 1 Introduction

The autoignition or self-ignition of a flammable mixture is a fundamental problem in combustion, where we deal with the transition from a slowly reacting state to a fully burning state corresponding to combustion at high temperature. A simple experiment on autoignition may be carried out where the initial temperature of a uniform stagnant fuel-air mixture in a close vessel is slowly raised until the mixture ignites at a temperature, which may be referred as the autoignition temperature. The time taken by the mixture to ignite is referred as the autoignition time or ignition delay time. Due to the increasing rate of chemical reactions with increasing temperature, even the very weak reaction starts to generate sufficient heat, which causes a further rise in temperature and thereby resulting in a faster heat release rate. Thermal runaway eventually takes place at an autoignition time that depends on the initial temperature, the concentrations of reactants, the nature of the fuel, and the pressure. The autoignition temperature of a fuel-air mixture provides useful information on how various compounds autoignite, but it does not provide any fundamental insights on the autoignition process and how it affects the flow behaviour in general.

Autoignition of turbulent fuel jets issued into a stagnant or coflowing oxidizer stream, and the subsequent flame stabilization is a well-known problem frequently encountered in Diesel engines, ramjets, and gas turbine combustors. The effect of turbulence on autoignition is not only a fundamental topic of research, but it is also crucial for the development of the next generation combustion systems for transportation and power generation such as homogeneous charge compression ignition (HCCI) engines and lean premixed pre-vaporized (LPP) gas turbines. Development of these industrial combustion systems is hindered by our capabilities to predict the interaction between turbulence and the slow chemistry leading to autoignition. It is needless to say that a better understanding and capability to predict the autoignition will facilitate the development of energy-efficient and environment-friendly combustors. In the practical combustors, the presence of considerable fluctuations of velocity, temperature, and mixture composition greatly impacts autoignition and the subsequent flame development process. With the ever-increasing requirement of emission regulation, alternative fuels or fuel additives are routinely being used and tested in Diesel engines and land-based gas turbines. The physical and chemical properties of the new fuels and their complex chemical kinetics are likely to affect their atomization, evaporation, mixing, auto-ignition, as well as combustion characteristics. A thorough understanding of the physical processes and their complex interactions are not available. Besides experimental capabilities, numerical models for these complex physical phenomena need to be developed in order to speed up the design, development, and optimization of the combustion chambers of these practical systems.

With the advancement of combustion diagnostics and numerical tools and the availability of more precise information on the chemical kinetics, auto-ignition of gaseous- and liquid fuels need to be revisited to address some of the fundamental aspects such as the development of ignition kernel, flame propagation and flame stability in the inhomogeneous mixtures. In the last decade, two different burners were developed to investigate auto-ignition in the inhomogeneous mixture, namely, vitiated co-flow burner [1] and preheated coflowing air burner [2, 3]. Besides revealing the vital information on autoignition under inhomogeneous mixture conditions, the experimental data from these burners have been routinely used for development of turbulent combustion models. A comprehensive database is now available [1–5] on autoigniting turbulent jet flames for different fuels. The prevailing auto-ignition mechanism is found to be consistent with those revealed by the direct numerical simulation (DNS) studies. In these burners, reaction kernels develop at some distance downstream of the fuel injector. These ignition kernels eventually initiate the flame development.

The book chapter amalgamates the knowledge on autoignition under laminar and turbulent conditions. In the subsequent sections, a review will be presented on fuel properties and its effects on autoignition, various experimental configurations employed to study autoignition under laminar and turbulent conditions, and modelling approaches for turbulent combustion. Towards the end of this chapter, major conclusions are drawn, and future recommendations are presented.

## 2 Fuel Properties

The properties of the fuel play an important role due to its applicability in different operating conditions. Due to the stringent emission regulations, combustion community is widely interested in replacing the conventional fuels with the alternative fuels. The fuel replacement in the internal combustion engines and gas turbine combustors also affects the combustion and emission characteristics. The physical and chemical properties of various fuels, e.g., dimethyl ether (DME), Diesel, butane, propane, hydrogen, and gasoline are shown in Table 1 [6–8]. In compression-ignition (CI) engines, ignition occurs due to self-ignition of the fuel. Autoignition temperature is one of the very important properties of the fuel in CI engines. The modern combustors are also being developed for alcohols, biofuels, synthetic and renewable fuels and different fuel mixtures.

**Table 1** Physical and chemical properties of different fuels [6–8]

| Property | DME | Diesel | Butane | Gasoline | Propane | Hydrogen |
|---|---|---|---|---|---|---|
| Chemical formula | $CH_3OCH_3$ | $C_8$ to $C_{25}$ | $C_4H_{10}$ | $C_7H_{16}$ | $C_3H_8$ | $H_2$ |
| Molecular weight | 46.07 | 96~ | 58.13 | 100.2 | 44.11 | 2.02 |
| Vapour pressure at 20 °C (bar) | 5.1 | <0.01 | 8.4 | – | 2.1 | – |
| Boiling temperature (°C) | −25 | ≈150–380 | −0.5 | 38–204 | −42.1 | −253 |
| Liquid density at 20 °C (kg/m$^3$) | 660 | 800–840 | 610 | 719.7 | 501 | 70.8 |
| Liquid viscosity at 25 °C (kg/ms) | 0.12–0.15 | 2–4 | 0.2 | – | 0.2 | – |
| Gas specific gravity (vs air) | 1.59 | – | 2.01 | – | 1.52 | – |
| Lower heating value (MJ/kg) | 28.43 | 42.5 | 45.74 | 44.5 | 46.36 | 119.93 |
| Cetane number | 55–60 | 40–55 | – | 80–100 | 5 | 130 |
| Stoichiometric A/F ratio (kg/kg) | 9.0 | 14.6 | 14.8 | 14.5 | 15.7 | 34.3 |
| Enthalpy of vaporization at NTP[a] (kJ/kg) | 460(−20 °C) | 250 | 390 | – | 426 | 448.69 |
| Auto-ignition temperature (K) | 508 | 523 | 638 | 501–744 | 743 | 858 |
| Lower explosion limit (vol.%) | 3.0–3.4 | 0.6 | 1.9 | 1.0 | 2.1 | 4.0 |
| Upper explosion limit (vol.%) | 17–18.6 | 6.5–7.5 | 8.4 | 7.6 | 9.4 | 75 |
| Minimum ignition energy (mJ) | 0.29 | – | 0.305 | 0.24 | 0.26 | 0.02 |

The physical and chemical properties of these new fuels determine whether they are suitable for different industrial applications. As an example, DME is a promising alternative fuel in Diesel engines and is being considered as a possible substitute for natural gas in power-generation applications. It is a clean-burning, non-toxic fuel with low soot formation potentials and PM emissions, which can be synthesized from natural gas, coal, and renewable fuels such as biomass and ethanol. DME exhibits Diesel-like performance due to its high Cetane number (55–60) [8] at the same time it offers propane-like handling properties due to its low boiling point (–25 °C) at atmospheric conditions. It has a lower autoignition temperature than Diesel, and it has a higher flammability range compared to other hydrocarbons due to the presence of the large oxygen content. However, it has a lower heating value compared to Diesel. Therefore, to produce the same output, we need to supply more DME in the combustion chamber. Also, very little is known about the fundamental combustion processes of DME-fuelled systems under highly turbulent conditions and at different operating pressures. Hence, fundamental understanding of the complex interplay between chemistry and turbulence is essential for speedy implementation of DME as an alternative fuel for internal combustion engines and gas turbines.

## 3 Experimental Approaches

In this section, fundamental findings and experimental approaches for autoignition in laminar and turbulent non-premixed flows are discussed. Experimental facilities to study autoignition are very limited till date. Most of the experiments are usually conducted under the ambient pressure conditions [1, 2]. Practical systems are very complex for experimental investigations. Replication of these complex systems in a laboratory is not only difficult and expensive, but rather a kind of impossible task. To gain fundamental insights on autoignition, simple lab scale setups have been designed and developed with optical access and good illumination. These experimental setups are designed with well-defined inlet boundary conditions to facilitate numerical model development. The various burner configurations presented in this section (e.g., lifted jet flames in the vitiated coflow, MILD combustion in the jet in hot coflow configuration, confined turbulent hot coflow burner and counterflow flame setup) demonstrate aspects of autoignition.

High spatial and temporal resolutions are also needed to study the transient phenomenon of turbulence-flame interactions. With the state-of-the-art combustion diagnostics systems, simultaneous measurements of the flow- and scalar fields are now possible to achieve complete physical insights of the problem. Various experimental techniques such as laser Doppler velocimetry (LDV), particle imaging velocimetry (PIV), phase Doppler anemometry (PDA) are used for flow field measurements and Mie scattering, laser absorption spectroscopy (LAS), planar laser-induced fluorescence (PLIF), 1D Raman/Rayleigh scattering, coherent anti-Stokes Raman spectroscopy (CARS) are used for scalar field measurements. Until early 1990's, experiments on autoignition had been limited to shock tubes, flow reactors and rapid compression machine (RCM). While each of these devices has own merits, but their application is restricted to certain specific conditions. As an example, flow reactors have limitations of maximum achievable pressure and temperature limits, while shock tubes are limited by short ignition time. On the other side, RCM is suitable for high-temperature and pressure applications with comparatively higher ignition time compared to the shock tubes.

## 3.1 Autoignition in Rapid Compression Machine

A schematic for the RCM is shown in Fig. 1. RCM operates at a compression ratio of 10:1. Primarily it is used to investigate chemical kinetics [9]. Carlier et al. [9] conducted consecutive experiments on autoignition of butane both at low and high pressures in RCM. They have observed autoignition as a two-stage process at low temperature, and their results indicate the appearance of negative temperature coefficient (NTC) by the autoignition delay times at high pressure. Also, they compared the measurements with the numerical predictions using available reaction mechanisms [11] by adjusting the rate constants (shown in Fig. 2). Improvement in the available reaction mechanisms was suggested in order to capture autoignition within the broad temperature and pressure range of extremely varying experimental conditions. Minetti et al. [12] investigated the oxidation and autoignition of the

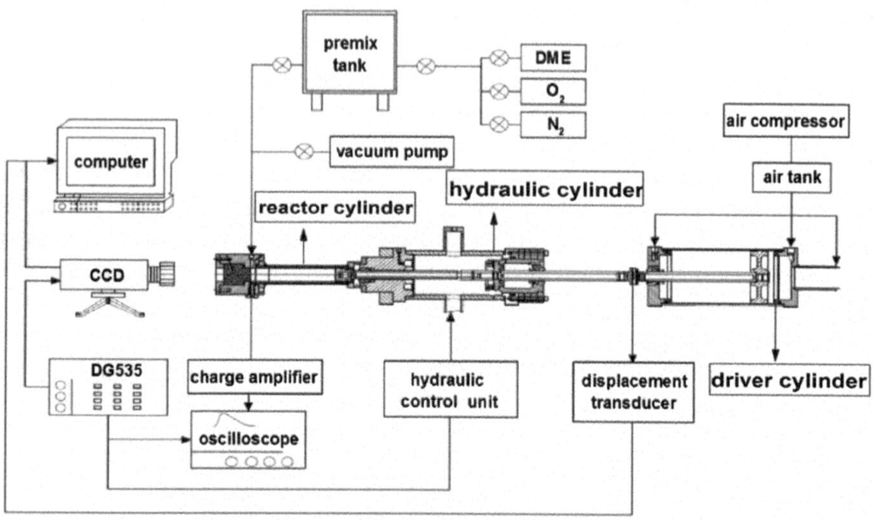


**Fig. 1** A schematic of rapid compression machine [10]

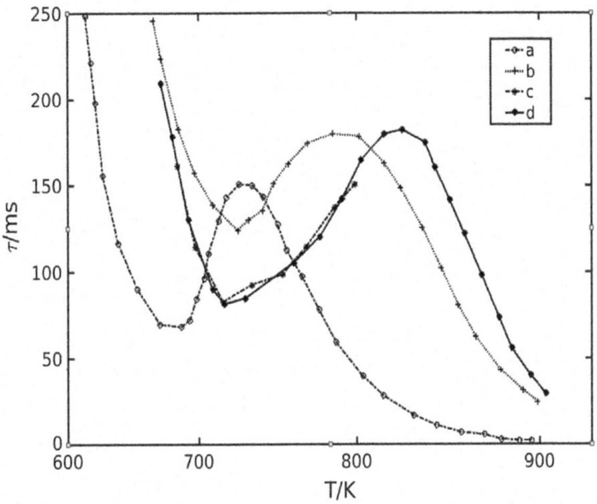


**Fig. 2** Ignition delay in RCM [9] for a butane oxygen-inert mixture: **a** experimental predictions, butane-oxygen (1 N2, 1 Ar), **b** butane-oxygen-Ar, **c** numerical results using the mechanism of [11], and **d** numerical results using a revised version of [11]

butane-air mixture in RCM at 700–900 K temperature and 9–11 bar pressure. A transition from the two-stage to a single-stage autoignition was observed under certain experimental conditions. Figure 3a shows two-stage autoignition within a temperature range of 700–760 K. In the first stage, an initial delay of $t_1$ was recorded, followed by an increase in pressure along with the significant emission of light in the second stage with a delay time $t_2$. However, for temperatures above 850 K, ignition occurred in single-stage only (Fig. 3c) with a decrease in the total delay time. In a recent study [13], the autoignition of tetralin/$O_2$/$N_2$ mixture in RCM failed to produce any evidence of NTC behaviour.

## 3.2 *Autoignition in Flow Reactor*

Flow reactors work based on the principle of direct mixing between fuels and flowing stream of hot oxidizer. The combustible mixture ignites over a long distance from the point of injection. Flow reactors have long ignition delay and particularly suitable for constant pressure applications. Also, the application is somewhat restricted to low-temperature $T < 1200$ K and intermediate pressure range of 1–20 bar. The flow reactors offer the advantage of having relatively simple measuring techniques due to the large spread of the reaction zone. Plenty of literature is available on the experimental investigations of autoignition using flow reactors [14–18]. Pressurized flow reactor was used to study autoignition of hydrogen in air using OH chemiluminescence technique [15]. The results revealed that the surface reaction has a serious impact on autoignition delay.

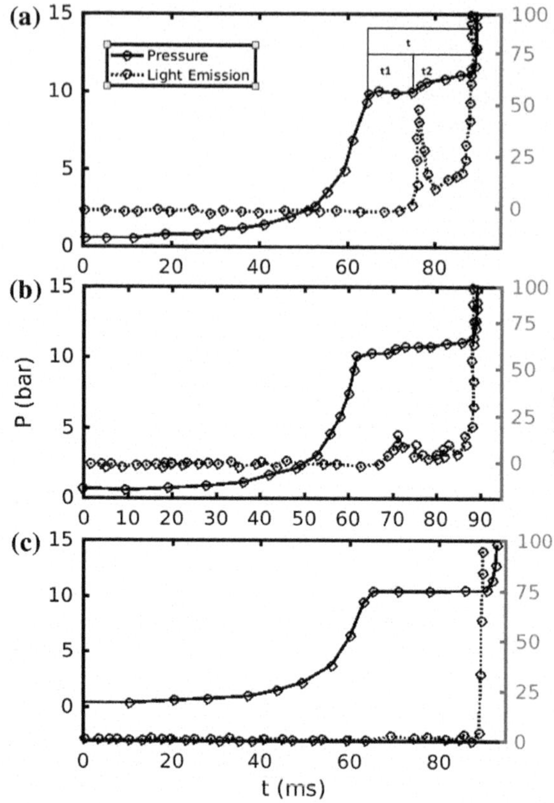


**Fig. 3** Diagram showing pressure and light emission during the compression and autoignition for different stoichiometric ratio. Core temperature: **a** $T_c = 749$ K, **b** $T_c = 796$ K, **c** $T_c = 818$ K [12]

## 3.3 *Autoignition in Shock Tube*

As the name suggests, shock tube uses the shock wave compression for bringing the reactants to a reactive state. Shock tubes are ideal for determining shorter ignition delay of the order of few milliseconds at high temperature. Autoignition of iso-cetane was investigated in a shock tube [19] and a new reaction mechanism was developed without changing the rate constants. The results with the newly developed reaction mechanism are found to be in reasonably good agreement with the experimental measurements. Autoignition of methyl decanoate [20], demonstrated the NTC behaviour for ignition delay at high pressure, which is comparable to that of *n*-decane. However, at low pressure, ignition delay time was found to be shorter than that of *n*-decane. Also, methyl decanoate demonstrates different oxidation mechanisms at low- and high temperatures. With the decrease in temperature, reactivity of methyl decanoate also reduces similar to that of *n*-decane.

### 3.4 Autoignition in Laminar Non-premixed Flows

The counterflow- or opposed-jet flow burner setup is the most widely used configuration for investigation of autoignition in laminar non-premixed flames (Fig. 4). Several experiments on autoignition have been performed in the counterflow configuration for a wide variety of fuels and at different operating conditions (such as temperature, pressure, and dilution). The results demonstrate the existence of a critical strain rate for autoignition. Autoignition is favoured for laminar non-premixed flames with low scalar dissipation rate. The same trend had also been observed in numerical simulations [22]. Autoignition of the $H_2/N_2$ mixture in an opposed flame configuration was examined with heated air stream as the oxidizer within a pressure range 0.1–6.0 atm [23]. It was found that under all ambient pressure conditions, the ignition temperature monotonically increases with an increase in the flow strain rates. However, under atmospheric pressure condition, the ignition temperature was found to be insensitive to changes in the flow strain rate. This is clearly shown in Fig. 5.

When hydrogen was added in methane, two-stage ignition took place, where the first stage is dominated by radical runaway followed by a thermal feedback mechanism in the second stage [24]. The two-stage ignition improves the ignition of methane significantly. The effect of $N_2$ dilution on ignition temperature was also investigated for DME jets in a counter-flow flame configuration with heated air as the oxidizer [25]. At constant pressure, nitrogen dilution in DME decreases the ignition temperature. Further, the ignition temperature also decreases with an increase in pressure. The results were found to be consistent with the previous experimental observations reported in previous literature [23, 24]. The flame structure of laminar $CH_4/N_2$ jet under moderate and intense low-oxygen-dilution (MILD) conditions in hot coflow was examined by Sepman et al. [26]. The flame height decreased with hydrogen addition. However, the maximum flame temperature and the thickness of combustion zone remained almost unaffected by the

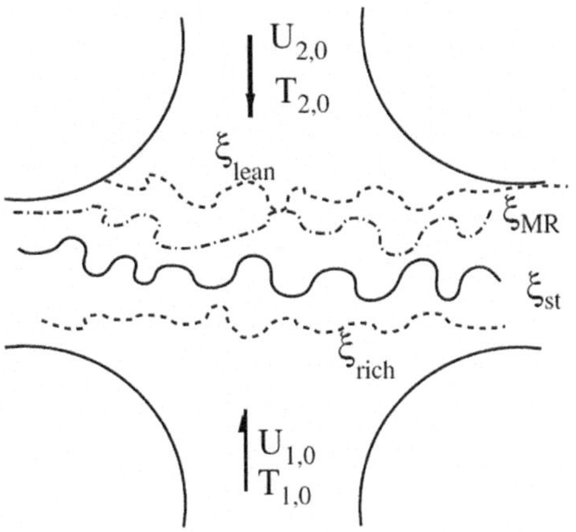


**Fig. 4** A schematic for counter-flow diffusion flame [21]

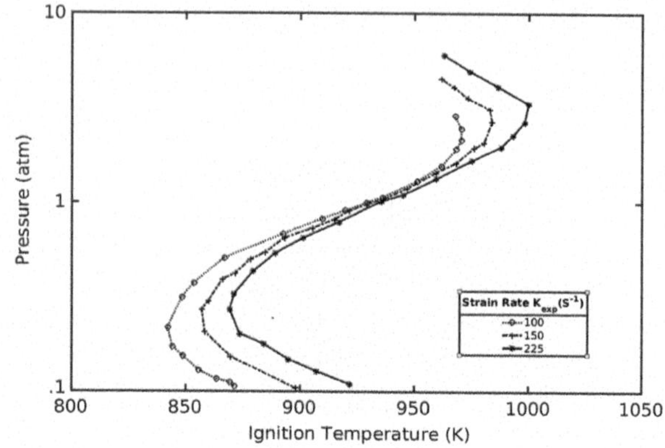


**Fig. 5** Ignition temperatures versus pressure at different strain rate [23]

addition of hydrogen. The flame structure and NO formation at different preheat conditions were also investigated [27]. Under the MILD conditions, NO formation decreases considerably.

## *3.5 Autoignition in Turbulent Non-premixed Flows*

Autoignition in turbulent nonpremixed flows is important from both industrial and academic perspective. The industrial burners are complex in nature due to their complex geometry and flow features. In order to gain physical insights of autoignition in inhomogeneous mixtures, simple experimental burner configurations have been developed in the past couple of decades. The Berkley burner was developed [1] to understand the physics behind the stabilization of lifted flames. They replicated a complicated process of exhaust gas recirculation with a very simple laboratory-based experimental setup in which the central jet is stabilized by hot products of lean $H_2$/air combustion. The schematic for the same is shown in Fig. 6. Their results concluded autoignition along with flame propagation as the relevant stabilization mechanism for lifted flames. However, they recommended further studies to confirm this phenomenon. The open flow configuration of the Berkley burner [1] provides a good optical access, which facilitates point and planar diagnostics for visualization of pre-ignition regions, autoignition spots, and the ensuing flamelets.

Mastorakos and co-workers [2, 3] experimentally investigated auto-ignition of diluted $H_2$ and prevaporized *n*-heptane jets in a confined turbulent hot co-flow (CTHC) burner. The schematic for the same is shown in Fig. 7. This burner has optical access and is well-suited for autoignition experiments under turbulent conditions. The experimental database from this burner has been used for development of turbulent combustion models.

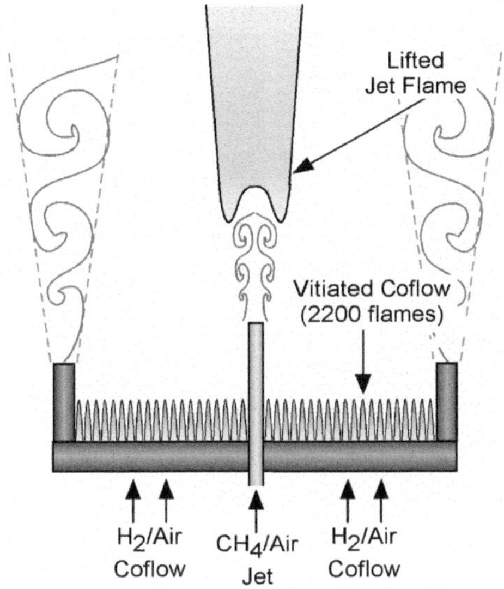


**Fig. 6** Schematic of Berkley vitiated coflow burner [28]

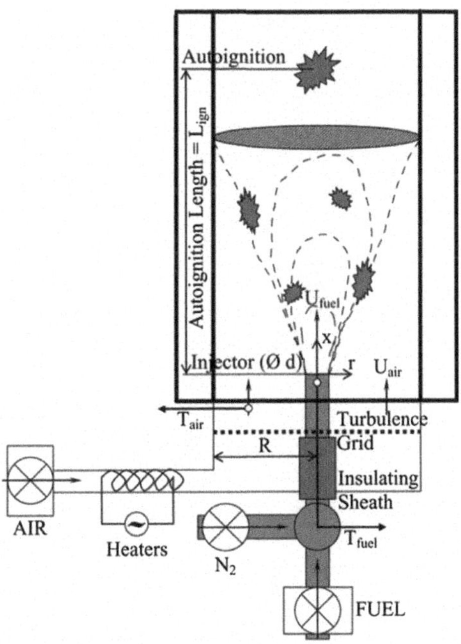


**Fig. 7** Schematic burner configuration used by Markides and Mastorakos [2]

During the experiments, auto-ignition kernels were identified using optical techniques, and their axial locations were measured by OH- radical chemiluminescence images. The auto-ignition length (and hence ignition delay time) increases with a decrease in the co-flow temperature (as shown in Fig. 8) and with an increase

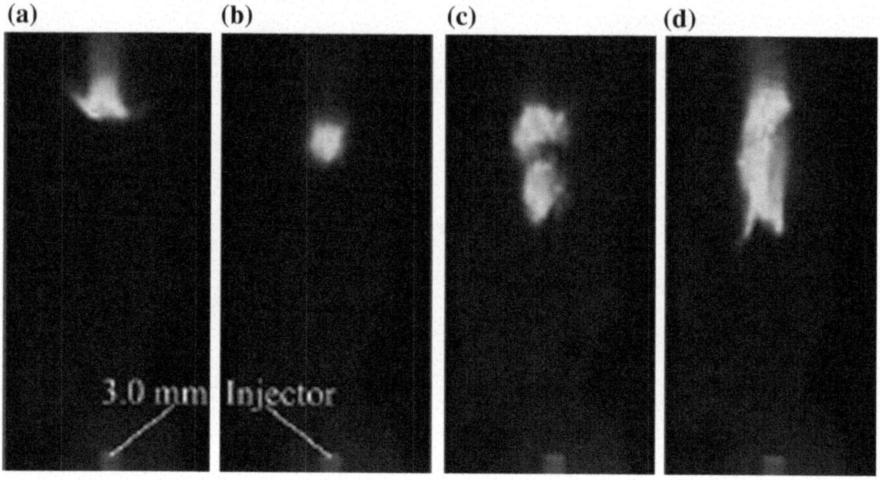


**Fig. 8** Instantaneous raw OH chemiluminescence snaps for the autoigniting $H_2$ jet: **a–c** for $T_{air}$ = 1000 K, whereas **d** for $T_{air}$ = 1020 K [2]

in the jet and co-flow velocity. Also, the results showed that the turbulent mixing delays auto-ignition. Four different regimes, namely "no ignition", "random spots", "flashback" and "lifted flame" were observed under different operating conditions. The optical imaging technique used by Mastorakos and co-workers [2, 3] was later used to explore turbulent mixing and auto-ignition of acetylene jets in inhomogeneous mixtures [29]. A parametric study was conducted wherein the effects of turbulent intensity, turbulent length scale, and injector diameter are considered. By retaining a constant level of turbulent intensity, an increase in the injector diameter resulted in an increase in the auto-ignition length. However, a simultaneous increase in the length scale and a decrease in the turbulent intensity shortened the auto-ignition length. Further, the frequency of appearance of auto-ignition increases with an increase in the air velocity.

Roubaud et al. [30] investigated autoignition and oxidation mechanisms of alkylbenzenes in the range of 600–900 K at pressure above 14 bars. Their results showed the double peroxidation and low-temperature branching phenomena during isomerization step when benzylic radicals combine with an oxygen molecule. Using a burner similar to the one used by Cabra [1] and high-speed Schlieren video imaging technique, Johannessen et al. [31] investigated unsteady $H_2/N_2$ flames in a vitiated co-flow. It was seen that nitrogen dilution increases the frequency of ignition up to a certain maximum value, beyond which the frequency of ignition decreases. The experimental results also established that re-ignition is primarily controlled by chemical kinetics and turbulent mixing. Also, the high-speed planar images obtained using Rayleigh scattering technique for a pulsed turbulent jet flame in a vitiated coflow [32] established that auto-ignition is most likely to occur at the periphery of the fuel jet, where the mixture is very lean and scalar dissipation rate is low. By using high-speed shadowgraphy, a two-stage ignition was reported during the transient injection of DME under high-pressure conditions [33] for reported operating conditions. Oldenhof et al. [34] experimentally investigated the ignition of impulsively started jets of natural gas in heated co-flow using particle image velocimetry (PIV) and Planar Laser Induced Fluorescence (PLIF) techniques. The jet leaving the nozzle was laminar initially, which suddenly transitioned to a turbulent jet. During the laminar phase, combustion was not observed and large ignition delay time was reported during the process of transition to turbulent phase. It is believed that the presence of the large-scale inhomogeneities increases the ignition delay time.

Chen and co-workers [35] performed an experimental study on the combustion characteristics in an HCCI engine by diluting the natural gas with DME. Their investigations predicted that when DME was mixed with the natural gas in small proportions, the engine can run over a comparatively large load range with higher thermal efficiency compared to the Diesel engine and also reduces the ignition delay and NOx emissions. Venkatesan et al. [36] conducted experiments using DME as an ignition improver in a hydrous methanol-fuelled HCCI engine. The combustion characteristics of methanol and ethanol were also investigated under HCCI mode of operation [37]. It was found that methanol auto-ignites faster than ethanol in HCCI conditions.

## 4 Turbulent Combustion Models

In the Reynolds-Averaged-Navier-Stokes (RANS) and large eddy simulations (LES) approach, the transport equations for the temperature and species mass fractions contain mean/filtered chemical source terms. Since the chemical source term is a nonlinear function of species mass fractions and temperature, accurate evaluation of the mean chemical source term becomes difficult. The so-called 'direct closure' approach is known to introduce erroneous values of the mean/filtered chemical source terms. This 'closure of mean/filtered chemical term' has been the central difficulty in RANS/LES of reacting flows.

Turbulent combustion may be modelled either assuming it is controlled by turbulent mixing (e.g., eddy breakup (EBU), and eddy dissipation models (EDM), etc.) or via statistical approaches, where the statistical properties of intermediate states are described using a PDF [38]. With the joint composition PDF, $\tilde{P}(\rho, Y_1, \ldots\ldots Y_N, T; x, t)$, the mean burning rate may be estimated as

$$\dot{\omega}_{Y_i}(x,t) = \int\limits_{\rho, Y_1, \ldots Y_N, T} \dot{\omega}_{Y_i}(\rho, Y_1, \ldots, Y_N, T)\tilde{P}(\rho, Y_1, \ldots Y_N, T; x, t)\, d\rho dY_1, \ldots dY_N dT \tag{1}$$

The PDF may be determined either by presumed PDF method [39] or transported PDF method [38]. In the presumed PDF method, an empirical expression for the PDF is chosen. Since the shape of the PDF usually depends on the local physical conditions, a few parameters of the PDF are computed at each physical location based on the balance equations of the first several moments, usually the mean and variance. The presumed PDF method is often used to model one variable. It becomes difficult to model more than one variable with a presumed joint PDF. For such problem, assumption is made that the variables are statistically independent and model the single variable with presumed PDF:

$$\tilde{P}(\psi_1, \psi_2, \ldots\ldots \psi_N) \approx \tilde{P}_1(\psi_1)\tilde{P}_2(\psi_2)\ldots\ldots\ldots\ldots\ldots\tilde{P}_N(\psi_N) \qquad (2)$$

Alternatively, significant efforts have been taken to simplify the expression of the joint PDF, $\tilde{P}(\rho, Y_1, \ldots\ldots Y_N, T; x, t)$. For instance, the composition space $Y_1, \ldots\ldots, Y_N$ is replaced with a conserved scalar; the joint PDF is presumed to have an empirical expression. Another class of combustion models assumes that the chemical reaction occurs in thin layers separating fuel and oxidizer (high Damköhler number Da $\gg$ 1 limit). The reaction zone is viewed as a collection of laminar flamelets. Coherent flamelet and flame surface density models are introduced for premixed turbulent combustion [40]. For non-premixed flames, hypotheses are formulated to construct different models such as (1) infinitely fast chemistry (IFC) model, (2) flamelet models and (3) conditional moment closure (CMC) method. First two models are applicable in the limit of Da $\gg$ 1. For non-premixed combustion, the joint PDF $\tilde{P}(\rho, Y_1, \ldots\ldots Y_N, T; x, t)$ is greatly simplified using a chemistry-independent conserved scalar, namely mixture fraction $\xi$.

## 4.1 Laminar Flamelet Model

The flamelet assumption takes into account the non-equilibrium effects. This model has been developed both for non-premixed [39] as well as premixed [41] combustion. It is derived based on the assumption that the thickness of the reaction zone is sufficiently small compared to the Kolmogorov length scale, and molecular diffusion can determine its internal structure without the influence of turbulence. The flamelet model views the turbulent diffusion flame as an ensemble of thin, laminar, locally one-dimensional structures called flamelets embedded within the turbulent flow field. The ensemble is weighted by a presumed PDF to produce the turbulent ‘flame brush’. Instead of calculating the reaction rate, the composition space is determined from a pre-calculated laminar flamelet library. The coupling between non-equilibrium chemistry and turbulence is achieved by the statistical description of two parameters: the mixture fraction $\xi$ and the scalar dissipation rate $\chi$. The mean value of any scalar $\Psi$ (temperature and compositions) is calculated by

$$\tilde{\psi}(x,t) = \int_0^1 \int_0^\infty \psi(\xi,\chi)\tilde{P}(\xi,\chi;x,t)d\chi d\xi \tag{3}$$

Here $\tilde{P}(\xi,\chi;x,t)$ is the joint PDF of the mixture fraction and the scalar dissipation rate, and these two parameters are assumed to be statistically independent

$$\tilde{P}(\xi,\chi;x,t) = \tilde{P}_\xi(\xi;x,t)\tilde{P}_\chi(\chi;x,t) \tag{4}$$

The steady flamelet equation can be given as

$$\rho\frac{N_{st}}{2}\frac{\partial^2 Y_K}{\partial Z^2} + \dot{\omega}_K = 0 \tag{5}$$

where, $N_{st}$ is the scalar dissipation at the stoichiometric value.

## 4.2 *Transported PDF Model*

In the transported PDF method, the turbulent flow is described by the transport equation of joint PDF of the velocity and scalars in the flow field namely mass fractions of species and enthalpy. This equation can be derived from Navier-Stokes equations and conservation equation for scalars [38]. In this formulation, the nonlinear chemical source term does not require any closure. However, terms involving molecular mixing in reactive scalar space, fluctuating pressure term and PDF transport in velocity space by viscous stresses require closure models. The qualitative performance of the PDF method largely depends on the accuracy of submodels used to close these terms [42]. None of the molecular mixing models namely, interaction by exchange with the mean (IEM), linear mean-square estimation (LMSE), mapping closure (MC) and Euclidean minimum spanning tree (EMST) [43] includes a physically realistic representation of scalar dissipation rate to capture phenomenon such as local extinction, re-ignition [44]. Theoretically, the transported PDF method can treat the arbitrary complex chemistry without any assumption. The transport equation of joint scalar PDF may be written as

$$\begin{aligned}\frac{\partial}{\partial t}(\rho P) + \frac{\partial}{\partial x_i}(\rho u_i P) + \frac{\partial}{\partial \psi_\alpha}(\omega_\alpha \rho P) = &\frac{\partial}{\partial \psi_\alpha}\left(\langle\frac{\partial}{\partial x_i}J_i^\alpha|\underline{\psi}\rangle\rho P\right) \\ &- \frac{\partial}{\partial x_i}\left(\langle u_i''|\underline{\psi}\rangle\rho P\right)\end{aligned} \tag{6}$$

In the above, $P$ is the joint scalar PDF, $u$ is the velocity, $\rho$ is the density, $\psi$ is the scalar space. All terms in the equation are closed except the first term on the right hand side, which is the micromixing term that needs mixing model for its closure.

### *4.3 Conditional Moment Closure Method*

CMC model may be derived either from joint PDF method or decomposition method [45]. The major assumption made in deriving the CMC equations is that most of the fluctuations in the scalar quantities of interest could often be associated with the fluctuation of one key quantity. In non-premixed flows, where there is mixing between two bodies or streams of fluid (fuel and oxidizer), the reactive scalars within the mixing field depend strongly on the local and instantaneous value of mixture fraction $\xi(x,t)$. Conditional expectation $Q(\eta;x,t)$ of a scalar variable, conditioned on the associated value $\xi(x,t)$ taking a specific value $\eta$ is defined as, $Q(\eta,x,t)=\langle\phi(x,t)|\xi(x,t)=\eta\rangle$. Here, the notation of vertical bar indicates that the average is taken of those members of the ensemble which satisfy the condition to the right of the bar. Then the local and instantaneous value of a scalar can be decomposed as,

$$\phi(x,t)=Q(\xi[x,t];x,t)+\phi''(x,t) \tag{7}$$

The transport equation for conditional scalars may be written as,

$$\begin{aligned}&\frac{\partial Q}{\partial t}+\langle u|\xi=\eta\rangle.\nabla Q+\frac{1}{P(\eta)\rho_\eta}\nabla.(\rho_\eta\langle u''\varphi''|\xi=\eta\rangle P(\eta)\\&-\langle N|\xi=\eta\rangle\frac{\partial^2 Q}{\partial\eta^2}=\langle\dot{\omega}_F'''|\xi=\eta\rangle\end{aligned} \tag{8}$$

Here, $Q=\{Q_h,Q_\alpha\}$, and $Q_h$ is the conditional mean enthalpy and $Q_\alpha$ is the conditional mean species mass fraction. In the above equation, $\eta$ is the sample space variable for the mixture fraction $\xi$. Angular brackets denote ensemble average subject to the condition $\xi(\mathbf{x},t)=\eta$. Also, $\rho_\eta$ is the conditional density of the gaseous mixture. If closure of the conditional scalars appearing in reaction rate term which needs to be modelled is achieved at the first-moment level by neglecting the conditional fluctuations, the model is said to be 'first-order CMC' model. The reaction rate term in first-order CMC model with single step reaction is given by,

$$\langle\dot{\omega}_F'''|\xi=\eta\rangle=-A_F Q_F Q_o\exp\left(-\frac{E}{RQ_T}\right) \tag{9}$$

For cases where the fluctuations about conditional means of a scalar are present in significant proportions, e.g., flames with extinction and re-ignition, the closure is necessary at the second-moment level, which is known as 'second-order CMC' model [46]. Sometimes, another variable such as sensible enthalpy is used as the second conditioning variable [47]. CMC equations with added dimension (mixture fraction space) are solved in tandem with flow field in case of fully coupled equations. The flow solver provides the mixing field information to the CMC equations, while the CMC solver provides the information of mean density field to the flow solver. However, CMC incurs huge computational efforts mainly due to the added dimension, namely the mixture fraction space.

## 5 Numerical Approaches

In this section, various modelling approaches that are used to capture autoignition are presented. In the early stage, most of the numerical work done on autoignition under turbulent conditions is based on DNS, where the problem is greatly simplified by assuming the homogeneous isotropic decaying turbulent flow like conditions.

### *5.1 Numerical Simulations of Autoignition in Laminar Flames*

Detailed chemical mechanisms for fuel oxidation with accurate reaction rate coefficients are essential for prediction of ignition delay time using numerical simulations. Sensitivity analysis revealed that modification of reaction rate constants did not improve the prediction of ignition delay time. To improve the accuracy of the available reaction mechanism, we must include a comprehensive model that describes low-temperature reactions adequately. This allows us to predict pre-ignition heat release and ignition delay time accurately.

Numerical investigations were performed for autoignition in laminar flows to determine the effects of a periodic strain rate on autoignition. The results revealed that autoignition may be still likely to occur depending on the amplitude and frequency of the oscillation, even if the time-averaged scalar dissipation rate ($\bar{N}$) exceeds the critical scalar dissipation rate, i.e. $\bar{N} > N_{crit}$. In this case, the flow remains in a Low-$N$ regime, which allows autoignition to take place. Converse of this situation may also happen, when $\bar{N} < N_{crit}$ moreover, the flow spends most of the time above the critical value, resulting in no autoignition.

### *5.2 Numerical Simulations of Autoignition in Turbulent Flames*

For engineering applications, numerical approaches based on RANS and LES are being used. Due to the limitations of computational resources and the availability of reliable sub-models, the numerical modelling of autoignition under turbulent conditions has yet to achieve widespread applications.

Sreedhara and Lakshmisha [22] performed a two-dimensional DNS using a single step chemistry to study autoignition of *n*-heptane in a hot oxidizer medium.

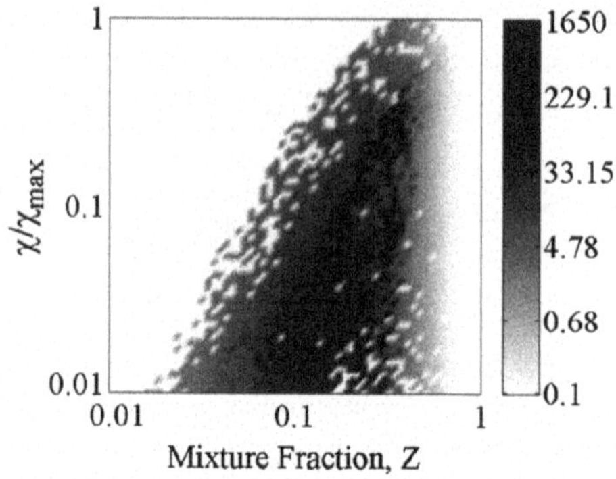


**Fig. 9** Diagram showing joint-conditional mean reaction rate during the first appearance of autoignition spots [22]

The DNS results led to a conclusion that ignition spots are favourable within some specific range of mixture fraction called the "most reactive mixture fraction" ($\xi_{MR}$) and also the region should have the low scalar dissipation rate, which is evident in Fig. 9 [22]. An increase in initial turbulence intensity promotes turbulent mixing and delays auto-ignition. Further, they investigated the difference between the effects of auto-ignition in two and three-dimensional flow fields using a four-step chemical kinetics. The discrepancies observed in earlier two-dimensional DNS with the experimental observations later got resolved due to the presence of vortex stretching in three-dimensional flows that are absent in two-dimensional flows [48]. Also, the first appearance of an ignition site is found to be independent of the turbulent time scale and with the increase in inhomogeneities in the mixture, the autoignition time also increases [49].

Further, they reported that the conditional scalar dissipation rate controls the ignition time and partial premixing decreases the delay in ignition time. In another DNS study [50], autoignition of DME was observed as a complex three stage process. Each stage evolves with a different chemical reaction pathway. Figure 10 shows a time history of temperature and heat release rate (HRR) for two different initial temperature conditions [50]. In the case of initial low temperature, three peaks were observed in HRR of approximately equal magnitude, and with every peak, a corresponding increase in temperature was observed. However, for the case with high initial temperature, time history of HRR showed a negligible heat release in the first stage and maximum heat release in the third stage with a comparatively very low value in the second stage.

Autoignition of hydrogen was investigated [51] in a heated co-flow using an LES—Eulerian PDF approach to model the turbulence-chemistry interaction and a detailed chemical mechanism involving 9 species and 19 reversible reactions. The model replicated the experimental results without tuning the model constants and the observed auto-ignition length matched reasonably well with the predicted auto-ignition length. The results showed a strong sensitivity to the variations in the co-flow temperature. Domingo et al. [52] investigated the lifted methane jet flames in a vitiated coflow using LES-CMC model, which established autoignition and premixed flame propagation as stabilization mechanism of flame liftoff. Further,

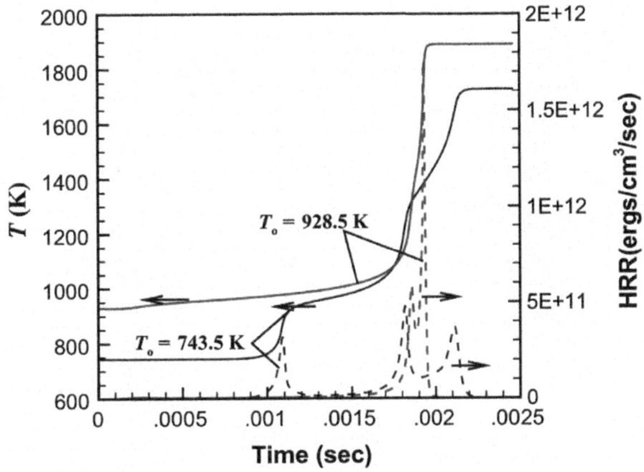


**Fig. 10** Change in temperature and heat release rate (HRR) with time for two different initial temperatures [50]

they reported that molecular diffusion effects ignition. Recently, the effect of differential diffusion on autoignition has also been investigated [53] in a vitiated coflow with LES/PDF modelling approach. It was found that differential diffusion decreases both lift-off height and thickness of autoignition zone. Further their results showed a reasonably good prediction with experimental results and no autoignition event were observed below the liftoff height. Out of auto-ignition and turbulent flame propagation, one of the phenomenon acts as the stabilization mechanism depending upon the location across the flame length. A numerical study was carried out in [54] using analytically reduced chemistry along with the thickened flame model for turbulence-chemistry interactions. Their results showed that autoignition was the dominant flame stabilization mechanism at the flame base. Autoignition pockets were observed in the lean mixture corresponding to the "most reactive mixture fraction". Around the centre line where the HRR is relatively high, flame propagation was found to be dominant in that region. Figure 11 shows the snapshot from the simulation, which supports the above discussion [54].

Autoignition of laminar non-premixed DME/air in a heated co-flow has been investigated [55] by varying the coflow temperature in the range of 700–1100 K. Autoignition serves as a stabilization mechanism when the coflow temperature is in the range of 700–900 K. Determination of stabilization point was carried out by the negative temperature coefficient chemistry in the mixture fraction space. At a low coflow temperature (about 700 K), the temperature was high enough to auto-ignite the gaseous mixture. However, the incoming flow velocity was not large enough to balance the flame propagation velocity. Due to this, pure kinetically stabilized flames were observed as shown in Fig. 12 [55]. On the other side, above a coflow temperature of 1100 K, the kinematic balance between the incoming flow velocity and the premixed flame propagation velocity acts as a stabilization mechanism. Figure 12 shows the change in the flame structure with respect to the change in coflow boundary temperature.

Further studies were conducted on the flame structure by varying the inlet flow velocity while keeping the coflow temperature constant [56]. When the flow

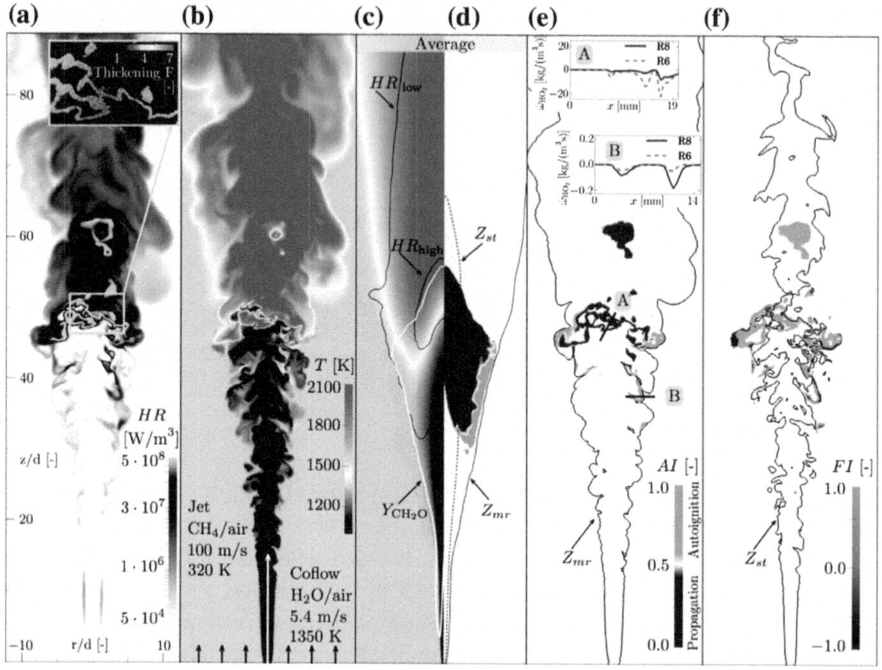


**Fig. 11** Snapshots at same instant of time **a** heat release rate, **b** temperature, **c** mean temperature field. **d** The conditioned mean Autoignition field. **e** Most reactive mixture fraction superimposed on autoignition index (AI) **f** Stoichiometric mixture fraction superimposed on the flame index (FI) [54]

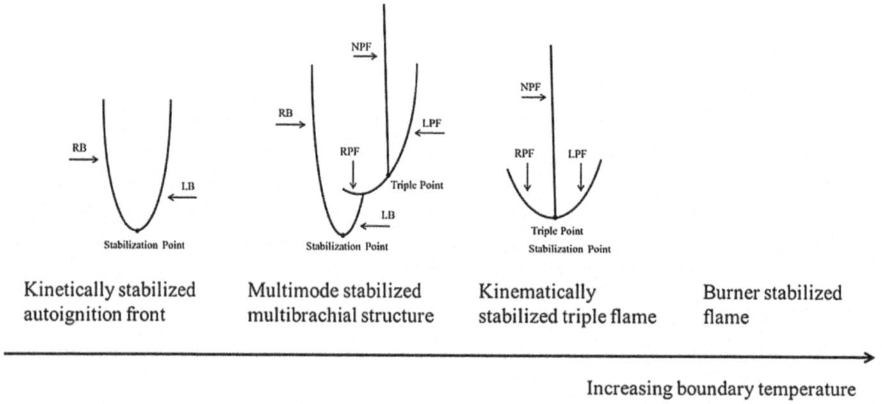


**Fig. 12** Change in stabilization mechanism with a change in coflow boundary temperature [55]

velocity is 2.4 m/s, the classical triple flame structure was observed at elevated temperature due to the kinematic balance between incoming flow velocity and premixed flame propagation. With an increase in the inlet velocity, the same

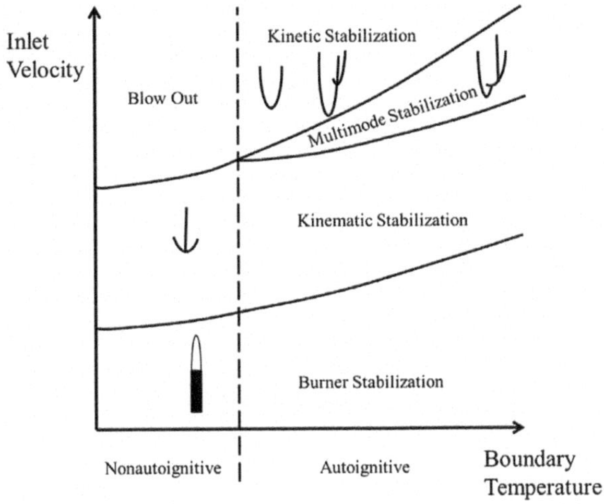


**Fig. 13** A regime diagram for the stabilization mechanism [56]

stabilization mechanism was not observed anymore at a certain mixture fraction values. Autoignition of inhomogeneous mixtures becomes the dominant stabilization mechanism, and a multimode stabilized flame was observed. With further increase in the inlet flow velocity, the kinematic balance was lost, and a pure kinetically balanced flame front was found. In Fig. 13, different stabilization regimes are shown as per the discussion presented here.

The applicability of the conditional moment closure (CMC) method was analysed for autoignition of *n*-heptane in a hot oxidizer medium [46]. Using DNS, a correction factor was introduced in CMC formulation for species fluctuations, which yielded the similar trend as shown by experimental results and matched well with the DNS data. The results for conditionally averaged temperature during autoignition for different eddy turnover time are shown in Fig. 14. Despite using a first order closure results were found to show an excellent agreement.

Patwardhan and Lakshmisha [57] numerically investigated auto-ignition of turbulent $H_2/N_2$ jets in a preheated co-flow using a RANS based first order CMC approach. They reported that modifications of $k$-$\varepsilon$ model constants would not alter the auto-ignition length significantly shown in Fig. 15. A slight increase in autoignition length was observed due to modification of empirical constants in the turbulence model, which resulted in reduced variance and increased mixing. This suggests autoignition process is primarily controlled by chemical kinetics rather than mixing. They concluded that first order CMC was unable to produce random spots that were observed in experimental results. The time sequences of autoignition and subsequent flame development are shown in Fig. 16 for different coflow temperature. Flashback is observed at all coflow temperature and random spots are not found due to the incapability of RANS because it averaged out the scalar field and unable to reproduce it.

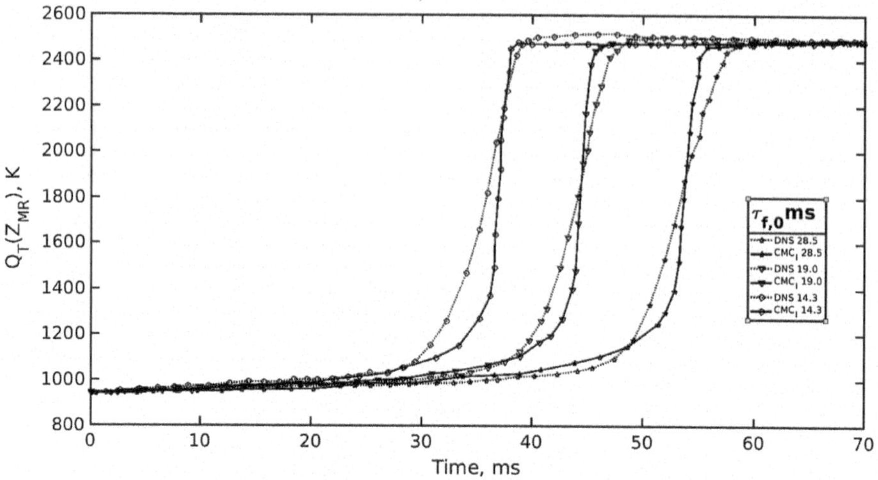


**Fig. 14** Time history of conditionally averaged temperature $Q_T$ ($Z_{MR}$) during autoignition comparison against DNS data for different initial eddy-turnover times [46]

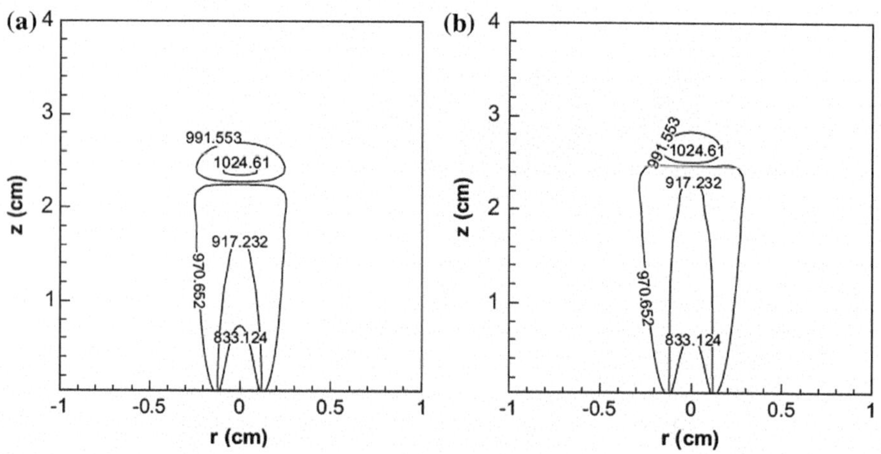


**Fig. 15** Contours of mean temperature at the time of autoignition: **a** standard $k$-$\varepsilon$ at $t = 0.938$ ms, and **b** modified $k$-$\varepsilon$ at $t = 1.019$ ms [57]

The effects of fluctuations of conditional scalar dissipation rate have also been included in the numerical model of [49] and found the autoignition time, and that shows a reasonably good agreement with DNS results. Wu et al. [58] investigated numerically ethanol and DME lifted flames in a hot vitiated co-flow. They used an eddy dissipation concept (EDC) model in the RANS context for turbulence-chemistry interaction. They concluded that value of the mixture fraction for the two

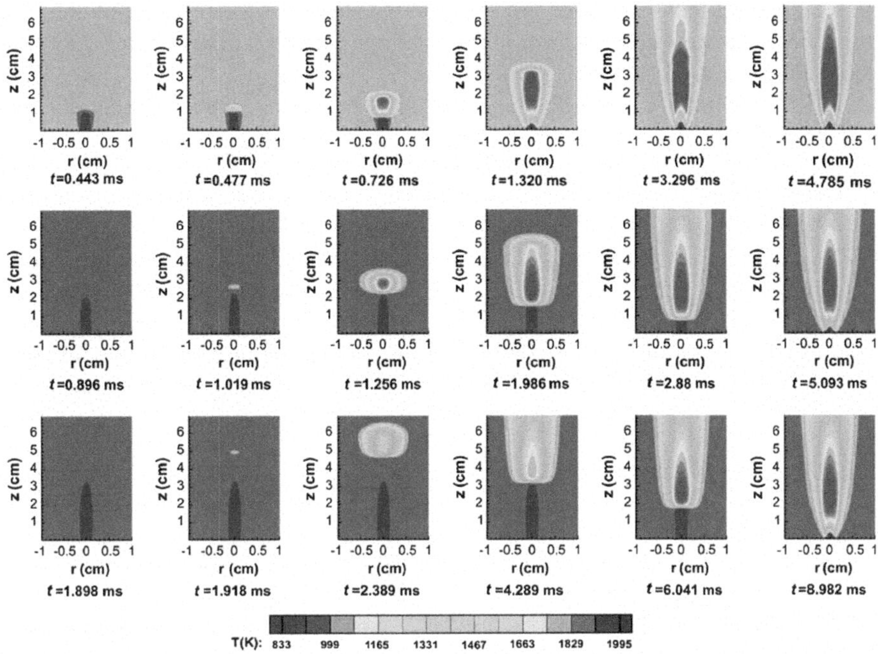


**Fig. 16** Contours of mean temperature at different time instances for different coflow temperature: $T_{\text{coflow}} = 1020$ K (*top row*), $T_{\text{coflow}} = 980$ K (*middle row*), and $T_{\text{coflow}} = 960$ K (*bottom row*) [57]

fuels is same at the stabilization point under identical conditions. Also, before auto-ignition, stabilization point was found to be affected simultaneously by ignition delay and reaction paths of the fuel under low co-flow temperature. However, at high co-flow temperature, lift off height is mainly controlled by ignition delay. Further, at high co-flow temperature flame stabilizes in the leaner mixture. The first-order spatially averaged CMC model was used to study auto-ignition of a gaseous *n*-heptane plume in a heated co-flow [3]. They concluded that the conditional scalar dissipation rate is much smaller value than the critical value for a perceptible span before auto-ignition takes place. After the critical value of conditional scalar dissipation rate, auto-ignition was observed.

Effect of DME dilution in natural gas was examined numerically in an HCCI engine [59] with detailed chemical kinetics to describe the low-temperature autoignition chemistry. The CHEMKIN chemistry solver was integrated with KIVA-3V CFD code for solving the multidimensional engine simulations. The numerical model used in [59] gave a reasonably good level of agreement with the available experimental results. A two-step ignition process was observed, where initially DME controls the ignition by self-oxidation before the major oxidation of natural gas could take place. With an increase in DME concentration in the natural gas, the ignition delay time decreases due to auto-ignition of DME at a relatively low temperature than methane. With an increase in natural gas concentration, the engine becomes unstable because operating range becomes narrower for very low excess air-fuel ratio of natural gas.

## 6 Conclusions

This chapter seeks to provide an overview on autoignition in laminar and turbulent nonpremixed flames. Various burners used for experimental investigations are reviewed along with the available numerical approaches with a specific attention for autoignition under turbulent conditions. Although substantial progress has been made in this area, we are yet to address several unresolved issues. The major findings and future research directions are summarized below:

- Autoignition of the fuel occurs at the specific temperature depending upon its physical and chemical properties for the given value of the strain rate. For a specific temperature, autoignition does not occur above a critical value of strain rate. Further, autoignition temperature increases with an increase in strain rate.
- The autoignition in non-premixed systems occurs at some specific value of mixture fraction usually away from the stoichiometric value of mixture fraction, referred as "most reactive mixture fraction". Autoignition is also found to be a dominant stabilization mechanism at the flame base of a lifted flame.
- Experimental data on autoignition in turbulent non-premixed flames are limited and only available for very simple flow configurations. Most of the work done on turbulent autoignition is specifically focused on the Berkley vitiated coflow burner and Cambridge confined turbulent hot coflow burner. More detailed experiments should be planned to investigate the direct effect of turbulence on autoignition in enclosed flow geometry.
- More experiments should be directed to investigate autoignition and subsequent flame development at the high-pressure environment using the heated oxidizer/product stream. Further, autoignition characteristics of synthetic fuels, alcohols, and different fuel mixtures under turbulent conditions should be investigated, before introducing them into the practical energy conversion systems.
- On the modelling side, new combustion models are needed to develop which consider the nature of flame propagation in mixtures with large mixture fraction fluctuations. Applications of higher order conditioning methods may be useful for the flames undergoing large mixture fraction fluctuations.
- New reaction mechanisms are needed to be developed, which are valid in a wide range of temperature and pressure. Moreover, suitably reduced chemical mechanisms for different fuel and fuel mixtures needs to be explored for their applications in numerical codes using advanced turbulent combustion models, namely, transported or stochastic PDF approach, and CMC method.

## References


1. Cabra R, Myhrvold T, Chen JY, Dibble RW, Karpetis AN, Barlow RS (2002) Simultaneous laser Raman-Rayleigh-Lif measurements and numerical modeling results of a lifted turbulent H-2/N-2 jet flame in a vitiated coflow. Proc Combust Inst 29:1881–1888
2. Markides CN, Mastorakos E (2005) An experimental study of hydrogen autoignition in a turbulent co-flow of heated air. Proc Combust Inst 30:883–891
3. Markides CN, De Paola G, Mastorakos E (2007) Measurements and simulations of mixing and autoignition of an n-heptane plume in a turbulent flow of heated air. Exp Therm Fluid Sci 31:393–401
4. Gordon RL, Masri AR, Mastorakos E (2008) Simultaneous Rayleigh temperature, OH- and CH2O-LIF imaging of methane jets in a vitiated coflow. Combust Flame 155:181–195
5. Gordon RL, Masri AR, Mastorakos E (2009) Heat release rate as represented by [OH] x [CH2O] and its role in autoignition. Combust Theory Model 13:645–670
6. Park SH, Lee CS (2014) Applicability of dimethyl ether (DME) in a compression ignition engine as an alternative fuel. Energy Convers Manag 86:848–863
7. Das LM (1990) Fuel induction techniques for a hydrogen operated engine. Int J Hydrog Energy 15:833–842
8. Semelsberger TA, Borup RL, Greene HL (2006) Dimethyl ether (DME) as an alternative fuel. J Power Sources 156:497–511
9. Carlier M, Corre C, Minetti R, Pauwels JF, Ribaucour M, Sochet LR (1991) Autoignition of butane: a burner and a rapid compression machine study. Symp (Int) Combust 23:1753–1758
10. Shi Z, Zhang H, Liu H, Lu H, Li J, Gao X (2015) Effects of buffer gas composition on autoignition of dimethyl ether. Energies 8:10198
11. Pitz WJ, Wilk RD, Westbrook CK, Cernansky NP (1988) Western States Section of the Combustion Institute
12. Minetti R, Ribaucour M, Carlier M, Fittschen C, Sochet LR (1994) Experimental and modeling study of oxidation and autoignition of butane at high-pressure. Combust Flame 96:201–211
13. Kukkadapu G, Weber BW, Sung CJ (2015) Autoignition study of tetralin in a rapid compression machine at elevated pressures and low-to-intermediate temperatures. Fuel 159:436–445
14. Yetter RA, Dryer FL, Rabitz H (1991) Flow reactor studies of carbon-monoxide hydrogen oxygen kinetics. Combust Sci Technol 79:129–140
15. Schonborn A, Sayad P, Konnov AA, Klingmann J (2014) OH*-chemiluminescence during autoignition of hydrogen with air in a pressurised turbulent flow reactor. Int J Hydrog Energy 39:12166–12181
16. Schonborn A, Sayad P, Konnov AA, Klingmann J (2014) Autoignition of dimethyl ether and air in an optical flow-reactor. Energy Fuel 28:4130–4138
17. Schonborn A, Sayad P, Konnov AA, Klingmann J (2013) Visualisation of propane autoignition in a turbulent flow reactor using OH* chemiluminescence imaging. Combust Flame 160:1033–1043
18. Beerer DJ, McDonell VG (2008) Autoignition of hydrogen and air inside a continuous flow reactor with application to lean premixed combustion, J Eng Gas Turb Power 130
19. Oehlschlaeger MA, Steinberg J, Westbrook CK, Pitz WJ (2009) The autoignition of iso-cetane at high to moderate temperatures and elevated pressures: shock tube experiments and kinetic modeling. Combust Flame 156:2165–2172
20. Wang WJ, Oehlschlaeger MA (2012) A shock tube study of methyl decanoate autoignition at elevated pressures. Combust Flame 159:476–481
21. Mastorakos E (2009) Ignition of turbulent non-premixed flames. Prog Energy Combust 35:57–97
22. Sreedhara S, Lakshmisha KN (2000) Direct numerical simulation of autoignition in a non-premixed, turbulent medium. Proc Combust Inst 28:25–33

23. Fotache CG, Kreutz TG, Zhu DL, Law CK (1995) An experimental study of ignition in nonpremixed counterflowing hydrogen versus heated air. Combust Sci Technol 109:373–393
24. Fotache CG, Kreutz TG, Law CK (1997) Ignition of hydrogen-enriched methane by heated air. Combust Flame 110:429–440
25. Zheng XL, Lu TF, Law CK, Westbrook CK, Curran HJ (2005) Experimental and computational study of nonpremixed ignition of dimethyl ether in counterflow. Proc Combust Inst 30:1101–1109
26. Sepman A, Abtahizadeh E, Mokhov A, van Oijen J, Levinsky H, de Goey P (2013) Experimental and numerical studies of the effects of hydrogen addition on the structure of a laminar methane-nitrogen jet in hot coflow under MILD conditions. Int J Hydrog Energy 38:13802–13811
27. Sepman AV, Abtahizadeh SE, Mokhov AV, van Oijen JA, Levinsky HB, de Goey LPH (2013) Numerical and experimental studies of the NO formation in laminar coflow diffusion flames on their transition to MILD combustion regime. Combust Flame 160:1364–1372
28. Cabra R, Chen JY, Dibble RW, Karpetis AN, Barlow RS (2005) Lifted methane-air jet flames in a vitiated coflow. Combust Flame 143:491–506
29. Markides CN, Mastorakos E (2011) Experimental investigation of the effects of turbulence and mixing on autoignition chemistry. Flow Turbul Combust 86:585–608
30. Roubaud A, Lemaire O, Minetti R, Sochet LR (2000) High pressure auto-ignition and oxidation mechanisms of o-xylene, o-ethyltoluene, and n-butylbenzene between 600 and 900 K. Combust Flame 123:561–571
31. Johannessen B, North A, Dibble R, Lovas T (2015) Experimental studies of autoignition events in unsteady hydrogen-air flames. Combust Flame 162:3210–3219
32. Papageorge MJ, Arndt C, Fuest F, Meier W, Sutton JA (2014) High-speed mixture fraction and temperature imaging of pulsed, turbulent fuel jets auto-igniting in high-temperature, vitiated co-flows, Exp Fluids 55
33. Fast G, Kuhn D, Class AG, Maas U (2009) Auto-ignition during instationary jet evolution of dimethyl ether (DME) in a high-pressure atmosphere. Combust Flame 156:200–213
34. Oldenhof E, Tummers MJ, van Veen EH, Roekaerts DJEM (2012) Transient response of the Delft jet-in-hot coflow flames. Combust Flame 159:697–706
35. Z. Chen, M. Konno, M. Oguma, T. Yanai, Experimental Study of CI Natural-Gas/DME Homogeneous Charge Engine, SAE International, 2000
36. Venkatesan M, Moorthi NSV, Karthikeyan R, Manivannan A (2014) Experimental study on hydrous methanol fuelled hcci engine using air pre-heater assisted controlled autoignition. Trans Famena 38:53–66
37. Maurya RK, Agarwal AK (2014) Experimental investigations of performance, combustion and emission characteristics of ethanol and methanol fueled HCCI engine. Fuel Process Technol 126:30–48
38. Pope SB (1985) Pdf methods for turbulent reactive flows. Prog Energ Combust 11:119–192
39. Peters N (1984) Laminar diffusion flamelet models in non-premixed turbulent combustion. Prog Energy Combust 10:319–339
40. Veynante D, Vervisch L (2002) Turbulent combustion modeling. Prog Energy Combust 28:193–266
41. Peters N (1999) The turbulent burning velocity for large-scale and small-scale turbulence. J Fluid Mech 384:107–132
42. Subramaniam S, Pope SB (1998) A mixing model for turbulent reactive flows based on Euclidean minimum spanning trees. Combust Flame 115:487–514
43. Subramaniam S, Pope SB (1999) Comparison of mixing model performance for nonpremixed turbulent reactive flow. Combust Flame 117:732–754
44. Bilger RW (2000) Future progress in turbulent combustion research. Prog Energy Combust 26:367–380
45. Klimenko AY, Bilger RW (1999) Conditional moment closure for turbulent combustion. Prog Energy Combust 25:595–687

46. Sreedhara S, Lakshmisha KN (2002) Assessment of conditional moment closure models of turbulent autoignition using DNS data. Proc Combust Inst 29:2069–2077
47. Kronenburg A (2004) Double conditioning of reactive scalar transport equations in turbulent nonpremixed flames. Phys Fluids 16:2640–2648
48. Sreedhara S, Lakshmisha KN (2002) Autoignition in a non-premixed medium: DNS studies on the effects of three-dimensional turbulence. Proc Combust Inst 29:2051–2059
49. Mastorakos E, Baritaud TA, Poinsot TJ (1997) Numerical simulations of autoignition in turbulent mixing flows. Combust Flame 109:198–223
50. Bansal G, Mascarenhas A, Chen JH (2015) Direct numerical simulations of autoignition in stratified dimethyl-ether (DME)/air turbulent mixtures. Combust Flame 162:688–702
51. Jones WP, Navarro-Martinez S (2007) Large eddy simulation of autoignition with a subgrid probability density function method. Combust Flame 150:170–187
52. Domingo P, Vervisch L, Veynante D (2008) Large-eddy simulation of a lifted methane jet flame in a vitiated coflow. Combust Flame 152:415–432
53. Han W, Raman V, Chen Z (2016) LES/PDF modeling of autoignition in a lifted turbulent flame: analysis of flame sensitivity to differential diffusion and scalar mixing time-scale. Combust Flame 171:69–86
54. Schulz O, Jaravel T, Poinsot T, Cuenot B, Noiray N (2016) A criterion to distinguish autoignition and propagation applied to a lifted methane–air jet flame, P Combust Inst, doi:10.1016/j.proci.2016.08.022
55. Deng SL, Zhao P, Mueller ME, Law CK (2015) Autoignition-affected stabilization of laminar nonpremixed DME/air coflow flames. Combust Flame 162:3437–3445
56. Deng SL, Zhao P, Mueller ME, Law CK (2015) Stabilization of laminar nonpremixed DME/air coflow flames at elevated temperatures and pressures. Combust Flame 162:4471–4478
57. Patwardhan SS, Lakshmisha KN (2008) Autoignition of turbulent hydrogen jet in a coflow of heated air. Int J Hydrog Energy 33:7265–7273
58. Wu ZJ, Zhang Q, Bao TT, Li LG, Deng J, Hu ZJ (2016) Experimental and numerical study on ethanol and dimethyl ether lifted flames in a hot vitiated co-flow. Fuel 184:620–628
59. Kong SC (2007) A study of natural gas/DME combustion in HCCI engines using CFD with detailed chemical kinetics. Fuel 86:1483–1489

# Langevin Simulation of Turbulent Combustion

**A.G. Nouri, S. Sammak, P.H. Pisciuneri and P. Givi**

**Abstract** A review is presented of modern developments in Langevin simulation of turbulent reacting flows. The filtered density function (FDF) methodology is the subject of main focus as it has proven very effective for stochastic simulations in turbulent combustion. The most recent contributions in mathematical modeling of FDF, its numerical solution, and its optimization for large scale simulations are reviewed.

## 1 Introduction

Combustion has propelled the modern society into the twenty-first century, but not without its negative impact on air quality and climate change. Even though significant effort is underway to develop alternative and sustainable resources of energy, combustion is likely to keep playing a major role for decades to come, in both developed and developing parts of the world. Concerns of air pollution, climate change, and dwindling petroleum resources have kept combustion research a high priority. In most cases, combustion is accompanied by turbulence, which provides the means of enhanced fuel-air mixing. The physical and chemical processes involved in turbulent reactive flows are extremely complex due to the intricacies of the interactions between chemistry and hydrodynamics.

Modern combustion systems require investigation of physio-chemical processes for a wide range of scales to understand their impact on overall system behavior. Access to the multi-scale and multi-physics phenomena is more easily and

A.G. Nouri (✉) · S. Sammak · P.H. Pisciuneri · P. Givi
Department of Mechanical Engineering and Materials Science, University of Pittsburgh, Pittsburgh, PA, USA
e-mail: arn36@pitt.edu

S. Sammak
e-mail: shs159@pitt.edu

P.H. Pisciuneri
e-mail: php8@pitt.edu

P. Givi
e-mail: pgivi@pitt.edu



A.K. Agarwal et al. (eds.), *Combustion for Power Generation and Transportation*, DOI 10.1007/978-981-10-3785-6_3

conveniently obtained via computer simulations. Such simulations, now conducted at very large scales, are taking an ever greater role in scientific discovery. This is enabled by high performance computing (HPC) environments and novel numerical algorithms designed for massively parallel computations. However, in order to utilize computer modeling as a reliable tool for simulations and development of combustion systems, it is necessary to obtain quantitatively accurate predictions, not just qualitative trends. Predictive simulations of highly transient turbulent combustion phenomena such as cycle-to-cycle variations in internal combustion engines, flame kernel development in the turbulent environment of a spark-ignited engine, instabilities in a gas-turbine combustor, and others cannot be performed using crude models.

Recent progress in HPC has resulted in significant progress for turbulence simulation [1]. Specifically, large eddy simulation (LES) is gradually replacing the traditional Reynolds averaged Navier-Stokes (RANS) methods. In its explicit implementations, the LES physical variables are passed through a (low-pass) filter with the goal to predict their transport at the large scales. Langevin simulations have proven effective in modeling the effects of the small scales. A very effective way of conducting such simulations is via the filtered density function (FDF). In this approach the effects of small scales are modeled in a probabilistic manner.

In this chapter, we present some of the most recent developments in Langevin simulations of turbulent reacting flows. After a review of the most recent FDF closures, we show that merging of the *discontinuous Galerkin* (DG) flow solver with the Monte Carlo (MC) simulation of the FDF is very effective for LES. The former allows LES with *both h and p*-refinements. The latter provides a convenient means of solving the FDF. The numerics and the physical model are bundled together in a computational package which allocates the processors in an *adaptive* manner. This facilitates massively parallel large scale simulations up to petascale, and towards future exascale. This new generation of LES tool facilitates predictions of complex turbulent flames.

## 2 Langevin Modeling

Langevin modeling goes back to the early work of Albert Einstein [2] (and in different forms by others prior to that). It provides a very powerful tool for the "stochastic" description of physical phenomena. Its generality is the reason for its wide application to many areas of science, in addition to other fields such as economics, weather prediction, Wall Street, etc. Stochastic modeling has also been very popular for turbulence description, going back to old school teachings of the subject [3, 4]. In fact, some of the powerful theories in stochastic processes have been developed for description of "random fields" with a specific focus on turbulence [5]. At times, the validity of the approach has been questioned for Navier-Stokes fluid dynamics. This has made the case for direct treatment of turbulence; e.g. direct numerical simulation (DNS) or other model free methods [6]. But the fact remains it will take a while before DNS can be used for prediction of practical flows, especially in the case of reacting and/or compressible flows [7].

The Langevin description of turbulence usually has been in the form of probability density function (PDF) modeling (although it certainly does not have to be in that way) [8, 9]. This is simply because the Fokker-Planck equation associated with the Langevin description is a (transitional) PDF evolution equation, albeit a deterministic one. The underlying idea with the Langevin description is to express everything we know and do not know via a *model*. A significant amount of science and art goes into the construction of this model. The hope is that with careful assessments of all of the sub-elements, the corresponding Fokker-Planck equation will provide a good closure for the "physical phenomena." The use of RANS/PDF methods in the U.S. was pioneered by Lundgren [10, 11], and has experienced significant growth over the past 50 years [8, 9, 12]. Equivalent to PDF in RANS is the FDF methods in LES [6, 13–17]. In this case, the Langevin model is to mimic the physics of the cut-off subgrid scale (SGS) field. LES via FDF has also proven very effective; particularly for prediction of reacting flows [18–22].

A large majority of LES methodologies are via closures for the SGS correlations. These are *second-order* moments which appear in the filtered transport equations of the first order moments [23, 24]. With the Langevin descriptor, we consider the *entire* stochastic variations of the transport variable, not just the first few moments. All of the SGS moments will be by-products of the Langevin model.

In the mathematical description of compressible turbulent flows involving $N_s$ species, the primary transport variables are the density $\rho(\mathbf{x}, t)$, the velocity vector $u_i(\mathbf{x}, t)$ $(i = 1, 2, 3)$, the pressure $p(\mathbf{x}, t)$, the temperature $T(\mathbf{x}, t)$, the internal energy $e(\mathbf{x}, t)$, and the species mass fractions $\phi_\alpha(\mathbf{x}, t)$ $(\alpha = 1, \ldots, N_s)$. We use the perfect gas assumption for Newtonian fluid with the viscous stress tensor $\tau_{ij}$, the heat flux $q_j$, and the species diffusive mass flux vector $J_j^\alpha$. Implementation of LES involves the use of the spatial filtering operation [23, 25] $\langle Q(\mathbf{x}, t)\rangle = \int_{-\infty}^{+\infty} Q(\mathbf{x}', t)\mathcal{H}(\mathbf{x}', \mathbf{x})d\mathbf{x}'$, where $\mathcal{H}$ denotes the filter function of width $\Delta_{\mathcal{H}}$, and $\langle Q(\mathbf{x}, t)\rangle$ represents the filtered value of the transport variable $Q(\mathbf{x}, t)$. In compressible and/or reacting flows, it is convenient to consider the Favré filtered quantity, $\langle Q(\mathbf{x}, t)\rangle_L = \langle \rho Q\rangle / \langle \rho \rangle$.

The SGS Langevin descriptor is best described in terms of modeled stochastic differential equations (SDEs) [26, 27] for each of the transport variables. The starting point is Albert Einstein's random walk equation:

$$dx_i^+ = u_i^+ dt + \sqrt{\frac{2\mu}{\langle \rho \rangle}} dW_i, \tag{1}$$

where $x_i^+$, $u_i^+$ are probabilistic representations of position, and velocity vectors, respectively. $\mu$ denotes the fluid dynamic viscosity, and $W$ denotes the Wiener-Lévy processes in the physical space.

The most sophisticated forms of the FDF to-date are the joint pressure-energy-velocity-scalar filtered mass density function (PEVS-FMDF) [28], the joint frequency-velocity-scalar filtered mass density function (FVS-FMDF) [29], and

VS-FMDF which does not include the SGS frequency [30–32]. Hydrodynamic closure in incompressible, non-reacting flows has been achieved via the marginal velocity-FDF (V-FDF) [33], and the one which considers only the species mass fraction field is the scalar-FDF (S-FDF and S-FMDF). This is the most elementary form of FDF when it was first introduced, [6, 13, 14, 16, 17] and has experienced widespread usage, especially in combustion [28, 30, 34–60]. See Refs. [18–22, 61–63] for reviews.

The most sophisticated form of the FDF is in a "self-contained" manner. That is when the FDF accounts for SGS modeling of all of the pertinent transport variables. The central part of the formulation is the "pressure" term, which provides the coupling between hydrodynamics and thermochemistry. This term is coupled with the internal energy, the fluid velocity and the species mass fractions. Consistent with established terminology, the resulting self-contained model is termed PEVS-FMDF. This is denoted by $P_L$ and is formally defined as [13]:

$$P_L(\boldsymbol{v},\boldsymbol{\psi},\theta,\eta,\mathbf{x};t) = \int_{-\infty}^{+\infty} \rho(\mathbf{x}',t)\zeta\Big(\boldsymbol{v},\boldsymbol{\psi},\theta,\eta;\boldsymbol{u}(\mathbf{x}',t),\boldsymbol{\phi}(\mathbf{x}',t), e(\mathbf{x}',t),p(\mathbf{x}',t)\Big)G(\mathbf{x}'-\mathbf{x})d\mathbf{x}', \quad (2)$$

where

$$\zeta\Big(\boldsymbol{v},\boldsymbol{\psi},\theta,\eta;\boldsymbol{u}(\mathbf{x},t),\boldsymbol{\phi}(\mathbf{x},t),e(\mathbf{x},t),p(\mathbf{x},t)\Big) = \left(\prod_{i=1}^{3}\delta\left(v_i - u_i(\mathbf{x},t)\right)\right) \times \left(\prod_{\alpha=1}^{\sigma=N_s}\delta\left(\psi_\alpha - \phi_\alpha(\mathbf{x},t)\right)\right)\times\delta\left(\theta - e(\mathbf{x},t)\right)\times\delta\left(\eta - p(\mathbf{x},t)\right). \quad (3)$$

Here $\delta$ denotes the Dirac delta function, and $\boldsymbol{v}$, $\boldsymbol{\psi}$, $\theta$ and $\eta$ are the velocity vector, the scalar array, the internal energy and pressure in the sample space. The term $\zeta$ is the "fine-grained" density [64, 65]. Equation (2) defines the PEVS-FMDF as the spatially filtered value of the fine-grained density. With the condition of a positive filter kernel [66], $P_L$ has all of the properties of a mass density function [9]. This function is governed by:

$$\begin{aligned}\frac{\partial P_L}{\partial t} + \frac{\partial v_j P_L}{\partial x_j} &= \frac{\partial}{\partial v_i}\left(\left\langle \frac{1}{\rho}\frac{\partial p}{\partial x_i}\middle| \boldsymbol{v},\boldsymbol{\psi},\theta,\eta\right\rangle_\ell P_L\right) - \frac{\partial}{\partial v_i}\left(\left\langle \frac{1}{\rho}\frac{\partial \tau_{ij}}{\partial x_j}\middle| \boldsymbol{v},\boldsymbol{\psi},\theta,\eta\right\rangle_\ell P_L\right)\\ &+ \frac{\partial}{\partial \psi_\alpha}\left(\left\langle \frac{1}{\rho}\frac{\partial J_j^\alpha}{\partial x_j}\middle| \boldsymbol{v},\boldsymbol{\psi},\theta,\eta\right\rangle_\ell P_L\right) + \frac{\partial}{\partial \theta}\left(\left\langle \frac{1}{\rho}\frac{\partial q_i}{\partial x_i}\middle| \boldsymbol{v},\boldsymbol{\psi},\theta,\eta\right\rangle_\ell P_L\right)\\ &- \frac{\partial}{\partial \theta}\left(\left\langle \frac{1}{\rho}\tau_{ij}\frac{\partial u_i}{\partial x_j}\middle| \boldsymbol{v},\boldsymbol{\psi},\theta,\eta\right\rangle_\ell P_L\right) + \frac{\partial}{\partial \theta}\left(\left\langle \frac{1}{\rho}p\frac{\partial u_j}{\partial x_j}\middle| \boldsymbol{v},\boldsymbol{\psi},\theta,\eta\right\rangle_\ell P_L\right)\end{aligned}$$

$$+(\gamma-1)\frac{\partial}{\partial\eta}\left(\left\langle\frac{\partial q_i}{\partial x_i}\middle|\boldsymbol{v},\boldsymbol{\psi},\theta,\eta\right\rangle_\ell P_L\right)+\gamma\frac{\partial}{\partial\eta}\left(\left\langle p\frac{\partial u_j}{\partial x_j}\middle|\boldsymbol{v},\boldsymbol{\psi},\theta,\eta\right\rangle_\ell P_L\right)$$
$$-(\gamma-1)\frac{\partial}{\partial\eta}\left(\left\langle\tau_{ij}\frac{\partial u_i}{\partial x_j}\middle|\boldsymbol{v},\boldsymbol{\psi},\theta,\eta\right\rangle_\ell P_L\right),\tag{4}$$

where $\gamma$ is the ratio of specific heats. This is the exact PEVS-FMDF transport equation in which the effect of convection, the second term on the left-hand side, appears in a closed form. The conditional terms at the right-hand side are unclosed. These terms are modeled by considering the general diffusion process, given by a system of stochastic differential equations (SDEs), and implementing the simplified Langevin model (SLM) and the linear mean-square estimation (LMSE) [67]. The internal energy can be modeled according to the first law of thermodynamics:

$$\mathrm{d}E^+=\left(-C_e\omega\left(E^+-\langle e\rangle_L\right)+\frac{\epsilon}{\rho^+}\right)\mathrm{d}t-P^+\mathrm{d}\xi^+,\tag{5}$$

where $C_e$ is the model constant, and $\epsilon$ and $\omega$ are the dissipation rate and the SGS mixing frequency, respectively. In Eq. (5), $\xi^+$ refers to the specific volume of the stochastic particle, related to $E^+$ and $P^+$ through the equation of state. The model for pressure is in the general form:

$$\mathrm{d}P^+=P^+\left(A\mathrm{d}t+B\mathrm{d}W_p\right).\tag{6}$$

The coefficients $A$ and $B$ must be specified so that the modeled transport equation of energy becomes identical to the exact equation. These equations are augmented by model SDEs for position, velocities, and scalars [16, 32, 33, 68, 69]. The Fokker-Planck equation corresponding to these SDEs [70] determine the FDF:

$$\frac{\partial P_L}{\partial t}+\frac{\partial v_iP_L}{\partial x_i}=\frac{1}{\langle\rho\rangle_\ell}\frac{\partial\langle p\rangle_\ell}{\partial x_i}\frac{\partial P_L}{\partial v_i}-\frac{2}{\langle\rho\rangle_\ell}\frac{\partial}{\partial x_j}\left(\mu\frac{\partial\langle u_i\rangle_L}{\partial x_j}\right)\frac{\partial P_L}{\partial v_i}$$
$$-\frac{1}{\langle\rho\rangle_\ell}\frac{\partial}{\partial x_j}\left(\mu\frac{\partial\langle u_j\rangle_L}{\partial x_i}\right)\frac{\partial P_L}{\partial v_i}+\frac{2}{3}\frac{1}{\langle\rho\rangle_\ell}\frac{\partial}{\partial x_i}\left(\mu\frac{\partial\langle u_j\rangle_L}{\partial x_j}\right)\frac{\partial P_L}{\partial v_i}$$
$$-\frac{\partial\left(G_{ij}\left(v_j-\langle u_j\rangle_L\right)P_L\right)}{\partial v_i}+\frac{\partial}{\partial x_i}\left(\mu\frac{\partial(P_L/\langle\rho\rangle_\ell)}{\partial x_i}\right)$$
$$+\frac{\partial}{\partial x_i}\left(\frac{2\mu}{\langle\rho\rangle_\ell}\frac{\partial\langle u_j\rangle_L}{\partial x_i}\frac{\partial P_L}{\partial v_j}\right)+\frac{\mu}{\langle\rho\rangle_\ell}\frac{\partial\langle u_k\rangle_L}{\partial x_j}\frac{\partial\langle u_i\rangle_L}{\partial x_j}\frac{\partial^2P_L}{\partial v_k\partial v_i}$$
$$+\frac{1}{2}C_0\frac{\epsilon}{\langle\rho\rangle_\ell}\frac{\partial^2P_L}{\partial v_i\partial v_i}+C_\phi\omega\frac{\partial\left(\left(\psi_\alpha-\langle\phi_\alpha\rangle_L\right)P_L\right)}{\partial\psi_\alpha}$$
$$+\frac{C_e\omega}{\gamma}\frac{\partial\left(\left(\theta-\langle e\rangle_L\right)P_L\right)}{\partial\theta}-\frac{\gamma-1}{\gamma}(\epsilon)\frac{\partial}{\partial\theta}\left(\frac{\theta}{\eta}P_L\right)-\frac{\gamma-1}{\gamma}\frac{\partial\left(\theta AP_L\right)}{\partial\theta}$$

$$+\frac{\gamma-1}{\gamma^2}\frac{\partial\left(\theta B^2 P_L\right)}{\partial\theta}-\frac{\partial\left(\eta A P_L\right)}{\partial\eta}+\frac{1}{2}\frac{(\gamma-1)^2}{\gamma^2}\frac{\partial^2\left(\theta^2 B^2 P_L\right)}{\partial\theta\partial\theta}$$
$$+\frac{\gamma-1}{\gamma}\frac{\partial^2\left(\theta\eta B^2 P_L\right)}{\partial\theta\partial\eta}+\frac{1}{2}\frac{\partial^2\left(\eta^2 B^2 P_L\right)}{\partial\eta\partial\eta}, \tag{7}$$

where

$$G_{ij}=\left(\frac{\Pi_d}{2k\left\langle\rho\right\rangle_\ell}-\omega\left(\frac{1}{2}+\frac{3}{4}C_0\right)\right)\delta_{ij}. \tag{8}$$

The parameters $C_0$ and $C_\phi$ are model constants and need to be specified [31, 71], $\Pi_d$ is the pressure dilatation, and $k$ is the SGS kinetic energy. The transport equations for the SGS moments are obtained by integration of Eq. (7) over the sample space domain.

## 3 Numerical Solution of the Langevin Equation

It is now widely established that an effective means of solving the Langevin equation is via Lagrangian Monte Carlo methods [72, 73]. In these methods, the stochastic equations are represented by an ensemble of *particles* which describe the physical field. For stable fluid dynamics simulations, the MC solver must be coupled with an Eulerian base flow solver; e.g. finite difference (FD), finite volume (FV), finite element (FE), spectral, etc. The coupling must be done in such a way that the overall accuracy of the solver is maintained. For LES, it is desired that the influence of the SGS quantities decrease with the increase of the resolution, and/or the order of accuracy of the discretization procedure. It is also desired that the LES predictions are *independent* of the grid size used in the base flow solver.

The most recent development in stochastic FDF simulation is via the use of the discontinuous Galerkin (DG) methodology [60, 74–76]. This methodology combines the versatility of FV discretization with the accuracy of spectral approximations and is particularly suitable for coupling with the MC simulator. The DG flow solver provides high-order approximations of the filtered transport equations with the flexibility to deal with complex geometries. To implement this solver, the domain is discretized into a number of structured or unstructured elements. Each of these elements are mapped from the physical space to the computational space ($\mathbf{x} \to \chi$). Within this space, the transport variables are represented via polynomial approximations of order $p$.

The Langevin model is represented by an ensemble of MC particles, each carrying the physical information about the SGS quantities. This information is updated via temporal integration of the modeled SDEs. By doing so, the position and the composition of the MC particles are updated. The DG-MC simulator exhibits several attractive features. A significant advantage is that DG allows convergence to the DNS limit via *both h* and *p*-refinements. Based on the close to exponential

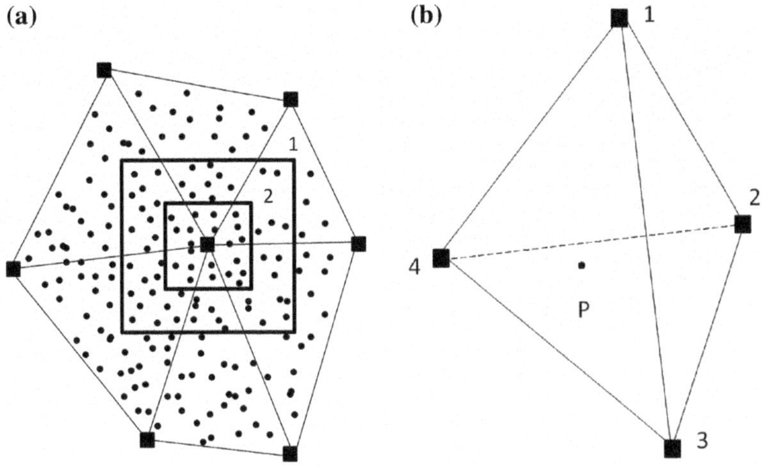


**Fig. 1** (**a**) Ensemble averaging on an unstructured mesh. (*1*) hexahedral with $\Delta_E = \Delta$. (*2*) hexahedral with $\Delta_E = \Delta/2$. (**b**) Schematic distribution of an MC particle within a tetrahedral cell. A *solid circle* denotes an MC particle, and the *solid squares* denote the quadrature points. Reprinted from [60] with permission of authors

convergence of the latter, the procedure is much more efficient than the conventional approach of refining the grid (reducing $h$). Another advantage is that the DG variables can readily be evaluated at the MC particle's locations since these variables are represented by polynomials within each element. Hence, there is no loss of accuracy due to the use of a lower order interpolation method as is typically used in conventional approximations.

The MC statistics are constructed by consideration of an ensemble of $N_p$ particles within a hexahedral domain of size $\Delta_E$. This is illustrated in Fig. 1 for a domain discretized by unstructured tetrahedral elements. For reliable statistics with minimal numerical dispersion, it is desired to minimize the size of the ensemble domain and maximize the number of the MC particles $N_p$.

Because the high-order DG approach involves multiple degrees of freedom within each mesh element, the coupling of the DG and MC solvers requires careful attention regarding the matching of the Langevin and DG local cell/element resolutions. This is accomplished by subdividing the DG elements into several smaller sampling cells that will be used to calculate local averages over the ensemble of MC particles that reside in each sampling cell. The equally spaced sampling cell averages will then be used to reconstruct the high-order polynomial distribution of these same quantities within each cell, as required by the DG discretization. Although initially this interface will be implemented for static discretization, the interface will be subsequently extended to support dynamic $h$–$p$ adaptive strategies for optimal accuracy.

Similar to that observed in other hybrid approaches, with the coupled DG-MC algorithm, several of the transport variables are calculated repeatedly [20]. This redundancy is very important to establish consistency. This is demonstrated in the

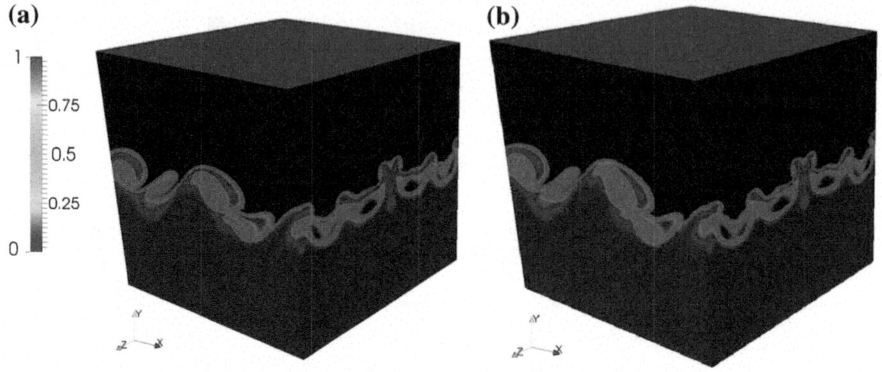


**Fig. 2** Contour plots of the filtered scalar field via DG-MC FDF simulations. (**a**) FDF, (**b**) DG. Reprinted from [60] with permission of authors

sample LES of a passive scalar in a temporally developing mixing layer [60]. Figure 2 shows the instantaneous contour plots of a filtered scalar field as computed via both DG and MC, and it provides a visual demonstration of the consistency of the Langevin simulation.

## 4 Langevin Simulation

It is sometimes argued that Langevin (or other MC) solvers are overly expensive as compared to other (conventional) methods. There could be some validity to this argument, but only if the computations are not mapped to the computational resources appropriately. With parallel algorithms designed to exploit petascale supercomputers of today and the exascale supercomputers of tomorrow, it is important to take advantage of modern computer science in any large scale simulations [1, 77].

Modern LES typically requires order of several billions of elements with an order of magnitude more MC particles. In reacting flows, when finite-rate kinetics are to be captured, the computational requirements become *monstrous*. To deal with such problems, the traditional HPC-CFD is typically via a temporally invariant uniform decomposition where the mesh is partitioned during the preprocessing phase as (more or less) equal sized partitions. For unsteady and inhomogeneous flows, this would most likely lead to *a poor load distribution.* In such cases, the computational requirements vary greatly depending on the flow composition.

The load imbalance problem can be effectively resolved by portioning the domain *adaptively* using a weighted graph partitioner [78]. Using the elapsed wall time to perform the integrations as the weights results in highly irregular (in terms of shape and size), but load-balanced, partitions. This approach is demonstrated via a very straightforward chemical kinetics simulation in an inhomogeneous field [54].

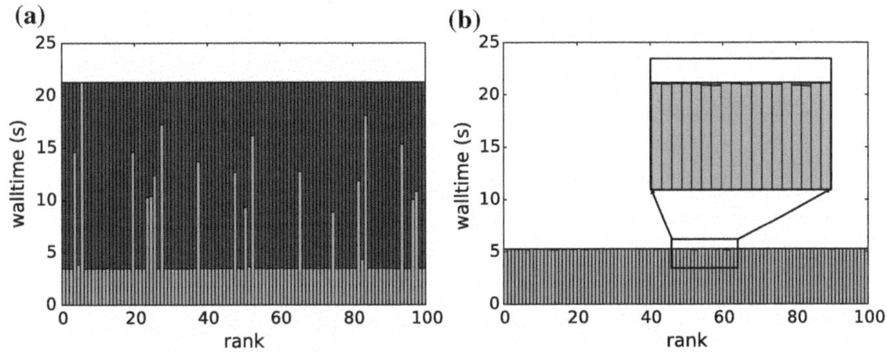


**Fig. 3** CPU times for a given time step for each rank. *Gray bars* denote time spent on computation. *Blue bars* correspond to idle time. (**a**) Uniform allocation of the processors. (**b**) Adaptive allocation of the processors

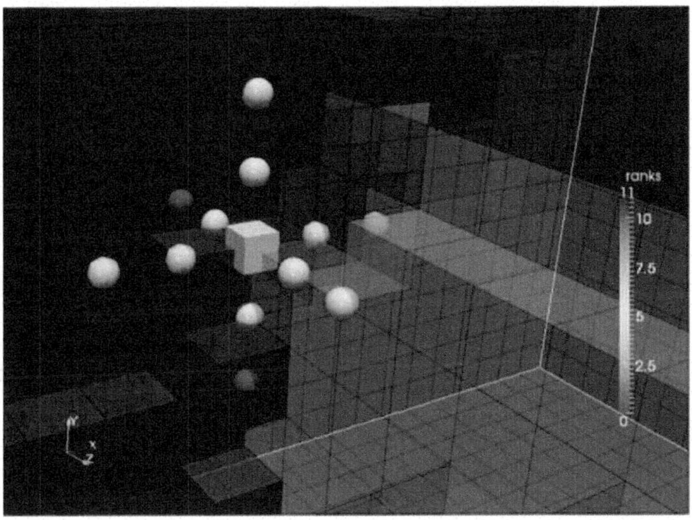


**Fig. 4** A typical finite difference stencil situated at the intersection of various partitions (ranks)

In conventional methods, the idle time is as much as half of the total time (Fig. 3a)! With the adaptive allocation, the idle time is virtually insignificant (Fig. 3b).

Figure 4 demonstrates that due to the irregularity of the domain decomposition, there is no *a priori* way of determining the number of, or boundaries of neighboring partitions. In this regard, the irregular and adaptive approach is already very suitable for unstructured solvers. It also facilitates highly parallel simulations as shown via sample results in Fig. 5.

Some sample results are shown of the Langevin FDF simulations in Figs. 6 and 7. Figure 6 [53] shows the instantaneous contour of *RT*, where *R* is the difference in specific heats, in LES of a premixed flame [79] and the underlying domain

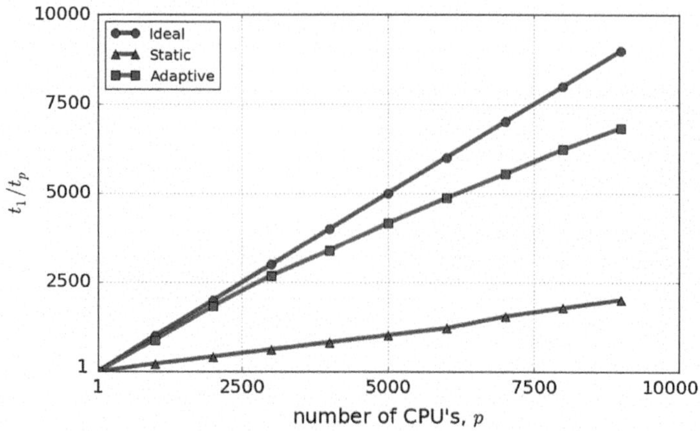


**Fig. 5** Strong scaling results for an adaptively load balanced and statically balanced simulation of chemical kinetics in an inhomogeneous field. Replotted from [53] with permission of authors

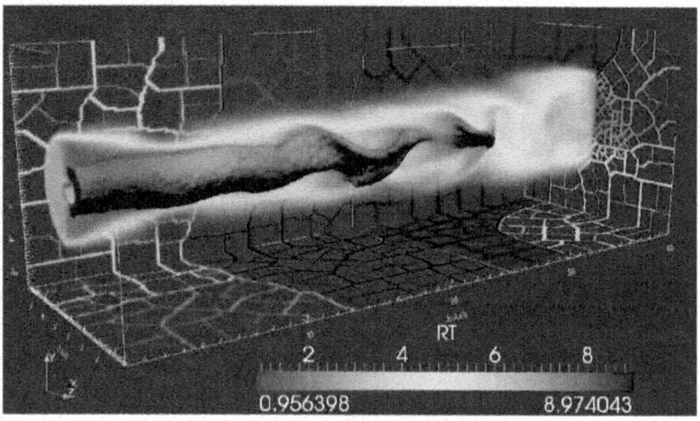


**Fig. 6** Instantaneous contour of the energy, *RT*, and the underlying domain decomposition outlined along the domain boundaries. This single frame is reprinted from the video in the supplementary material of [53] with permission of authors

decomposition (outlined regions along the domain boundaries). Figure 7 [58] compares the scatter plots of the temperature versus the mixture fraction at several transverse planes of a partially premixed flame [80]. In both of these simulations, and several others not shown here, the good agreement with the experimental data indicates that the compositional structure of the flame is very nicely captured by the Langevin simulations.

In order to make efficient and *complete* usage of existing petascale and future exascale systems, additional enhancements to the parallel flow solver should be considered. One such consideration is overdecomposition. This is the concept of decomposing the problem into more partitions than computational cores. This allows for computation and communication to overlap, reducing some of the idle time associated with communication [81]. Another consideration is to combine distributed and shared memory approaches to parallelism. Typically MPI [82, 83], used at the distributed (internode) level, is combined with OpenMP [84], used at the intranode level. This approach abstracts nicely to the actual architecture of modern

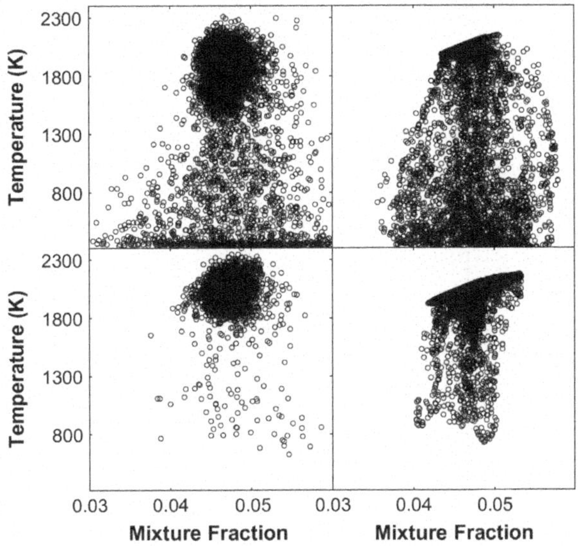


**Fig. 7** Scatter plot of the temperature versus the mixture fraction at $x = 10$ mm (*top*) and $x = 40$ mm (*bottom*). Experimental measurements are on *left* and LES/FDF results are on the *right*. Reprinted from [58] with permission of authors

supercomputers and will more efficiently utilize memory on each node compared to a flat MPI approach. An additional consideration is the utilization of accelerators attached to each node, such as graphics processing units (GPUs) and coprocessors (i.e. Intel Phis). Accelerators are integral components of existing supercomputers and will remain critical components to future supercomputers due to power constraints. Software such as CUDA [85] and OpenACC [86] can be used to port existing code for use on accelerators. A final consideration is the efficient handling of adapting an existing domain decomposition while minimizing data migration in order to keep the computational load balanced among partitions during the duration of a simulation. The issue of efficient data migration required for such dynamic load balancing applications has been investigated by many; e.g. see Ref. [87]. Softwares like Zoltan [88] and Charm++ [89] are tools dedicated to load balancing, repartitioning, data migration and even overdecomposition.

## 5 Conclusions

The capability of Langevin simulation via FDF is now widely recognized within the turbulent combustion community [19, 21, 62]. This is not surprising as such simulations account for the effects of SGS chemical reactions in an exact manner. Future utilization of the methodology will depend on the rate of progress in its numerical and computational implementations. We feel that the DG discretization coupled with the irregular and adaptive portioning as described here would make future FDF simulations routine for engineering predictions.

The superior numerical efficiency and scalability of the DG method can be leveraged by using large polynomial approximation orders $p \sim 4 - 6$ and/or higher. Such large polynomial orders result in very low numerical dissipation [90] which is very desired (in fact essential) in LES. A major advantage of performing LES in this way is that the spectral content of the solution is controlled by the filter and thus the performance of the SGS models is improved since the smallest scales are not polluted by numerical errors [91].

With irregular portioning of the partitions to ensure a reasonable load balance, stochastic LES/FDF can be applied to complex flow geometries with Reynolds/Damköhler numbers that are currently intractable with DNS. As an example, LES of a small size industrial combustor with $10^{10}$ elements and $\mathcal{O}(10^{11})$ MC particles requires of order of several days on $10^4$ processors. DNS of this flame would take order of months on the world's largest supercomputers, if dedicated to such simulations.

**Acknowledgements** This work is sponsored by AFOSR under Grant FA9550-12-1-0057, and by NSF under Grant CBET-1609120 and Grant CBET-1603131. Computational resources are provided by the University of Pittsburgh Center for Research Computing.

## References

1. Slotnick J, Khodadoust A, Alonso J, Darmofal D, Gropp W, Lurie E, Mavriplis D (2013) CFD Vision 2030 study: A path to revolutionary computational aerosciences. Tech. rep., NASA Langley Research Center, Hampton, VA. NASA/CR-2014-218178
2. Einstein A (1956) Investigations on the theory of the brownian movement, 2nd edn. Dover Publications
3. Monin AS, Yaglom AM (1975) Statistical fluid mechanics, vol. 2. MIT Press
4. Lumley JL (1970) Stochastic tools in turbulence. Academic Press, New York
5. Panchev S (1970) Random functions and turbulence. Pergamon Press, Oxford, England
6. Givi P (1989) Prog Energy Combust 15(1):1. doi:10.1016/0360-1285(89)90006-3
7. Poinsot T, Veynante D (2011) Theoretical and numerical combustion, 3rd edn. R.T.Edwards Inc., Philadelphia
8. Heinz S (2003) Statistical mechanics of turbulent flows. Springer, New York
9. Pope SB (2000) Turbulent flows. Cambridge University Press, Cambridge
10. Lundgren TS (1967) Phys Fluids 10(5):969. doi:10.1063/1.1762249
11. Lundgren TS (1972) Statistical models and turbulence, Rosenblatt M, van Atta C (eds) Lecture Notes in Physics, vol. 12. Springer, pp 70–100. doi:10.1007/3-540-05716-1_5
12. Fox RO (2003) Computational models for turbulent reacting flows. Cambridge University Press, Cambridge
13. Pope SB (1990) Proc Combust Inst 23(1):591. doi:10.1016/S0082-0784(06)80307-3
14. Gao F, O'Brien EE (1993) Phys Fluids A 5(6):1282. doi:10.1063/1.858617
15. Madnia CK, Givi P (1993) Large eddy simulation of complex engineering and geophysical flows. In: Galperin B, Orszag SA (eds) Cambridge University Press, Cambridge, England, chap 15, pp 315–346
16. Colucci PJ, Jaberi FA, Givi P, Pope SB (1998) Phys Fluids 10(2):499. doi:10.1063/1.869537
17. Jaberi FA, Colucci PJ, James S, Givi P, Pope SB (1999) J Fluid Mech 401:85. doi:10.1017/S0022112099006643
18. Givi P (2006) AIAA J 44(1):16. doi:10.2514/1.15514

19. Haworth DC (2010) Prog Energy Combust 36(2):168. doi:10.1016/j.pecs.2009.09.003
20. Ansari N, Jaberi FA, Sheikhi MRH, Givi P (2011) Engineering applications of computational fluid dynamics. In: Maher ARS (ed), vol 1. International Energy and Environment Foundation, chap. 1, pp 1–22
21. Pope SB (2013) Proc Combust Inst 34(1):1. doi:10.1016/j.proci.2012.09.009
22. Foster JW, Miller RS (2015) 53rd AIAA aerospace sciences meeting. AIAA, Kissimmee, FL, pp 1–21. doi:10.2514/6.2015-1379, (AIAA-2015-1379)
23. Sagaut P (2010) Large eddy simulation for incompressible flows, 3rd edn. Springer, New York
24. Lesieur M (1990) Turbulence in fluids, fluid mechanics and its applications, 2nd edn. Kluwer Academic Publishers, Boston
25. Geurts BJ (2004) Elements of direct and large-eddy simulation. R. T. Edwards Inc, Philadelphia
26. Gikhman II, Skorokhod AV (1972) Stochastic differential equations. Springer, New York
27. Karlin S, Taylor HM (1981) A second course in stochastic processes. Academic Press, New York
28. Nouri AG, Givi P, Nik MB, Pope SB (2015) Bulletin of the American physical society, 68th annual meeting of the APS division of fluid dynamics, vol. 60. American Physical Society, Boston, MA, p 565
29. Sheikhi MRH, Givi P, Pope SB (2009) Phys Fluids 21(7):075102. doi:10.1063/1.3153907
30. Nik MB, Yilmaz SL, Givi P, Sheikhi MRH, Pope SB (2010) AIAA J 48(7):1513. doi:10.2514/1.50239
31. Sheikhi MRH, Givi P, Pope SB (2007) Phys Fluids 19(9):095106. doi:10.1063/1.2768953
32. Sheikhi MRH, Drozda TG, Givi P, Pope SB (2003) Phys Fluids 15(8):2321. doi:10.1063/1.1584678
33. Gicquel LYM, Givi P, Jaberi FA, Pope SB (2002) Phys Fluids 14(3):1196. doi:10.1063/1.1436496
34. Zhou XY, Pereira JCF (2000) Flow Turbul Combust 64(4):279. doi:10.1023/A:1026595626129
35. Heinz S (2003) Flow Turbul Combust 70(1–4):115. doi:10.1023/B:APPL.0000004933.17800.46
36. Raman V, Pitsch H, Fox RO (2005) Combust Flame 143(1–2):56. doi:10.1016/j.combustflame.2005.05.002
37. Raman V, Pitsch H (2005) Combust Flame 142(4):329. doi:10.1016/j.combustflame.2005.03.014
38. Van Vliet E, Derksen JJ, van den Akker HEA (2005) AIChE J 51(3):725. doi:10.1002/aic.10365
39. Carrara MD, DesJardin PE (2006) Int J Multiph Flow 32(3):365. doi:10.1016/j.ijmultiphaseflow.2005.11.003
40. Sammak S, Nouri AG, Ansari N, Givi P (2015) Mathematical modeling of technological processes, Danaev N, Shokin Y, Darkhan AZ. Springer International Publishing, chap. 13, pp 124–132. doi:10.1007/978-3-319-25058-8_13
41. Mustata R, Valiño L, Jiménez C, Jones W, Bondi S (2006) Combust Flame 145(1–2):88. doi:10.1016/j.combustflame.2005.12.002
42. Jones WP, Navarro-Martinez S, Röhl O (2007) Proc Combust Inst 31(2):1765. doi:10.1016/j.proci.2006.07.041
43. Drozda TG, Sheikhi MRH, Madnia CK, Givi P (2007) Flow Turbul Combust 78(1):35. doi:10.1007/s10494-006-9052-4
44. Jones WP, Navarro-Martinez S (2007) Combust Flame 150(3):170. doi:10.1016/j.combustflame.2007.04.003
45. James S, Zhu J, Anand MS (2007) Proc Combust Inst 31(2):1737. doi:10.1016/j.proci.2006.07.160
46. McDermott R, Pope SB (2007) J Comput Phys 226(1):947. doi:10.1016/j.jcp.2007.05.006
47. Raman V, Pitsch H (2007) Proc Combust Inst 31(2):1711. doi:10.1016/j.proci.2006.07.152
48. Afshari A, Jaberi F, Shih T (2008) AIAA J 46(7):1576. doi:10.2514/1.25467

49. Drozda TG, Wang G, Sankaran V, Mayo JR, Oefelein JC, Barlow RS (2008) Combust Flame 155(1–2):54. doi:10.1016/j.combustflame.2008.06.012
50. Yilmaz SL, Nik MB, Givi P, Strakey PA (2010) J Propuls Power 26(1):84. doi:10.2514/1.44600
51. Ansari N, Goldin GM, Sheikhi MRH, Givi P (2011) J Comput Phys 230(19):7132. doi:10.1016/j.jcp.2011.05.015
52. Yilmaz SL, Nik MB, Sheikhi MRH, Strakey PA, Givi P (2011) J Sci Comput 47(1):109. doi:10.1007/s10915-010-9424-8
53. Pisciuneri PH, Yilmaz SL, Strakey PA, Givi P (2013) SIAM, J Sci Comput 35(4):C438. doi:10.1137/130911512
54. Pisciuneri PH, Yilmaz SL, Strakey PA, Givi P (2015) Stochastic equations for complex systems: theoretical and computational topics. In: Heinz S, Bessaih H (eds) Mathematical engineering. Springer International Publishing, chap. 8, pp 175–192. doi:10.1007/978-3-319-18206-3_8
55. Nik MB, Yilmaz SL, Sheikhi MRH, Givi P (2010) Flow Turbul Combust 85(3–4):677. doi:10.1007/s10494-010-9272-5
56. Ansari N, Pisciuneri PH, Strakey PA, Givi P (2012) AIAA J 50(11):2476. doi:10.2514/1.J051671
57. Banaeizadeh A, Afshari A, Schock H, Jaberi F (2013) Int J Heat Mass Transf 60:781. doi:10.1016/j.ijheatmasstransfer.2012.12.065
58. Ansari N, Strakey PA, Goldin GM, Givi P (2015) Proc Combust Inst 35(2):1433. doi:10.1016/j.proci.2014.05.042
59. Esmaeili M, Afshari A, Jaberi FA (2015) Int J Heat Mass Transf 89:1239. doi:10.1016/j.ijheatmasstransfer.2015.05.055
60. Sammak S, Brazell MJ, Givi P, Mavriplis DJ (2016) Comput Fluids 140:158. doi:10.1016/j.compfluid.2016.09.003
61. Yilmaz SL, Ansari N, Pisciuneri PH, Nik MB, Otis CC, Givi P (2013) J Appl Fluid Mech 6(3):311
62. Kuo KK, Acharya R (2012) Fundamentals of turbulent and multiphase combustion. John Wiley and Sons Inc., Hoboken
63. Almeida T, Afshari A, Jaberi FA (2005) Advances in combustion and noise control, Roy GD, Yu KH, Whitelaw JH, Witton JJ (eds) Cranfield University Press, chap. 35, pp 551–564
64. O'Brien EE (1980) Turbulent reacting flows, Libby P, Williams F (eds) Topics in Applied Physics, vol 44. Springer Berlin, chap. 5, pp 185–218. doi:10.1007/3540101926_11
65. Pope SB (1985) Prog Energ Combust 11(2):119. doi:10.1016/0360-1285(85)90002-4
66. Vreman B, Geurts B, Kuerten H (1994) J Fluid Mech 278:351. doi:10.1017/S0022112094003745
67. Dopazo C, O'Brien EE (1976) Combust Sci Technol 13(1–6):99. doi:10.1080/00102207608946731
68. Haworth DC, Pope SB (1986) Phys Fluids 29(2):387. doi:10.1063/1.865723
69. Dreeben TD, Pope SB (1997) Phys Fluids 9(1):154. doi:10.1063/1.869157
70. Risken H (1989) The Fokker-Planck equation, methods of solution and applications. Springer, New York
71. Martin MP, Piomelli U, Candler GV (2000) Theor Comput Fluid Dyn 13(5):361. doi:10.1007/s001620050005
72. Grigoriu M (1995) Applied non-Gaussian processes. Prentice-Hall, Englewood Cliffs
73. Kloeden PE, Platen E, Schurz H (1997) Numerical solution of stochastic differential equations through computer experiments, 2nd edn. Springer, Berlin
74. Blaise S, St-Cyr A, Mavriplis D, Lockwood B (2013) J Comput Phys 232(1):416. doi:10.1016/j.jcp.2012.08.022
75. Brazell M, Mavriplis DJ (2013) 21st AIAA CFD conference. AIAA, San Diego, CA. (AIAA-2013-2855)
76. Glasby RS, Burgess NK, Anderson WK, Wang L, Mavrplis DJ, Allmaras SR (2013) 51st AIAA aerospace sciences meeting and exhibit. AIAA, Grapevine, TX. (AIAA-2013-0691)

77. Advanced Scientific Computing Advisory Committee (ASCAC) Subcommittee, The opportunities and challenges of exascale computing. Tech. rep., U.S. Department of Energy (2010)
78. Karypis G, Kumar V (1998) METIS: A software package for partitioning unstructured graphs, partitioning meshes and computing fill-reducing orderings of sparse matrices, version 4.0. University of Minnesota, Minneapolis, MN
79. Chen YC, Peters N, Schneemann GA, Wruck N, Renz U, Mansour MS (1996) Combust Flame 107(3):223. doi:10.1016/S0010-2180(96)00070-3
80. Meier W, Weigand P, Duan XR, Giezendanner-Thoben R (2007) Combust Flame 150(1–2):2. doi:10.1016/j.combustflame.2007.04.002
81. Valiant LG (1990) Commun ACM 33(8):103. doi:10.1145/79173.79181
82. Gropp W, Lusk E, Skjellum A (1999) Using MPI: portable parallel programming with the message-passing interface, 2nd edn., Scientific and Engineering Computation. MIT Press, Cambridge
83. Gropp W, Lusk E, Thakur R (1999) Using MPI-2: advanced features of the message-passing interface., Scientific and Engineering Computation. The MIT Press, Cambridge
84. The OpenMP API specification for parallel programming (2016). http://www.openmp.org
85. NVIDIA CUDA parallel programming and computing platform (2016). http://www.nvidia.com/object/cuda_home_new.html
86. OpenACC directives for accelerators (2016). http://www.openacc-standard.org
87. Devine KD, Boman EG, Karypis G (2006) Frontiers of scientific computing, Heroux M, Raghavan A, Simon H (eds) SIAM, Philadelphia, PA, chap. 1, pp 1–29
88. Boman E, Devine K, Heaphy R, Hendrickson B, Leung V, Riesen LA, Vaughan C, Catalyurek U, Bozdag D, Mitchell W, Teresco J (2007) Zoltan 3.0: parallel partitioning, load balancing, and data-management services; user's guide. Tech. Rep. SAND2007-4748W, Sandia National Laboratories, Albuquerque, NM
89. Kale LV, Bhatele A (2013) Parallel science and engineering applications: the charm++ approach, Series in Computational Physics. CRC Press, chap 1
90. Gassner GJ, Beck AD (2013) Theor Comput Fluid Dyn 27(3):221. doi:10.1007/s00162-011-0253-7
91. Gullbrand J, Chow FK (2003) J Fluid Mech 495:323. doi:10.1017/S0022112003006268

# Numerical Modeling of MILD Combustion at High Pressure to Predict the Optimal Operating Conditions

V. Mahendra Reddy and William L. Roberts

**Abstract** This Chapter presents numerical simulation on MILD combustion operating at high pressure. Influence of preheat and dilution of oxidizer and operating pressure on stabilization of MILD combustion are presented. Three different preheat temperatures (1100, 1300 and 1500 K) and three different dilution levels (3, 6 and 9% $O_2$) are simulated over an operating pressure variation from 1 atm to 16 atm. A classical jet in hot coflow burner is considered for this study. Total of 45 cases are simulated and analyzed. Essential characteristics of MILD combustion, i.e., maximum temperature ($T_{max}$), temperature rise ($\Delta T$) and temperature distributions, are analyzed. The distribution of emissions OH and CO are also studied and presented. Well-stabilized MILD combustion is observed for all cases except for two cases with high preheated (1500 K). Peak temperature is observed to decrease with increasing operating pressure for a given level of preheat and dilution. OH mass faction is reduced with increasing pressure. The CO emissions show little sensitivity to operating pressure. However, CO mass fraction is slightly higher at 1 atm operating pressure as compared to 4 to 16 atm. Since the residence time of reactants increases as the operating pressure increases, well-stabilized MILD combustion is observed for all highly diluted and low temperature preheat cases (3% $O_2$ and 1100 K).

**Keywords** MILD combustion · High pressure · Auto-ignition

## 1 Introduction

Enhancement of thermal efficiency with ultralow emissions is most challenging goal of combustion researchers. After many techniques have been investigated for this concern, moderate or intense low-oxygen dilution (MILD) combustion has

---

V. Mahendra Reddy · W.L. Roberts (✉)
Clean Combustion Research Center, Physical Sciences and Engineering Division,
King Abdullah University of Science and Technology, Thuwal 23955-6900,
Kingdom of Saudi Arabia
e-mail: William.Roberts@kaust.edu.sa



A.K. Agarwal et al. (eds.), *Combustion for Power Generation and Transportation*, DOI 10.1007/978-981-10-3785-6_4

been identified as a promising concept to achieve these goals [1–8]. This technique is also known as high temperature air combustion (HiTAC) [9–11] or flameless combustion (FLOX) [1, 7, 11–15] or colorless distributed combustion (CDC) [16, 17] in the literature. MILD combustion is defined by the reactants temperature $T_r$, being higher than reactants mixture's autoignition temperature $T_{Ign}$, and the temperature rise in the combustor $\Delta T = (T_p - T_r)$, is smaller than autoignition temperature $(T_{Ign})$ of reactants under typical MILD conditions [2, 3, 5, 6]. In MILD combustion the efficiency increases because the excess heat of the products is recirculated and preheats the reactants. The emissions decrease because the maximum temperature in the combustion chamber is reduced as a result of distributed reactions. In addition, the combustion takes place uniformly, without emitting any sound, and there is no visible flame emissions in the MILD combustion regime.

Evaluation of chemical kinetic times is an essential step to define the operating parameters for the well stabilized MILD combustion. Autoignition delay time of the fuel-oxidizer mixture implies the crucial influence on the rate of reaction [18–22]. Distinctive characteristics of the MILD combustion, e.g., lower temperature gradients and low reactants concentration, maintain longer kinetic characteristic times than conventional combustion. Inlet temperature of the reactants, operating pressure of the chamber, dilution level of the oxidizer and C/O ratio of the local fuel-air mixture are the major operating parameters of the MILD combustion. These parameters have a critical influence on the autoignition delay time. Autoignition delay time for MILD combustion is greater than the conventional combustion $(\tau_{igd,MILD} \gg \tau_{igd,Conv})$. The autoignition delay time is in inversely proportional to the inlet temperature of the reactants and operating pressure $\left(\tau_{igd} \propto \frac{1}{P_{comb}}; \tau_{igd} \propto \frac{1}{T_{reactants}}\right)$. Level of reactants dilution and autoignition delay time are proportional to each other $(\tau_{igd} \propto R_{dil})$. The C/O ratio does not follow a simple relation with autoignition delay time. The C/O ratio is inversely and proportional to autoignition temperature at higher and lower inlet temperatures respectively. Along with these operational conditions, jet velocities or Reynolds number of fuel and oxidizer also have crucial role in autoignition delay time and subsequent stabilization of MILD combustion.

The optimal operating conditions of well-stable MILD combustion operating at a particular pressure will fall within a certain range of inlet temperature and dilution level of oxidizer [20]. Hence, a number of experimental or computational runs are required to predict the optimal operating conditions. These optimal operating conditions of MILD combustion operating at ambient pressure can be predicted either from experimental or numerical study. However, in case of high pressure MILD combustion, it becomes very difficult and expensive to run a wide range experimental conditions to predict the optimal operating parameters. The number of experimental runs necessary at high pressure exclude a trial and error approach. Hence, computational simulations are very helpful in predicting the optimal operating conditions for well-stabilized MILD combustion at high pressure.

The present study is an attempt to simulate the MILD combustion operating at high pressure. The Adelaide Jet in Hot Coflow (JHC) burner configuration of [23] is considered for numerical simulations, since this JHC burner has received much attention from the modelling community due to the comprehensive set of measured data available. For the present study the basic model (RANS) along with assumptions and parameters of [24] are considered. The effect of operating pressure, preheat temperature and dilution levels on sustainability of MILD combustion is studied. Peak temperature, average temperature of the MILD combustion zone, the temperature difference $\Delta T$ and reaction spread rate are analyzed of all the cases.

## *1.1 Literature on High Pressure MILD Combustion*

Studies of [25, 26, 27] on MILD combustion operates at high pressure are presented.

a. Ye et al. [25]: Experimental study on high pressure MILD combustion with pre evaporated fuels [25]

Ye et al. [25] have developed a reverse flow combustor to stabilize MILD combustion at elevated pressures. Experimental setup of [25] is shown in Fig. 1. The reverse-flow design enhances the internal recirculation of exhaust gases, which dilutes the reactants before combustion takes place. In their study, three different liquid fuels of Ethanol, acetone and n-heptane are pre vaporised and passed to the combustor with carrier gas. The influence of fuel type, equivalence ratio, carrier gas, operating pressure and air jet velocity on the combustion stability and emissions are investigated. The combustor developed for a thermal input of 4.7 kW, with the thermal intensity of 7 $MW/m^3$. Fuel and carrier gas are injected into the main combustion chamber from a central fuel nozzle as shown in Fig. 1. The fuel nozzle and the air nozzle are in coaxial position. Two different sizes of air nozzles were used to study the effect of air jet momentum flow rate while the fuel nozzle diameter is kept constant. The combination of air preheating and strong recirculation of hot products produce MILD combustion conditions.

The experiment of [25] has observed many interesting results. The lean extinction limit of fuels is shifted to leaner conditions under elevated pressures. NOx emission increases as pressure increases. It is suspected that the combustion instability and higher NOx emissions observed under elevated pressure are partially due to the inversely proportional relationship between pressure and volumetric flow rate according to the ideal gas law. The decrease in volumetric flow rate under elevated pressure results in a considerable increase in residence time, thereby increasing the available time for NOx to be produced. CO emission does not show a clear dependence on pressure. Due to a decreased volumetric flow rate, the air and fuel jet velocities under 5 bar are four times lower than those under 1 bar, which can affect the mixing process, thereby affecting NOx emissions. [25] also have calculated the ignition delays of all three fuels in the Closed Homogeneous Batch Reactor (CHBR) model in CHEMKIN software. The calculations were performed for a stoichiometric fuel–air mixture with inlet preheat temperature in the range of

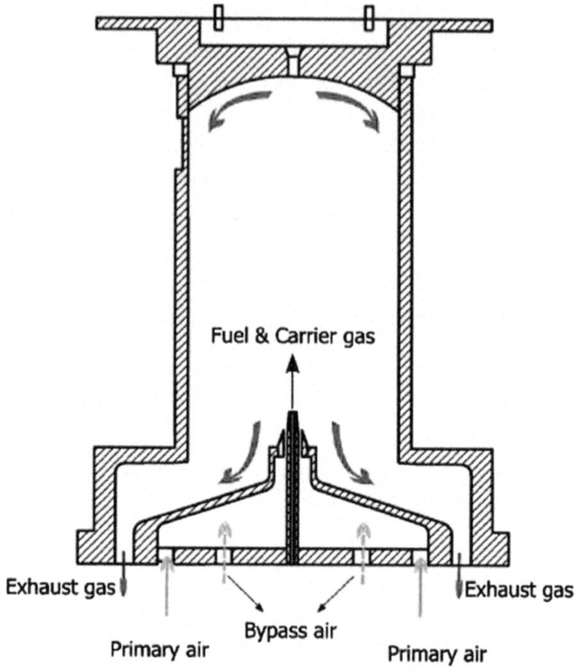


**Fig. 1** Reverse flow MILD combustor developed by [25]

823 K to 923 K. The calculated ignition delays are function of the inverse of the inlet temperature and pressures. Thc ignition delays are reduced with pressure for all fuels (shown in Fig. 2). The chemistry timescale decreases greatly under elevated pressure. This indicates that ignition could occur before effective mixing. The early ignition leads to combustion instability under elevated pressure.

b. Lückerath et al. [26]: High pressure MILD combustion [26]

Lückerath et al. [26] have investigated FLOX combustion at high pressure in order to attribute its applicability for gas turbine combustors. A FLOX burner was equipped with a combustion chamber with 12 single nozzles were positioned on a reference circle with a diameter of 80 mm as shown in Fig. 3. Each single nozzle comprised a fuel nozzle surrounded by an air nozzle. The length of the combustion chamber was 200 mm and the inner diameter of the hexagon was 100 mm. The air was supplied at T = 600–735 K and fuel at room temperature. The mean air velocity was in the range of $v_{air}$ = 40–160 m/s. The burner was operated at a pressure of P = 20 bar with a maximum thermal power of $P_{therm}$ = 475 kW. The penetration depth of the injected jets of reactants and the products recirculation depend on the inlet velocity of the fuel/air mixtures. Therefore, Lückerath et al. [26] have considered the air and fuel injection velocities are parameters for the FLOX combustor.

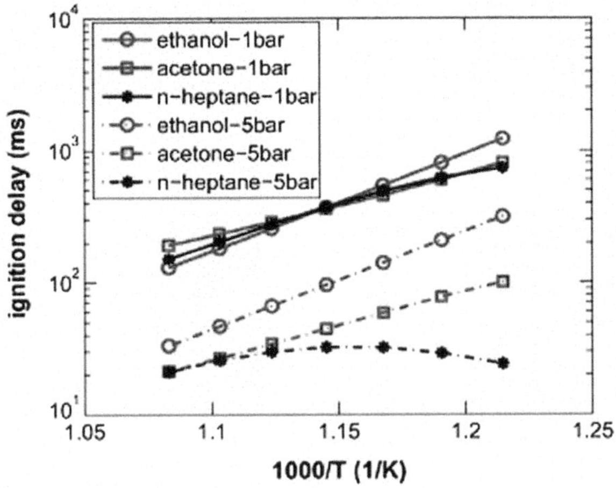


**Fig. 2** Calculated ignition delays for three fuels at pressures of 1 bar and 5 bar [25]

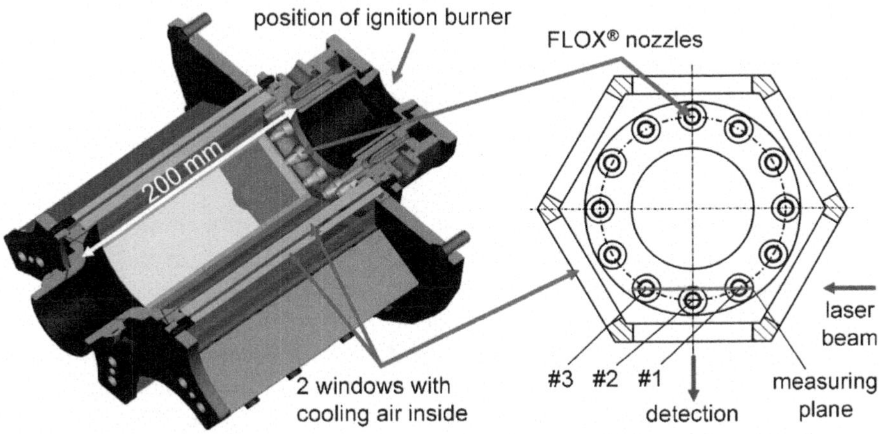


**Fig. 3** The hexagonal combustion chamber with FLOX® burner: (*left*) longitudinal section and (*right*) cross section view of 12 single nozzles [26]

In Fig. 4, the OH—PLIF and chemiluminescence distributions are displayed for the inlet velocities $v_{air}$ = 40, 85, 128, and 160 m/s. Considering that low OH—PLIF signals reflect cold regions, it can be seen that the penetration depth of the relatively cold jets in front of the nozzles varies with $v_{air}$ from ~25 mm at 40 m/s to more than 55 mm at higher velocities. Accordingly, the positions of the flame zones, characterized by high $OH^*$ chemiluminescence intensities, move downstream with an increase in the air injection velocity. At 160 m/s, the flame zone covers the downstream region z ~ 30–90 mm almost homogeneously and combustion is completed at z ~ 100–110 mm. In the OH—PLIF images, the relatively

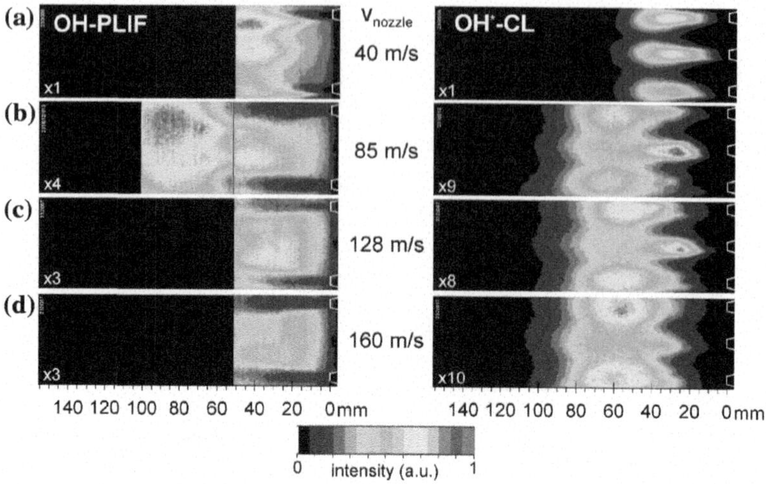


**Fig. 4** Averaged OH—PLIF and OH* chemiluminescence images for different inlet velocities [26]

high intensities between the jets reflect a high temperature level. This region is, thus, completely filled with hot recirculated flue gas.

c. Mingming et al. (2014): Chemkin analysis on High pressure MILD combustion [27]

Mingming et al. (2014) have studied the characteristics of high pressure MILD combustion in the operating pressure range of 1–19 atm. Hydrogen and syngas are used as fuels and studied to evaluate the effects of fuel type and pressure on MILD combustion. Numerical simulation was carried out in Chemkin package by considering closed homogeneous batch reactor (CHBR) and perfectly stirred reactor (PSR) model as shown in Fig. 5. Autoignition delay times from Mingming et al. is in conflicting trend with [25] results. However, in both the studies the autoignition delay time is calculated by Chemkin package by considering closed homogeneous batch reactor (CHBR).

As pressure increases from 1 atm to 19 atm, autoignition delay, $\tau_{ig}$ of MILD mixture increased from 25 ms to 180 ms. Variations autoignition and OH with pressure is shown in Fig. 6. With pressure increased from 1 atm to 19 atm, maximum reaction rate rises. MILD mixture ignition process is still delayed for 19 atm system compared to 1 atm system. OH radicals content in 19 atm MILD mixture is extremely low in camparison with 1 atm system. In reality, OH intermediate species is essential for inducing MILD mixture ignition. OH mole fraction of MILD mixture at p = 1 atm is obviously higher than that of other pressures. Higher OH mole fraction for atmospheric pressure promotes MILD mixture autoignition and a relatively lower $\tau_{ig}$ is required for 1 atm system.

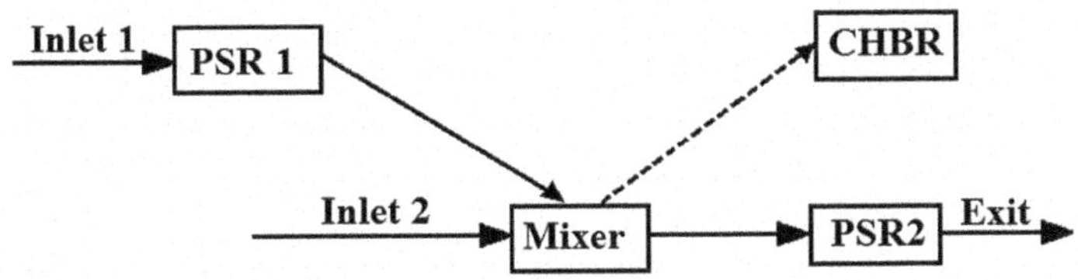


**Fig. 5** MILD combustion model of Chemkin (Redrawn) [27]

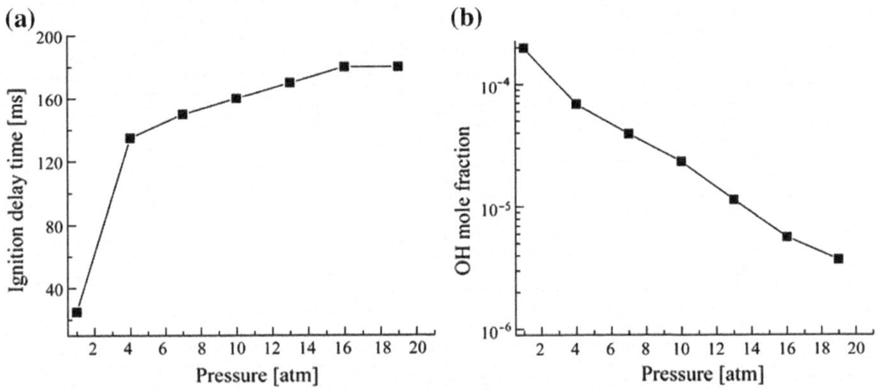


**Fig. 6** **a** $\tau_{ig}$ profile as a function of pressure **b** OH mole fraction MILD mixture for pressures of 1–19 atm [27]

## 2 RANS Governing Equations and Model

Many researchers have developed various numerical models to study MILD combustion and compared their results with experimental results of Dally et al. (2002) [23]. Christo and Dally [24] compared measurements with simulation results using conserved scalar-based (e.g. flamelet approaches) and Eddy Dissipation Concept (EDC) models. Aminian et al. [28] further explored the capabilities of EDC models by modifying the chemical time constant. Ihme et al. [29] modelled the burner as a three-stream problem using Large Eddy Simulation (LES) with a flamelet/progress variable (FPV) formulation. Lee et al. [30] used a Multi-environment Probability Density Function (MPDF) approach for the JHC configuration. Recently, De and Dongre [31] extensively compared various models used for MILD combustion including the steady flamelet model, transported PDF models as well as EDC based models. Chen et al. [32] have developed a novel concept of perfectly stirred reactor model to simulate the MILD combustion. In the present study the basic model (RANS) along with assumptions and parameters of Christo and Dally [24] are considered.

In the present study, RANS equations are derived by averaging the instantaneous governing equations, given in Eq. (1) to (4). The standard two-equation $k-\varepsilon$ model is used for the turbulent jet flow. Christo and Dally [24] have shown that the standard $k-\varepsilon$ model with a modified constant $C_{\varepsilon 1}=1.6$, instead of the more commonly used 1.44, in the dissipation rate equation is the most suitable model among different $k-\varepsilon$ models for the numerical simulation of MILD combustion. Governing equations are [33, 34]:

$$\frac{\partial \bar{\rho}}{\partial t}+\frac{\partial}{\partial x_i}(\bar{\rho}\tilde{u}_i)=0 \tag{1}$$

$$\frac{\partial \bar{\rho}\tilde{u}_i}{\partial t}+\frac{\partial}{\partial x_i}(\bar{\rho}\tilde{u}_i\tilde{u}_j)+\frac{\partial \bar{P}}{\partial x_j}=\frac{\partial}{\partial x_i}\left(\overline{\tau_{ij}}-\bar{\rho}\overline{u_i^{''}u_j^{''}}\right)+\bar{\rho}g \tag{2}$$

$$\frac{\partial \bar{\rho}\tilde{Y}_k}{\partial t}+\frac{\partial}{\partial x_i}(\partial\bar{\rho}\tilde{Y}_k)=\frac{\partial}{\partial x_i}\left(\mu\frac{\partial \tilde{Y}_k}{\partial x_j}-\bar{\rho}\overline{u_i^{''}Y_k^{''}}\right)+\bar{\dot{\omega}}_k \quad k=1,\ldots\ldots,N \tag{3}$$

$$\frac{\partial \bar{\rho}\tilde{h}}{\partial t}+\frac{\partial}{\partial x_i}(\bar{\rho}\tilde{u}_i\tilde{h})=\frac{\overline{Dp}}{Dt}+\frac{\partial}{\partial x_i}\left(\frac{\mu}{Pr}\frac{\partial \tilde{h}}{\partial x_j}-\bar{\rho}\overline{u_i^{''}h^{''}}\right) \tag{4}$$

In above Equations $\bar{\rho}\overline{u_i^{''}Y_k^{''}}$ and $\bar{\rho}\overline{u_i^{''}h^{''}}$ are closed using a classical gradient assumptions of

$$\bar{\rho}\overline{u_i^{''}Y_k^{''}}=-\frac{\mu_t}{Sc_{kt}}\frac{\partial \tilde{Y}_k}{\partial x_j} \qquad \bar{\rho}\overline{u_i^{''}h^{''}}=-\frac{\mu_t}{Pr_t}\frac{\partial \tilde{h}}{\partial x_i}$$

It is also assumed that $Sc_{kt}$ (turbulent Schmidt number), and $Pr_t$ (turbulent Prandtl number) are unity. The turbulent viscosity is estimated as:

$$\mu_t=\bar{\rho}C_\mu\frac{k^2}{\varepsilon}$$

Following the turbulence viscosity model proposed by Boussinesq, the turbulent Reynolds stresses $\bar{\rho}\overline{u_i^{''}u_j^{''}}$ are described using the viscous tensor $(\tau_{ij})$ expression that is obtained for Newtonian fluids.

$$\bar{\rho}\overline{u_i^{''}u_j^{''}}=\bar{\rho}\overline{u_i^{''}u_j^{''}}=-\mu_t\left(\frac{\partial \tilde{u}_i}{\partial x_j}+\frac{\partial \tilde{u}_i}{\partial x_i}-\frac{2}{3}\delta_{ij}\frac{\partial \tilde{u}_k}{\partial x_k}\right)+\frac{2}{3}\bar{\rho}k$$

**Turbulence-chemistry interaction model for non-premixed combustion**

The assumed-shape Probability Density Function (PDF) approach is used for the non-premixed combustion model in the present numerical simulations. The PDF written as $p(f)$, is the fraction of time that the fluid spends in the vicinity of the state $f$. Probability Density Function is represented as [34].

$$p(f)\Delta f = \lim_{T\to\infty}\frac{1}{T}\sum_i \tau_i \tag{6}$$

where $T$ is the time scale and $\tau_i$ is the amount of time that f spends in the $\Delta f$ band. The shape of the assumed PDF, $p(f)$, is described by one of two mathematical functions: the double-delta function (two-mixture-fraction cases only) and the β -function (single- and two- mixture-fraction cases). The double-delta function is the most easily computed, while the β-function most closely represents experimentally observed PDFs. The shape produced by this function depends solely on the mean mixture fraction $\bar{f}$, and its variance, $\overline{f'^2}$. In the present numerical study β–function is used.

The b-function PDF shape defined the mean mixture fraction $\bar{f}$, and its variance, $\overline{f'^2}$ as:

$$p(f) = \frac{f^{\alpha-1}(1-f)^{\beta-1}}{\int f^{\alpha-1}(1-f)^{\beta-1}df} \quad (1) \tag{7}$$

where

$$\alpha = \bar{f}\left[\frac{\bar{f}(1-\bar{f})}{\overline{f'^2}} - 1\right] \quad \beta = (1-\bar{f})\left[\frac{\bar{f}(1-\bar{f})}{\overline{f'^2}} - 1\right]$$

## 3 Numerical Setup

In this study, the Jet in Hot Coflow (JHC) burner configuration of Dally et al. [23] is considered for numerical simulations. A schematic of the numerical flow field of the present study and experimental setup of Dally et al. [23] is shown in Fig. 7a and b respectively. The fuel jet with internal diameter of 4.25 mm is arranged centrally in the annulus hot coflow of 82 mm diameter. The unheated fuel jet consists of 50% methane and 50% hydrogen based on mole fraction and the mean jet velocity of the fuel is 62 m/s corresponding to a Reynolds number of approximately 14,000. The velocity of hot coflow is 3.2 m/s at 1300 K preheat temperature. The burner was placed in a wind tunnel where the cold air entrainment velocity is the same as the coflow. In the present numerical study, three different preheat temperatures (1500, 1300 and 1100 K) are considered. The operating pressure varied from 1 atm to 16 atm in 4 atm increments. Three different dilution levels of oxidizer (3, 6 and 9% $O_2$) are considered. A total of 45 cases are simulated to find the optimal operating

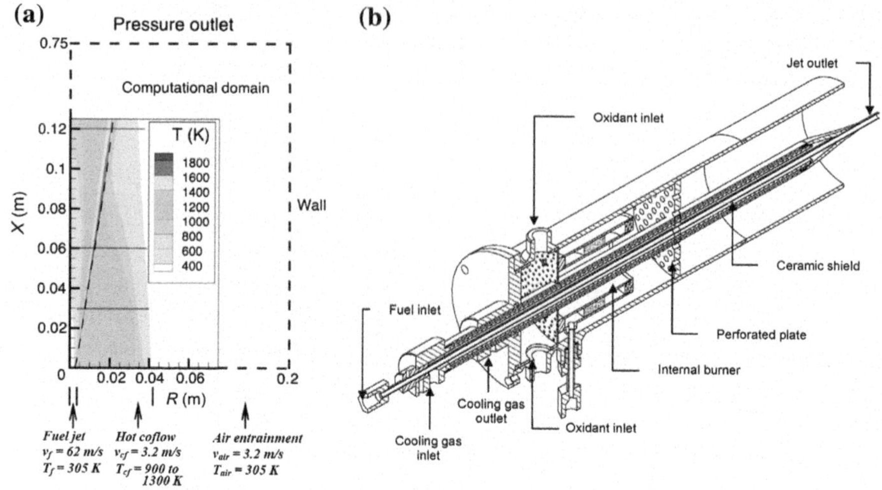


**Fig. 7** Geometry and flow condition of the MILD burner [23]

**Table 1** Composition of Hot coflow dilution ($Y_i$ *in %*)

| Dilution level, $Y_{O_2}$ | $Y_{N_2}$ | $Y_{CO_2}$ | $Y_{H_2O}$ |
|---|---|---|---|
| 3 | 85 | 5.5 | 6.5 |
| 6 | 82 | 5.5 | 6.5 |
| 9 | 79 | 5.5 | 6.5 |

conditions to achieve well-stabilized MILD combustion. The composition of hot coflow oxidizer for these three different dilution levels is given in the Table 1.

Flow conditions of coflow preheat temperature of 1300 K at atmospheric pressure are considered as the base flow parameters. The mass flux of fuel and hot coflow are 0.000562 and 0.0043 kg/s respectively. A constant mass flux condition is maintained to scale the flow parameters for higher operating pressures and different preheat conditions. The Reynolds number of the fuel jet, $Re_f$, is around 14000. The Reynolds numbers of hot coflow is 1410 at 1300 K preheat temperature. The flow parameters for all flow conditions are shown in Table 2. The composition of diluted hot coflow is not sensitive to Reynolds number, since the $CO_2$ and $H_2O$ mass fractions are maintained constant and the dilution composition is varied by exchanging $O_2$ with $N_2$. A simplified axisymmetric computational domain used in this study is shown in Fig. 2a along with the boundary conditions. The domain starts from the burner exit plane, and extends to 1000 mm downstream in the axial direction $X$ and 200 mm in the radial direction. A grid independence study was carried out using three different mesh sizes and the final grid with 141,000 cells is chosen based on velocity and temperature comparisons [32].

**Table 2** Flow parameters of Coflow and fuel

| Preheat temperature: 1100 K | | | |
|---|---|---|---|
| $P$ (atm) | $v$ (m/s) | $\dot{V}$ $(m^3/s)$ | $\tau_R$ (s) |
| 1 | 2.492 | 0.0136 | 0.368 |
| 4 | 0.623 | 0.0034 | 1.474 |
| 8 | 0.311 | 0.0017 | 2.94 |
| 12 | 0.207 | 0.0011 | 4.424 |
| 16 | 0.155 | 0.0008 | 5.899 |
| Preheat temperature: 1300 K | | | |
| $P$ (atm) | $v$ (m/s) | $\dot{V}$ $(m^3/s)$ | $\tau_R$ (s) |
| 1 | 3.2 | 0.0161 | 0.311 |
| 4 | 0.799 | 0.0040 | 1.247 |
| 8 | 0.399 | 0.0020 | 2.495 |
| 12 | 0.266 | 0.0013 | 3.743 |
| 16 | 0.199 | 0.0010 | 4.991 |
| Preheat temperature: 1500 K | | | |
| $P$ (atm) | $v$ (m/s) | $\dot{V}$ $(m^3/s)$ | $\tau_R$ (s) |
| 1 | 4.007 | 0.0185 | 0.270 |
| 4 | 1.002 | 0.0046 | 1.081 |
| 8 | 0.501 | 0.0023 | 2.163 |
| 12 | 0.334 | 0.0015 | 3.244 |
| 16 | 0.250 | 0.0011 | 4.326 |
| *Fuel* | | | |
| $P$ (atm) | $v$ (m/s) | | |
| 1 | 62 | | |
| 4 | 15.484 | | |
| 8 | 7.742 | | |
| 12 | 5.161 | | |
| 16 | 3.871 | | |

# 4 Results and Discussion

## 4.1 *Residence Time of Reactants*

The residence time of reactants in the flow filed is analyzed and listed in Table 2. The flow velocity decreases with increased operating pressure for the constant mass flux of reactants. Hence the residence time of reactants increase as the operating pressure increases. The density of the oxidizer is inversely proportional to preheat temperature. For instance, the residence time of hot coflow with preheat temperature of 1300 K is increased from 0.31 s to 4.99 s for pressure rise from 1 atm to 16 atm. At the operating pressure of 16 atm, residence times for the hot coflow with preheat temperature of 1100, 1300 and 1500 K, are respectively, 5.89, 4.99 and 4.33 s. The variation of residence time for all cases with pressure is plotted in Fig. 8.

**Fig. 8** Residence time of reactants

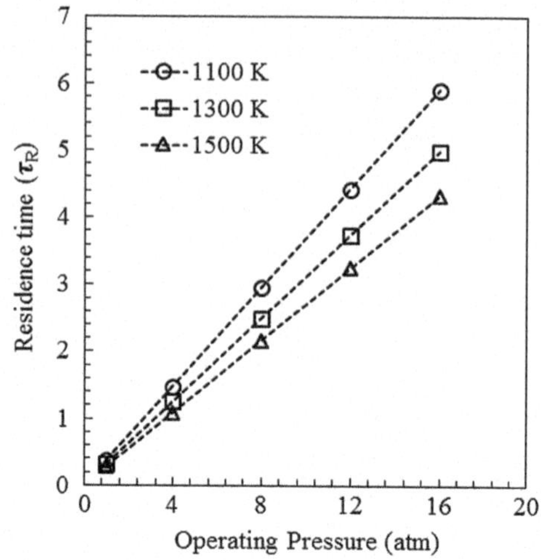


## 4.2 *Temperature Distribution*

Maximum temperature ($T_{max}$) and temperature rise ($\Delta T$) are determining characteristics of MILD combustion. $T_{max}$ and $\Delta T$ are noted from temperature contours of each case of simulation and plotted in Fig. 9a and b. For all sets of preheat and dilution cases simulated in this work, the maximum temperature is observed at 1 atm pressure, thus $T_{max}$ decreases mildly with increasing pressure from 1 atm to 16 atm. However this reduction in peak temperature is in the range of 0.5 to 1.4% of $T_{max}$ at P = 1 atm. Temperature contours for 1300 K preheat and 9% $O_2$ level dilution is shown in Fig. 10a. For this case temperatures at 1 atm and 16 atm are 1672 and 1651 K respectively, the difference is 1.25%.

At a constant operating pressure, $T_{max}$ is proportional to the preheat temperature and inversely proportional with the dilution level. Figure 10b shows the axial distribution of mean mixture fraction ($\xi$) for two different operating pressures, 1 atm and 16 atm. Increasing the pressure leads to a rapid decrease in the axial distribution of mixture fraction. This implies that the fuel is consumed more rapidly with increasing pressure and that the reactions occur over a shorter length scale as the pressure increases. Since the residence time increases with operating pressure, thermal diffusion in the lateral direction is slightly increased. Hence $T_{max}$ is decreased slightly with increased operating pressure.

Another distinctive characteristic of MILD combustion is that the temperature rise in the reaction zone is lower than the autoignition temperature of fuel. Temperature rise, $\Delta T$, is calculated as $T_{max} - T_{hotcoflow}$. The $\Delta T$ for all cases simulated in this work are plotted in Fig. 9b. It can be seen that a maximum and minimum temperature rise of 429 K and 101 K are observed for 1500 K preheat, 9% $O_2$ level operating at 1 atm pressure and 1100 K preheat, 3% $O_2$ level operating at 16 atm,

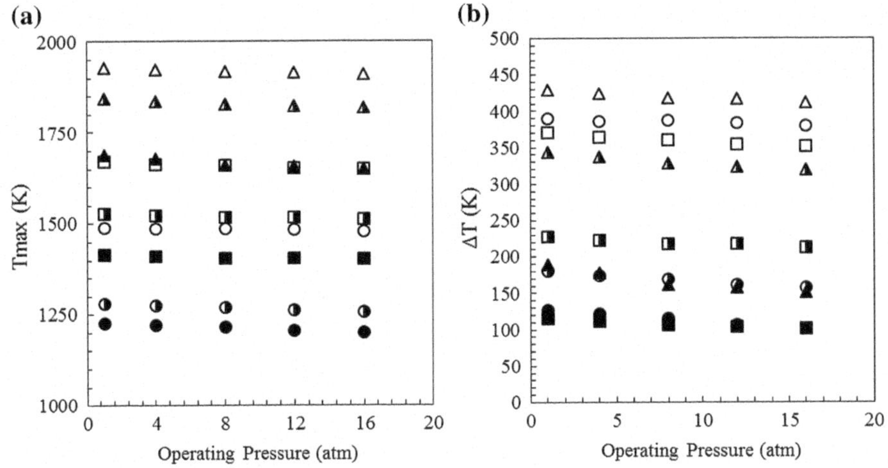


**Fig. 9** **a** Maximum temperature **b** Temperature rise. (Symbols: Δ–1500 K, □—1300 K and ○—1100 K preheat conditions. No fill—9% $O_2$, Gradient fill—6% $O_2$ and Solid fill—3% $O_2$ dilution levels)

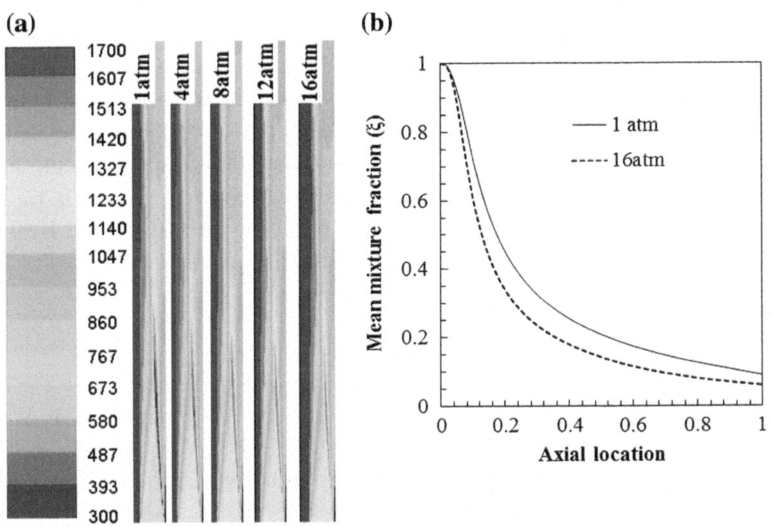


**Fig. 10** **a** Temperature contour (1 atm to 16 atm pressure) **b** Mean mixture fraction variation, for 1300 K preheat and 9% $O_2$ dilution level

respectively. An interesting observation is that the temperature rise is higher for lower dilution levels (9% $O_2$) and lower for stronger dilution (3% $O_2$).

The extent of the hot zone, area where the temperature is greater than 1800 K, $A_{T \geq 1800K}$, are calculated and plotted in Fig. 11.

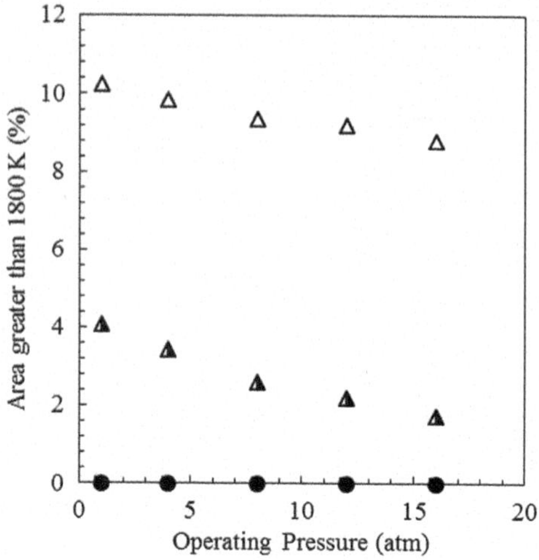


**Fig. 11** Area of the domain the temperature greater than 1800 K. (Symbols: Δ–1500 K, □—1300 K and O—1100 K preheat conditions. No fill—9% $O_2$, Gradient fill—6% $O_2$ and Solid fill—3% $O_2$ dilution levels)

$$A_{T \geq 1800K} = \frac{\sum_{i=1}^{i=n} f_i dA_i}{A}, f_i = \begin{cases} 0, & T_i < 1800K \\ 1, & T_i \geq 1800K \end{cases}$$

At all operating pressures (1 atm to 16 atm) with 9 and 6% $O_2$ dilution level at 1500 K preheat, a significant hot zones of $A_{T \geq 1800K}$ are observed. The area of this hot zone decreases as the dilution increases. For the 9% $O_2$ dilution condition, a larger hot zone is observed when compared with 6% $O_2$ dilution. The hot zone decreases in size with increasing operating pressure. For example, with 1500 K preheat and 9% dilution, the hot zone area decreases from 10.2% to 8.4% as the pressure increases from 1 atm to 16 atm. Except these two cases (1500 K, 9% and 6% $O_2$ levels), for remaining all cases the temperature in the reaction zone is well below 1800 K.

## 4.3 *OH Distribution*

OH mass faction is directly proportional to mass flow rates of the reactants. However, throughout all the cases the total mass flow rate is fixed. Hence the OH variation due to mass flow rate is negligible. OH chemiluminescence decreases with increasing pressure due to increased collisional quenching at higher pressures. Since the mass flux is maintained constant, the residence time of reactants increases with increasing operating pressure. These simulations do not calculate excited state OH, but rather ground state OH. The consequence of increasing pressure is higher third body collision efficiency and hence conversion of OH to water. Figure 12a shows the contours of OH mass fraction with operating pressure from 1 atm to 16 atm for 1300 K preheat and 6% $O_2$ cases. The OH mass fraction decreases with pressure from 1 atm to 16 atm due to higher conversion rates to primarily water.

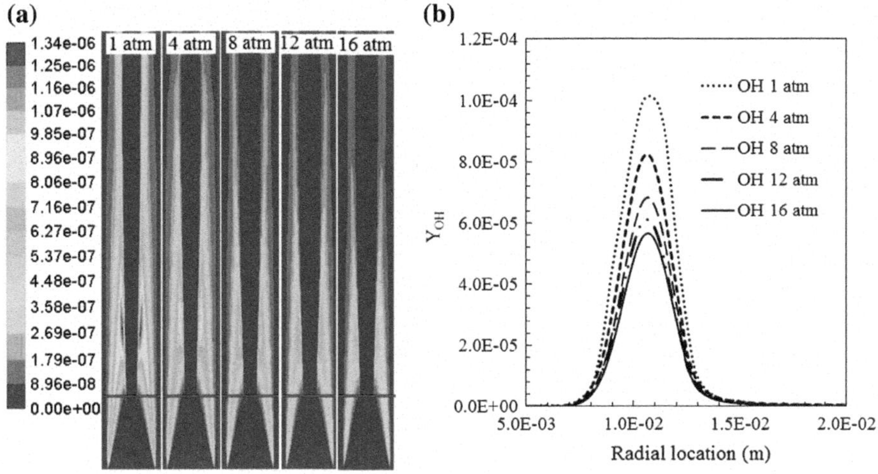


**Fig. 12** OH mass fraction variation for 1300 K prehat, 6% $O_2$ for 1 atm to 16 atm pressure, **a** Contours **b** Massfraction variation at 120 mm axial location

The variation of OH in the reaction zone is analysed at an axial location of 120 mm, results are shown in Fig. 12b. The peak OH mass fraction of $1.02 \times 10^{-6}$ occurs at one atm and decreases to $5.75 \times 10^{-7}$ as the pressure increases to 16 atm, at 120 mm axial location.

The maximum OH mass fraction for all cases simulated are plotted in Fig. 13. It is observed that the maximum OH occurs at the lower dilution case of 9% $O_2$, for all three preheat temperature. At 1 atm, higher levels of OH are observed for all cases simulated. The OH mass fraction is observed to decrease with increasing pressure In $CH_4 - O_2$ combustion the OH formation takes place as [35, 36]

**Fig. 13** Maximum OH mass fraction for all cases (Symbols: Δ–1500 K, □—1300 K and ○—1100 K preheat conditions. No fill—9% $O_2$, Gradient fill—6% $O_2$ and Solid fill—3% $O_2$ dilution levels)

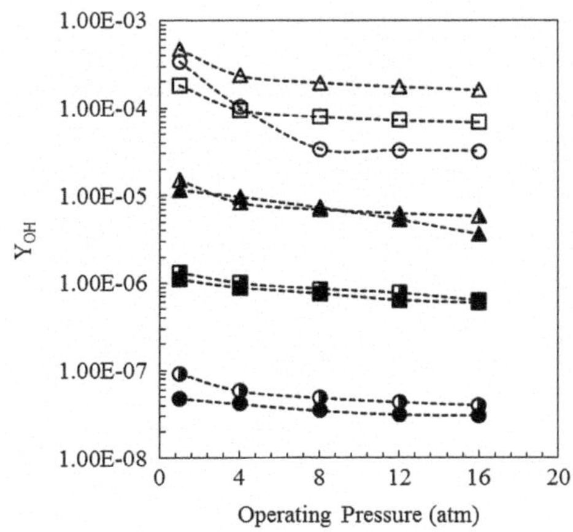

$$CH + O_2 \leftrightarrows CH + OH^*$$
$$OH^* + M \leftrightarrows M + OH$$

The rate of formation of $OH^*$ dependence on the concentrations of $CH$ and $O_2$. Further formation of $OH$ from $OH^*$ purely depends on intermittent reactant partner ($M$). The reactant partners can be among products or reactants ($CO_2$, $O_2$, $H_2O$, CO, $CH_4$, $H_2$, and $N_2$). The $O_2$ concentration in hot coflows with 3% and 6% $O_2$ level dilution is lower as compared to 9% $O_2$ diluted hot coflow. Hence the OH mass fraction is higher for 9% $O_2$ diluted cases and lower for 3% and 6%.

## 4.4 CO Distribution

Contours of CO mass fraction for 1500 K preheat with 9% $O_2$ and 1300 K preheat with 6% $O_2$ are shown in Fig. 14. The CO mass fraction is observed to decrease as the pressure increases from 1 atm and 4 atm. The variation in CO mass fraction with pressure is very much less sensitive for pressures above 4 atm, as shown in

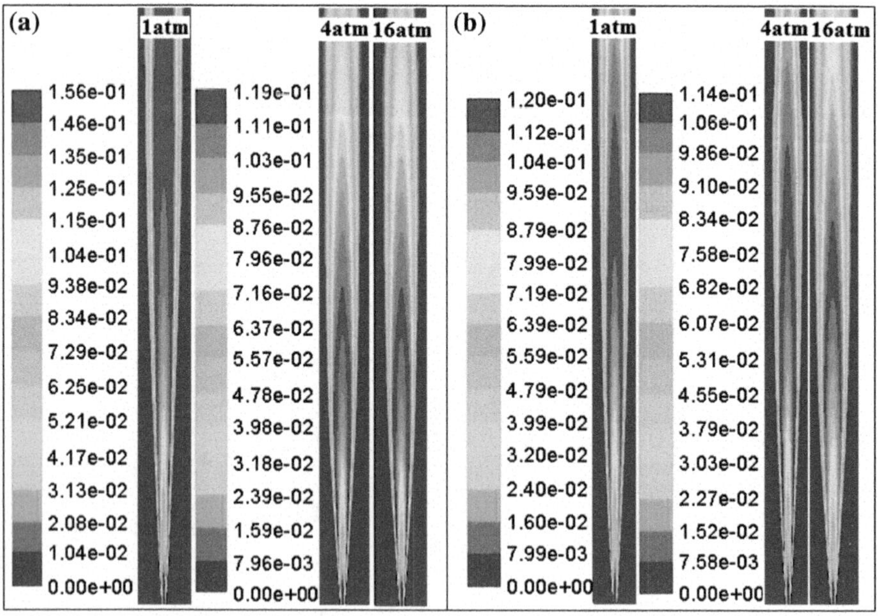


**Fig. 14** Contours of $Y_{CO}$ for preheat temperature and dilution levels of **a** 1500 K and 9% $O_2$ and **b** 1300 K and 6% $O_2$

**Fig. 15** Maximum OH mass fraction for all cases (Symbols: Δ–1500 K, □—1300 K and ○—1100 K preheat conditions. No fill – 9% $O_2$, Gradient fill—6% $O_2$ and Solid fill—3% $O_2$ dilution levels)

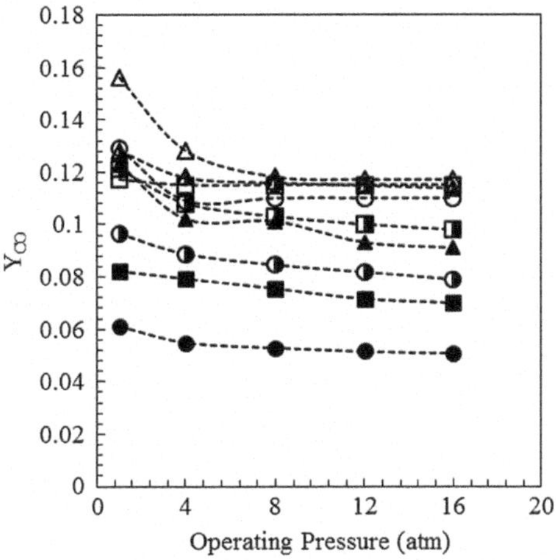


Fig. 15. For all the cases simulated, $Y_{CO}$ is high at 1 atm and almost flat for 4 atm to 16 atm. Higher and lower $Y_{CO}$ are observed for lower (9% $O_2$) and higher (9% $O_2$) dilution levels respectively. For a given dilution level, a higher mass fraction of CO is observed for higher preheat temperatures, as shown in Fig. 15.

## 4.5 *Ignition Delay*

Numerical simulations are carried out by CHEMKIN software to find out ignition delay time for mixtures of all cases (Table 2). Closed Homogeneous Batch Reactor (CHBR) model is considered for a stoichiometric fuel–air mixture. It is observed that ignition delay decreased with increase in inlet reactants temperature and operating pressure and increased with increase in dilution. Ignition delay for all cases plotted in Fig. 16. When the preheat temperature reduced from 1500 to 1100 K, the ignition delay difference is maximum and minimum for higher (3% $O_2$) and lower dilution levels (9% $O_2$). It is $4.1e^{-2}$ and $7.75e^{-5}$ for 3% and 9% $O_2$ levels respectively, at 4 bar operating pressure. The ignition delay variation due to pressure rise is maximum at lower preheat temperature (1100 K) and minimum at higher preheat temperature (1500 K). The ignition delay difference between 1 atm and 16 atm operating pressure are 2.72e-4 and 5.4e-3 s respectively, for 1500 and 1100 K for 9% $O_2$ level. The case of preheat temperature of 1100 K and dilution level of 3% $O_2$ operates at 1 atm pressure has highest ignition delay of 4.8e-2 s. On contrary, the case of preheat temperature 1500 K and dilution level of 9% $O_2$ operates at 16 atm has lowest ignition delay of 6.15e-05 s.

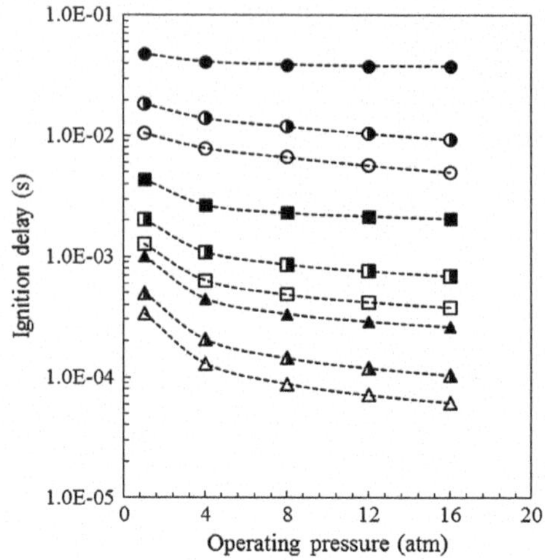


**Fig. 16** Calculated ignition delay time (CHBR model) for all cases (Symbols: Δ–1500 K, □—1300 K and O—1100 K preheat conditions. No fill—9% $O_2$, Gradient fill—6% $O_2$ and Solid fill—3% $O_2$ dilution levels)

## 5 Discussion and Scope of Research in MILD Combustion at High Pressure

Many researchers have conducted various studies to investigate the role of autoignition delay of reactants on MILD combustion stabilization. Autoignition delay of different mixtures with range of dilution and preheat are reported in the literature extensively. This present study is initiated to understand the high pressure MILD combustion with different mixtures of reactants. At the initial step, it is considered that the residence time of the reactants in the computational domain is maintained constant. Therefore, the inlet velocity of reactants maintained constants for all operating pressures. The density of reactants (fuel jet and oxidizer coflows) increased with increase in operating pressure. Hence, the Reynolds number (*Re*) of fuel jet increased in same multiplication with rise in pressure and the mass flux of the fuel jet is increased extensively. At low pressures up to 4 atm, complete combustion is achieved and MILD combustion is well stabilized. When the operating pressure is above 4 atm, the fuel jet is not completely decayed. The mean mixture fraction at the axis of the fuel jet remains higher and fuel jet persists till the exit of the domain. The mean mixture fraction variation for 1 atm and 20 atm operating pressure of 1500 K preheat and 9% dilution is shown in Fig. 17a. Due to high Reynolds number of the fuel jet, wrinkled reaction zone is observed around the fuel jet. The temperature contours for 1and 20 atm cases are shown in Fig. 17b, the wrinkled reaction zone in 20 atm case is clearly observed.

In the next step of this study, as discussed in the numerical setup, mass flux of the reactants are maintained constant. The flow velocities of fuel and oxidizer are decreased while increasing the operating pressure. The residence time at the high pressures is many times higher than the autoignition delay time. Hence, complete reaction is observed and well-stabilized MILD combustion is achieved for all the

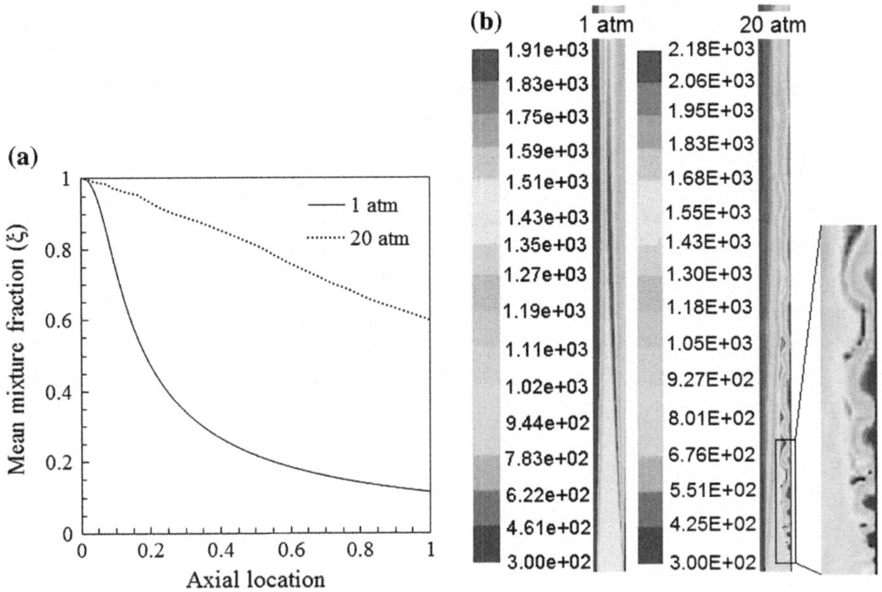


**Fig. 17** Mean mixture fraction variation and temperature contours at 1 atm and 20 atm operating pressure for 1500 K preheat and 9% dilution (constant jet velocity cases)

cases, except two cases of 1500 K preheat with 9% and 6% $O_2$ dilution level. To understand the effect of ignition delay due to operating parameters (dilution, preheat temperature and operating pressure) on MILD combustion, the resident time of reactants should be just higher than the autoignition delay time of that particular case. Such crucial situation is not existed in this study due to large residence time at high pressure conditions.

Detailed basic study is required to understand the MILD combustion at high pressures. Transition of MILD combustion from conventional combustion is to be studied at high pressures. The lifted flame in hot coflow helps in calculating autoignition delay for diluted and preheated mixtures. The autoignition delay measurement of fuel in coflow with different $O_2$ levels generate the data for autoignition delay time at high pressures. Jet in hot coflow setup designed and developed for high pressure operating conditions helps for this purpose.

Based on constant residence time, set of combustors to be designed and developed computationally and experimentally. The volume of the combustor decreases with increased pressure. The diameter of inlet ports to be scaled to maintain jet momentum for better stirring and recirculation along with maintain the constant residence time at all pressures. This study helps in analyze the role of autoignition delay on MILD combustion stabilization with certainly diluted and preheated mixtures for a particular operating pressure.

## 6 Conclusions

The present study is an attempt to simulate the MILD combustion operates at high pressure. The Jet in Hot Coflow (JHC) burner configuration is considered for numerical simulations. The effect of operating pressure, preheat temperature and dilution levels on sustainability of MILD combustion is studied. Essential characteristics of MILD combustion i.e. maximum temperature ($T_{max}$), temperature rise ($\Delta T$) and temperature distributions are analyzed. Distribution of emissions OH and CO are studied and presented. In the present numerical study, three different preheat temperatures of 1500, 1300 and 1100 K are considered. The operating pressure varied from 1 atm to 16 atm in the step variation of 4 atm. Three different dilution levels of oxidizer with 3, 6 and 9% $O_2$ are considered. Flow conditions at preheat temperature of 1300 K, operates at atmospheric pressure are considered as base flow parameters. Constant mass flux condition is maintained to scale the flow parameters for higher operating pressures and different preheat conditions. The Reynolds number of fuel jet, $Re_f$ is around 14000. The Reynolds numbers of hot coflow is 1410. Total of 45 cases are simulated to find optimal operating conditions for well stabilized MILD combustion.

The flow velocity decreases with increased operating pressure for the constant mass flux of reactants. Hence the residence time of reactants is increases as the operating pressure increases. For all sets of preheat and dilution cases, maximum temperature is observed at 1 atm pressure. The maximum temperature is decreased with increased pressure from 1 atm to 16 atm. However this decrement is in the range of 0.5–1.4% of maximum temperature at P = 1 atm. $T_{max}$ is in proportional with preheat temperature and inverse relation with dilution level. Increase of pressure lead to a rapid decrease in the axial distribution of mean mixture fraction ($\xi$). It means that the fuel is consumed more rapidly with increasing pressure and that the reaction occurs in a smaller distance with the increase of pressure. The temperature rise is higher for lower dilution levels (9% $O_2$) and lower for stronger dilution (3% $O_2$). The area of hot zone decreases as dilution increases. The area of hot zone decreased with increased operating pressure. Except these two cases (1500 K, 9 and 6% $O_2$ levels), for remaining all cases the temperature in the reaction zone is well below 1800 K.

OH decreases with increasing pressure due to increased collisional quenching at higher pressures. Since the mass flux maintained constant, the residence time of reactants is increases with increased operating pressure. That causes for increase in probability of collision quenching and results OH reduces. Intensity of OH mass-fraction decreased with pressure from 1 atm to 16 atm. It is observed that maximum OH is occurred for lower dilution case of 9% $O_2$, for all three preheat temperatues. At 1 atm operating pressure higher level of OH is observed for all cases of three preheating and three dilution cases. Then the OH is reduced with increased operating pressure. The $O_2$ concentration in hot coflows with 3 and 6% $O_2$ level dilution is lower as compared to 9% $O_2$ diluted hot coflow. Hence the OH mass fraction is higher for 9% $O_2$ diluted cases and lower for 3 and 6%. For all the cases the $Y_{CO}$ is high at 1 atm and almost flat for 4 atm to 16 atm. Higher and lower $Y_{CO}$ are observed for lower (9% $O_2$) dilution levels. For certain dilution level, higher mass fraction of CO is observed for higher preheat temperatures.

## References

1. Wünning JA, Wünning JG (1997) Flameless oxidation to reduce thermal NO-formation. Prog Energy Combust Sci 23(1):81–94
2. Minamoto Y, Swaminathan N (2014) Scalar gradient behaviour in MILD combustion. Combust Flame 161(4):1063–1075
3. Minamoto Y, Swaminathan N, Cant RS et al (2014) Morphological and statistical features of reaction zones in MILD and premixed combustion. Combust Flame 161(11):2801–2814
4. Cavaliere A, de Joannon M (2004) Mild combustion. Prog Energy Combust Sci 30:329–366
5. de Joannon M, Cavaliere A, Faravelli T et al (2005) Analysis of process parameters for steady operations in methane mild combustion technology. Proc Combust Inst 30(2):2605–2612
6. de Joannon M, Matarazzo A, Sabia P et al (2007) Mild combustion in Homogeneous Charge Diffusion Ignition (HCDI) regime. Proc Combust Inst 31(2):3409–3416
7. Kumar S, Paul PJ, Mukunda HS (2002) Studies on a new high intensity low-emission burner. Proc Combust Inst 29(1):1131–1137
8. Szego GG, Dally BB, Nathan GJ (2008) Scaling of NOx emissions from a laboratory-scale mild combustion furnace. Combust Flame 154(1–2):281–295
9. Katsuki M, Hasegawa T (1998) The science and technology of combustion in highly preheated air. Proc Combust Inst 27:3135–3146
10. Gupta AK (2006) Developments in high temperature air combustion (Flameless Oxidation) and fuel reforming. In: 44th AIAA aerospace sciences meeting and exhibit 9-12 Jan 2006, Reno, Nevada
11. Blasiak W, Yang WH, Narayanan K et al (2007) Flameless oxyfuel combustion for fuel consumption and nitrogen oxides emissions reductions and productivity increase. J Energy Inst 80:3–11
12. Gupta AK, Bolz S (1999) Effect of air preheat temperature and oxygen concentration on flame structure and emission. J Energ Resour-ASME 121:209
13. Levy Y, Sherbaum V, Arfi P (2004) Basic thermodynamics of FLOXCOM, the low-NOx gas turbines adiabatic combustor. Appl Therm Eng 24:1593–1605
14. Reddy VM, Sawant D, Trivedi D et al (2013) Studies on a liquid fuel based two stage flameless combustor. Proc Combust Inst 34:3319–3326
15. Reddy VM, Katoch A, Roberts WL et al (2015) Experimental and numerical analysis for high intensity swirl based ultra-low emission flameless combustor operating with liquid fuels. Proc Combust Inst 35(3):3581–3589
16. Arghode VK, Gupta AK, Bryden KM (2012) High intensity colorless distributed combustion for ultra low emissions and enhanced performance. Appl Energ 92:822–830
17. Arghode VK, Gupta AK (2011) Development of high intensity CDC combustor for gas turbine engines. Appl Energ 88:963–973
18. Sabia P, Lavadera ML, Giudicianni P et al (2015) $CO_2$ and $H_2O$ effect on propane auto-ignition delay times under mild combustion operative conditions. Combust Flame 162:533–543
19. Sabia P, de Joannon M, Picarelli A et al (2013) Methane auto-ignition delay times and oxidation regimes in MILD combustion at atmospheric pressure. Combust Flame 160:47–55
20. Sabia P, de Joannon M, Lavadera ML et al (2014) Autoignition delay times of propane mixtures under MILD conditions at atmospheric pressure. Combust Flame 161:3022–3030

21. Sabia P, de Joannon M, Sorrentino GR et al (2015) Effects of mixture composition, dilution level and pressure on auto-ignition delay times of propane mixtures. Chem Eng J 277: 324–333
22. de Joannon M, Cavaliere A, Donnarumma R et al (2002) Dependence of autoignition delay on oxygen concentration in mild combustion of high molecular weight paraffin. Proc Combust Inst 29:1139–1146
23. Dally BB, Karpetis AN, Barlow RS (2002) Structure of turbulent non-premixed jet flames in a diluted hot coflow. Proc Combust Inst 29:1147–1154
24. Christo FC, Dally BB (2005) Modeling turbulent reacting jets issuing into a hot and diluted coflow. Combust Flame 142:117–129
25. Ye J, Medwell PR, Varea E, Kruse S et al (2016) An experimental study on MILD combustion of prevaporised liquid fuels. Appl Energ 151:93–101
26. Lückerath R, Meier M (2008) Aigner M (2008) FLOX Combustion at high pressure with different fuel compositions. ASME J Eng Gas Turbines Power 130:011505
27. Huang M, Zhang Z, Shao W et al (2014) MILD combustion for hydrogen and syngas at elevated pressures. J Therm Sci 23:96–102
28. Aminian J, Galletti C, Shahhosseini S et al (2012) Numerical investigation of a MILD combustion burner: analysis of mixing field, chemical kinetics, and turbulence-chemistry interaction. Flow Turbul Combust 88:597–623
29. Ihme M, Zhang J, He G et al (2012) Large-eddy simulation of a jet-in-hot-coflow burner operating in the oxygen-diluted combustion regime. Flow Turbul Combust 89:449–464
30. Lee J, Jeon S, Kim Y (2015) Multi-environment probability density function approach for turbulent $CH_4/H_2$ flames under the MILD combustion condition. Combust Flame 162: 1464–1476
31. De A, Dongre A (2015) Assessment of turbulence-chemistry interaction models in MILD combustion regime. Flow Turbul Combust 94:439–478
32. Chen Z, Reddy VM, Ruan S et al (2017) Simulation of MILD combustion using perfectly stirred reactor model. Proc Combust Inst. doi: 10.1016/j.proci.2016.06.007
33. Poinsot T, Veynante D (2005) Theoretical and numerical combustion. Edwards, USA
34. FLUENT 6.3 User's Guide
35. Dandy DS, Vosen SR (1992) Numerical and experimental studies of hydroxyl radical chemiluminescence in Methane-Air flames. Combust Sci Tech 82:131–150
36. Higgins B, McQuay MQ, Lacas F et al (2001) Systematic measurements of OH chemiluminescence for fuel-lean, high-pressure, premixed, laminar flames. Fuel 80:67–74

# Investigation of Forward and Reverse Flow CDC Combustors

**Vaibhav K. Arghode and Ashwani K. Gupta**

**Abstract** Colorless distributed combustion (CDC) is characterized by discrete and direct injection of air and fuel at high velocity in the combustor. Injected air mixes with the internally recirculated product gases to form hot and diluted oxidizer. Rapid mixing of the injected fuel with the oxidizer (prior to ignition) results in an overall lean combustion (near the global equivalence ratio) as compared to near stoichiometric combustion found in conventional diffusion flames. Hence, with CDC low pollutant emissions ($NO_x$, CO and UHC) can be achieved even in the non-premixed combustion mode. Different air and fuel injection configurations could be realized for a CDC combustor. In forward flow, air is injected from the opposite end of combustor exit and in reverse flow, air is injected from the same end of combustor exit. In both forward and reverse flow configurations, fuel can be injected from the same end of air injection or opposite end of air injection. Here, we will discuss the impact of various flow configurations on the overall flow field, gas recirculation, fuel/air mixing, residence time characteristics and combustion, emission behavior of CDC combustors.

**Keywords** Colorless distributed combustion · Forward flow · Reverse flow · Residence time characteristics · Emissions

**Nomenclature**

| | |
|---|---|
| *A* | Air |
| *B* | Burned gases |
| *C* | Tracer exit mass fraction |
| *CDC* | Colorless distributed combustion |

V.K. Arghode (✉)
Department of Aerospace Engineering, Indian Institute of Technology, Kanpur 208016, UP, India
e-mail: varghode@iitk.ac.in

A.K. Gupta
Department of Mechanical Engineering, University of Maryland, College Park, MD 20742, USA
e-mail: akgupta@umd.edu



A.K. Agarwal et al. (eds.), *Combustion for Power Generation and Transportation*, DOI 10.1007/978-981-10-3785-6_5

| | |
|---|---|
| *CFD* | Computational fluid dynamics |
| $D$ | Diameter of air injection |
| $D_{con}$ | Length scale of confinement |
| $F$ | Fuel |
| $F$ | Forward flow (air injection from same side of combustor exit) |
| *FLOX* | Flameless oxidation |
| *Hi-TAC* | High temperature air combustion |
| $L$ | Length scale of combustor |
| $m_{air}$ | Mass flow rate of air |
| $m_{fuel}$ | Mass flow rate of fuel |
| $m_{rec}$ | Mass flow rate of recirculated gases |
| *MILD* | Moderate and intense low oxygen dilution |
| $O$ | Oxidizer (air and burned gas mixture) |
| $O$ | Fuel injection from opposite side of air injection |
| $P$ | Premixed mode (air and fuel premixed prior to injection in combustor) |
| *psr* | Perfectly stirred reactor |
| $R$ | Reverse flow (air injection from opposite side of combustor exit) |
| $S$ | Fuel injection from same side of air injection |
| *SPRF* | Stagnation point reverse flow |
| $T$ | Temperature |
| *FO* | Air injection opposite to combustor exit, fuel injection opposite to air injection |
| *FP* | Air and fuel premixed, mixture injection opposite to combustor exit |
| *FS* | Air injection opposite to combustor exit, fuel injection on same side of air injection |
| *RO* | Air injection on same side of combustor exit, fuel injection on opposite of air injection |
| *RC* | Air injection on same side of combustor exit, fuel injection in cross flow to air injection |
| *RP* | Air and fuel premixed, mixture injection on same side of combustor exit |
| *RS* | Air injection opposite to combustor exit, fuel injection on same side of air injection |
| $V_i$ | Inlet velocity |
| $V_x$ | Axial velocity |
| $W$ | Length scale of combustor |
| $X$ | Distance along the combustor length |

# 1 Introduction

Recently, colorless distributed combustion (CDC) and related technologies have demonstrated very low pollutant emissions for gas turbine application [1–6] following its success in furnaces [7–9]. Note that the gas turbine combustors operate at significantly higher thermal intensity (25–50 MW/m$^3$-atm) as compared to furnaces (<1 MW/m$^3$-atm), thus posing additional challenges to achieve desirable behavior. Colorless combustion employs discrete and direct injection of air and fuel at high velocities (~50–100 m/s) into a combustion chamber. The injected air entrains the burned gases to form hot and diluted oxidizer and the injected fuel is expected to mix thoroughly with the oxidizer prior to ignition. Hence, in this case, lean combustion can be achieved (if the overall equivalence ratio is less than unity) even in the non-premixed mode of operation of the combustor. This and related technologies are known as, flameless oxidation (FLOX [3, 4, 8]), stagnation point reverse flow (SPRF [5, 6]), high temperature air combustion (HiTAC [7]), and moderate and intense low oxygen dilution (MILD [9]) combustion. Note that if the diluted oxidizer temperature is above the autoignition temperature of the fuel, spontaneous ignition of fuel is facilitated and reactions can occur in a distributed mode instead of a concentrated flame [10, 11]. In conventional non-premixed flames, the reactions are stabilized near unity equivalence ratio causing hot-spots, resulting in higher pollutant emissions (especially $NO_x$).

As there are three different ports, namely air, fuel and exit, various flow configurations can be realized by placing these ports in different combinations. Here the flow configurations are broadly classified as forward and reverse flow based on the positioning of the air jet with respect to the exit port and the same is depicted as schematics in Fig. 1. In the forward flow configuration ('F') air is injected from the opposite end of the exit and the bulk flow at any cross section is from the air injection towards exit. The injected air can entrain the product gases to support ignition and combustion, however, the net flow at any combustor cross section will correspond to the mass flow rate of the injected air. On the other hand, air can be injected from the same end as exit so that the bulk flow has to make a complete 'U'

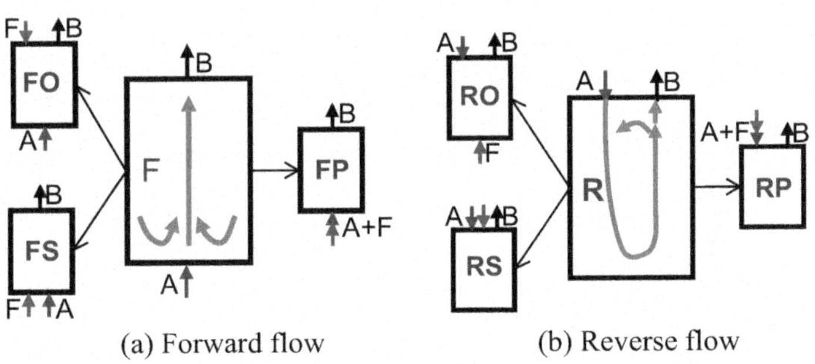


**Fig. 1** Different flow configurations for a CDC combustor

turn before escaping from the combustor, hence, this configuration is termed as reverse flow configuration, 'R'. Note that in this configuration also the product gases can be entrained and at any cross section of the combustor the net flow rate of the gases is zero as there is a complete reversal of the bulk flow.

The forward and reverse flow configurations can be further classified based on the location of fuel jet with respect to the air jet. Fuel can be injected from the same end of air injection ('S'), opposite end of air injection ('O'), or premixed with air prior to injection in the combustion chamber ('P'). The combination of these notations will give a specific flow configuration, for example, a combustor configuration in which air is injected from the same end as the exit and fuel is injected from the opposite end of air injection will be termed as 'RO', see Fig. 1b. Note that the momentum of the fuel jet is 1–2 orders of magnitude lower than that of the air jet primarily because of the high operating air/fuel ratios for hydrocarbon fuels. Hence, the bulk flow is expected to be governed by the dominant air jet and hence the classification of forward and reverse flow is based on the positioning of the air jet. The focus here is on investigation of different flow field configurations of CDC combustors to achieve low emissions of $NO_x$ and CO.

Figure 2 shows some of the forward and reverse flow configurations that have been reported in the literature for non-conventional combustors similar to CDC technology. These combustors operated in non-premixed mode wherein discrete injection of fuel and air is applied. In almost all the combustors a single flow configuration (either forward or reverse) has been reported, hence, the role of configuration is not clear on the combustion characteristics.

A summary of different non-conventional combustors as classified by the flow configurations is presented in Fig. 3. From the figure it can be noted that the configurations where fuel is injected from the same end as that of air (FS and RS) have been investigated by many researchers whereas the other configurations where fuel is injected from the opposite end of the air injection (FO and RO) have very few associated studies.

## 2 Background Calculations

Sample calculations are presented in this section to understand the effect of forward and reverse flow configurations in a combustion chamber. Forward and reverse flow configurations have significantly different bulk flow patterns, hence, simplified computations are presented to highlight the characteristics of a forward and revere flow configuration.

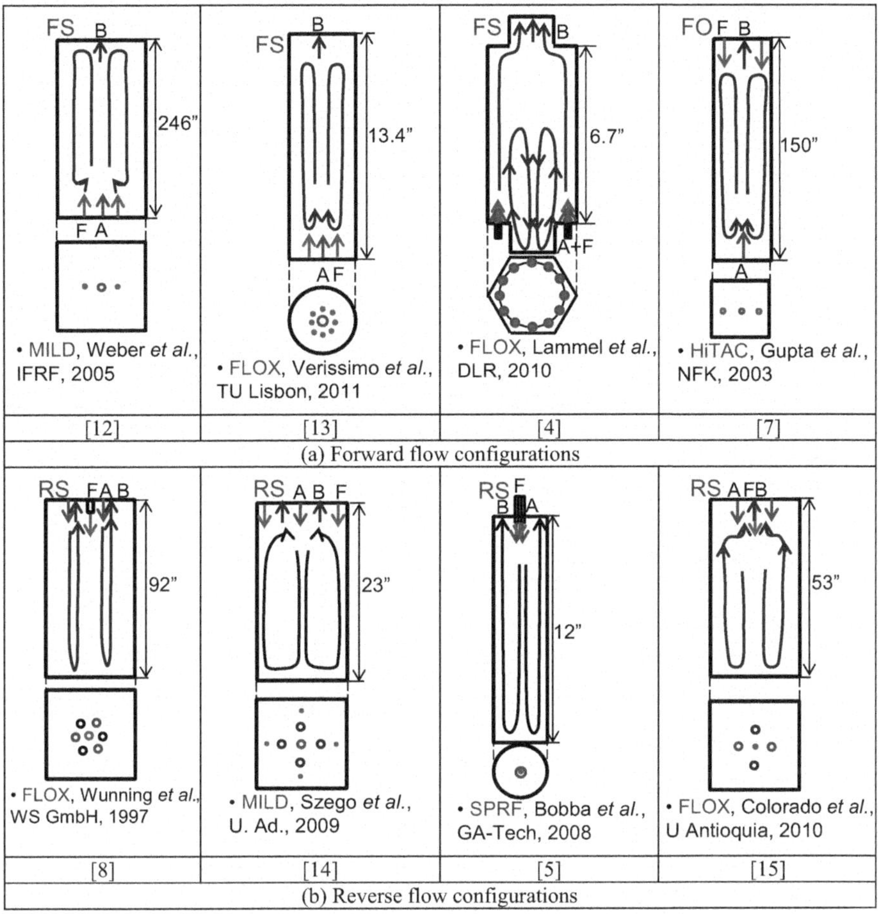


**Fig. 2** Schematics of different flow configurations reported in the literature

## 2.1 *Jet Decay and Recirculation Ratio*

To understand the effect of flow field configuration on gas flow and recirculation inside a confined chamber, CFD simulations were performed on an axisymmetric jet and results are shown in Fig. 4a, b. Both forward and reverse flow configurations were considered for the simulations, see Fig. 4c and d, respectively. The mesh used for simulations is shown in Fig. 4e. The air is injected with a velocity of 128 m/s and normal temperature and atmospheric pressure. For these CFD simulations commercial software FLUENT was used and standard k-ε model was used to model turbulence. Note that these are cold flow (non-reacting) simulations with only air as the fluid.

**Fig. 3** Summary of forward and reverse flow configurations reported in the literature

| | |
|---|---|
|  | 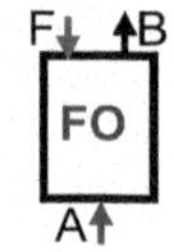 |
| Weber 2005 [12],<br>Gupta 2003 [7],<br>Sobiesiak 1998 [16],<br>Xing 2007 [17],<br>Verissimo 2011 [13],<br>Kumar 2002 [18],<br>Kumar 2005 [19],<br>Lukerath 2008 [3] | Gupta 2003 [7] |
| (a) Forward flow configurations | |
| 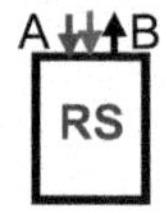 |  |
| Colorado 2010 [15],<br>Wunning 1997 [8],<br>Mi 2009 [20],<br>He 2008 [21],<br>Dally 2004 [22],<br>Szego 2009 [14],<br>Bobba 2008 [5],<br>Hsu 1998 [23] | Straub 2005 [24],<br>Melo 2009 [25] |
| (b) Reverse flow configurations | |

The results for the centerline air jet velocity decay are shown in Fig. 4a. The results are shown for different confinement sizes ($D/D_{con}$). It can be noted that for reverse flow configuration the jet velocity decay is faster as compared to the forward flow configuration. This may be because of higher shear experienced by the injected jet due to reversing flow from the annulus. The difference is escalated for smaller confinement size where space is limited for the gases to escape from the combustor.

Effect of the recirculation ratio is presented in Fig. 4b. Recirculation ratio is defined as the ratio of mass flow rate of entrained air with respect to mass flow rate of injected air. Note that the net mass flow rate across a combustor section will be equal to the mass flow rate of the injected air for the forward flow configurations and zero for the reverse flow configuration. For both forward and reverse flow configuration the mass flow rate of entrained air corresponds to the flow in the positive 'X' direction minus the mass flow rate of injected air. Note that the maximum recirculation ratio is limited by the confinement size, it increases along

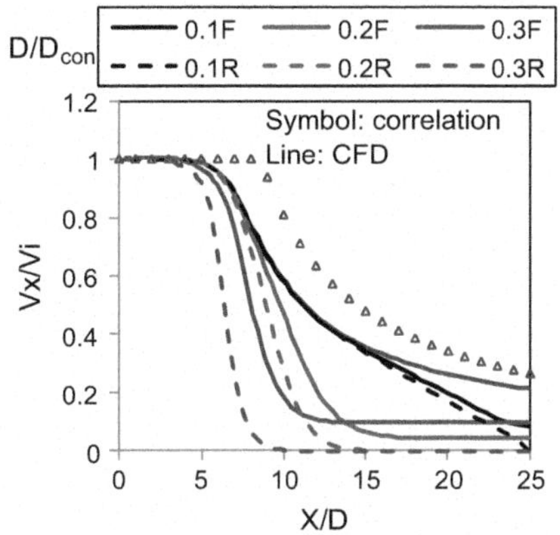


(a) Effect of flow configuration on centerline jet velocity decay (blue=free jet), corelation from [28]

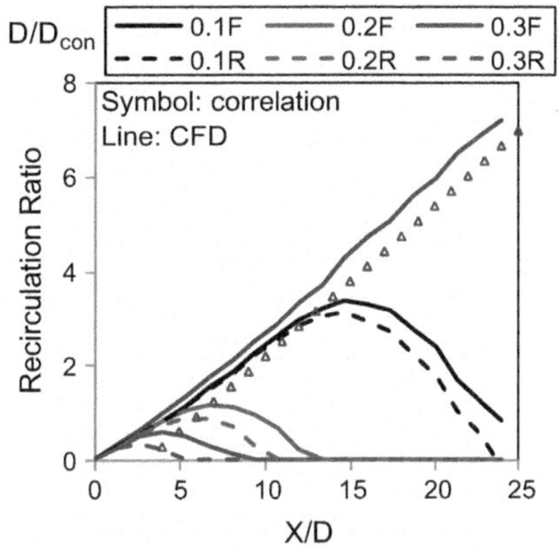


(b) Effect of flow configuration on recirculation ratio (blue=free jet) corelation from [27]

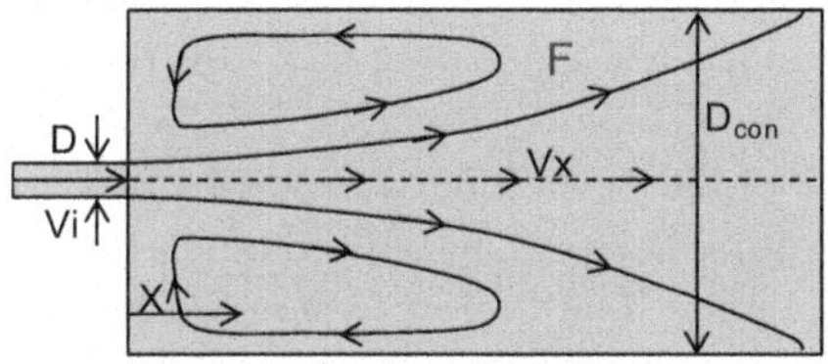


(c) Forward flow configuration

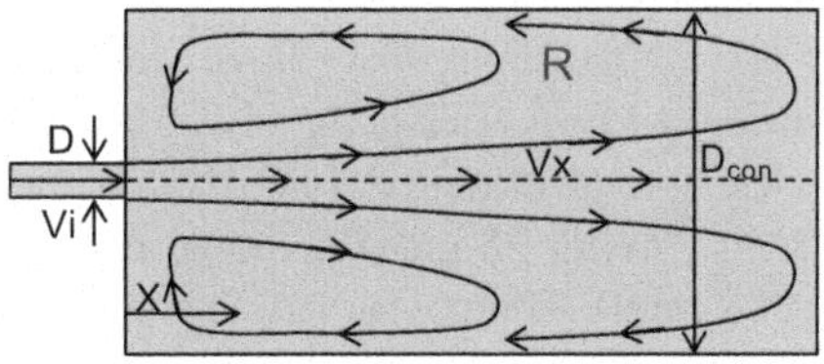


(d) Reverse flow configuration

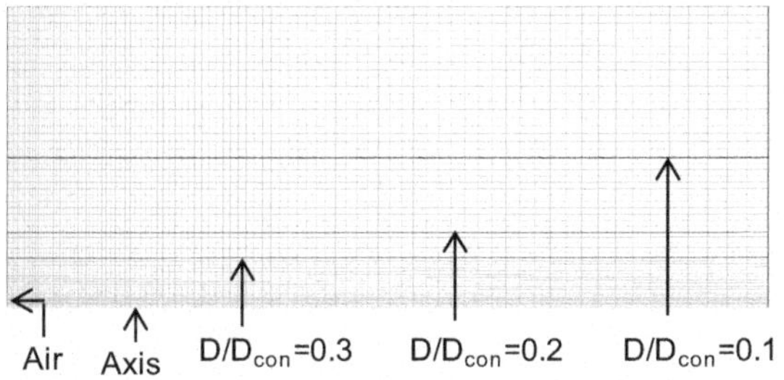


(e) Mesh used for simulations, length=25D, D=3/16 inch

**Fig. 4** Effect of flow configuration on centerline jet velocity decay and recirculation ratio

the length, reaches a peak value and then decreases. Reverse flow configuration results in slightly lower gas recirculation as compared to the forward flow configuration. This may be because the reversed gases occupy a certain portion (cross section) of the combustor thus effectively reducing the confinement size and it is clear from the figure that smaller confinement size leads to lower recirculation ratio, see Fig. 4b.

## 2.2 Residence Time Characteristics

Carbon monoxide oxidation is a slow process; hence, residence time has a significant effect on the CO emissions. To understand the effect of residence time on CO emissions, chemical kinetics simulations were performed assuming perfectly stirred reactor using the GRI-3.0 mechanism [26] and the results are shown in Fig. 5. The equivalence ratio used for the computations is 0.8 and fuel is methane and air is the oxidizer. The CO conversion rate (d[CO]/dt) with residence time is shown in Fig. 5a and from the figure it can be inferred that the CO conversion rate is much higher at low residence time, this means that if the gases are allowed to have more time before exiting the combustor it will result in lower CO emissions even though the average residence time is same.

The calculation for average residence time of 25 ms with different percentage of gases having low residence time is shown in Fig. 5b. From the figure it can be noted that if we have a higher percentage of gases having low residence time, CO emissions will increase even though the average residence time is same of all the cases. For example, if we have all the gases having residence time of 25 ms the CO emission from a perfectly stirred reactor will be 971 ppm. However, if 71% of the gases have a residence time of 5 ms and 29% of the gases have a residence time of 75 ms (giving 25 ms of average residence time) CO emission will be much higher (1788 ppm). Hence, with favorable residence time distribution, where we have a minimal fraction of gases having low residence time, higher conversion of CO to $CO_2$ and consequently lower CO emissions can be realized.

Residence time distribution was obtained for forward 'F' and reverse 'R' configurations by using a tracer having the same properties as that of air at confinement ratio $D/D_{con} = 0.2$ (see Fig. 4e). After obtaining a steady state solution of the flow field, the tracer concentration at the air inlet is set to 1 at time = 0, and only tracer

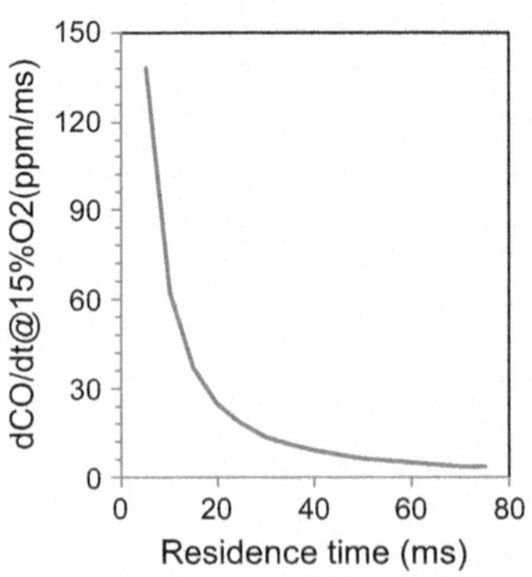


(a) Conversion rate of CO to $CO_2$, obtained for dt=5ms

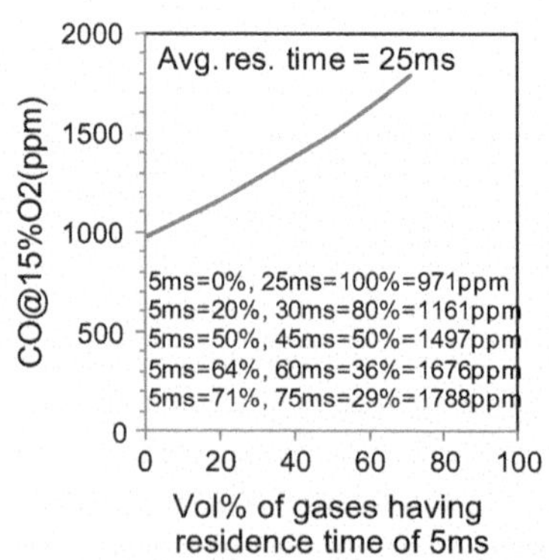


(b) Effect of gases having low residence time on CO emissions

**Fig. 5** Effect of residence time distribution on CO emissions conservation equation is solved in the transient mode with the time step size of 0.1 ms for both the configurations. The tracer is tracked from the air injection location to simulate the presence of reaction zone near the air injection location. The tracer concentration at the outlet is monitored with respect to time. The residence time is obtained from Eq. 1 given below. For these simulations, commercial software FLUENT was used. The standard k-ε model was used to model turbulence.

$$Residence\ time = t_{res} = \frac{1}{C_{mix}} c \int_{0}^{C_{mix}} t dC \tag{1}$$

$$C_{mix} = 1\,(for\ only\ air)$$

$$C = tracer\ mass\ fraction\ at\ exit$$

$$t = time$$

The residence time profile obtained from numerical simulations is also compared to the case of a perfectly stirred reactor. The exit tracer concentration profile with respect to time, for a perfectly stirred reactor corresponding to a step input of tracer at the inlet is given below in Eq. 2.

$$\frac{C}{C_{mix}} = 1 - exp\left(-\frac{t}{t_{res}}\right) \tag{2}$$

Residence time distribution for a perfectly stirred reactor is given by Eq. 3.

$$\frac{d(C/C_{mix})}{dt} = \frac{1}{t_{res}} exp\left(-\frac{t}{t_{res}}\right) \tag{3}$$

The results of tracer mass fraction contours, tracer exit mass fraction profile and residence time frequency distribution for both forward and reverse flow configurations is shown in Fig. 6. The average residence time is 23.8 ms based on the confinement volume and the flow rate. The residence time obtained from Eq. 1 for the forward flow configuration 'F' is 23.8 ms and that for the reverse flow mode "R" is 15.2 ms. Note that the residence time for the forward flow configuration is very close to that for a perfectly stirred reactor; however, for the reverse flow configuration residence time is lower. This may be due to insufficient penetration of the injected air jet, which results in large "dead" region near the closed end of the reverse flow configuration. This dead zone can be observed from the tracer mass fraction contour as shown in Fig. 6b, where very low concentration of tracer (mass fraction~0) is present. This suggests lower effective volume for the reverse flow configuration. The distribution of residence time is also important as the gases which escape much earlier can result in significant levels of CO emissions at the exit even though the average residence time is same in the combustor. From

(a) Tracer mass fraction for the forward flow configuration, $D/D_{con}=0.2$

(b) Tracer mass fraction for the reverse flow configuration, $D/D_{con}=0.2$

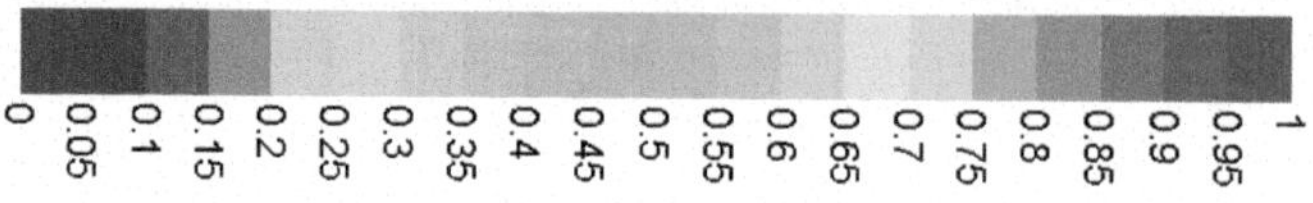


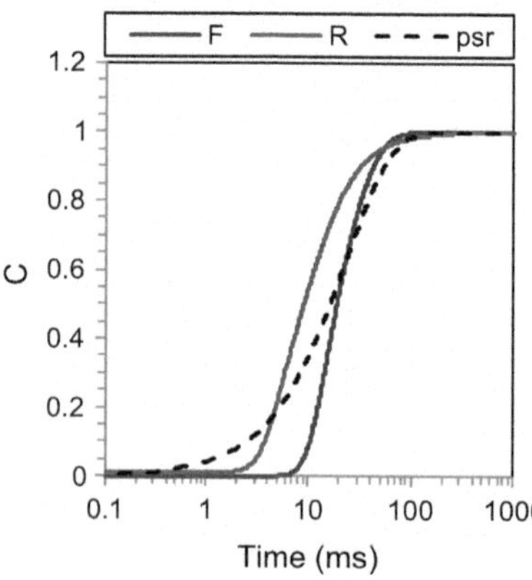


(c) Tracer mass fraction at the combustor exit.

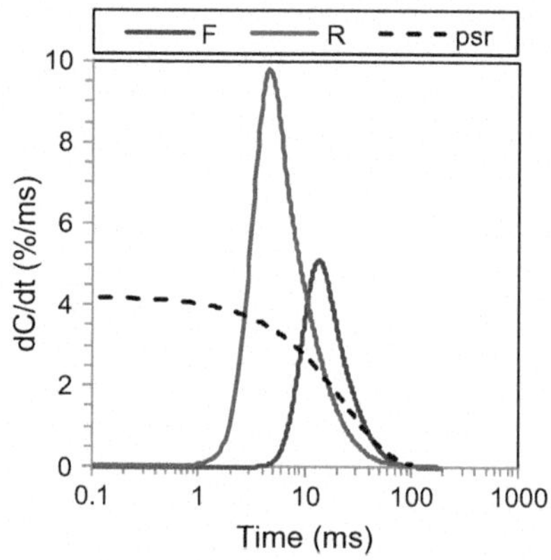


(d) Residence time distribution.

**Fig. 6** **a** Tracer mass fraction at the combustor exit and **b** residence time frequency distribution for configurations F and R (psr = perfectly stirred reactor), **c** Tracer mass fraction at the combustor exit, **d** Residence time distribution

frequency distribution plot given in Fig. 6d we can observe that for the reverse flow configuration more tracer has lower residence time of <10 ms as compared to the forward flow configuration. This may be due to the close vicinity of gas exit near the air injection location which results in early inception of tracer from the exit (see Fig. 6c) as well as lower overall residence time as mentioned above. Hence, for the investigated geometry, the reverse flow configuration suggests the undesirable residence time distribution of gases as compared to the forward flow geometry and this may result in an adverse pollutant emission profile of carbon monoxide.

It may be noted that these conclusions could not be generalized for all reverse and forward flow configurations and for other geometries we may obtain different results depending on the flow field inside the combustor. For certain cases reverse flow configuration could result in full penetration inside the confinement volume and due to reverse flow geometry the time required for the gases to escape can be higher as compared to the forward flow geometry resulting in favorable residence time and pollutant emission characteristics.

# 3 Experimental and Numerical Results

## *3.1 Operation at Low Thermal Intensity (5 MW/m³-atm)*

Initially a low thermal intensity combustor operating at 5 MW/m$^3$-atm was designed and examined. This combustor had four air and four fuel injection ports with the desired flexibility to operate in various air and fuel injection configurations. A photograph of this combustor and the different flow configurations investigated is shown in Fig. 7. Eight flow field configurations are reported and these include five forward flow configurations ('FS1', 'FS2', 'FO', 'FP1', 'FP2') and three reverse

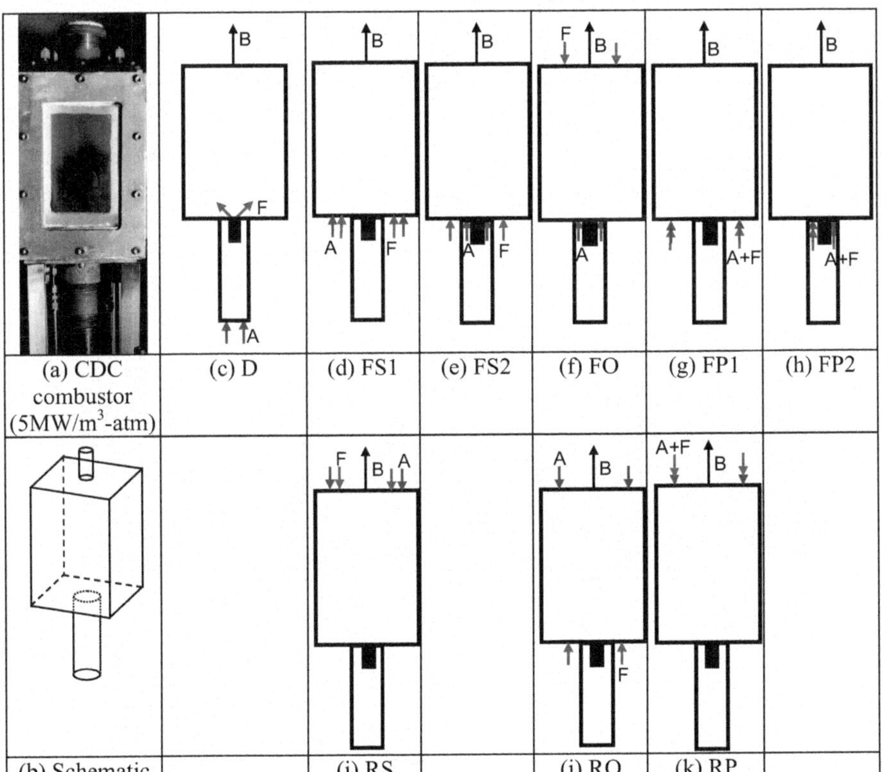


**Fig. 7** CDC combustor at thermal intensity of 5 MW/m$^3$-atm, **a** photograph, **b** schematic and **c**–**k** schematic diagrams of different flow configurations examined, heat load is 25 kW

flow configurations ('RS', 'RO', 'RP'). The combustor is started in diffusion flow mode ('D') and then transitioned to CDC mode. The combustor operates at a heat load of 25 kW with methane as the fuel and air as the oxidizer. Both air and fuel were injected at ambient temperature and the combustor was operated at atmospheric pressure. Fuel injection velocity was 97 m/s, and air injection velocity was 128 m/s at an equivalence ratio of 0.8.

Figure 8 shows velocity contour plots (from CFD simulations) for the nine configurations investigated here. Note that these are cold flow (non-reacting) simulations with air and fuel injection and the plane of visualization is along the

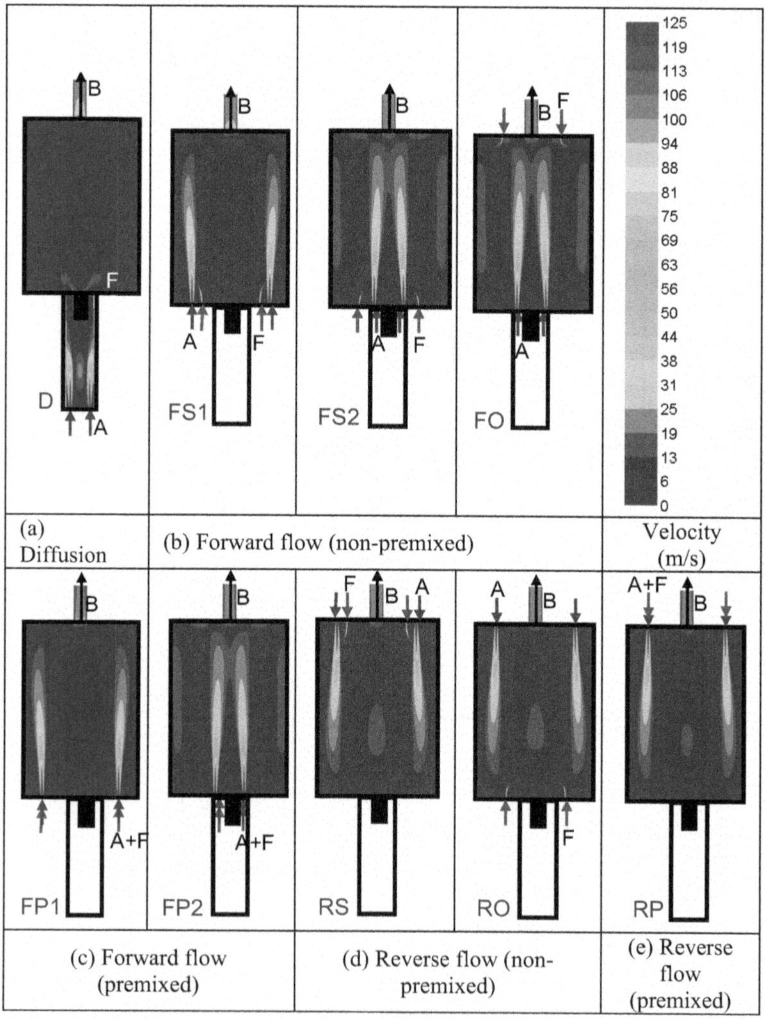


**Fig. 8** Velocity magnitude contour plots for different flow field configurations at low thermal intensity (5 $MW/m^3$-atm)

diagonal of the combustor cross section. For these simulations commercial software FLUENT was used. Standard k-ε model was used to model turbulence. In the diffusion flame mode (see, Fig. 8a) rapid air jet decay is observed due to confinement from the pipe. Uniform air flow is observed near the fuel injection location where fuel is injected at 45° angle. The flame is stabilized due to the fuel injector which acts as a bluff body. The recirculated hot product gases also aid in stabilization of the flame which recirculate inside the combustor.

Velocity contour plots for the forward flow configurations in the non-premixed mode are shown in Fig. 8b. Air jet decays along the length of the combustor and mostly complete decay is observed as the air jet reaches the top of the combustor. Due to high momentum of air jet, the fuel jet entrains into the air jet for both configurations 'FS1' and 'FS2'. In configuration 'FS2' due to closeness of air jets to each other the air jets bend towards each other, which makes the product gases even more likely to escape earlier from the combustor. In the opposed flow configuration 'FO', the fuel jet is sheared due to the mixture of recirculated gases and the air jet. In all the forward flow modes the fuel jet decay is fast suggesting better mixing. The velocity contour plots for the forward flow configurations in premixed mode ('FP1' and 'FP2') are shown in Fig. 8c, from the contour plots it can be observed that the absence of fuel jet does not affect the air jet profile significantly.

Figure 8d shows velocity contour plots for the reverse flow configurations in non-premixed mode 'RS', 'RO'. For configuration 'RS' the fuel jet gets entrained into the air jet due to the presence of recirculated gases. For configuration 'RO' the fuel jet decay is faster due to shearing by the recirculated gases. Similar to the forward flow cases the premixed mode in reverse flow also suggest that the air jet is the dominant jet and is unaffected by the presence of fuel jet in the non-premixed modes. Thus, these simulations highlight significant differences in the flow field and fuel/air mixing patterns based on the flow configurations.

Numerical calculations of residence time distribution were performed for configurations 'FP1', 'FP2' and 'RP'. Premixed cases are considered as air is the dominant jet and mostly governs the flow field for the geometries under investigation. First a steady state solution of non-reacting (only air) flow field was obtained and then a tracer, having the same property as that of air is introduced from the air injection location, and the exit mass fraction is tracked with respect to time, refer to Sect. 2.2 for more details. Residence time calculations for the three premixed modes ('FP1', 'FP2' and 'RP') are shown in Fig. 9. Figure 9a shows the exit mass fraction profile and Fig. 9b shows the residence time distribution. The average residence time for the three cases is 589 ms for 'FP1', 588 ms for 'FP2', 590 ms for 'RP' and 585 ms based on the combustor volume and the flow rate (the same is used for perfectly stirred reactor (psr)). This suggests that the average residence times are very similar irrespective of the flow configuration. Note that this is in contrast with the observations made in Sect. 2.2 where the residence time for the reverse flow configuration was lower than that in the forward flow configuration. This may be due to sufficient air jet penetration in the reverse flow configuration for the geometrical cases shown in Figs. 7 and 8 thus allowing complete utilization of the combustor volume.

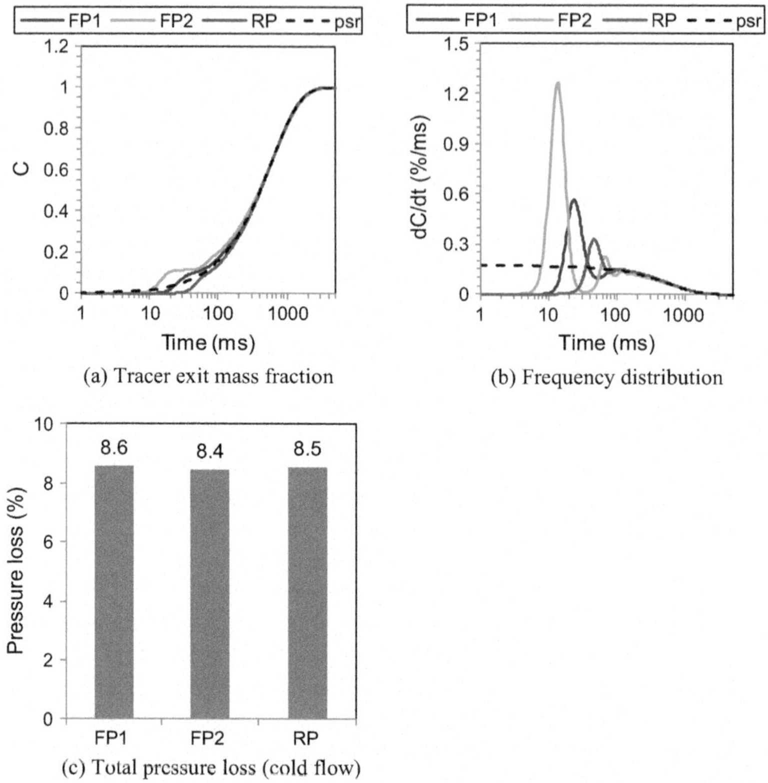


**Fig. 9** **a** Tracer exit mass fraction, **b** frequency distribution for configurations FP1, FP2 and RP and its comparison with perfectly stirred reactor (psr), **c** Total pressure loss (cold flow)

From the residence time distribution, configuration 'FP1' has the highest fraction of tracer particles having low residence time, whereas configuration 'RP' has lowest fraction of tracer particles having low residence time. This could also be observed from early inception of tracer at the exit for configuration 'FP2' as compared to configuration 'RP'. From Fig. 5b we learn that for the case with higher fraction of gases having lower residence time can lead to the overall high emission of CO in the exit gases even though the average residence time is same. Hence, the residence time calculation suggests that the reverse flow configuration can result in low CO emissions as compared to the forward flow configuration in the present case. It may be noted that these observations are in contrast with that noted from Fig. 6, where the reverse flow configuration resulted in unfavorable residence time characteristics. Thus the influence of confinement and flow configuration both play a role in flow field characteristics and generalizations are to be done with caution. These simple calculations suggest that one can have the benefit of favorable residence time distribution by tailoring the flow field by operating optimal configuration even though the average residence time is the same for different configurations for a given volume.

Total pressure loss from the cold flow condition was obtained for configurations ‘FP1’, ‘FP2’ and ‘RP’ from the CFD simulations. The pressure loss is shown in Fig. 9c. From the figure it may be noted that the pressure loss is about 8.5% of the inlet total pressure and is independent of the flow field configurations. The pressure loss calculated from simple calculations based on loss of dynamic heat at the combustor inlet is about 8.7%, which is very close to the values obtained from the CFD simulations. Note that the loss in total pressure can be reduced by lowering the air injection velocity so that the dynamic pressure loss is reduced.

The NO and CO emissions at an equivalence ratio of 0.8 are shown in Fig. 10. It can be observed that the premixed flow mode (‘FP1’, ‘FP2’, ‘RP’) produced lowest NO levels (about 1 ppm). Lowest CO emission (about 5 ppm) was obtained for the reverse flow configuration, operating in the premixed mode (‘RP’). This may be due

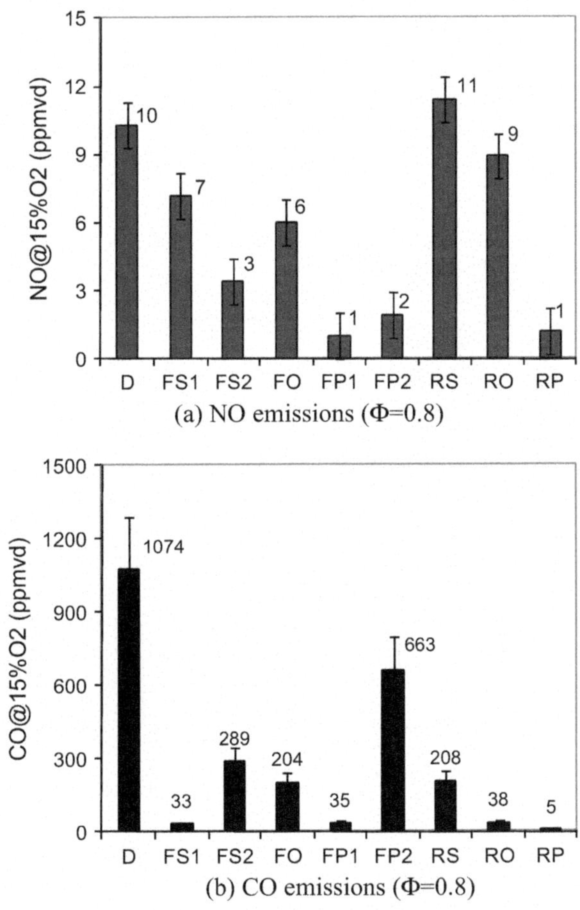


**Fig. 10** **a** NO and **b** CO emissions of CDC combustor at thermal intensity of 5 MW/m$^3$-atm and heat load of 25 kW, for different flow-field configurations

to higher fraction of gases having higher residence time associated for this configuration as compared to other forward flow configurations (‘FP1’ and ‘FP2’) as the burned gases have to reverse the direction before exiting from the combustor, also see Fig. 9b for residence time distribution for different configurations. Amongst the non-premixed modes, configuration ‘FS1’ resulted in low NO (about 7 ppm) and low CO (about 33 ppm) emission levels. As the air injection port was shifted towards the combustor centerline (‘FS2’, ‘FP2’), CO emissions increased significantly as compared to the configurations where the air injection port was near the combustor walls (‘FS1’, ‘FP1’). This may be due to higher fraction of gases having lower residence time for the configurations ‘FS2’ and ‘FP2’ due to faster escape of burned gases from the combustor exit which is located along centerline of the combustor, also see Fig. 9b. This highlights the importance of location of injection of the reactants into the combustor.

### 3.2 *Operation at High Thermal Intensity (28 MW/m³-atm)*

Different flow field configurations at higher thermal intensity of 28 MW/m$^3$-atm were investigated using a simple combustor geometry shown in Fig. 11a. For the higher thermal intensity combustor the number of air and fuel injection ports has been decreased to one from four and the length of the combustor was decreased to 3 in.

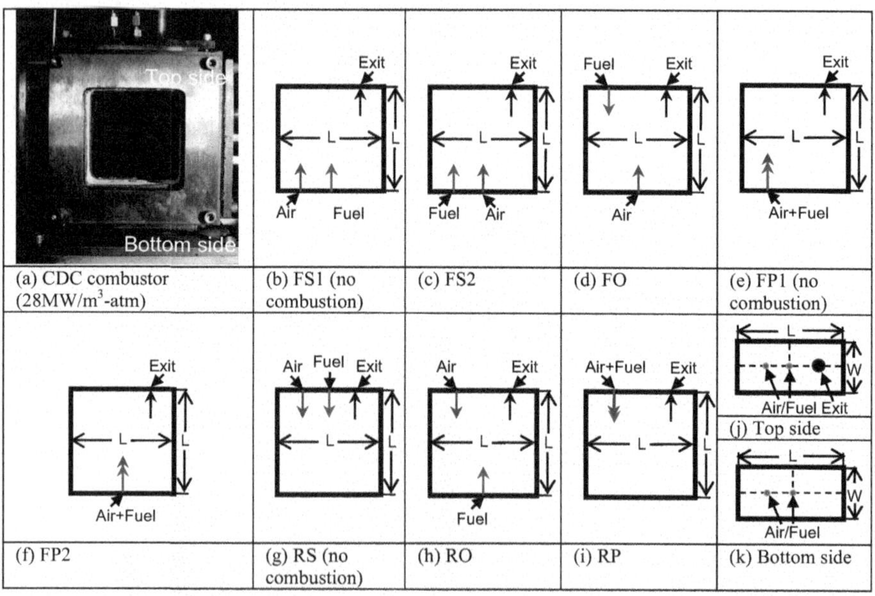


**Fig. 11** CDC combustor (28 MW/m$^3$-atm, 6.25 kW), **a** photograph and **b–i** schematic diagrams of different geometries examined

from 9 in. as compared to the lower thermal intensity (5 MW/m$^3$-atm) combustor discussed in Sect. 3.1. Note that the flow field, though expected to be similar for the analogous configurations at lower and higher thermal intensities, need not be exactly same as the confinement geometry size and shape and number of ports are different. The heat load was fixed at 6.25 kW and air and fuel was injected at normal temperature and combustor operated at atmospheric pressure for all the cases. The fuel injection velocity was 97 m/s and air injection velocity varied from 128–205 m/s to provide the variation in equivalence ratio from 0.8 to 0.5, respectively. The combustor was operated at atmospheric pressure.

Figure 12 shows velocity contour plots for the non-premixed configurations in cold flow (non-reacting) conditions. Commercial software FLUENT was used for the simulations and standard k-ε model was used to model turbulence. As observed from the contour plots, air jet decays along the length of the combustor; however, the combustor length appears to be insufficient for the complete decay of the air jet. One interesting feature that can be noted from the contour plots for configurations 'FS2' and 'FO' is the bending of air jet away from the exit location (see Fig. 12b, c). This suggests that a significant portion of the injected air turns away from the exit and recirculate inside the combustor from the left side of the air jet. The velocity contour plot for configuration 'FS1' is shown in Fig. 12a. It can be observed that the air jet does not bend as observed for configurations 'FS2' and 'FO', suggesting that the gases recirculate from both sides of the air jet. For configuration 'FS1' the fuel jet decay is delayed, suggesting weaker cross flow for this configuration as observed from the velocity contour plots (see Fig. 12a). The cross flow from the recirculating gases near the fuel injection for configurations 'FS1' and 'FO' is expected to be strong, hence the fuel jet decay is faster suggesting faster fuel/oxidizer mixing for these configurations. For configurations 'RS' and 'RO' the air jet splits in two parts and recirculate from both left and right side of the air jet. The contour plots also suggests that the cross flow of recirculating gases near the fuel injection is very weak for configuration 'RS', and hence the fuel jet decay is delayed for this configuration as observed from the velocity contour plots (see Fig. 12e). The cross flow from the recirculating gases near the fuel injection for configurations 'RO' is expected to be strong (see Fig. 12d) hence the fuel jet decay is faster, suggesting faster fuel/oxidizer mixing for these configurations, which is desirable to achieve reactions closer to distributed regime.

Figure 13 shows the centerline velocity decay for air jet and its comparison with a free jet decay (from correlation [28]) for the three forward flow configurations. It can be observed that the decay of air jet is much faster as compared to the free jet suggesting that the confinement has a significant effect on the air jet. Configuration 'FS1' suggests delayed decay of air jet as compared to the other two non-premixed configurations. Figure 13b shows the decay of fuel jet and its comparison with a free jet (from correlation [28]) for the three forward flow configurations. It can be observed that the fuel jet decay is significantly faster for configurations 'FS2' and 'FO' as compared to configuration 'FS1'. This reveals slower mixing for configurations 'FS1' and this may result in lower stability for this configuration.

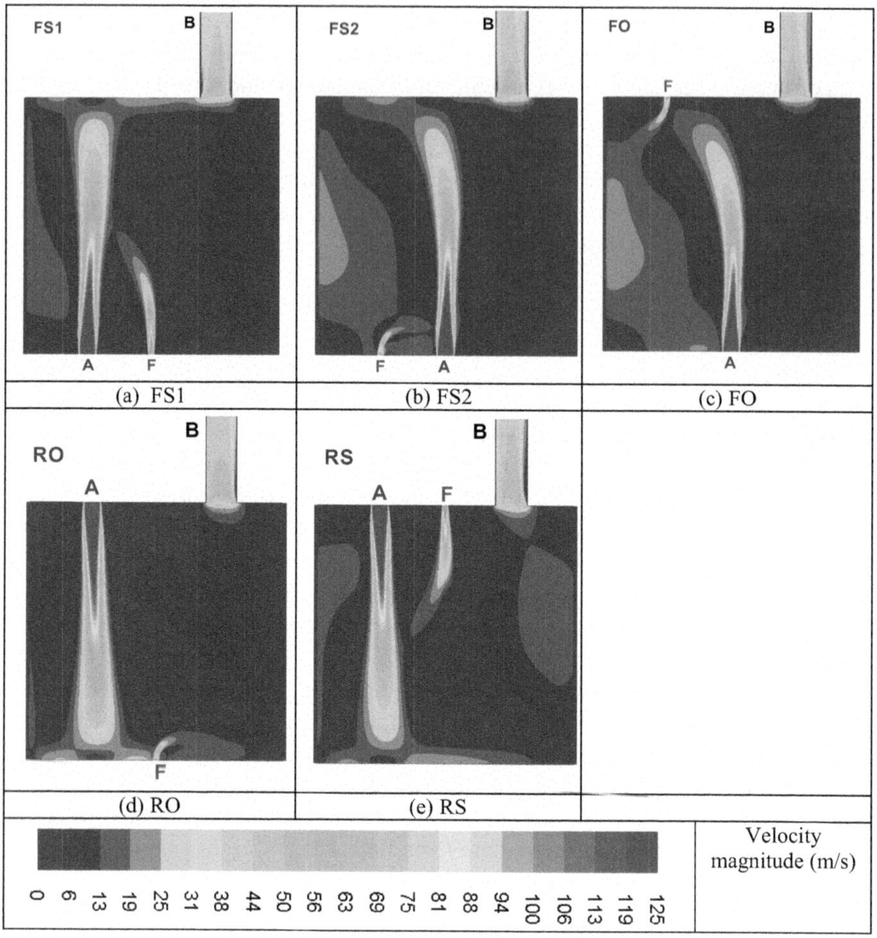


**Fig. 12** Velocity magnitude contour plots for different configurations

Variation of the recirculation ratio of gases inside the combustor along the length of the combustor as well as a comparison with the entrainment ratio in a free jet (from correlation [27]) is presented in Fig. 13c. The figure reveals that the variation of the recirculation ratio along the length of the combustor is significantly different for the confined chamber as compared to the entrainment ratio for the free jet case. The amount of recirculation near the air injection is much higher for all three configurations as compared to a free jet. This may be due to the presence of recirculating gases inside the combustor that turns towards the bottom side of the combustor after coming from combustor side walls. It may be noted that although the recirculation ratio is higher as compared to the entrainment in a free jet, the recirculation ratio increases initially and then decreases along the length of the combustor, suggesting that the maximum recirculation is limited by the

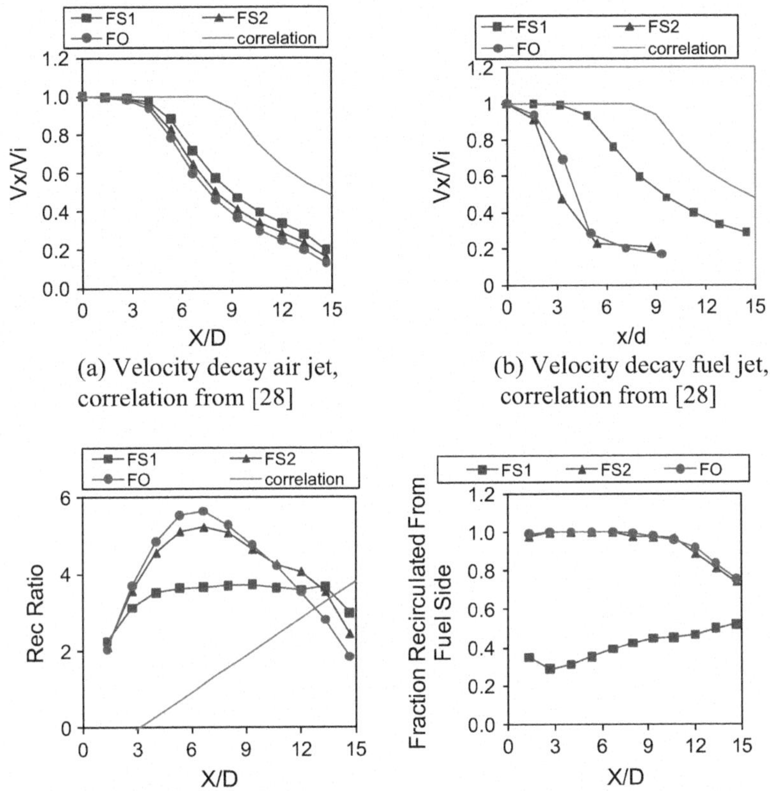


(a) Velocity decay air jet, correlation from [28]

(b) Velocity decay fuel jet, correlation from [28]

(c) Gas recirculation, correlation from [28]

(d) Fraction recirculated from fuel side

**Fig. 13** Velocity decay and gas recirculation for forward flow configuration

confinement size. It can also be observed that the recirculation ratio for configuration 'FS1' is significantly lower than the recirculation ratio for configurations 'FS2' and 'FO'.

The portion of recirculated gases from the fuel side of the combustor is shown in Fig. 13d. The fuel side is the portion of the combustor, when partitioned vertically based on the positioning of air jet, in which the fuel injection port is present. Hence, for the present investigation the fuel side for configurations 'FS2' and 'FO' is the left portion of the combustor w.r.t. the air injection port and for configuration 'FS1' it is the right portion of the combustor w.r.t. the air injection port, Figs. 11 and 12 for different flow field configurations. From Fig. 13d it can be noted that the gases recirculate almost entirely from the fuel side for configurations 'FS2' and 'FO'; however, only about 40% of the gases recirculate from the fuel side for configurations 'FS1'. Recirculation ratio calculations reveal that configuration 'FS1' may not facilitate conditions conducive for reaction zone stabilization. This was also observed from experiments and combustion could not be sustained for

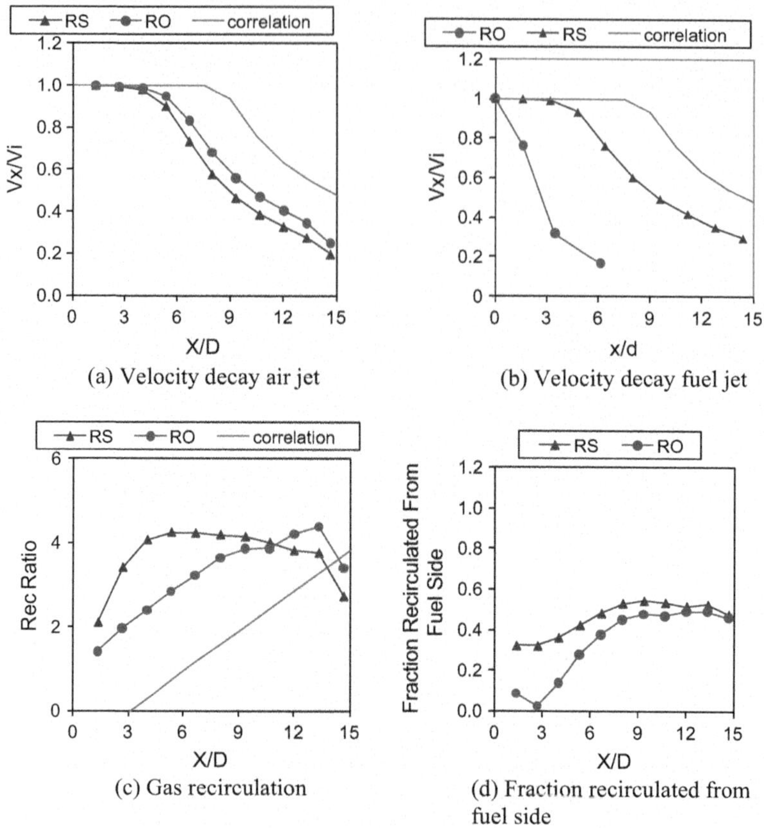


(a) Velocity decay air jet
(b) Velocity decay fuel jet
(c) Gas recirculation
(d) Fraction recirculated from fuel side

**Fig. 14** Velocity decay and gas recirculation for reverse flow

configuration 'FS1'. The premixed mode where air was injected from the side port as in 'FS1' also could not sustain the flame experimentally (configuration 'FP1').

Therefore, from now onwards, only results corresponding to configurations 'FS2', 'FO' and 'FP2' are presented for the forward flow configurations.

Figure 14a shows centerline velocity decay for air jet and its comparison with free jet decay (from correlation) for the reverse flow configurations 'RS' and 'RO'. It can be observed that the decay of air jet is much faster as compared to the free jet decay suggesting that the confinement has a significant effect on the air jet. Figure 14b shows the decay of fuel jet and its comparison with a free jet (from correlation [28]) for the reverse flow configurations 'RS' and 'RO'. It can be observed that the fuel jet decay is significantly faster for configurations 'RO' as compared to configuration 'RS'. This reveals slower mixing for configurations 'RS' and this may result in lower stability for this configuration.

Variation of the recirculation ratio of gases inside the combustor along the length of the combustor as well as comparison with entrainment ratio for free jet (from correlation [27]) is presented in Fig. 14c. The figure reveals that the variation of the recirculation ratio along the length of the combustor is significantly different for a confined chamber as compared to the entrainment ratio for a free jet. The amount of recirculation near the air injection is much higher for all three configurations as compared to the free jet similar to that observed or the forward flow configurations. It may be noted that although the recirculation ratio is higher as compared to the entrainment in free jet, the recirculation ratio increases initially and then decreases along the length of the combustor, suggesting that the maximum recirculation is limited by the confinement size. It can also be observed that the recirculation ratio for configuration 'RS' is significantly higher than the recirculation ratio for configuration 'RO', and this may be due to the effect of fuel injection jet which also will entrain the gases inside the combustor. The portion of recirculated gases from the fuel side of the combustor is shown in Fig. 14d. The fuel side is the portion of the combustor w.r.t. the air jet from where the fuel is injected, hence, for the present investigation fuel side for configurations 'RS' and 'RO' is on the right side of the air injection, see Figs. 11 and 12 for different flow field configurations. From the figure it can be noted that the gases recirculate from both sides of the air jet for both configurations 'RS' and 'RO'.

Experimentally, configuration 'RS' was not able to stabilize combustion hence onwards no results are presented for 'RS' configuration and only results corresponding to the configurations 'RO' and 'RP' are presented.

Experimentally, configurations 'FS1', 'FP1' and 'RS' could not sustain combustion due to lower recirculation of gases as well as slower decay of fuel jet as observed numerically. Note that configurations FS1, FP1 and RS were able to support combustion experimentally for the low thermal intensity combustor, as shown in Fig. 7. This further highlights the importance of the confinement size and shape on the flow field inside the combustor and reaction zone stabilization. Detailed investigations are required to analyze the combustor performance due to interrelations of many factors involved in stabilization of the reaction zone and generalizations are to be done with caution.

Experimentally measured NO and CO emissions are presented in Fig. 15. In the non-premixed mode, higher NO emissions were observed for both forward and reverse flow configurations (see Fig. 15a) with highest emissions observed for the reverse flow configuration 'RO'. Forward flow configuration 'FO' resulted in lower CO levels (about 200 ppm) as compared to the forward flow configuration 'FS2' but higher CO emissions as compared to the reverse flow configuration 'RO' (see Fig. 15b). This may be due to favorable residence time distribution for the opposed flow configuration ('FO') as compared to the configuration where both air and fuel are injected from the same side ('FS2'). This was also confirmed from the numerical simulations (not presented here). Reverse flow configuration 'RO' had lower CO emissions as compared to the forward flow configuration 'FO' and this further suggests a favorable residence time for reverse flow configuration as compared to the forward flow configuration.

The premixed flow mode in both forward ('FP2') and reverse ('RP') flow configuration resulted in the lowest NO levels (about 1 ppm); however, CO

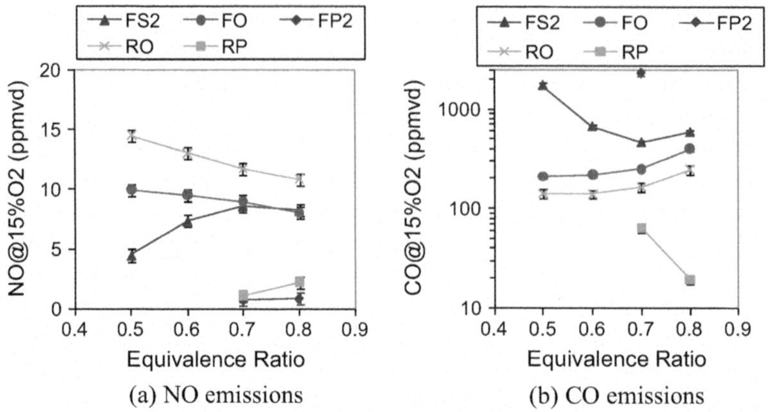


**Fig. 15** NO and CO emissions for the CDC combustor at thermal intensity of 28 MW/m$^3$-atm and heat load of 6.25 kW

emission was very high (about 2300 ppm) for the forward flow configuration but considerably low (about 50 ppm) for the reverse flow configuration. In the reverse flow geometry, the gases turn around and then exit the combustor, whereas in the forward flow geometry, there is a possibility of higher fraction of gases exiting the combustor quickly. This would lead to favorable residence time distribution (less fraction of gases having small residence time) for reverse flow configuration and hence lower CO emission as compared to the forward flow configuration. These observations are similar to that observed for the case with low thermal intensity combustor. Higher CO emissions also suggest that there may be higher unburned hydrocarbon (UHC) emission and hence, lower combustion efficiency for the forward flow configuration 'FP2'.

## 4 Conclusions

Amongst the investigated flow field configurations, the reverse flow configurations resulted in lower CO emissions (measured) possibly due to favourable residence time distribution (computed) as compared to the forward flow configurations. The bulk flow field inside the combustor was observed to be governed by the dominant air jet and the fuel jet had minimal influence on the flow field (computed). The location of fuel jet was found to have significant influence on the emission levels as placing fuel jet in strong cross flowing gases suggested better fuel/oxidizer mixing (computed) resulting in lower emissions (measured). The recirculation pattern of gases inside the combustor was found to be very sensitive to the air injection location with respect to the exit port, (computed). Higher gas recirculation from the fuel side suggested better reaction zone stabilization. The influence of confinement geometry was found to be very important in governing the flow field and the combustion behaviour as the flow field configurations which supported combustion at lower thermal intensity (larger confinement volume) could not sustain combustion at higher thermal intensity (smaller confinement volume). Thus, many factors need to be taken into account in order to design a low emission CDC combustor and the results presented here can serve as a guideline for development of CDC combustors for practical applications.

**Acknowledgements** This research was supported by ONR, Program Managers, Dr. Gabriel D. Roy and Dr. Clifford Bedford. The support is gratefully acknowledged. The authors also acknowledge the support provided by Reaction Design for the Chemkin-PRO code.

## References

1. Arghode VK, Gupta AK, Bryden KM (2012) High intensity colorless distributed combustion for ultra low emissions and enhanced performance. Appl Energy 92:822–830
2. Arghode VK, Gupta AK (2013) Role of thermal intensity on operational characteristics of ultra-low emission colorless distributed combustion. Appl Energy 111:930–956
3. Luckerath R, Meier W, Aigner M (2008) FLOX® combustion at high pressure with different fuel compositions. J Eng Gas Turbines Power 130:011505-1–011505-7
4. Lammel O, Schutz H, Schmitz G, Luckerath R, Stohr M, Noll B, Aigner M, Hase M, Krebs W (2010) FLOX® combustion at high power density and high flame temperatures. J Eng Gas Turbines Power 132:121503-1-10
5. Bobba MK, Gopalakrishnan P, Periagaram K, Seitzman JM (2008) Flame structure and stabilization mechanisms in a stagnation-point reverse-flow combustor. J Eng Gas Turbines Power 130:031505-1-8
6. Gopalakrishnan P, Bobba MK, Seitzman JM (2007) Controlling mechanisms for low NOx emissions in a non-premixed stagnation point reverse flow combustor. Proc Combust Inst 31:3401–3408
7. Tsuji H, Gupta AK, Hasegawa T, Katsuki M, Kishimoto K, Morita M (2003) High temperature air combustion: from energy conservation to pollution reduction. CRC Press
8. Wunning JA, Wunning JG (1997) Flameless oxidation to reduce thermal NO formation. Prog Energy Combust Sci 23:81–94
9. Cavaliere A, de Joannon M (2004) Mild combustion. Prog Energy Combust Sci 30:329–366
10. Ozdemir IB, Peters N (2001) Characteristics of the reaction zone in a combustor operating at mild combustion. Exp Fluids 30:683–695
11. Plessings T, Peters N, Wunning JG (1998) Laseroptical investigation of highly preheated combustion with strong exhaust gas recirculation. Proc Combust Inst 27:3197–3204
12. Weber R, Smart JP, vd Kamp W (2005) On the (MILD) combustion of gaseous, liquid, and solid fuels in high temperature preheated air. Proc Combust Inst 30:2623–2629
13. Verissimo AS, Rocha AMA, Costa M (2011) Operational, combustion and emission characteristics of a small scale combustor. Energy Fuels 25:2469–2480
14. Szego GG, Dally BB, Nathan GJ (2009) Operational characteristics of a parallel Jet MILD combustion burner system. Combust Flame 156:429–438
15. Colorado AF, Herrera BA, Amell AA (2010) Performance of a flameless combustion furnace using biogas and natural gas. Bioresour Technol 101:2443–2449
16. Sobiesiak A, Rahbar S, Becker HA (1998) Performance characteristics of the novel low-NOx CGRI burner for use with high air preheat. Combust Flame 115:93–125
17. Xing X, Wang B, Lin Q (2007) Structure of reaction zone of normal temperature air flameless combustion in a 2 ton/h coal-fired boiler furnace. J Power Energy 221:473–480

18. Kumar S, Paul PJ, Mukunda HS (2002) Studies on a new high-intensity low-emission burner. Proc Combust Inst 29:1131–1137
19. Kumar S, Paul PJ, Mukunda HS (2005) Investigations of the scaling criteria for a mild combustion burner. Proc Combust Inst 30:2613–2621
20. Mi J, Li P, Dally BB, Craig RA (2009) Importance of initial momentum rate and air-fuel premixing on moderate or intense low oxygen dilution (MILD) combustion in a recuperative furnace. Energy Fuels 23:5349–5356
21. He Y (2008) Flameless combustion of natural gas in the SJ/WJ furnace. PhD dissertation-Queen's University
22. Dally BB, Riesmeier E, Peters N (2004) Effect of fuel mixture on moderate and intense low oxygen dilution combustion. Combust Flame 137:418–431
23. Hsu KY, Goss LP, Roquemore WM (1998) Characteristics of a trapped-vortex combustor. J Propuls Power 14:57–65
24. Straub DL, Casleton KH, Lewis RE, Sidwell TG, Maloney DJ, Richards GA (2005) Assessment of rich-burn, quick-mix, lean-burn trapped vortex combustor for stationary gas turbines. J Eng Gas Turbines Power 127:36–41
25. Melo MJ, Sousa JMM, Costa M, Levy Y (2009) Experimental investigation of a novel combustor model for gas turbines. J Propuls Power 25:609–617
26. Smith GP, Golden DM, Frenklach M, Moriarty NW, Eiteneer B, Goldenberg M, Bowman CT, Hanson RK, Song S, Gardiner WC, Lissianski VV, Qin Z, http://www.me.berkeley.edu/gri_mech/
27. Ricou FP, Spalding DB (1961) Measurements of entrainment by axisymmetrical turbulent jets. J Fluid Mech 11:21–32
28. Hussein HJ, Capp SP, George WK (1994) Velocity measurements in a high-Reynolds-number, momentum-conserving, axisymmetric, turbulent jet. J Fluid Mech 258:31–75

# Global and Local Viewpoints to Analyze Turbulence-Premixed Flame Interaction

**Swetaprovo Chaudhuri**

**Abstract** Turbulence remains one of the most important unresolved problems in classical physics. The addition of exothermicity through a myriad of chemical reactions in fluid turbulence renders turbulent combustion a formidable challenge alongside its ubiquity in all chemical to mechanical energy conversion devices. Here, we review some recent advancements in theory, experiments, and computations of turbulent premixed combustion through two ostensibly decoupled: global and local viewpoints. Global viewpoints are essential to obtain useful statistics of turbulent flame propagation such as turbulent flame speed. Local viewpoints are crucial towards understanding detailed phenomena such as how local flame elements are stretched or annihilated. These fine grained understandings must culminate towards various sub-models of a successful global model. Essential features of recent theoretical developments towards turbulent flame speed using the level set formulation are reviewed here. The theoretical turbulent flame speed is shown to conform to the experimental expanding flame measurements over a wide range of fuels, pressure and turbulence intensity. Finally, recent advances to track flame elements and their local properties from detailed chemistry direct numerical simulations, are reviewed through a Lagrangian viewpoint.

## 1 Introduction

Turbulent flame speed: the global statistic of turbulent premixed flame propagation is a topic of high interest in turbulent combustion research, as evidenced by the large volume of analytical [1–8], experimental [9–15], computational [16, 17] and review literature [18–22] that has emerged. Its practical relevance can be readily appreciated by recognizing that, turbulent flame speed can be correlated to the global heat release rate in turbulent reacting flows. Furthermore, turbulent flame speed determines mean flame location in a premixed combustor and holds the key towards unraveling the

S. Chaudhuri (✉)
Department of Aerospace Engineering, National Center for Combustion Research and Development, Indian Institute of Science, Bengaluru, India
e-mail: schaudhuri@aero.iisc.ernet.in; swetaprovo@gmail.com



A.K. Agarwal et al. (eds.), *Combustion for Power Generation and Transportation*, DOI 10.1007/978-981-10-3785-6_6

deflagration to detonation transition (DDT) phenomena [23] in a supernova Ia. Fundamentally, the problem of turbulent flame speed is of very high complexity due to multiscale interaction of turbulence, transport and chemical reactions. The difficulty is further compounded by the disagreement between theories as well as the high degree of uncertainties in the experimental turbulent flame speed datasets and their sensitivity on the geometry and type of the burner used in the investigation [22]. Consequently, while understanding of the laminar flame speed is at such an advanced stage that it is now routinely used for validating reaction mechanisms; the complexities just mentioned have prevented unequivocal utilization of turbulent flame speed as a fundamental quantity in the study of turbulent reacting flows.

It is of high interest to seek a unified scaling description of turbulent flame speed as a function of few flow and flame parameters, at least under some special flow conditions such as those in nearly isotropic turbulence. Such a unified scaling description, if it exists, could for example be incorporated as a sub-grid scale model for Large Eddy Simulations (LES) or Reynolds Averaged Navier Stokes (RANS) modeling of turbulent premixed combustion. The apparent choice of the flame parameters for such a scaling would be the planar laminar flame speed $S_L$ and the corresponding laminar flame thickness $\delta_L$. Then to the leading order the problem of turbulent flame speed could be considered as a geometric one, where one has to estimate the flame surface area, in which the effect of turbulence is to stretch and wrinkle the flame over a wide range of length scales without disturbing the inner structure. Such a problem was recently considered in [24] for unity Lewis number (*Le*) statistically planar flames, propagating in homogeneous isotropic turbulence. In fact, it is also recognized that the results of [24] should be valid for any mixture with positive Markstein number (*Mk*) as long as the governing formulation of the *G*-equation remains well posed. To demonstrate this generalization, we note that the concept used to arrive at the normalized turbulent flame speed expression in [24] is the following: turbulence imposes the $k^{-5/3}$ dependence on the $g^2$ spectrum $\Gamma(k)$, (where $g$ at a point in space is the interface height from its mean position and assumes the role of a passive scalar) by wrinkling the interface at a multitude of wavenumbers $k$ for moderate to large turbulent Reynolds number $Re_T$. In analogy with the hydrodynamic theory of premixed flames, positive Markstein diffusivity $D_M = S_L.\delta_M$ ($S_L$ being the planar laminar flame speed and $\delta_M$ being the Markstein length) dissipates the flame surface fluctuations at large $k$, which results in exponential decaying terms on $\Gamma(k)$. If $D_M$ is negative, instead of dissipation it provides amplification, but in this condition the governing *G*-equation becomes ill-posed and hence the negative $D_M$ case is not addressed by the theory. For positive $D_M$, the dissipation effects become cumulative as the normalized turbulent flame speed can be expressed as a square root of the integral, for all $k$, over the $k^2\Gamma(k)$ spectrum.

In [24], the following was thus found to the leading order

$$S_T/S_L \sim \left[\left(u'_{rms}/S_L\right)\left(L_I/\delta_M\right)\right]^{1/2} \tag{1}$$

where $u'_{rms}$ is the root mean square of velocity fluctuations, $L_I$ the velocity integral length scale which was also assumed to be the hydrodynamic length scale of the flame, and $\delta_M = Mk.\delta_L$ the Markstein length. Brief review of this approach and a simpler derivation is presented in Sect. 3.

Primarily, non-equidiffusive transport in flames due to non-unity Lewis number (*Le*) results in variations of *Mk*. Hence, to arrive at a fuel independent scaling of turbulent flame speed it is crucial to develop a fundamental understanding of *Le* and *Mk* effects. In this review we will mainly focus on conditions that are free from hydrodynamic and diffusion thermal instabilities i.e. positive *Mk*, unless otherwise stated. All lean premixed flames of large hydrocarbon-air mixtures are characterized by $Le > 1$ and positive *Mk*, with hydrodynamic instability being only important at weak turbulence intensity as shown in [24]. Thus the interest to study such intrinsic instability-free condition is natural from a practical point of view as well.

In this chapter we also review recent experimental turbulent flame speed datasets measured from constant-pressure statistically spherical flames in near isotropic turbulence, for different values of *Mk* for various fuels. Examining the data over large range of conditions of pressure, $u'_{rms}$ and fuels: $H_2$, $C_1$-$C_4$ n-alkanes and dimethyl ether [25], it has been shown that for all conditions, the scaling of Eq. (1), slightly modified due to the evolving nature of the flames, approximately holds for all positive *Mk*, fuels, pressure and turbulence intensity conditions considered. As shown in [25], the recent iso-octane turbulent flame speed database from Leeds [26] also appears to follow this scaling and agrees well with the above data in the framework of Eq. (1).

Finally, we will discuss recent developments on Lagrangian analysis of turbulent premixed flames called Flame Particle Tracking (FPT) [27]. Rigorous Lagrangian formulations introduced by Pope [28] in non-reacting flows, has been successfully implemented in DNS datasets of reacting flows through FPT. Interesting insights on flame propagation and annihilation, obtained through FPT will be discussed.

## 2 Turbulent Premixed Combustion Regimes

Peters [29] used the level set approach to derive an expression for the turbulent flame speed in the thin reaction zone regime. According to his definition, such a regime is characterized by the Karlovitz Number: $Ka > 1$ such that the flame time scale is smaller than the Kolmogorov time scale, implying that Kolmogorov eddies penetrate the preheat zone to replace molecular transport with turbulent transport. Peters obtained the following expression of the normalized turbulent flame speed in the thickened flamelet regime given by

$$S_T/S_L = \sqrt{u'_{rms} L_I/D} \tag{2}$$

where $D$ is the mass diffusivity. Relating $D$ to $S_L$, $\delta_L$, and the thermal diffusivity through $Le$ [23], (2) can be expressed as,

$$S_T/S_L = \sqrt{Le\left(u'_{rms}/S_L\right)\left(L_I/\delta_L\right)} \quad (3)$$

If Peters expression is correct, in this regime characterized by moderate to large turbulent Reynolds number: $Re_{T,f} = \left(u'_{rms}/S_L\right)\left(L_I/\delta_L\right)$, the normalized turbulent flame speed increases with increasing $Le$ or is independent of $Le$ if $S_L$ and $\delta_L$ are assumed to scale with molecular diffusivity instead of thermal diffusivity. However, there is strong evidence from present and recent experimental studies, which suggests a different trend in such moderate to strong turbulent Reynolds numbers. The overwhelming observation is that with decrease in $Le$ or $Mk$, the normalized turbulent flame speed increases. It is to be carefully noted that this observation may not invalidate Eq. (2) for it is possible that statistically thickened, purely turbulence transport dominated flame elements may not be found in the so called "thin reaction zone regime" when the Kolmogorov length scale ($\eta$) is smaller than $\delta_L$. Evidence of decreasing normalized turbulent flame speed (at a constant $Re_{T,f}$) with increasing $Le$ or $Mk$, in such a regime suggests the presence of laminar flamelet mode of local flame propagation i.e. strong influence of molecular and thermal diffusion, at least in a statistical sense. The most crucial question that arises out of these observations is that, do eddies of Kolmogorov length scale ($\eta$) have sufficient kinetic energy to transfer thermal energy and mass, in and out of the preheat zone, noting that $\eta$ is precisely the smallest statistical length scale in continuum turbulence where dissipation rather than energy transfer dominates? Secondly, any activity around $\eta$ should be dissipation dominated and dissipation range intermittency is a well-known phenomenon in fluid turbulence. The fractal theories of turbulence that can explain the anomalous scaling of the higher order velocity structure functions (i.e. deviation from Kolmogorov scaling beyond the third order structure function) as a consequence of intermittency, suggest that the small scales around $\eta$ may not be space filling in nature. Thus even if they have sufficient energy to cause local mass and thermal transport, their sparsity may largely reduce their statistical significance for moderate turbulent Reynolds numbers. Of course, it is also well known that these intermittent events increase with turbulent Reynolds number. Pope [30] suggested using the characteristic length scale of $13\eta$ which corresponds to the centroid of the dissipation spectrum. While this can be considered as the characteristic length scale for the boundary between the thin reaction zone and the corrugated flamelet regimes, a less conservative estimate at moderate $Re_{T,f}$, for thickened flamelets, can be given by the integral length scale ($L_I$). Eddies of size $L_I$ carry a major fraction of the total turbulent kinetic energy and are also space filling such that the disruption of the preheat zone thickness by an eddy of size $L_I$, could be guaranteed almost over the entire flame surface. If the turnover time of this eddy is smaller than the flame time given by $\delta_L/S_L$, turbulent transport is expectedly predominant in the preheat zone. In recent direct numerical simulations by Poludnenko and Oran [31] and Aspden et al. [32] where $L_I$ was chosen to be $O(\delta_L)$, this is precisely the case. Thus the $Da = 1$

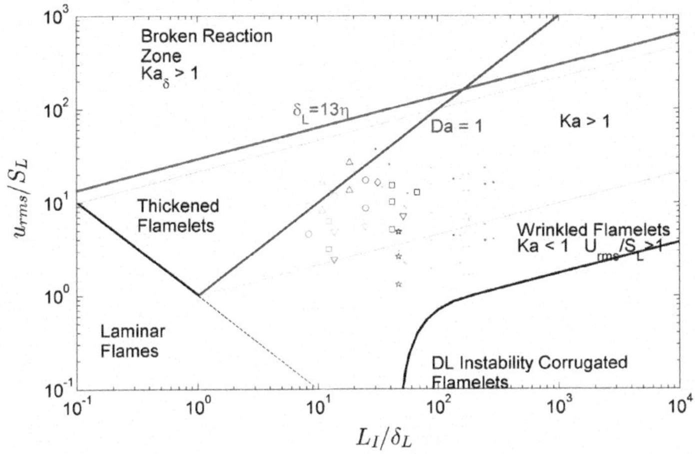


**Fig. 1** Regime diagram with conditions of present ($C_1$-$C_4$) experiments and those for iso-octane data from [26]. The legends for present experimental conditions (*symbols*) and corresponding laminar flame and turbulence parameters could be found in Table 1. The legends (*dots*) flame and flow parameters for the Leeds data could be found in Table 2. Reproduced with permission from [25]

line where

$$Da = t_I/t_F = L_I S_L u_{rms}^{-1} \delta_L^{-1} \tag{4}$$

can provide a better estimate of a statistically thickened flamelet regime above which all the statistical properties of combustion, such as turbulent flame speed and brush thickness are governed by turbulent transport. These boundaries, conditions of present experiments and recent Leeds data for iso-octane [26] to be discussed in Sects. 4 and 5, are presented in the regime diagram of Fig. 1. It seems logical that a combination of the $Da = 1$ line (for moderate $Re_{T,f}$ when $(u'_{rms}/S_L)_{Da=1} < (u'_{rms}/S_L)_{\delta_L=13\eta}$) and the $\delta_L = 13\eta$ line (for large $Re_{T,f}$ when $(u'_{rms}/S_L)_{Da=1} > (u'_{rms}/S_L)_{\delta_L=13\eta}$) could be a more prominent boundary between the laminar flamelet and thickened flamelet regime. As shown in [24], the regime diagram on Fig. 1 highlights the introduction of a new regime boundary which separates flame affected by Darrieus-Landau instability versus those in which turbulence prevents any intrinsic instability growth. Preliminary experimental validation of this regime boundary has been recently obtained in [33].

## 3 Eulerian Theory

Starting with the work of Kerstein et al. [5], the problem of premixed flame propagation in turbulent flows has often been analyzed using the level set formulation or the *G*-equation. For the turbulent flame speed, the *G*-equation and the flame surface density equation coupled with the progress variable approach have been used in the framework of Reynolds-Averaged Navier Stokes (RANS) equations formulation [4–6]. However, a RANS model with $k - \varepsilon$ type closure cannot explicitly consider the evolution of scale-dependent global properties. In general the multiscale nature of turbulence in premixed combustion has been considered only in very few analytical studies. In particular, Peters [34] proposed spectral closure of the $G$-equation on the basis of dimensional arguments and extended it in [35] to include effects of gas expansion. Collins and co-workers adopted a more rigorous but complex approach [36, 37] by using the Eddy Damped Quasi Normal Markovian (EDQNM) approximation. EDQNM is one of the statistical closure theories for homogenous turbulence. Dandekar and Collins [38] performed DNS of the $G$-equation including gas expansion effects on the flame surface by considering the Sivashinsky integral over different $Le$.

It was recognized by Oberlack et al. [39] that due to the generalized scaling symmetry of the $G$ equation, Reynolds averaging of $G$ cannot yield a unique mean flame position. This is due to the fact that the functional value of $G$ on the interface, being a level set, does not carry any unique physical information. This problem could be circumvented, as in [40], by using the following well known ansatz. We define a $z$-distance function $g(x, y, z, t) = G(x, y, z, t) - z = G_N - z$ at all points $(x, y, z)_N$ belonging to each of the interfaces given by $G = G_N$ required for the field formulation. The flame can be considered to be one of these interfaces, say given by $G = G_0$. For the homogeneity and isotropy of the scalar field $g(x, y, z, t)$ over a volume, it is necessary that $\langle g(x, y, z, t)\rangle = 0$, by averaging over each interface. This requires $\langle G\rangle = \langle z\rangle = G_N$ at each of the interfaces, constraining assignment of the level set values. A mean velocity in the $z$ direction equal to the turbulent flame speed of the interfaces is required to ensure that each interface remains statistically stationary. Of course, this also predefines the mean interface location at $z = G_N$. However, defining $g$ as the physical $z$-distance of the instantaneous interface position from the mean surface, Reynolds averaging and subsequent spectral closure by Peters [34] could be adopted for the $g^2$ transport equation as in [40]. As in any level set problem, the $g^2$ transport equation is defined within the entire cubic volume and is not limited by interface folding. Since the fluctuating velocity and $g$ field is homogenous and isotropic, we need not distinguish between the actual flame and other interfaces. As in the work of Kerstein et al. [5], volume averaging can be performed with ergodic hypothesis to yield global properties such as the surface area of the flame. All the interfaces, each of which can be considered as a realization of a flame surface is defined by its local laminar flame speed $\tilde{S_L}$

$$\tilde{S_L} = S_L - S_L\delta_M.\kappa - \delta_M.S \tag{5}$$

where $S_L$ is the laminar flame speed of a planar premixed flame with respect to the unburnt gas; $\kappa$ and $S$ are the local curvature and strain rate respectively. Here, only a linear correction to $S_L$ by $\kappa$ is considered which is ostensibly applicable only for small $\kappa$. Kelley et al. [41] showed that experimental laminar flame speed data from spherical flames can be exactly explained only by consideration of nonlinear stretch rate effects. However it also shows that the linear correction above, which was Markstein's original empirical correction, works reasonably well for large $\kappa$ as well, for both positive and negative $\delta_M$. Since the target of the work is to obtain a leading order scaling only, the linear correction is adopted in this work.

Following [34], the autocorrelation of the isotropic $g$ field given by $R(\mathbf{r}, t) = \langle g(\mathbf{x}, t) g(\mathbf{x} + \mathbf{r}, t) \rangle$ was Fourier transformed to yield the $g^2$ spectrum given by $\Gamma(k, t) = 4\pi k^2 \hat{R}(k, t)$, where $\mathbf{r}$ is the separation vector, $k$ the wavenumber and $\hat{R}$ the Fourier transform of $R$. The transport terms of the transport equation for $\Gamma(k, t)$, derived from the $G$ equation, was closed by Pao's gradient transfer hypothesis and the dissipation terms were closed on the basis of dimensional arguments to give the following form of the $g^2$ spectrum.

$$\begin{aligned}\Gamma(k) = BH\left(k - k_I\right) k^{-5/3} \exp\left[-3c_1 (2\pi)^{1/3} \left(\frac{u'_{rms}}{S_L}\right)^{-1} \left(\frac{k}{k_I}\right)^{1/3}\right] \times \\ \exp\left[-\frac{3}{4} (2\pi)^{4/3} c_2 Mk \left(\frac{u'_{rms}}{S_L}\right)^{-1} \left(\frac{k_I}{k_L}\right) \left(\frac{k}{k_I}\right)^{4/3}\right]\end{aligned} \tag{6}$$

where $B$ is the integration constant and $H\left(k - k_I\right)$ is a function determined from initial conditions. Here $H = 0$ if $k < k_I$ and $H = 1$ otherwise, in such a manner that $\Gamma$ is independent of time. Furthermore, $c_1, c_2$ are constants, $k_I = 2\pi / L_I$, and $k_L = 2\pi / \delta_L$. In $\Gamma(k)$, the term $k^{-5/3}$ arises from stretching and wrinkling of the interfaces due to the imposed isotropic turbulence followed by two exponentially decaying dissipation terms. Here, kinematic restoration is described by the first dissipation term while curvature dissipation by positive $Mk$ is captured by the second dissipation term. Peters [35] showed that gas expansion effects, through the Sivashinsky integral, results in a term identical to the one due to kinematic restoration albeit with a positive sign within the exponent. Hence in strong turbulence, the net result of gas expansion is a reduction of dissipation by Huygens propagation, leaving the curvature dissipation term dominant. It is important to recognize that the expression of $\Gamma(k)$ given by Eq. (6) is only valid for positive $Mk$. Clearly a negative $Mk$ would result in a blowup of the spectrum at large wavenumbers. It is well known that negative $Mk$ results in diffusion-thermal instability [42] manifested as amplification of flame surface fluctuations instead of dissipation (for positive $Mk$). The blowup of the spectrum for negative $Mk$ thus indicates similar phenomena, though with the unrealistic unlimited amplification at high wavenumbers. Physically, when a flame is stretched or curved at very large wavenumbers, eventually the stretch time scales will be shorter than the chemical time scales thereby causing extinction even for negative $Mk$. In the following, we will mainly focus on positive $Mk$ cases.

Kerstein et al. [5] showed that with the field formulation of the scalar $g$, the turbulent flame speed is given by the volume averaged functional of the absolute gradient of $G$ for initially planar flames.

$$S_T/S_L = \langle | \nabla G | \rangle \approx \left\langle \sqrt{1 + \nabla g . \nabla g} \right\rangle \tag{7}$$

It was shown in [24] that for the derivatives of $g$ following normal distribution or the gradient squared of following lognormal distribution, the following approximation holds, which allows determination of the turbulent flame speed directly from the spectrum given by Eq. (6),

$$\begin{aligned} S_T/S_L = \left\langle \sqrt{1 + \nabla g . \nabla g} \right\rangle \approx \sqrt{\langle 1 + \nabla g . \nabla g \rangle} &= \sqrt{1 + \int_0^\infty k^2 \Gamma(k) dk} \\ &= \sqrt{1 + \int_{k_I}^\infty k^2 \Gamma(k) dk} \end{aligned} \tag{8}$$

Dandekar and Collins [38] showed by DNS of the $G$-equation but including the Sivashinsky integral that the probability density function of $| \nabla G |$ is indeed lognormal. It is known that if a variable $X$ is lognormal with its natural logarithm having mean and variance as $\mu$ and $\sigma$ respectively, the variable $X^a$ is also lognormal with its natural logarithm having mean and variance as $a\mu$ and $\sigma^2 a^2$ for $a \neq 0$. Therefore the assumption of $\partial g / \partial x_i . \partial g / \partial x_i$ following lognormal distribution is justified. Equation (8) is further simplified to finally yield the following scaling to the leading order:

$$\frac{S_T}{S_L} \sim \sqrt{\left(\frac{1}{Mk}\right)\left(\frac{u'_{rms}}{S_L}\right)\left(\frac{L_I}{\delta_L}\right)} \tag{9}$$

Thus from the result and discussion above, it is clear that the normalized turbulent flame speed should decrease with increasing $Mk$ or $Le$ at a constant $Re_{T,f}$. The negative-$Mk$ turbulent flame speed has often been described with the "leading points" concept proposed by the Russian school and discussed in [18] which states that since the positive curvature leading points have maximum $\tilde{S_L}$, they locally propagate farthest out into the fresh mixture, thereby increasing the overall flame surface area and turbulent flame speed. Broadly, it is clear, that the spectral concepts discussed above arrive at an identical physical mechanism, to explain the effects of $Mk$ which controls the flame surface area by dissipating or amplifying the fluctuations depending on its sign and magnitude.

Instead of using the spectrum, the same result can be arrived at by using the dissipation rate scaling [29] which has been one of the most central concepts of turbulence for the past several decades. Though the approach is simpler, it lacks the detailed insight of scale dependent interaction of kinematic restoration, gas expansion and curvature. This is also the concept by which Peters [43] arrived at Eq. (4). But there are important conceptual differences in the ostensibly similar final expression for normalized turbulent flame speed. Since we assume a locally laminar flame structure, we assume dissipation by Markstein diffusivity rather than molecular diffusivity as assumed by Peters [29] for the thickened flamelets.

It is known that the mean turbulent kinetic energy dissipation rate $\langle \varepsilon \rangle = \nu . \left\langle \left( \partial u_i / \partial x_j + \partial u_j / \partial x_i \right)^2 \right\rangle$ follows the "dissipative anomaly" from moderate $Re_\lambda$ onwards, where $\lambda$ is the Taylor scale. The mean turbulent kinetic energy dissipation rate can be equated to the inertial range scale invariant turbulent kinetic energy transfer rate which is independent of kinematic viscosity. This is one of the central concepts of homogenous isotropic turbulence which implies: $\langle \varepsilon \rangle \sim u_l'^3 / l \sim u_{rms}'^3 / L_I$. Here $l$ could be any length scale in the inertial range. Similarly, for the scalar dissipation rate $\langle \chi \rangle = D \left\langle \left( \partial \phi / \partial x_i + \partial \phi / \partial x_j \right)^2 \right\rangle$ of the conserved scalar $\phi$ for unity Schmidt number, the following scaling law exists for $Re_\lambda > 50$: $\langle \chi \rangle \sim \left\langle \phi_l'^2 \right\rangle u_l'^2 / l \sim \left\langle \phi_l'^2 \right\rangle u_{rms}'^2 / L_I$ as shown by DNS of Donzis et al. [44].

Now $g$, which is the interface fluctuation, is like a passive scalar in the field formulation by Kerstein et al. [5], albeit much more complex due to the nonlinearity of its governing equation. However, if we assume that the leading-order dissipation is provided by $Mk$ by smearing out the small scale flame surface fluctuations, we can expect the following to hold by proposing the following $g$ dissipation rate scaling ($Mk > 1$) in the limit of moderate and strong turbulence.

$$D_M \langle \nabla g . \nabla g \rangle \sim \left\langle g_l^2 \right\rangle u_l' / l \sim \left\langle g^2 \right\rangle u_{rms}' / L_I \tag{10}$$

A very crucial question however emerges from this assumption. Physically, why Markstein diffusivity is considered as the dominant flame surface fluctuation dissipation mechanism, when we are aware of other dissipation mechanisms such as the nonlinear kinematic restoration? The answer to this question might be traced back to the hydrodynamic theory of premixed flames. The original Darrieus-Landau theory [45, 46] of premixed flames states that flames are unstable to all wavenumbers ($k$) of perturbation and the growth rate increases linearly with $k$. However, it was soon known that it is not the case in reality as large perturbations are unconditionally stable for $Le > 1$ flames. This was first empirically clarified by Markstein [47] who introduced the concept of Markstein length and rigorously explained later by the linear theory of Pelce and Clavin [48], Matalon and Matkowsky [49] and the nonlinear theory of Sivashinsky [50]. The important point to note here is that the $k^2$ term of which the Markstein length ($\delta_M$) is a dominant coefficient is the stabilizing term that prevents DL instability at large $k$. Therefore, Markstein diffusivity $D_M = S_L \delta_M$—an intrinsic physico-chemical property of the premixed flame is essentially a large wavenumber dissipation mechanism of flame surface fluctuations irrespective of the source of perturbation. When turbulence instead of DL instability is the flame surface fluctuation production mechanism, the role of $D_M$ is invariant and provides the dissipation to flame surface fluctuations at large $k$. The analogy of $D_M$ with viscosity or molecular diffusivity is also apparent when the Navier Stokes, species conservation and $G$ equation are compared. Kinematic viscosity, diffusivity, and Markstein diffusivity appear as coefficients of the Laplacian of velocity, species and $G$ respectively. Of course for $G$-equation there are additional terms that appear through expansion of the curvature term. Physically, molecular diffusion smears away scalar fluctuation at small scales. Markstein diffusion smoothens flame surface fluctuations by varying the local flame speed with curvature. It can also be easily shown that such dissipation is dominant at small scales, like viscous dissipation or molecular diffusion, as the curvature term scales as $k^2$ in wavenumber space, shown in Peters [34]. Furthermore, since both the kinematic restoration and gas expansion terms originating from the Sivashinsky integral [35] scale as $k$, they tend to cancel out due to similar scaling in wavenumber space leaving the Markstein diffusivity to provide the leading order dissipation at large $Re_T$. Such cancellation is only possible when the flame is strongly wrinkled at small scales as otherwise the nonlinear kinematic restoration has a quadratic to the wavenumber scaling, to the leading order due to Taylor expansion. Furthermore, the dominance of diffusion over kinematic restoration can also be verified from the curvature distribution, experimentally obtained by various groups. For small $u'_{rms}/S_L$, curvature pdf is typically skewed due to corner formation showing the importance of Huygens dissipation. But for large $u'_{rms}/S_L$ the curvature pdf is symmetric Gaussian, suggesting diffusive processes in action. Clearly any study on weak turbulence requires explicit consideration of kinematic restoration, as can be found in Refs. [1, 4].

Returning to Eq. (10), for a statistically planar flame one can assume that $\langle g^2 \rangle \approx \delta_T^2 \sim L_I^2$ i.e. the flame brush thickness scales as integral length scale. Thus

$$D_M \langle \nabla g . \nabla g \rangle \sim u'_{rms} L_I \tag{11}$$

as the integral length scale should be proportional to the hydrodynamic length scale. It should however be noted that a primary requirement for this scaling is moderate to large $Re_\lambda \sim O(10^2)$ and above.

It was shown in [24] that the turbulent flame speed normalized by the laminar flame speed is given by, $S_T/S_L = \left\langle \sqrt{1 + \nabla g . \nabla g} \right\rangle \approx \sqrt{\langle 1 + \nabla g . \nabla g \rangle} \sim \sqrt{\langle \nabla g . \nabla g \rangle}$ as all the scaling holds for large $Re_\lambda$ anyways. As reported above, this operation approximately holds for derivatives of $g$ following Gaussian and gradient squared of $g$ following lognormal distributions. For the lognormal distributions at higher $Re_\lambda$, there will be a pre-factor with some additional but small dependence on $Re_\lambda$ [24] which is neglected in the following. Therefore for statistically planar flames,

$$S_T/S_L \sim \sqrt{\langle \nabla g . \nabla g \rangle} \sim \sqrt{\frac{L_I u'_{rms}}{D_M}} = \sqrt{\frac{u'_{rms} L_I}{S_L \delta_M}} \tag{12}$$

This result for the statistically planar flame could be extrapolated to the spherical configuration. At an instantaneous realization of the turbulent expanding flame, a cube of the dimension $\lambda_i$ could be selected to contain a sector of a statistically isotropic distribution of $g$. At any $i$th instant of the flame expansion event, when the average flame radius is given by $\langle R \rangle$ and given the solid angle subtended by the sector remains constant: $\lambda_i \sim \langle R \rangle$. Thus $\langle R \rangle$ is a measure of the largest hydrodynamic length scale that the flame encounters. This is elucidated in Fig. 2. Now, the functional of

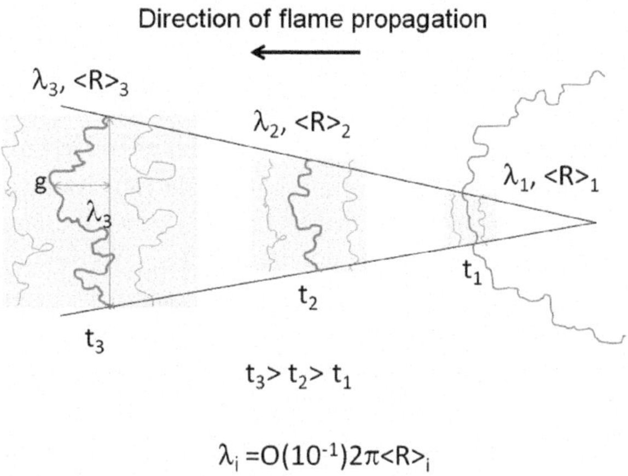


**Fig. 2** Schematic of the statistcially quasi planar assumption for an expanding flame to utilize the theoretical scaling of Eq. (5). The *shaded squares* are the *vertical* cross sections of cubes of dimension $\lambda_i$. Reproduced with permission from [25]

the field variable $g$ distributed over this cube could then be spatially integrated to obtain the normalized turbulent flame speed at each $\langle R \rangle$. Recalling Eq. (10) we had: $D_M \langle \nabla g.\nabla g \rangle \sim \langle g_l^2 \rangle u_l^{'}/l \sim \langle g^2 \rangle u_{rms}^{'}/L_I$. Consider the second term where $l$ represents any length scale in the inertial range and $u_l^{'2}$ is the turbulent kinetic energy contained in eddies of size $l$. For the expanding flame sector contained in the cubic box, replacing $l$ with $\langle R \rangle$ we can write:

$$D_M \langle \nabla g.\nabla g \rangle \sim \left\langle g_{\langle R \rangle}^2 \right\rangle u_{\langle R \rangle}^{'}/\langle R \rangle \sim \delta_{T,\langle R \rangle}^2 u_{\langle R \rangle}^{'}/\langle R \rangle \tag{13}$$

From [25] it was shown $\delta_{T,\langle R \rangle} \sim \langle R \rangle$ which yields:

$$\begin{aligned} &D_M \langle \nabla g.\nabla g \rangle \sim u_{\langle R \rangle}^{'} \langle R \rangle \\ &\Rightarrow S_T/S_L \sim \langle \nabla g.\nabla g \rangle^{1/2} \sim \left[ \left( u_{\langle R \rangle}^{'} \langle R \rangle \right) /D_M \right]^{1/2} \sim \left[ \left( u_{\langle R \rangle}^{'} \langle R \rangle \right) / \left( S_L \delta_M \right) \right]^{1/2} \end{aligned} \tag{14}$$

Physically, this implies that eddies of size equal or smaller than length scale $\langle R \rangle$ wrinkles the flame surface and contributes to turbulent flame speed. Whereas, eddies of size greater than $\langle R \rangle$ only convect the overall flame ball and do not contribute to enhancing the flame surface area and turbulent flame speed. The best way to test this assumption is to measure the normalized turbulent flame speed itself as a function of $\langle R \rangle$ or perform direct numerical simulations or three-dimensional tomographic measurements. Renou et al. [51] performed two-dimensional laser sheet tomography measurements of the flame brush thickness in such expanding configurations but that study is restricted to $u_{rms}^{'}/S_L < 2$ and small $Re_\lambda$, hence cannot be applied in this context. For unity-*Le*, methane-air flames at $\phi = 0.9$ [52], it was found that all the experimental data obtained over different $u_{rms}^{'}$ and pressure could be scaled as:

$$S_T/S_L \sim \left[\left(u'_{eff}/S_L\right)\left(\langle R\rangle/\delta_L\right)\right]^{1/2} \tag{15}$$

In this case, variation of $Mk$ is small and hence it does not appear as a variable in the above scaling. $u'_{eff}$ or $u'_{\langle R\rangle}$ is the effective r.m.s. of fluctuating velocity that the flame experiences at a length scale $\langle R\rangle$. As first pointed out by Abdel Gayed et al. [9] and also clarified in [25, 52], since the largest length scale of the flame: $\langle R\rangle$ is increasing, $u'_{eff}$ is not equal to $u'_{rms}$. We find $u'_{eff}$ approaches $u'_{rms}$ as approximately $\langle R\rangle^{0.23}$. From PIV measurements it was found that in the domain of the reported measurements, $(0.012\,\text{m} \leq \langle R\rangle \leq 0.022\,\text{m})$, and fan speed ranging from 2000 to 8000 rpm, the $u'_{eff}$ can be represented (approximately) by the correlation below: $u'_{eff} = (0.0018875 \times rpm + 0.06622) \times \langle R\rangle^{0.2333}$. Here the $u'_{eff}$ was based on radial velocity. For a slightly updated correlation based on $u_x$ and $u_y$, see Ref. [53].

Working with non-unity $Le$ flames however requires additional considerations. The local laminar flame speed and temperature are not constant over the flame surface, as a result of which the normalized area is not exactly equal to the normalized turbulent flame speed. This discrepancy could be accounted for, by defining the normalized turbulent flame speed as an $\tilde{S}_L$ weighted averaging of normalized area. Kerstein et al. [5] showed:

$$S_T \approx \left\langle \tilde{S}_L \mid \nabla G \mid \right\rangle = S_L \langle \mid \nabla G \mid \rangle - S_L \delta_M \langle \kappa \mid \nabla G \mid \rangle \tag{16}$$

While the first term on the RHS is given by Eq. (14), the second term concerns with the joint distribution of curvature and flame surface density which is highly non-trivial in general. To circumvent this problem one can safely assume the well-known quasi-symmetric distribution of curvature although with a non-zero mean. Then, the second term could be decomposed as:

$$\begin{aligned}\langle \kappa \mid \nabla G \mid \rangle = \langle -\kappa_n \mid \nabla G \mid_n - \kappa_{n-1} \mid \nabla G \mid_{n-1} \ldots\ldots \kappa_0 \mid \nabla G \mid_0 \ldots\ldots \\ + \kappa_{n-1} \mid \nabla G \mid_{n-1} + \kappa_n \mid \nabla G \mid_n \rangle\end{aligned}$$

where the $n$th bin of the negative or positive side of the joint pdf is denoted by the subscript $n$. Now, the negative side of the pdf is cancelled by the positive part of the distribution if the pdf is symmetric. This is because $\mid \nabla G \mid_n$ is always positive and independent of the sign of $\kappa$. Hence, to the leading order the only term retained would be: $\langle \kappa \rangle \langle \mid \nabla G \mid \rangle$. Hence, $\langle \kappa \mid \nabla G \mid \rangle \sim \langle \kappa \rangle \langle \mid \nabla G \mid \rangle$. Thus for the expanding flame for any positive $Mk$, Eq. (13) is revised to

$$S_T/S_L \sim \left(1 - 2\delta_M/\langle R\rangle\right)\left[\left(u'_{eff}/S_L\right)\left(\langle R\rangle/\delta_M\right)\right]^{1/2} \tag{17}$$

The pre-factor can be called the stretch factor and will be represented by

$$I_0 = 1 - 2\delta_M/\langle R\rangle \tag{18}$$

For most conditions, contribution by $I_0$ is only a small correction to the scaling laws of Eq. (14) as the dominant contribution to the normalized turbulent flame speed evolves from the normalized area.

We note that Lipatnikov and Chomiak [7] attempted to explain the acceleration of expanding premixed flames by extrapolating laminar flame stretch concepts, defining a turbulent Markstein number as a coefficient of turbulent flame brush thickness over average radius. The two approaches are fundamentally different as [7] apparently explains the acceleration as a manifestation of mean curvature $\sim 1/\langle R \rangle$ since their turbulent Markstein number is a constant. In contrast we consider the integrated effects of curvature in dissipating flame surface fluctuations from $O(1/\delta_L)$ to $O(1/\langle R \rangle)$ as detailed in the spectral approach.

## 4 Recent Expanding Flame Experiments

The experiments to be reviewed here, were conducted in a constant-pressure spark ignited expanding flame apparatus which has been extensively employed for measurement of laminar flame speeds [41]. This review, by no means cover all experiments on turbulent expanding flames which could be found in [9, 18, 22, 54] and the references thereof. Rather, here we focus on recent experimental results primarily reported over [25, 26, 52, 55], in light of the discussions in previous sections. The fan stirred turbulent expanding flame apparatus consists of an inner chamber situated within a much larger outer chamber. Detailed description of the apparatus could be found in the above references and is not repeated here for brevity. Important point to note here is that, in this apparatus, the flame propagation event upon ignition is essentially isobaric as the expanding gases could be absorbed in the outer chamber. This prevents any perturbation on the local flame structure due to pressure rise. Therefore experiments can be conducted under high initial pressures, up to 30 bars as in the studies of Refs. [41, 56], without damaging the quartz windows. Four orthogonally positioned fans generated isotropic turbulence as in [9] and the flow was characterized by high-speed particle image velocimetry. Detailed flow-field statistics and quantification of the small but unavoidable anisotropy could be found in [52]. The experiments reviewed here, were conducted at pressures of 1, 2, 3 and 5 bars and with $u'_{rms}$ between 1.34 and 5.33 m/s. Here the $u'_{rms}$ refers to the mean $u'_{eff}$ in the domain of interest, which was chosen to be $0.21 \leq \langle R \rangle / R_{chamber} \leq 0.38$, identified from laminar flame speed experiments. By resorting to the range above, the influence of ignition and wall on flame propagation could be minimized. Also measurements were initiated after an integral time scale $\tau_I = L_I / u'_{rms}$. $L_I \sim 4mm$ for all conditions in the present experiments.
In this section we review turbulent flame speed data reported in [25] from experiments performed with the following mixtures: $H_2$-air $\phi = 4.0$; $C_2H_4$-15%, $O_2$-85%, $\phi = 1.3$; $C_2H_4$-air, $\phi = 1.3$; n-$C_4H_{10}$-air, $\phi = 0.8$ and $C_2H_6O$-air $\phi = 1.0$; $N_2$ was the inert for all the mixtures. In addition, the $CH_4$-air, $\phi = 0.9$ data from Ref. [52]

are also used for comparison purposes. All the required flame properties and the symbols corresponding to each condition of the experiments in relevant figures are given in Table 1. The values of $\delta_{M,b}$ for all conditions were obtained from laminar stretched flame speed measurements using the approach detailed in [41]. The turbulent flame speed data were extracted from the time derivative of the average radius: $d\langle R\rangle/dt$, where $\langle R\rangle$ is defined as $\langle R\rangle = \sqrt{A/\pi}$, $A$ is the area circumscribed by the flame edge tracked from the high-speed Schlieren imaging. First, $d\langle R\rangle/dt$ normalized by $S_{L,b}$ for pressure varying from 1 to 5 atm and $u'_{rms}$ between 1.43 and 5.33 m/s are presented in Fig. 3. Clearly each data set corresponding to a particular mixture approximately collapsed on a line, suggesting the possible validity of the $S_{L,b}^{-1} d\langle R\rangle/dt \sim \left[\left(u'_{eff}/S_L\right)\left(\langle R\rangle/\delta_L\right)\right]^{1/2}$ scaling irrespective of the fuel-air mixtures. It has been shown in [25] that for all fuel-air mixtures under consideration the effect of stretch factor $I_0$ is small, hence the data is presented without $I_0$. However, for larger $Mk_b$, effect of $I_0$ could be significant. It is found that in Fig. 3, for different mixtures, the data sets collapse on lines with different slopes indicating the influence of $Mk$. As shown in Table 1 for a given mixture such as $CH_4$-air, $\phi = 0.9$, the variation in $Mk_b$ over pressure is small; hence including $Mk_b$ in the scaling above will not result in any significant deviation of the data from a single line for a given mixture. The subscript $b$ implies measurement with respect to the burnt gas. As can be observed from Table 1, the variation of $Mk_b$ over different mixtures is however significant. It is to be noted that in Fig. 3 the largest $S_{L,b}^{-1} d\langle R\rangle/dt$ corresponds to negative $Mk_b$. Even without considering this case, focusing solely on the positive $Mk_b$ cases suggests a factor of 4 variation of $S_{L,b}^{-1} d\langle R\rangle/dt$ at a constant $Re_{T,f,\langle R\rangle}$. Hence, dominance of

**Table 1** Legends, laminar flame properties and turbulent intensity for all present ($C_0$-$C_4$) experimental conditions. Reproduced with permission from [25]

| Symbol | Mixture | $\phi$ | $p/p_0$ | rpm | $u_{rms}$ (m/s) | $S_L$ (m/s) [1] | $S_{L,b}$ (m/s) [1] | $\delta_L$ (m) [1] | $\delta_{M,b}$ (m) [2] | $S_{L,b}\delta_L$ | $S_{L,b}\delta_{M,b}$ | $Mk_b$ | $Mk_u$ | $\Theta$ |
|---|---|---|---|---|---|---|---|---|---|---|---|---|---|---|
| ☆ | H2-air | 4.00 | 5 | 2000 | 1.43 | 1.13 | 5.45 | 8.55E-05 | 3.50E-04 | 4.66E-04 | 1.91E-03 | 4.10 | 5.67 | 4.82 |
| ☆ | H2-air | 4.00 | 5 | 4000 | 2.85 | 1.13 | 5.45 | 8.55E-05 | 3.50E-04 | 4.66E-04 | 1.91E-03 | 4.10 | 5.67 | 4.82 |
| ☆ | H2-air | 4.00 | 5 | 7500 | 5.33 | 1.13 | 5.45 | 8.55E-05 | 3.50E-04 | 4.66E-04 | 1.91E-03 | 4.10 | 5.67 | 4.82 |
| ○ | $CH_4$-air | 0.90 | 1 | 2000 | 1.43 | 0.32 | 2.31 | 4.80E-04 | 5.64E-04 | 1.11E-03 | 1.30E-03 | 1.18 | 3.15 | 7.17 |
| ○ | $CH_4$-air | 0.90 | 1 | 4000 | 2.85 | 0.32 | 2.31 | 4.80E-04 | 5.64E-04 | 1.11E-03 | 1.30E-03 | 1.18 | 3.15 | 7.17 |
| ○ | $CH_4$-air | 0.90 | 2 | 4000 | 2.85 | 0.25 | 1.83 | 2.90E-04 | 3.08E-04 | 5.31E-04 | 5.65E-04 | 1.06 | 3.05 | 7.32 |
| ○ | $CH_4$-air | 0.90 | 3 | 4000 | 2.85 | 0.22 | 1.57 | 2.20E-04 | 2.09E-04 | 3.45E-04 | 3.29E-04 | 0.95 | 2.92 | 7.14 |
| ○ | $CH_4$-air | 0.90 | 5 | 2000 | 1.43 | 0.17 | 1.22 | 1.60E-04 | 1.17E-04 | 1.96E-04 | 1.43E-04 | 0.73 | 2.70 | 7.20 |
| ○ | $CH_4$-air | 0.90 | 5 | 4000 | 2.85 | 0.17 | 1.22 | 1.60E-04 | 1.17E-04 | 1.96E-04 | 1.43E-04 | 0.73 | 2.70 | 7.20 |
| △ | $C_2H_4$-15%$O_2$-85%$N_2$ | 1.30 | 2 | 2000 | 1.43 | 0.18 | 1.15 | 3.67E-04 | -4.04E-04 | 4.22E-04 | -4.64E-04 | -1.10 | 0.75 | 6.39 |
| △ | $C_2H_4$-15%$O_2$-85%$N_2$ | 1.30 | 2 | 4000 | 2.85 | 0.18 | 1.15 | 3.67E-04 | -4.04E-04 | 4.22E-04 | -4.64E-04 | -1.10 | 0.75 | 6.39 |
| △ | $C_2H_4$-15%$O_2$-85%$N_2$ | 1.30 | 5 | 2000 | 1.43 | 0.11 | 0.75 | 2.17E-04 | -1.80E-04 | 1.63E-04 | -1.35E-04 | -0.83 | 1.09 | 6.82 |
| △ | $C_2H_4$-15%$O_2$-85%$N_2$ | 1.30 | 5 | 4000 | 2.85 | 0.11 | 0.75 | 2.17E-04 | -1.80E-04 | 1.63E-04 | -1.35E-04 | -0.83 | 1.09 | 6.82 |
| ▽ | $C_2H_4$-air | 1.30 | 1 | 2000 | 1.43 | 0.61 | 5.01 | 2.92E-04 | 4.87E-04 | 1.46E-03 | 2.44E-03 | 1.67 | 3.77 | 8.22 |
| ▽ | $C_2H_4$-air | 1.30 | 1 | 4000 | 2.85 | 0.61 | 5.01 | 2.92E-04 | 4.87E-04 | 1.46E-03 | 2.44E-03 | 1.67 | 3.77 | 8.22 |
| ▽ | $C_2H_4$-air | 1.30 | 2 | 4000 | 2.85 | 0.53 | 4.36 | 1.61E-04 | 2.77E-04 | 7.02E-04 | 1.21E-03 | 1.72 | 3.83 | 8.23 |
| ▽ | $C_2H_4$-air | 1.30 | 3 | 2000 | 1.43 | 0.48 | 3.95 | 1.16E-04 | 2.06E-04 | 4.59E-04 | 8.12E-04 | 1.77 | 3.88 | 8.25 |
| ▽ | $C_2H_4$-air | 1.30 | 5 | 4000 | 2.85 | 0.41 | 3.42 | 7.85E-05 | 1.47E-04 | 2.68E-04 | 5.03E-04 | 1.87 | 3.98 | 8.26 |
| ◇ | $nC_4H_{10}$-air | 0.80 | 5 | 4000 | 2.85 | 0.18 | 1.18 | 1.27E-04 | 3.49E-04 | 1.50E-04 | 4.12E-04 | 2.75 | 4.63 | 6.56 |
| □ | $C_2H_6O$-air | 1.00 | 1 | 2000 | 1.43 | 0.47 | 3.92 | 3.24E-04 | 8.71E-04 | 1.27E-03 | 3.41E-03 | 2.69 | 4.81 | 8.33 |
| □ | $C_2H_6O$-air | 1.00 | 1 | 4000 | 2.85 | 0.47 | 3.92 | 3.24E-04 | 8.71E-04 | 1.27E-03 | 3.41E-03 | 2.69 | 4.81 | 8.33 |
| □ | $C_2H_6O$-air | 1.00 | 5 | 2000 | 1.43 | 0.29 | 2.44 | 9.70E-05 | 2.60E-04 | 2.37E-04 | 6.35E-04 | 2.68 | 4.81 | 8.42 |
| □ | $C_2H_6O$-air | 1.00 | 5 | 4000 | 2.85 | 0.29 | 2.44 | 9.70E-05 | 2.60E-04 | 2.37E-04 | 6.35E-04 | 2.68 | 4.81 | 8.42 |
| □ | $C_2H_6O$-air | 1.00 | 5 | 6000 | 4.29 | 0.29 | 2.44 | 9.70E-05 | 2.60E-04 | 2.37E-04 | 6.35E-04 | 2.68 | 4.81 | 8.42 |
| □ | $C_2H_6O$-air | 1.00 | 10 | 4000 | 2.85 | 0.23 | 1.95 | 5.97E-05 | 6.95E-09 | 1.16E-04 | 1.35E-08 | 2.78 | 4.92 | 8.47 |

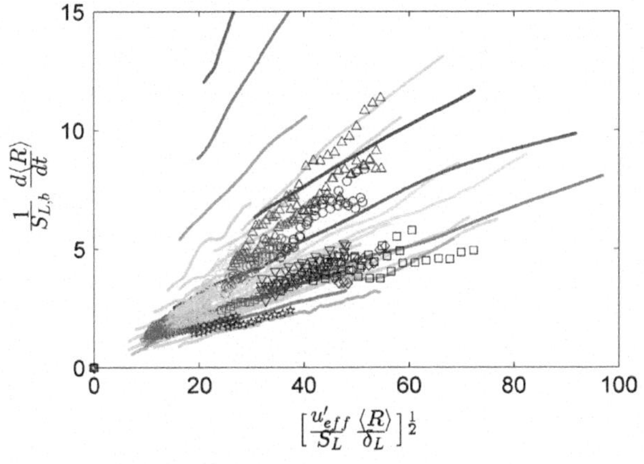


**Fig. 3** Normalized $d\langle R\rangle/dt$ versus $\sqrt{\left(u'_{eff}/S_L\right)\left(\langle R\rangle/\delta_L\right)}$ for $C_2H_4$-15% $O_2$- 85% $N_2$, $\phi = 1.3$; $CH_4$-air $\phi = 0.9$, $C_2H_4$-air, $\phi = 1.3$, n-$C_4H_{10}$-air, $\phi = 0.8$ and $C_2H_6O$-air $\phi = 1.0$ mixture from present experiment. The *lines* represent the iso-$C_8H_{18}$ data from [26]. The *symbols* for present experimental condition and corresponding laminar flame and turbulence parameters could be found in Table 1. The symbols, flame and flow parameters for the Leeds data could be found in Table 2. Reproduced with permission from [25]

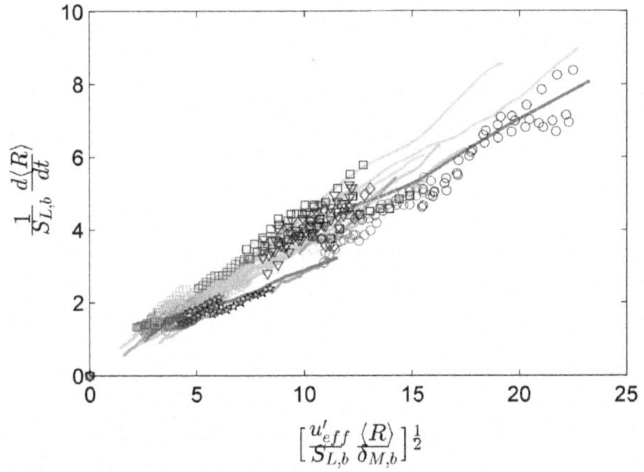


**Fig. 4** Normalized $d\langle R\rangle/dt$ versus $\sqrt{\left(u'_{eff}/S_{L,b}\right)\left(\langle R\rangle/\delta_{M,b}\right)} = Re^{0.5}_{T,fb}$ for $C_2H_4$-15% $O_2$- 85% $N_2$, $\phi = 1.3$; $CH_4$-air $\phi = 0.9$, $C_2H_4$-air, $\phi = 1.3$, n-$C_4H_{10}$-air, $\phi = 0.8$ and $C_2H_6O$-air $\phi = 1.0$ mixture from present experiment. The *lines* represent the iso-$C_8H_{18}$ data from [26]. The *symbols* for present experimental condition and corresponding laminar flame and turbulence parameters could be found in Table 1. The *symbols*, flame and flow parameters for the Leeds could be found in Table 2. Reproduced with permission from [25]

molecular diffusion based transport rather than turbulent transport within the local flame structure is strongly suggested.

Thus, using $C_0$, $C_1$, $C_2$, $C_4$ and $C_8$ hydrocarbon flames the following observations emerge: (a) the scaling of Eq. (15) appears valid for these hydrocarbons, (b) for a constant $Re_{T,f}$, $S_{L,b}^{-1}d\langle R\rangle/dt$ decreases with increasing $Mk$. This trend is also predicted by the theoretical result of Eq. (14) although with the restriction of positive $Mk$. (c) Therefore, from (a) and (b) a further degree of validation of the physical mechanism of the theoretical results beyond the general scaling is provided. It is however noted that $S_{L,b}^{-1}d\langle R\rangle/dt \neq S_T/S_L$ due to gas expansion effects and could be shown that $S_{L,b}^{-1}d\langle R\rangle/dt = c\left(S_T/S_L\right)$ where $c$ is a constant, weakly depending on the density ratio $\Theta = \rho_u/\rho_b$ Recently, in [55] the above the scaling has been extended for $C_5$-$C_8$ alkanes in strongly burning turbulent expanding flames. Thus, the scaling seems to explain turbulent flame propagation rates of $C_0$-$C_8$ alkanes.

The trend (b) was also qualitatively observed by Fairweather et al. [57] and in many other works reviewed by Lipatnikov [21] where it was found that all other flow and flame parameters being equal, turbulent flame speed increases with decreasing $Le$. In the present work, we have utilized the concept of dissipation by $D_M$ coupled with the statistical turbulence theory to arrive at an expression for normalized turbulent flame speed.

As mentioned above, the scaling of Eq. (13) appears valid for individual positive $Mk$ flames. If it is indeed true, then all positive $Mk$ data can be collapsed on one line irrespective of fuel, pressure and turbulent intensity. The positive $Mk$ data for the $H_2$-air, $CH_4$-air, $C_2H_4$-air, $C_4H_{10}$-air and $C_2H_6O$-air mixtures are plotted in Fig. 4. In addition, the ensemble averaged iso-$C_8H_{18}$-air data of Ref. [26], is also plotted with $\left[\left(u'_{eff}\langle R\rangle/S_L\right)\left(\langle R\rangle/\delta_{M,b}\right)\right]^{1/2}$ on the $x$-axis in Fig. 4. Here, we choose $\delta_M = \delta_{M,b}$ and coherent collapse of the data is achieved. While most of the data sets from both present experiments and Leeds data, collapse well within a band and are continuous across turbulence intensities, the $u'_{rms} = 1$ and 4 m/s of $\phi = 1.0$, $p = 10$ atm cases of the Leeds iso-octane data does not fall on a single line. This discrepancy could be attributed to the leading order nature of the scaling, where dissipation by kinematic restoration was neglected and/or uncertainty in the scaling of $u'_{eff}$ with $\langle R\rangle$ or uncertainties in the measurement itself. Overall, Figs. 3 and 4 are the most important results of [25] which holds the possibility of a unified, fuel invariant scaling of normalized turbulent flame speed with a Reynolds number based on turbulence intensity, flame hydrodynamic length scale and Markstein diffusivity at least for the positive $Mk$ flames. Explicit consideration of kinematic restoration and thermal expansion from the hydrodynamic instability theory might yield a stronger result with Markstein length refined towards a DL cutoff scale.

It is expected that this is a generalized scaling irrespective of fuels, for expanding flames and could be easily extrapolated to other flame configurations such as the Bunsen flame, as long as the cold flow turbulence is nearly isotropic. It is impossible to ascertain if this scaling is absolutely unique to produce the data collapse. However, given the absence of any empirical constants, use of well-defined parameters and theoretical background, it could be considered a strong candidate for predicting turbulent flame speeds.

Flame surface area generation and destruction are the key processes that determine the turbulent flame speed. Next we proceed to the local viewpoints of turbulence-flame interaction to enhance our much required understanding of local flame surface propagation and annihilation.

# 5 Local Lagrangian Viewpoints

It has been observed in the previous section that global viewpoints of turbulent premixed flame propagation in terms of flame geometry can provide useful insights and scaling laws that could explain turbulent flame speed variation of expanding flames over a range of pressure, fuels and turbulent intensity. It was shown in Ref. [55] that strong turbulence can lead to local extinction events which results in deviation from the above scaling laws meant for strongly burning flames. Above all, the turbulent flame speed scaling relationship obtained above is based on the assumption that flame surface fluctuations follow the same dissipation rate anomaly as that by passive scalar fluctuations. To prove or disprove this hypothesis by experiments or DNS is not straightforward and must be arrived at from the local behavior of flame elements that are stretched, wrinkled and annihilated by turbulence. A Lagrangian viewpoint which will allow tracking individual flame elements is most suitable for such a purpose. In reference [28], Pope defined: "A *surface point* is defined, first, by its location on the surface at a reference initial time $t_0$; and, second, by the specification that it remains on the surface by moving relative to the fluid (if at all) in the direction of the local normal to the surface". Pope and co-workers studied [58, 59], unconnected, propagating surface elements that were randomly distributed in non-reacting, uniform temperature, isotropic turbulence. Important insights were obtained from the surface point statistics generated in these studies. However, these surface elements did not belong to a premixed flame and they were not reactive. For the first time, in [27] the time history of the straining rate and curvature of surface elements belonging to iso-surfaces of statistically planar premixed $H_2$-air flames simulated with detailed chemistry interacting with oncoming isotropic turbulence, were studied. To delineate between surface points in nonreacting flows versus those embedded on and comoving with the reacting isosurfaces inside a real flame, the later were named *flame particles*. Flame particles thus follow the flame surface where fluid particles follow a fluid element. The basic concept of flame particle tracking (FPT) is the following. Say, flame particle $P^F$'s position vector at a time $t$ is denoted by $\mathbf{X}^F(t)$ where $\mathbf{X}^F(t_0) = \mathbf{X}(x_0, y_0, z_0, t_0)$. Here $(x_0, y_0, z_0)$ is the position of the flame particle at an initial time $t = t_0$. Then motion of $P^F$ is given by:

$$\frac{d}{dt}\mathbf{X}^F(t) = \mathbf{U}(\mathbf{X}^F[t], t) + S_d(\mathbf{X}^F[t], t)\mathbf{n}^F(\mathbf{X}^F[t], t) \qquad (19)$$

Here, $\mathbf{U}(\mathbf{X}^F[t], t)$ is the local flow velocity at $\mathbf{X}^F(t)$; $S_d(\mathbf{X}^F[t], t)$ is the local displacement flame speed and $\mathbf{n}^F(\mathbf{X}^F[t], t)$ is the local surface normal at $\mathbf{X}^F(t)$. This equation

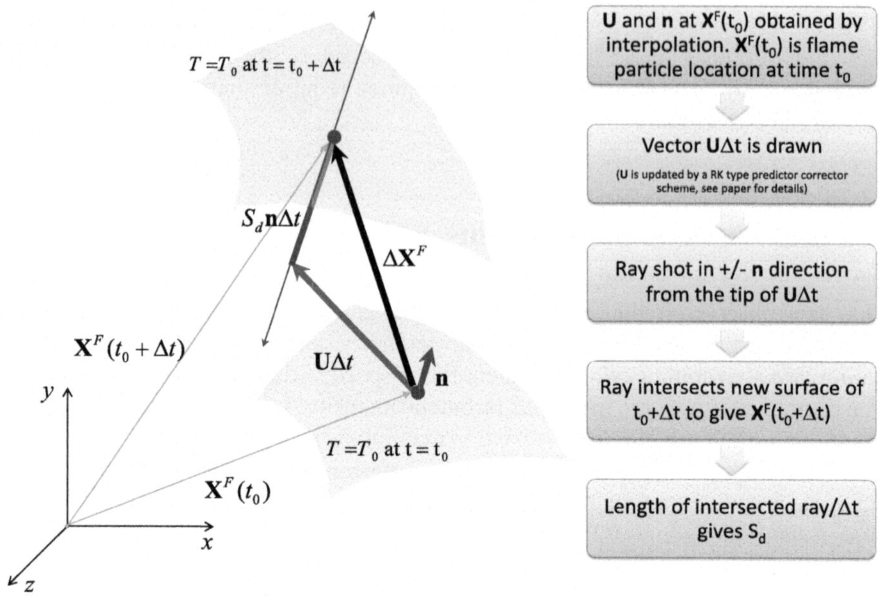


**Fig. 5** Schematic of the flame particle tracking algorithm. Reproduced with permission from [27]

was solved numerically in the DNS datasets of statistically planar, turbulent premixed flames, using the algorithm shown in Fig. 5. Lagrangian analysis of turbulence has gained recent attention due to rapid development of high performance computing allowing realistic Reynolds number DNS computations [60] alongside PDF modeling [61] and advent of high speed detectors [62]. However, motion of flame particles and their similarity, if any with fluid particles, species tracers or surface points has not been studied. Furthermore, strong dependence of on the local strain rate and curvature alongside the gas expansion caused by heat release can greatly complicate such possible analogy. Flame particles motion needs to be studied for the following reasons, which includes but are not limited to:

1. Time evolution of particular positions and corresponding properties of a reacting surface. The time history of the properties of individual flame particles would quantify their intermittent/persistent nature.
2. Connect turbulence physics with combustion through a Lagrangian viewpoint to enhance fundamental understanding.
3. How is flame surface generated and destructed? This will enhance our understanding and explain observations related to turbulent flame speed.
4. How are flame elements extinguished in turbulence? This will have strong bearing on phenomena of practical relevance like flame blowoff.
5. Use Pope's stochastic flamelet model [63] to provide sub grid scale modeling of small scale limiting phenomena.

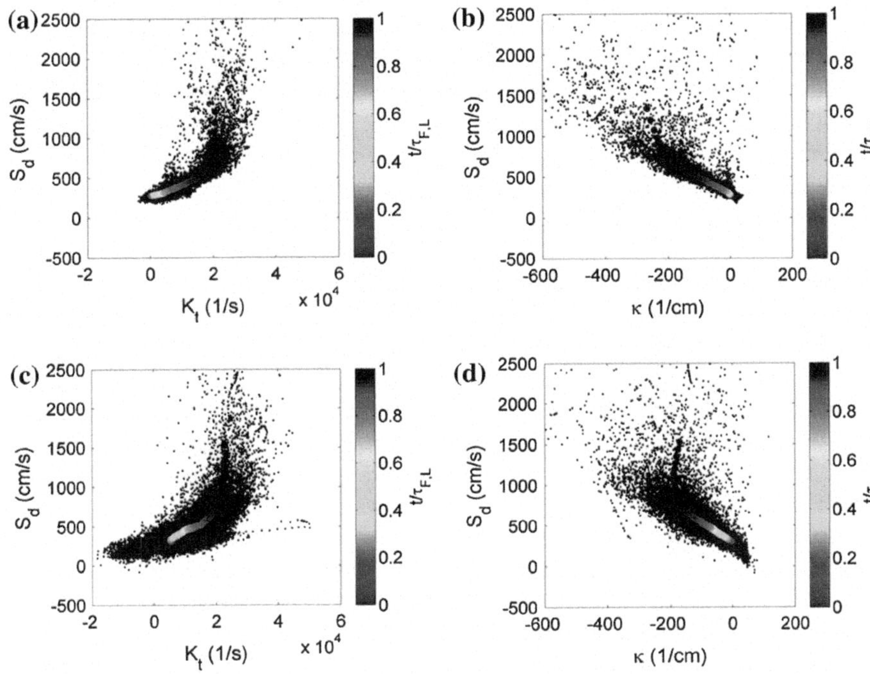


**Fig. 6** From Case 1: **a** $S_d$ versus $K_t$, **b** $S_d$ versus $\kappa$. From Case 2: **c** $S_d$ versus $K_t$, **d** $S_d$ versus $\kappa$. In all plots the phase space location of individual flame particles are shown with *black dots* while the ensemble averaged time history is shown by filled *circles color coded* with the $t/\tau_{F,L}$ value at which the averaging was performed. Reproduced with permission from [27]

In this section we will attempt to address few of the above using FPT analysis on DNS data of statistically planar premixed $H_2$-air flames interacting with near isotropic turbulence. Figure 6a–d shows that for such turbulent premixed flames, on ensemble averaging, $S_d$ and $\kappa$ for a set of flame particles are negatively correlated. However, unlike conventional averaging, here the averaging is performed at constant values of $t/\tau_{F,L}$. The lifetime $\tau_{F,L}$ introduced in [27] represents the lifetime of individual flame particles. Unlike fluid particles, the flame particles have a finite lifetime as the resident surface of these flame particles could be annihilated by different mechanisms like kinematic restoration, island formation and Markstein diffusion. Thus, following the flame particles in the normalized time of $t/\tau_{F,L}$ allows fundamental understanding of these flame surface annihilation processes in turbulence. This is critical due to the following: turbulence continuously stretches and wrinkles a premixed flame to enhance flame surface area and turbulent flame speed. However, the above mentioned flame surface dissipation mechanisms destroy flame surface to prevent unbounded growth of flame surface area, turbulent flame speed and maintain statistical stationarity of the same. In Fig. 6a–d the $\langle K_t \rangle$ - $\langle S_d \rangle$ and $\langle \kappa \rangle$ - $\langle S_d \rangle$ trajectories are thus color coded by the corresponding $t/\tau_{F,L}$. $t/\tau_{F,L} =$ 0 represents the time instant at which the flame particles were injected on the T = 500 K isosurface, whereas $t/\tau_{F,L} = 1$ represents the time instant at which the flame particle and its corresponding local surface was annihilated. It is noted that for the first Case (Fig. 6a–b) for $t/\tau_{F,L} \rightarrow 0$; $\langle K_t \rangle \rightarrow 0, \langle \kappa \rangle \rightarrow 0$, and $\langle S_d \rangle \rightarrow S_{L,0,T=500\,K}$. However, with progress in the normalized time i.e. towards surface annihilation, the individual flame particles and their ensemble get progressively stretched. This results in rapid increase of $S_d$. As the flame particles approach the end of their lifetime denoted by $t/\tau_{F,L} \rightarrow 1$ (colored by dark red) they attain extreme values of $\langle K_t \rangle \rightarrow +20000\,\text{s}^{-1}, \langle \kappa \rangle \rightarrow -200\,\text{cm}^{-1}$, and $\langle S_d \rangle \rightarrow +1500\,\text{cm/s}$. It is to be noted that except for $t/\tau_{F,L} \rightarrow 0$ the entire isosurface is never represented by the ensemble of flame particles. Rather the ensemble of flame particles represents newer states of their initial positions that once constituted the initial isosurface. However, this ensemble represents the unified behavior of all flame particles approaching annihilation and thereby unraveling the physics of the same. In a recent work [64], the question of increasing $S_d$ has been attributed to the effects of increasing negative curvature of the flame surface and eventual homogenization of temperature gradients. This is because, near the end the flame particles reside on reactant islands detached from the main flame surface and these shrink due to flame propagation and turbulent mixing. However, question arise how generic is this observation of increasing $S_d$ with curvature and tangential strain rate. As suggested by Eq. (5), a near linear laminar flame speed response to stretch rate is well known for laminar flames. Unfortunately, for turbulent flames, local flame speed and local stretch rate often appeared uncorrelated in Eulerian viewpoint. In [64] characterization of the effect of turbulence intensity on $S_d$ is attempted by analyzing the local Karlovitz number determined at the flame particle positions by employing the FPT technique. As in [65], here the stretch Karlovitz number is defined as

$$Ka_S = \frac{\delta_L}{S_{L,T}} \left( K_t + S_d \nabla.\mathbf{n} \right) = Ka_T + Ka_C \tag{20}$$

with $K_t$ being the tangential strain rate. The first term on the RHS of Eq. (20) is referred to as the tangential Karlovitz number and the second term as curvature Karlovitz number, respectively.

The variation of ensemble averaged quantities $\left\langle S_d/S_{L,T} \right\rangle$ versus $\langle Ka_T \rangle$, $\langle Ka_C \rangle$, and $\langle Ka_S \rangle$ for four different cases of slightly lean $H_2$-air, statistically planar turbulent premixed flames, are shown in Figs. 7 and 8 respectively. It is seen that for all four cases, the curvature Karlovitz number dominates the overall non-dimensional stretch rate, overwhelmingly. More importantly the $\left\langle S_d/S_{L,T} \right\rangle$ versus $\langle Ka_S \rangle$ for all cases collapse to a strong correlation curve, irrespective of the Reynolds, Damköhler or the turbulent Karlovitz numbers considered. A slight non-linearity in $\left\langle S_d/S_{L,T} \right\rangle$ versus $\langle Ka_S \rangle$ is observed, though the slopes of these curves apparently do not correspond to the laminar Markstein number. The origin of this generic dependence of flame speed on curvature Karlovitz number in the Lagrangian viewpoint and the physical interpretation of the resulting slope or its connection with the laminar Markstein number are open questions so far. Further investigation of $S_d$ versus $Ka_S$ in the FPT viewpoint could open up new avenues in modeling turbulent premixed flame dynamics.

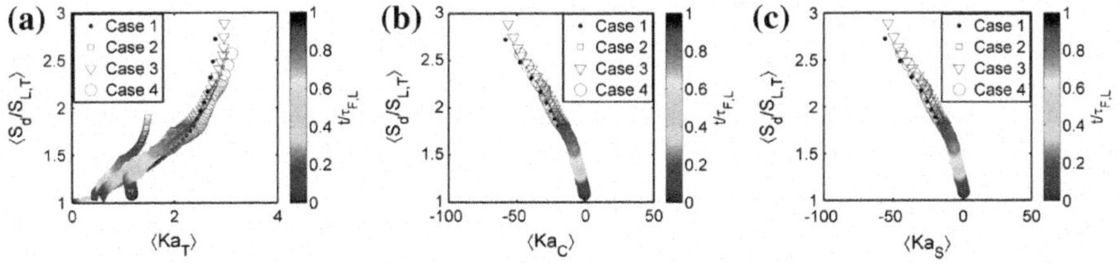


**Fig. 7** Time evolution of $\langle S_d/S_{L,T}\rangle$ versus **a** $\langle Ka_T\rangle$, **b** $\langle Ka_C\rangle$, and **c** $\langle Ka_S\rangle$, for the T = 665 K isosurface. Reproduced with permission from [64]

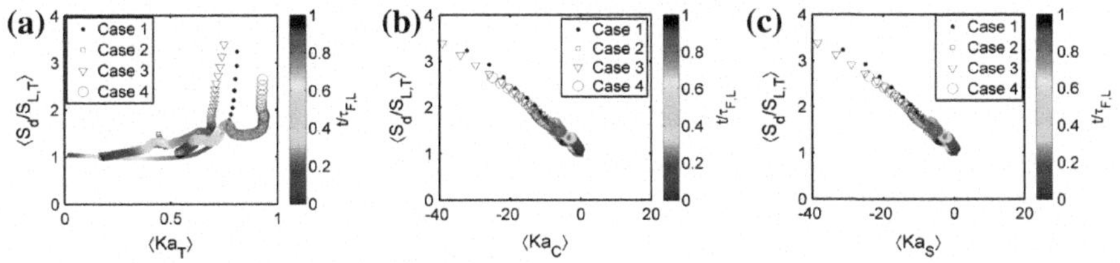


**Fig. 8** Time evolution of $\langle S_d/S_{L,T}\rangle$ versus **a** $\langle Ka_T\rangle$, **b** $\langle Ka_C\rangle$, and **c** $\langle Ka_S\rangle$, for the T = 1321 K isosurface. Reproduced with permission from [64]

**Table 2** Legends, laminar flame properties and turbulent intensity for all experimental conditions of iso-octane (i-$C_8H_{18}$) data from Leeds [26]. The $S_{L,b}$ and $\delta_{M,b}$ are obtained from experiments in our group such that the nonlinear extrapolation technique used for other data sets in Table 1 is consistent throughout. Same colors are assigned for the different $u'_{rms}$ cases corresponding to same $\phi$ and $p$ as $u'_{rms}$ variation can be identified from $Re_T$, different variants of which are used as abscissa of most of the figures. Reproduced with permission from [25]

| Symbol | $\phi$ | $p/p_0$ | $\Theta$ | $S_L$ (m/s) [1] | $S_{L,b}$ (m/s) [1] | $\delta_{M,b}$ (m) [2] | $\delta_L$(m) [1] | Le | $Mk_b$ | $Mk_u$ |
|---|---|---|---|---|---|---|---|---|---|---|
| • | 0.80 | 1.00 | 6.09 | 0.31 | 1.89 | 1.42E-03 | 4.27E-04 | 2.98 | 3.33 | 5.13 |
| • | 1.00 | 1.00 | 6.84 | 0.43 | 2.94 | 1.13E-03 | 3.72E-04 | 1.43 | 3.04 | 4.96 |
| • | 1.20 | 1.00 | 6.96 | 0.41 | 2.85 | 6.00E-04 | 3.83E-04 | 0.93 | 1.57 | 3.51 |
| • | 1.40 | 1.00 | 6.80 | 0.25 | 1.73 | -8.00E-04 | 5.81E-04 | 0.90 | -1.38 | 0.54 |
| • | 0.80 | 5.00 | 6.11 | 0.21 | 1.28 | 5.20E-04 | 1.31E-04 | 2.98 | 3.96 | 5.77 |
| • | 0.90 | 5.00 | 6.59 | 0.25 | 1.65 | 3.55E-04 | 1.17E-04 | 2.94 | 3.03 | 4.92 |
| • | 1.00 | 5.00 | 6.95 | 0.28 | 1.95 | 3.10E-04 | 1.12E-04 | 1.43 | 2.77 | 4.71 |
| • | 1.10 | 5.00 | 7.05 | 0.29 | 2.04 | 2.60E-04 | 1.11E-04 | 0.94 | 2.33 | 4.29 |
| • | 1.20 | 5.00 | 6.99 | 0.28 | 1.96 | 2.30E-04 | 1.23E-04 | 0.93 | 1.88 | 3.82 |
| • | 1.40 | 5.00 | 6.80 | 0.24 | 1.63 | NA | 2.29E-04 | 0.90 | NA | NA |
| • | 1.60 | 5.00 | 6.60 | 0.15 | 0.99 | NA | 3.61E-04 | 0.88 | NA | NA |
| • | 1.80 | 5.00 | 6.39 | 0.07 | 0.47 | NA | 5.08E-04 | 0.86 | NA | NA |
| • | 2.00 | 5.00 | 6.16 | 0.05 | 0.31 | NA | 6.62E-04 | 0.84 | NA | NA |
| • | 0.80 | 10.00 | 6.12 | 0.16 | 0.98 | 3.10E-04 | 8.19E-05 | 2.98 | 3.78 | 5.60 |
| • | 1.00 | 10.00 | 6.99 | 0.23 | 1.61 | 1.73E-04 | 6.93E-05 | 1.43 | 2.49 | 4.44 |
| • | 1.20 | 10.00 | 7.00 | 0.24 | 1.68 | NA | 8.14E-05 | 0.93 | NA | NA |
| • | 1.40 | 10.00 | 6.81 | 0.21 | 1.43 | NA | 1.48E-04 | 0.90 | NA | NA |

## 6 Conclusions

In this article we have reviewed some recent developments in theory, experiments and computations of turbulent premixed combustion through global and local viewpoints. Global viewpoints are essential to obtain overall useful statistics of turbulent flame propagation such as turbulent flame speed. Local viewpoints are crucial towards understanding fine grained detailed phenomena such as how local flame elements are stretched or annihilated. These local observations and understandings must build up the numerous sub-models that should feed into the global viewpoints. Essential features of the turbulent premixed combustion regime diagram are reviewed here and few essential modifications/shortcomings are highlighted. Some recent theoretical advances towards turbulent flame speed using the level set formulation are considered. Here, for the first time, an explicit connection is established between statistically planar turbulent premixed flames and turbulent expanding flames using the inertial range scale invariant scalar fluctuation flux transfer rate and small scale scalar dissipation rate. Using this, the turbulent flame speed scaling of Reynolds number to the power of half is obtained which is shown to stand the turbulent expanding flame experiments using $C_0$-$C_8$ alkane hydrocarbons. Finally, recent advances to track flame elements and their local properties from DNS, detailed chemistry solutions, through a Lagrangian viewpoint are reviewed. Such local viewpoints would further refine the global models to allow their pervasive use in combustor CFD calculations alongside unraveling the unknowns of turbulent combustion physics.

## References

1. Clavin P, Williams F (1979) J Fluid Mech 90:589–604
2. Victor Y (1988) Combust Sci Technol 60:191–214
3. Kerstein AR (1991) Phys Rev A 44
4. Kerstein AR, Ashurst WT (1992) Phys Rev Lett 68:934–937
5. Spalding DB (1977) Development of the eddy-break-up model of turbulent combustion In: Symposium (International) on combustion. Elsevier, pp 1657–1663
6. Pocheau A (1994) Phys Rev E 49:1109–1122
7. Lipatnikov A, Chomiak J (2007) Proc Combust Inst 31:1361–1368
8. Kolla H, Rogerson JW, Chakraborty N, Swaminathan N (2009) Combust Sci Technol 181:518–535
9. Abdel-Gayed RG, Bradley D, Lawes M (1987) Proceedings of the royal society A: mathematical. Phys Eng Sci 414:389–413
10. Bradley D, Lawes M, Mansour MS (2011) Combust Flame 158:123–138
11. Filatyev SA, Driscoll JF, Carter CD, Donbar JM (2005) Combust Flame 141:1–21
12. Kwon S, Wu MS, Driscoll JF, Faeth GM (1992) Combust Flame 88:221–238
13. Kobayashi H, Tamura T, Maruta K, Niioka T, Williams FA (1996) Symp (International) Combust 26:389–396
14. Kobayashi H, Seyama K, Hagiwara H, Ogami Y (2005) Proc Combust Inst 30:827–834
15. Venkateswaran P, Marshall A, Shin DH, Noble D, Seitzman J, Lieuwen T (2011) Combust Flame 158:1602–1614
16. Bell JB, Day MS, Shepherd IG, Johnson MR, Cheng RK, Grcar JF, Beckner VE, Lijewski MJ (2005) Proc Natl Acad Sci 102:10006–10011

17. Shim Y, Tanaka S, Tanahashi M, Miyauchi T (2011) Proc Combust Inst 33:1455–1462
18. Lipatnikov AN, Chomiak J (2002) Prog Energy Combust Sci 28:1–74
19. Lipatnikov AN, Chomiak J (2010) Prog Energy Combust Sci 36:1–102
20. Peters N (2000) Turbulent combustion. Cambridge university press
21. Lipatnikov AN, Chomiak J (2005) Prog Energy Combust Sci 31:1–73
22. Driscoll J (2008) Prog Energy Combust Sci 34:91–134
23. Akkerman V, Chaudhuri S, Law CK (2013) Phys Rev E 87:023008
24. Chaudhuri S, Akkerman V, Law CK (2011) Phys Rev E 84
25. Chaudhuri S, Wu F, Law CK (2013) Phys Rev E 88:033005
26. Lawes M, Ormsby MP, Sheppard CGW, Woolley R (2012) Combust Flame 159:1949–1959
27. Chaudhuri S (2015) Proc Combust Inst 35:1305–1312
28. Pope SB (1988) Int J Eng Sci 26:445–469
29. Peters N (1999) J Fluid Mech 384:107–132
30. Pope S (1987) Ann Rev Fluid Mech 19:237–270
31. Poludnenko AY, Oran ES (2011) Combust Flame 158:301–326
32. Aspden AJ, Day MS, Bell JB (2011) J Fluid Mech 680:287–320
33. Troiani G, Creta F, Matalon M (2015) Proc Combust Inst 35:1451–1459
34. Peters N (1992) J Fluid Mech 242:611–629
35. Peters N, Wenzel H, Williams FA (2000) Proc Combust Inst 28:235–243
36. Collins LR, Ulitsky M (1996) Symp (International) Combust 26:315–322
37. Ulitsky M, Ghenai C, Gokalp I, Wang L-P, Collins LR (2000) Combust Theor Modell 4:241–264
38. Dandekar A (1995) Combust Flame 101:428–440
39. Oberlack M, Wenzel H, Peters N (2001) Combust Theor Modell 5:363–383
40. Chaudhuri S, Akkerman V, Law CK (2011) Phys Rev E 84:026322
41. Kelley AP, Law CK (2009) Combust Flame 156:1844–1851
42. Law CK (2006) Combustion physics. Cambridge University Press, New York
43. Peters N (1999) Combust Flame 116:675–676
44. Donzis DA, Sreenivasan KR, Yeung PK (1999) J Fluid Mech 532:199–216
45. Darrieus G (1938) La Technique Moderne and Congrés de Mécanique Appliquée Paris
46. Landau L (1944) Acta Physicochim
47. Markstein GH (1964) N.A.T.O.A.G.f.A. Research, development, nonsteady flame propagation, published for and on behalf of advisory group for aeronautical research and development, North Atlantic treaty organization by Pergamon press
48. Pelce P, Clavin P (1982) J Fluid Mech 124:219
49. Matalon M, Matkowsky BJ (1982) J Fluid Mech 124:239
50. Sivashinsky GI (1977) Acta Astronaut 4:1177–1206
51. Renou B, Mura A, Samson E, Boukhalfa A (2002) Combust Sci Tech 174:143–179
52. Chaudhuri S, Wu F, Zhu D, Law CK (2012) Phys Rev Lett 108:044503
53. Chaudhuri S, Saha A, Law CK (2015) Proc Combust Inst 35:1331–1339
54. Bradley D, Sheppart CGW, Woolley R, Greenhalgh DA, Lockett RD (2000) Combust Flame 122:195–209
55. Wu F, Saha A, Chaudhuri S, Law CK (2015) Proc Combust Inst 35:1501–1508
56. Rozenchan G, Zhu DL, Law CK, Tse SD (2002) Proc Combust Inst 29:1461–1470
57. Fairweather M, Ormsby MP, Sheppard CGW, Woolley R (2009) Combust Flame 156:780–790
58. Yeung PK, Girimaji SS, Pope SB (1990) Combust Flame 79:340–365
59. Girimaji SS, Pope SB (1992) J Fluid Mech 234:247–277
60. Yeung PK (2002) Ann Rev Fluid Mech 34:115–142
61. Pope SB (1985) Prog Energy Combust Sci 11:119–192
62. Toschi F, Bodenschatz E (2009) Ann Rev Fluid Mech 41:375–404
63. Pope SB, Cheng WK (1989) Symp (International) Combust 22:781–789
64. Uranakara HA, Chaudhuri S, Dave HL, Arias PG, Im HG (2016) Combust Flame 163:220–240
65. Chen JH, Im HG (1998) Correlation of flame speed with stretch in turbulent premixed methane/air flames. In: Symposium (International) on combustion. Elsevier, pp 819–826

# Part III
# Gas Turbines

# Pulse Combustor Driven Pressure Gain Combustion for High Efficiency Gas Turbine Engines

**Joel C. Lisanti and William L. Roberts**

**Abstract** The gas turbine engine is an essential component of the global energy infrastructure which accounts for a significant portion of the total fossil fuel consumption in transportation and electric power generation sectors. For this reason there is significant interest in further increasing the efficiency and reducing the pollutant emissions of these devices. Conventional approaches to this goal, which include increasing the compression ratio, turbine inlet temperature, and turbine/compressor efficiency, have brought modern gas turbine engines near the limits of what may be achieved with the conventionally applied Brayton cycle. If a significant future step increase in gas turbine efficiency is to be realized some deviation from this convention is necessary. The pressure gain gas turbine concept is a well established new combustion technology that promises to provide a dramatic increase in gas turbine efficiency by replacing the isobaric heat addition process found in conventional technology with an isochoric process. The thermodynamic benefit of even a small increase in stagnation pressure across a gas turbine combustor translates to a significant increase in cycle efficiency. To date there have been a variety of methods proposed for achieving stagnation pressure gains across a gas turbine combustor and these concepts have seen a broad spectrum of levels of success. The following chapter provides an introduction to one of the proposed pressure gain methods that may be most easily realized in a practical application. This approach, known as pulse combustor driven pressure gain combustion, utilizes an acoustically resonant pulse combustor to approximate isochoric heat release and thus produce a rise in stagnation pressure.

## 1 Introduction to Pressure Gain Combustion for Gas Turbine Applications

As conventional gas turbine technology has reached maturity it has become increasingly apparent that future gains in the efficiency of these devices will be both small and expensive. Despite this, the gas turbine engine continues to be an important part

J.C. Lisanti (✉) · W.L. Roberts
King Abdullah University of Science and Technology, Thuwal, Saudi Arabia



A.K. Agarwal et al. (eds.), *Combustion for Power Generation and Transportation*, DOI 10.1007/978-981-10-3785-6_7

of the global energy infrastructure and presently accounts for a large of portion of the consumption of fossil fuels and thus greenhouse gas and pollutant emissions. The combined importance of the need to reduce fossil fuel emissions and the key role of the gas turbine engine in transportation and electric power generation has created a strong interest in further improving the efficiency of these heat engines. In order to circumvent the high cost, low reward challenges associated with further iterating on existing gas turbine components, significant research efforts have been carried out to develop what is known as pressure gain combustion technology. This is a new combustion concept that has the potential to dramatically increase efficiency by deviating from the conventional gas turbine cycle, the Brayton cycle, and instead harnessing the Humphrey thermodynamic cycle. The Humphrey cycle replaces the isobaric heat addition process found in conventional gas turbine technology with an isochoric process. The motivation for this is given as follows.

It is known that the conventional steady flow combustor is the source of the largest loss of thermodynamic availability in the gas turbine cycle [32] and this loss is characterized by a reduction in the stagnation pressure of the working fluid. In a conventional gas turbine engine the dimensionless pressure loss ratio, a key design parameter, is defined as the drop in stagnation pressure across the combustor divided by the stagnation pressure of the working fluid as delivered by the compressor. These ratios, commonly referred to the overall pressure loss, are generally reported to be on the order of 4–8% of the stagnation pressure delivered by the compressor and given values do not typically include the fundamental loss in pressure due to isobaric combustion [15]. This overall pressure loss can largely be attributed to complex fluid flow paths necessary for flame stabilization, increasing turbulence, and dilution schemes for cooling of the combustion products and combustor itself. The typically unaccounted fundamental loss in stagnation pressure due to isobaric ('constant pressure') combustion occurs as the density of the working fluid is decreased and velocity increased due to heating. The additional loss is on the order of 20–30% of the complete measurable loss in stagnation pressure across the combustion chamber [20] and this reduction in the stagnation pressure of the working fluid represents an inherently large efficiency penalty. Pressure gain gas turbine technology seeks to not only eliminate this loss but also replace it with a net gain in stagnation pressure. The American Institute of Aeronautics and Astronautics (AIAA) program committee for pressure gain combustion has developed the following definition:

> Pressure gain combustion (PGC) is a periodic process, in a fixed volume, whereby gas expansion by heat release is constrained, causing a rise in stagnation pressure and allowing work extraction by expansion to the initial pressure [29].

Effectively a pressure gain combustion gas turbine utilizes an isochoric combustor to perform work (compress) the working fluid and is thus able to increase the stagnation pressure above that delivered by the turbine. This gain replaces the loss

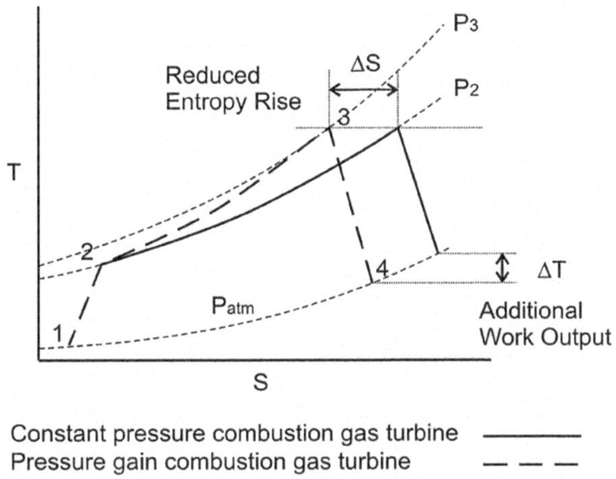


**Fig. 1** Thermodynamic benefit of pressure gain gas turbine [34]

in stagnation pressure found in a conventional isobaric combustor, thereby significantly increasing the thermodynamic availability of the fluid as it reaches the turbine and thus the overall cycle efficiency.

Conventional gas turbine operation is described thermodynamically by the Brayton cycle, for which the idealized cycle consists of four processes: isentropic compression, isobaric heat addition, isentropic expansion and isobaric heat rejection. The Humphrey thermodynamic cycle is a modification of the Brayton cycle such that the isobaric heat addition has been replaced by isochoric heat addition.

The thermodynamic benefit of modifying gas turbine engines to operate on the Brayton cycle is easily visualized in temperature entropy space as shown in Fig. 1. In this diagram, the isochoric heat addition process results in a reduced entropy rise compared with the isobaric case, thus allowing more work to be extracted at the turbine and reducing the necessary fuel input for a constant work output.

## 2 Proposed Methods of Achieving Pressure Gain in Gas Turbine Engine

At the current stage of development in pressure gain combustion technology there are primarily two pathways through which forms of isochoric combustion are being developed for gas turbine applications. These two approaches coincide with the different modes at which combustion can occur, detonation and deflagration. The deflagration process is characterized by a subsonic combustion wave that propagates through the reactive mixture decreasing the pressure and density of the fluid. Conversely, the detonation combustion process is characterized by the presence of a strong shock wave that is trailed by a reaction front and travels through the detonatable mixture at sonic velocity. As opposed to the deflagration process, the shock wave in detonative combustion increases the pressure and density of the fluid. This alludes to the key difference between these two distinct combustion modes and their application to pressure gain combustion for gas turbines. Detonation based approaches harness the compressive capability of the detonation wave to achieve what is effectively constant volume combustion without full geometric constraint. A deflagration based technique, however, must use a carefully designed geometry to produce an increase in stagnation pressure. This may be accomplished through direct confinement, as in the case of a wave rotor engine, or through a coupling with acoustic waves as with a pulse combustor based system.

A survey of the literature on the topic of pressure gain combustion indicates that, especially in recent history, detonation based approaches have received significantly more attention than technologies that utilize deflagrations. This is likely attributed to the fundamental understanding that the theoretical thermodynamic gains in stagnation pressure achievable with a detonation are substantially higher than what may be achieved with a deflagration. However, this benefit does not come without cost. Despite their potential performance advantages, stemming from an inherently more efficient combustion process, it is difficult to effectively harness the energy released by these technologies with conventional gas turbine components. Inefficiencies resulting from poor coupling between an isochoric combustor and the turbomachinery in a gas turbine could easily eliminate any potential gain in cycle efficiency that may have been possible with an increase in stagnation pressure at the turbine inlet. The velocity of a detonation front (order of km/s) is significantly higher than that of a deflagration (cm/s or m/s) and in the case of a pulse detonation engine is inherently unsteady. This results in extraordinary difficulty when attempting to couple a detonative combustor with conventional gas turbine components designed for use with a subsonic, steady flow device. Furthermore, detonations produce higher burned gas temperatures for the same initial mixture as a deflagrations, this may result in difficulty maintaining low turbine inlet temperatures and drive higher rates of $NO_x$ formation. Beyond these implementation challenges, significant difficulty remains in maturing detonation based technology to the level necessary for practical application to gas turbine engines, this will be discussed in more detail in the sections that follow [35]. At present no net gain in the stagnation pressure of the working fluid has been demonstrated using detonative device in gas turbine conditions. Therefore significant further research must be carried before current production gas turbine engines may benefit from detonation based pressure gain combustion technology.

## 2.1 Detonation Based Approaches

The pressure gain combustion concepts that have received the most attention and investment in recent years are those that harness a detonation wave to produce a rise in stagnation pressure. These approaches may be generally classified as either pulse detonation engines (PDEs) or rotational detonation engines (RDEs), both of which are distinctly different in physical form and function but operate on the same ther-

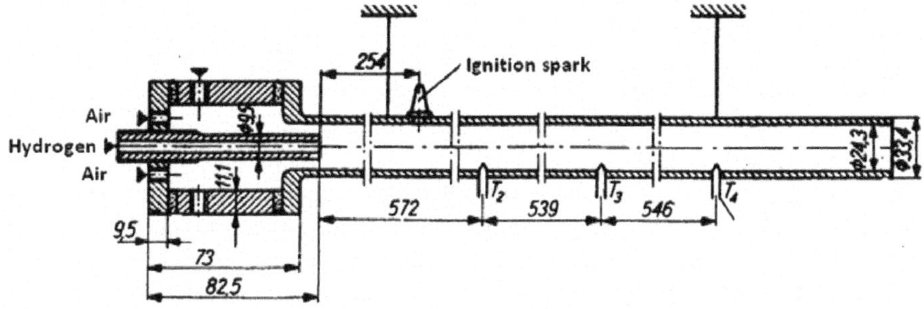


**Fig. 2** First PDE designed by Nicholls et al. [21]

modynamic principles. Beyond replacing the combustion chamber of a gas turbine engine both PDEs and RDEs have found potential application as standalone propulsion devices. For the purposes of this text, the overview of these systems will focus upon the application to pressure gain gas turbine engines. Other applications are well described in the literature [8, 35].

Pulse Detonation Engines

Perhaps the most well developed of the detonation based approaches is the pulse detonation engine (PDE). This device is generally composed of a tube filled with a combustible mixture that is detonated and subsequently purged and filled with another mixture to complete the cycle. Most research systems are composed of a single tube which is used to contain and direct the detonation but a full, practical implementation of this technology in a gas turbine engine would likely include multiple tubes firing in sequence. An example of a single experimental detonation tube is shown in Fig. 2.

The pulse detonation engine cycle is composed of three main phases. Initially, a fuel-oxidizer mixture is injected into the tube. This mixture is then typically ignited as deflagration and transitions into a detonation that propagates from the ignition source to the open end. As the detonation travels down and out of the tube a rarefaction wave forms behind the detonation and partially purges the tube of combustion products. In the final phase the tube is purged with an injected gas (typically air) to remove the remainder of the combustion products, thereby completing the combustor cycle. At present PDEs still face a number of challenges that impede implementation, mainly the difficulty valving of the upstream oxidizer and fuel injectors and minimizing the inefficiency of the deflagration to detonation transition. Many suitable reviews of the development and current state of the art in pulse detonation technology may be found in the literature [1, 4, 9, 22].

Rotating Detonation Engines

The rotating detonation engine (RDE) provides an alternative format to harness the benefits of detonative combustion while addressing some of the fundamental difficulties associated with PDEs. Typical RDE configurations consist of an annular combustion chamber with choked fuel and oxidizer injectors at the inlet and a diffuser at the outlet. A detonable gas mixture is injected into the inlet of the annulus and a periodic detonation wave propagates radially around the annulus consuming these fresh reactants. As the detonation wave travels through the chamber the pressure gradient aids in drawing fresh reactants into the chamber while purging the exhaust products. While RDE operation is still inherently unsteady the detonation wave moves at such a high velocity that the exhaust flow is continuous for practical purposes. This alleviates many of the challenges associated with integrating a PDE system into a gas turbine engine, in some ways making RDE technology far better suited for pressure gain gas turbine engines. At the time of writing, however, RDE technology is still far from seeing practical application as a fundamental understanding of how to establish and sustain operation as well as how to overcome the loss in stagnation pressure inherent of a choked inlet has not yet been achieved. Reviews of RDE technology are also present in the literature [18, 35] (Fig. 3).

## 2.2 *Deflagration Based Approaches*

In addition to the denotative systems, deflagration based devices have also been developed for pressure gain combustion gas turbine applications. These approaches are further subdivided into two primary categories, wave rotor engines and acousti-

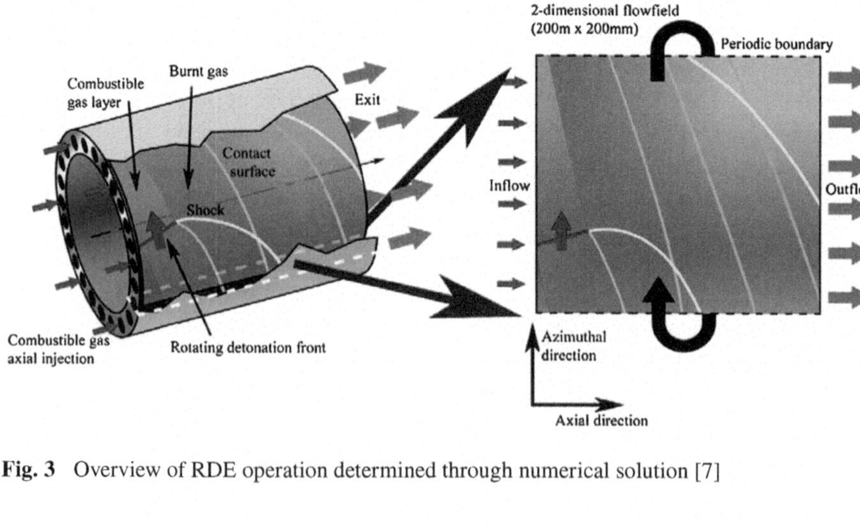


**Fig. 3** Overview of RDE operation determined through numerical solution [7]

cally resonant pulse combustors. Unlike systems utilizing a detonation wave to compress the working fluid the wave rotor and pulse combustor use confined, rapid heat release or coupling with acoustic waves to achieve a stagnation pressure rise. While this introduces additional mechanical complication for lower theoretical stagnation pressure gains, the absence of shock waves and the vibrations inherent of detonations may allow for deflagration based approaches to easily integrate into current gas turbine technology.

Wave Rotor Engines

A wave rotor combustor is composed of a set of channels that rotate between two fixed plates with ports allowing the channels to periodically open and close for filling, isochoric combustion, and expansion. These systems directly confine the heat release event within a fixed volume producing a pressure rise as the combustion products expand. The rotating drum would fit naturally between the compressor and turbine present in gas turbine engines and is an elegant solution to the pressure gain problem. Even so, while this technology displays several favorable characteristics (mechanically simple, low rotational rates, self-cooling) and a significant development effort has been conducted, unresolved challenges have thus far impeded the implementation of these combustors into commercial gas turbine engines. These challenges are primarily mechanical and associated with sealing of the rotating channels between the fixed plates and thermal expansion of the drum. The sealing difficulty is inherent of the concept and the associated losses offset the potential effiency gains achieved with isochoric heat release. Research efforts into wave rotor engines is ongoing and a suitable review is provide by Akbari et al. [2] (Fig. 4).

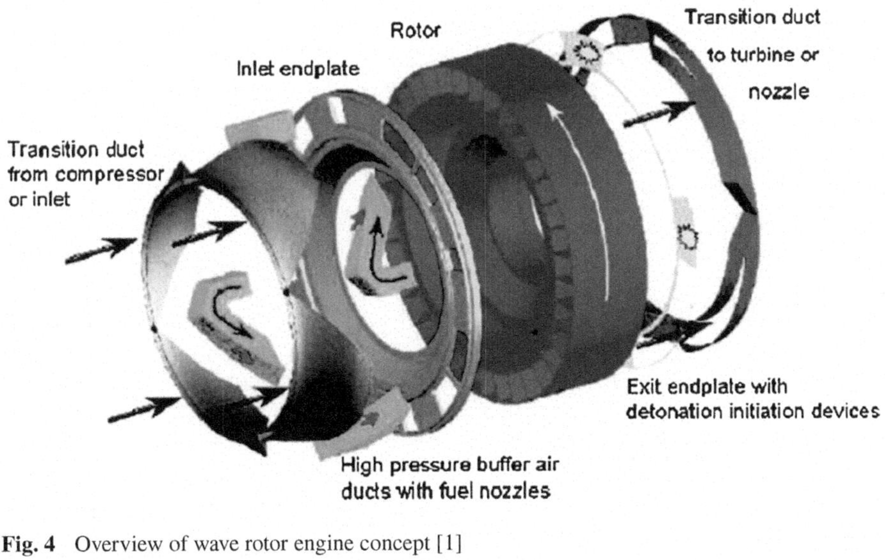


**Fig. 4** Overview of wave rotor engine concept [1]

*Pulse Combustors*

The second deflagration based approach harnesses a thermoacoustic device known as a pulsejet or pulse combustor to achieve heat release with self compression. These devices, initially made infamous by the World War II German V1 "buzzbomb", use a carefully designed geometry that couples unsteady heat release with acoustic fluctuations to create a naturally aspirated, self compressing combustor. Pulse combustors typically provide several advantages over their steady flow counterparts including lower emissions, higher heat transfer rates, and higher efficiency. Furthermore, several studies have demonstrated their suitability for pressure gain combustion applications. Most of these systems couple a pulse combustor with an unsteady fluidic device and bypass flow to cool and smooth the high enthalpy, unsteady flow exiting the combustor to levels that are acceptable for turbine inlets. Experimental apparatus of this design have been used to demonstrate stagnation pressure gains on the order of 4 to 8% of the stagnation pressure delivered by the compressor. The ability to demonstrate a meaningful stagnation pressure gain with practical components and relatively benign turbine inlet conditions make this approach highly favorable to provide a short term dramatic increase in gas turbine efficiency through pressure gain combustion technology.

## 3 Historical Overview of Pulse Combustor Based Pressure Gain Combustion Systems

The use of an acoustically resonant pulse combustor to increase the efficiency of gas turbine engines was first proposed by Reynst [31] in a series of publications with predictions on the potential increase in efficiency that may be possible. They failed, however, to include any experimental evidence to prove viability of the concept. Experimental investigation followed shortly after in the 1950s and has continued slowly to the present day. The following is an overview of the investigations that have sought to understand and develop pulse combustor based pressure gain combustion technology.

### *3.1 Literature Review*

The pulse combustor based pressure gain combustion concept was investigated for gas turbine applications with some success as early as 1958 in a study by Porter [28]. In this work a valveless pulse combustor was developed and tested at one to three atmospheres and successfully demonstrated a net gain in stagnation pressure across the combustor. It was reported in this study that the influence of operating pressure on the performance of the pulse combustor as a pressure gain device was largely

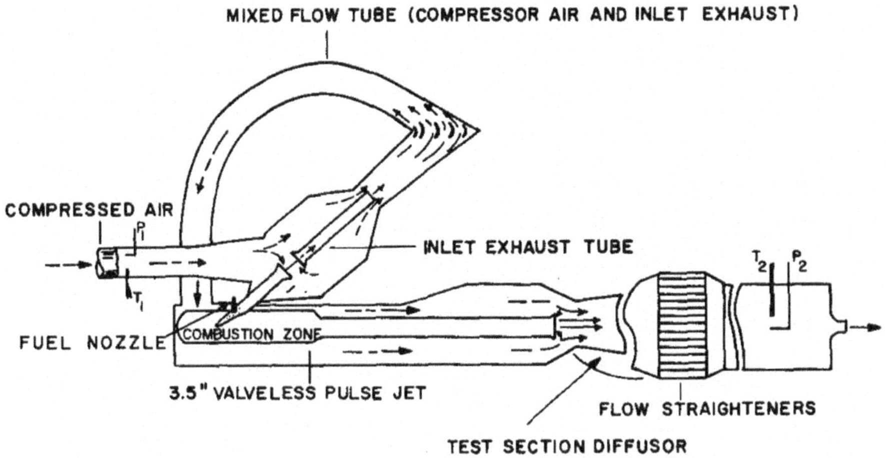


**Fig. 5** Early liquid fueled pressure gain combustion system used in study by Porter 1958 [28]

due to the modification of the delay in onset of heat release. Thus, for the combustor design utilized for this study the performance was retarded for gaseous fuels, which were demonstrated to operate well at atmospheric pressure, and improved for liquid fuels. This was likely due to the additional delay required to allow for liquid fuels to vaporize. This early study provides some indication that proper design may produce a valveless pulse combustor that maintains operation allowing stagnation pressure gains at elevated pressures (Fig. 5).

In the 1971 study of Muller [20] a complex system of actively valved resonant pulse combustors in an annular arrangement were used to demonstrate a 1.015% pressure ratio for temperature rise values between 400 and 500 °C. This is the only example available in the literature of an actively driven rotary valved pulse combustor. This work, however, provides little data on the operation or performance characteristics of the actively valved resonant pulse combustors themselves and is instead focused on the overall pressure gain combustion system performance. One of the key conclusions of this study was that the improvements in gas turbine performance associated with resonant combustion depends primarily on the design of the combustor and the pressure loss of the combustion chamber that would be replaced and the compressor outlet turbine inlet temperatures. It was also shown in this study that the pressure ratio across the resonant combustion process was a strong function of the temperature ratio, with pressure ratio increasing with increasing temperature rise (Fig. 6).

Further significant efforts to develop pulse combustor based pressure gain combustion for gas turbine applications were carried out in the series of experimental and theoretical works by Kentfield. These studies utilized a Lockwood type [17] valveless pulse combustor with a purpose built fluidic device termed a thrust-augmenter flow-rectifier. This device catches the backflow from the forward facing inlet of the valveless combustor and redirects it rearwards while simultaneously entraining (through

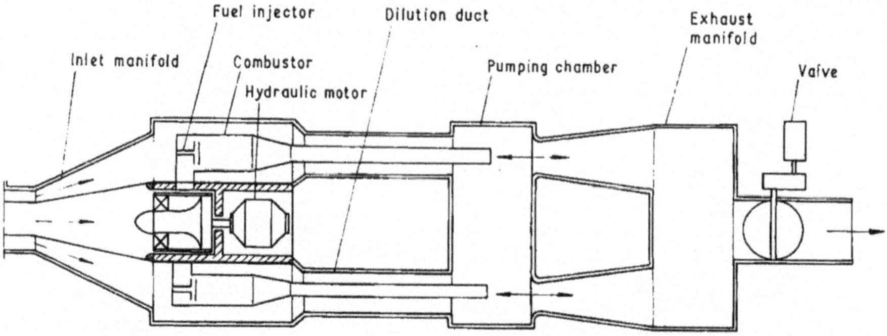


**Fig. 6** Rotary valved pulse combustor based pressure gain combustion system [20]

pulsating-flow ejector action) the secondary stream which bypasses the combustion zone. In their preliminary study [10] it was found that the combustor configuration with added fluidic device would provide a satisfactorily steady flow at the inlet of a turbine and represented a good configuration for combustor start up and sustained operation. It was demonstrated that the proposed configuration produced a maximum stagnation pressure gain of approximately five percent at a stagnation temperature ratio of three to one. Additionally, emissions measurements indicated that emission of unburnt hydrocarbons, carbon monoxide, and nitric oxides is would not likely be more of a problem with valveless, pulse combustors than for conventional combustors. In a second study Kentfield et al. [11] modified a small, educational gas turbine such that the conventional constant pressure burner could be replaced with their valveless pulse combustor and thrust-augmenter flow rectifier. The primary purpose of this study was to investigate the compatibility of the valveless pulsating combustor with the compressor-turbine unit. Through this work they were able to conclude that a resonant combustor has sufficient operational flexibility to serve as a gas turbine combustor and is capable of eliminating combustor pressure losses and producing a small combustion generated stagnation pressure gain (Fig. 7).

A study was carried out by Gemmens [5] with the objective of further evaluating the feasibility of a pulse combustor based pressure gain combustion system for gas turbine applications. In this work the pressure gain performance and pollutant emissions were characterized with a pulse combustor geometry based on that of Kentfield [12]. A pressure gain of approximately 1% of the stagnation pressure and $NO_x$ and CO emissions of about 30 ppmv and 8 ppmv respectively were reported. This provided evidence that such a system may improve the efficiency of gas turbine engines while reducing the pollutant emissions.

More recently, in the 2005 study Paxson [26] utilized an off the shelf, passively valved pulse jet to demonstrate a 3.7% stagnation pressure-gain at total temperature ratios relevant to gas turbine engines (ratios of 2–3). The pulse combustor, ejector, shroud system used in this study, though largely adapted instead of being purpose built, provided some of the most promising evidence of the potential of this technology (Fig. 8).

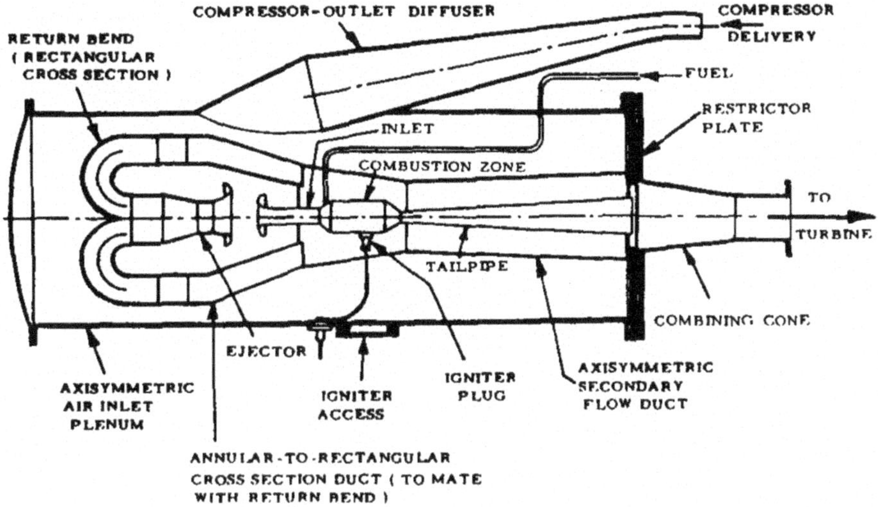


**Fig. 7** Valveless pulse combustor based pressure gain combustion system used in studies by Kentfield et al. [13]

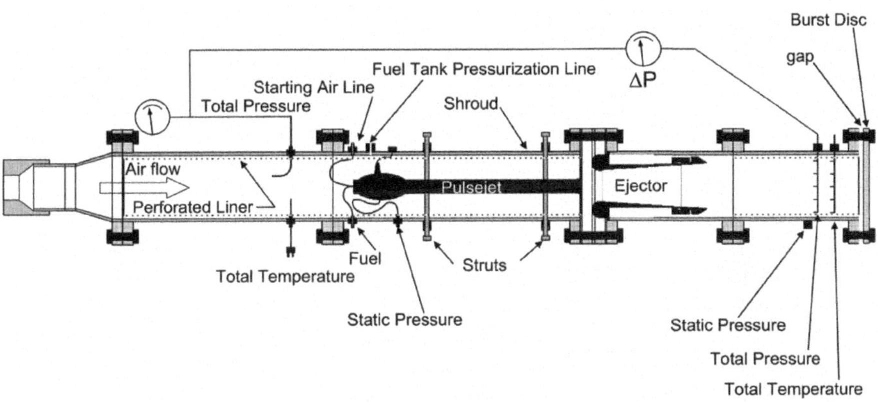


**Fig. 8** Pressure gain combustion system of Paxson et al. [26]

Some work has been carried out [16] to develop pulse combustor technology that would be adaptable to practical gas turbine technology. This work was focused on further understanding rotary inlet valves and their influence on the operation and performance of the acoustically resonant pulse combustor. It was found that the "ball" type rotary valve design, shown in Fig. 9, was capable of operating over a wide range of frequencies and fuel flow rates without large changes in the performance. These preliminary results indicate a rotary valve may be a strong contender for pulse combustor based pressure gain combustion systems.

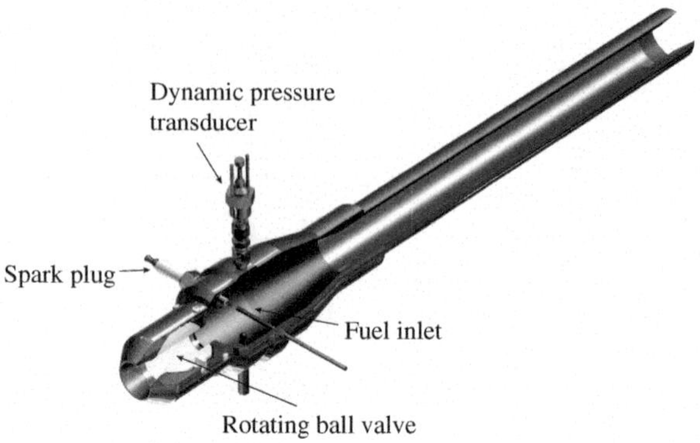


**Fig. 9** 50 cm acoustically resonant pulse combustor with "ball" type rotary valve [16]

## *3.2 Knowledge Gaps*

Despite the large potential to provide a dramatic increase in gas turbine efficiency, thereby reducing the cost, pollutant and greenhouse gas emissions per unit of work output, pressure gain combustion technology has not yet been integrated into production gas turbine engines in any form. As the topic of this chapter is pulse combustor based pressure gain combustion systems, the following is an overview of the aspects of pulse combustor based system for which our current lack of understanding is impeding the implementation of this technology.

*Pulse Combustor Design for Pressure Gain Combustion*

One of the key features of these systems is the acoustically resonant pulse combustor that approximates isochoric heat release. At this time little research has been carried out to develop this technology specifically for pressure gain combustion gas turbines, thus leaving large uncertainty in regards to the proper design approach for these applications. Furthermore, these complex devices become even more difficult to purposely design when attempting to account for acoustic interaction between the pulse combustor and other gas turbine components. It is highly important to understand these interactions as they can manifest as destructive interference that renders the acoustics driven combustor inoperable. In order to design a practical pulse combustor driven system an improved understanding of the key pulse combustor parameters that influence the operation and performance on the pressure gain system as a whole is necessary.

Design of a pulse combustor for a pressure gain combustion system requires the choice between an aerodynamic or mechanical valve for the inlet located at the front of the combustion chamber. This valve allows fresh air to enter the combustor for the portion of the cycle during which the combustion chamber pressure is below the inlet pressure and conversely closes the chamber inlet during compression and the subsequent heat release. This reduces back flow through the inlet and directs combustion products through the exhaust tube of the device and strongly impacts the operability and performance of the device.

Valveless pulse combustors provide the distinct advantage of requiring no moving parts but introduce further complication in difficulty establishing resonant operation and the presence of significant amounts of reverse flow through the inlet. Mechanical valves, found in three forms: reed, flapper, and rotary, alleviate many of these difficulties but also introduce further complications. Reed and flapper valve motion is passively driven by the combustion chamber pressure fluctuations and thus must have a low enough mass such that the valve can accelerate in the time necessary to open in a fraction of the operating frequency of the combustor (typically sub millisecond timescale). The low mass requirement combined with the presence of high temperatures and cyclic loads results in significant difficulty designing a mechanically robust passive valve. This was shown to be an important limitation in the pressure gain combustion study of Paxson et al. [27] where failure of passive reed valves of the pulse combustor in the study limited experimental run times to the extent that thermal equilibrium could not be achieved. The third form, rotary valves, are actively driven at a fixed frequency and therefore not mass limited as passive valves are. This alone allows for a significantly more robust mechanical design. Furthermore, in the case of gas turbine applications the necessary rotating components are already present in conventional technology. Thus, in summary, valveless inlets create significant operation difficulty, passively valved inlets are effective and simple but are not robust enough to be suitable for a practical application, but rotary valves may be both robust and effective.

Despite their apparent suitability for pressure gain combustion gas turbine applications, little in the literature exists describing the operational or performance characteristics of an actively driven, mechanically valved pulse combustor. To the authors knowledge all proposed active valve designs have been of the rotary type as described in [19] with the exception of the pressure gain study by [20] in which little detail of the pulse combustor operation was provided. No study exist in which the operational and performance characteristics of an actively driven, rotary valved, acoustically resonant pulse combustor are described. A recent review by Meng et al. [19] of pulse combustor technology provides insight to the extensive research efforts that have been carried out to understand valveless and passively valved pulse combustors, thereby highlighting the apparent gap in knowledge of the operation of actively valved devices.

In order for rotary valve, acoustically resonant pulse combustors to be effectively utilized in a pressure gain combustion system, some understanding of the influence of the rotating valve on the operation and performance of the device is desired. Furthermore, to achieve meaningful stagnation pressure gains optimization of the valve combustor design may be necessary. Information regarding these challenges is absent from current literature.

*Pulse Combustor Based Pressure Gain Combustion System Emissions*

Before any new combustion concept may be considered for practical systems it must be expected meet or exceed current emissions standards. Although most standalone pulse combustors are shown to have lower $NO_x$ and $CO$ emissions than their steady combustion counterparts [19], the influence of integrating these combustors into full pressure gain combustion systems operating at elevated pressure and temperature on their operation, and thus emissions, is poorly understood. This is because, as discussed in the previous section, the majority of pulse combustor based pressure gain research studies have failed to include quantitative emissions measurements. Solid experimental evidence suggesting that the emissions of a pulse combustor based pressure gain gas turbine would scale below or at the pollutant emissions levels of a conventional gas turbine would greatly aid in the adaptation of this technology into commercial applications.

*Pulse Combustor Based Pressure Gain Combustion System Performance in Gas Turbine Conditions*

A typical gas turbine combustion chamber has an inlet condition that includes gas temperatures on the order of 500 °C and static pressures ranging from 16–40 bar. These conditions are exceedingly difficult to simulate within a laboratory environment, especially while allowing access for sufficient instrumentation to characterize the performance and emissions characteristics of such a system. For this reason many of the research efforts into the development of pulse combustor based pressure gain combustion systems have been limited to low inlet pressures (1–4 bar) and temperatures (300 K). Although the results of these studies are invaluable in the development of pulse combustor based pressure gain combustion technology, it is difficult to extrapolate the performance and emissions characteristics to the conditions present in a practical gas turbine.

For this reason one of the key knowledge gaps that must be bridged before this technology will see practical implementation is experimental evidence providing insight into how the performance and emissions characteristics of a pulse combustor based system will scale with increasing inlet pressures and temperatures. The current literature has examples of a variety of systems, all with different flow geometry, combustors, instrumentation for characterizing performance, and operating conditions. These studies are scattered across a such wide variety of different configurations that it is impossible to draw conclusions outside of the independent studies. What is required is a set of consistent experimental measurements utilizing one combustor and geometric configuration across a range of operating conditions, thereby providing direct information on performance and emissions scaling.

## 4 Current State of the Art in Pulse Combustor Based Technology

At time of publishing, pulse combustor based pressure gain combustion systems remain a niche research topic conducted by only a few groups internationally. Despite this, the most modern experiments carried out to date have been able to demonstrate meaningful stagnation pressure gains at total temperature ratios relevant to gas turbine engines [26] (indicative of realistic bypass ratios) and include a demonstration of the systems ability to perform work by coupling the outlet of one such device with the inlet of a modified turbocharger [27]. This represents a substantial achievement as few pressure gain concepts can boast the experimentally demonstrated ability to produce a meaningful stagnation pressure gain much less perform work of any form.

Current research pulse combustor based pressure gain combustion systems are composed of three key components: an acoustically resonant pulse combustor, an unsteady ejector, and a shroud to enclose these component thus allowing for the presence of bypass flow in the device. Although in modern systems each of these components maintains some key characteristics, large variation exists in their physical realizations and in the level of performance they are able to achieve. As such, the state of the art in pulse combustor based pressure gain technology is defined, for the purposes of this text, as the most recent and complete technology. In discussion of the state of the art in these devices it is useful to separate the pulse combustor technology from the full systems. The motivation for this being that pulse combustor technology is better established for applications (i.e. pulse combustor drying) outside of pressure gain combustion gas turbines.

*Acoustically Resonant Pulse Combustors*

The pulse combustor is a relatively old technology, it was first invented and documented in the early 19 century with an indepth theoretical description provided by Rayleigh in 1976 [30]. The subsequently defined Rayleigh criterion provided a basis for the fundamental understanding of the operation of an acoustically resonant pulse combustor. A simple geometric design and unique capability to run naturally aspirated inspired significant interest in evaluating the operational capability of these devices and experiments quickly demonstrated an ability to run with a variety of different geometries and high fuel flexibility (gaseous, liquid, and pulverized solid all demonstrated) [3]. Perhaps the most well known implementation of pulse combustor technology was as a form of propulsion for the World War II V-1 buzz-bomb. After its success as a propulsion device, the pulse combustor found further application in residential heating and commercial drying [3, 19].

While many of these applications are unrelated to pressure gain combustion gas turbines, this technology has benefited from the extensive efforts put forth to understand the many complex processes dictating their operation and performance. Suitable reviews detailing the development of pulse combustor technology are provided by Bellest et al. [3] and more recently Meng et al. [19] (Fig. 10).

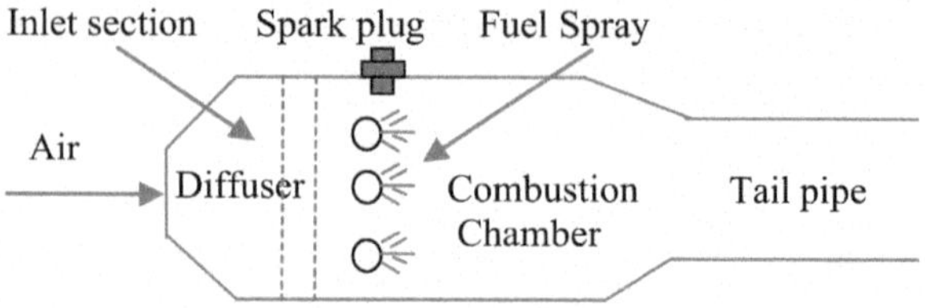


**Fig. 10** Typical pulse combustor geometry [19]

*Principle of Operation*

In general, all pulse combustors operate on the same principle with only slight variations resulting from differences in configuration. Acoustically resonant pulse combustors operate on the Rayleigh criterion which states that if unsteady heat release occurs in a system with fluctuating pressure that the point of heat release must occur in phase with the pressure oscillation for that pressure wave to be amplified [30]. This criterion is described mathematically with the following equation where $q$ is the unsteady heat release, $p$ is the oscillating pressure, and $t$ represent time.

$$\int_{cycle} \dot{q}pdt > 0 \tag{1}$$

If the heat release occurs directly at the point of maximum pressure then the amplitude of the oscillation is increased with no other alteration to the pressure fluctuations. Conversely if the heat release occurs directly at the point of minimum pressure then the amplitude is decreased with no other effect. If the heat release occurs at the mean value of the oscillation before the peak pressure, then the frequency of the oscillation will increase with no effect on the amplitude. Likewise if the heat release occurs at the mean value after the peak in pressure then the frequency will decrease with no effect on pressure. These represent the extreme cases with heat release being approximated as a point energy release. In practice the unsteady heat release will influence both the amplitude of pressure oscillation and frequency of operation.

The steady state operation of an acoustically resonant pulse combustor is described as follows with a graphical representation of this operation occurring in a passively valved Helmholtz pulse combustor in Fig. 11.

- Phase 1—Air is drawn into the combustion chamber through the forward inlet as fuel is injected and mixes with the air. Residual exhaust gases from the previous cycle mix with the fresh reactants raising the temperature of the mixture and causing the reactants to ignite. This results in a rapid rise in pressure within the combustion chamber.
- Phase 2—As the combustion products expand they are channeled down the exhaust tube and the pressure rise is followed by an over-expansion caused by the momentum of the fluid exiting the combustor. This results in a drop in pressure in the combustion chamber below that of the inlet pressure.

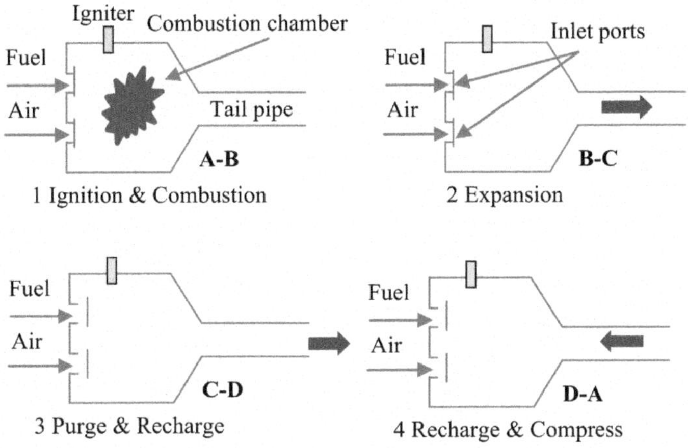


**Fig. 11** Cycles of acoustically resonant, passively valved pulse combustor operation [19]

- Phase 3—As the pressure in the combustion chamber drops below the inlet pressure a portion of the combustion products is drawn back towards the combustion chamber and fresh reactants drawn through the air and fuel inlets.
- Phase 4—As the fresh reactants mix with the hot products and the oscillating pressure again exceeds the inlet pressure the mixture is ignited and the cycle is completed.

In general a spark, and in some cases starting air, is required to initiate the cycle. Once steady state resonant operation is achieved a properly designed pulse combustor can operate indefinitely with only a continuous fuel supply.

*Pulse Combustor Advantages*

Energy conversion through pulse combustion provides several important advantages over steady combustion devices. In general, they have higher combustion efficiency and intensities while maintaining lower $NO_x$ and CO pollutant emissions. They are compact and capable of operating on wide range of fuels without significant modification. Futhermore, they are relatively simple and do not require ducting that results in high aerodynamic losses to ensure stable operation.

*Pulse Combustor Variations*

Three key components are characteristic of all pulse combustion devices relevant to pressure gain applications. These components are the air inlet, combustion chamber, and an exhaust tube extending from the combustion chamber. The size of these components relative to each other influence the fundamental operation of the device and general guidelines for selection of these parameters may be found in the literature [3].

Pulse combustor technology is generally categorized by its overall geometric and operational characteristics and most typically fall into one of the following three categories: Schmidt, Helmholtz, or Rijke. A Schmidt-type pulse combustor operates on the principle of a quarter-wave sound resonator and is characterized by an open/closed inlet followed by a combustion chamber and exhaust tube (which may be the same diameter as the combustion chamber). In these systems the heat release event occurs in the combustion chamber near the inlet and the mixing and combustion processes must occur within half a period of the pulsation. Helmholtz type pulse combustors operate on the acoustic Helmholtz resonator principle and are characterized by an closed/open or open/open inlet, followed by a separate combustion chamber and a long, small diameter exhaust tube. For this device the frequency of operation is determined by both the combustion chamber volume, overall length, and the cross-sectional area of the exhaust tube. The Rijke tube operates on the half wave principle and is open on both ends with the heat release event occurring at the quarter length and the frequency of operation being determined by the length of the tube. Choice of a resonant pulse combustor for pressure gain gas turbine applications typically falls in the quarter-wave or Helmholtz category as these devices have clearly defined pressure zones allowing for a robust resonant system.

The pulse combustors that would typically be consider for application in a pressure gain gas turbine can be further subdivided into two main categories, those with a mechanical valve at the forward end of the device and those without a mechanical valve, commonly known as valveless pulse combustors. Mechanical valves are typically of the reed, flapper, or rotary types as described in Fig. 12 of which the reed and flapper are the most popular and are passively driven by the pressure differential resulting from pressure oscillations in the combustor. In the case of a rotary valve the operational frequency can be precisely controlled by the rotational speed of the valve assuming the rotational rate is sufficiently close to the resonant frequency of the geometry. Mechanically valved pulse combustors have the ability to achieve unidirectional flow during the pressure rise and exhaust phases of the cycle as the

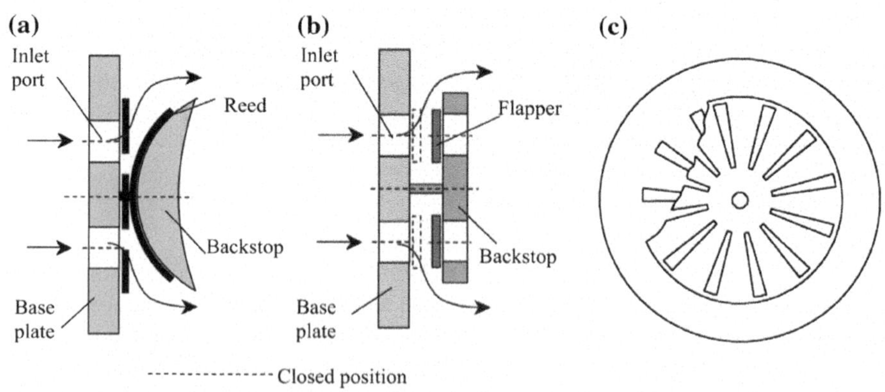


**Fig. 12** Reed (**a**), flapper (**b**) and rotary valves (**c**) [19].

mechanical vales are able to effectively seal the front air inlet. Aerodynamic valves provide the convenient advantage of requiring no moving parts, this is accomplished by using a carefully chosen geometry to create fluid-mechanic effects to reduce back flow through the inlet [19].

Theory

One of the most important parameters in the design of a pulse combustor for a pressure gain combustion system is the resonant frequency of the device. This frequency determines the fuel flexibility and is important for specifying the mass flow rate of the working fluid that is ingested into the combustor. As such, the following are provided as approximate methods of calculating the natural frequencies of Schmidt and Helmholtz type acoustically resonant pulse combustors.

*Resonant Frequency*

For the Schmidt type combustor the frequencies of all possible longitudinal modes are given in the following equation [36]. For practical purposes only the fundamental mode is relevant.

$$f_{Schmidt} = \frac{(2k-1)c_e}{4(l_e + l_c)} \tag{2}$$

where $f_{Schmidt}$ is the frequency, $k$ is the mode number, $c_e$ is the average speed of sound in the exhaust tube, $l_e$ is the exhaust tube length, and $l_c$ is the combustion chamber length.

The resonant frequency of a Helmholtz type combustor is given as follows [36].

$$f_{Helmholtz} = \frac{c_e}{2\pi}\left(\frac{A_e}{A_c}\right)^{\frac{1}{2}}\left(\frac{1}{l_e l_c}\right)^{\frac{1}{2}} \tag{3}$$

where $A_e$ is the cross-sectional area of the exhaust tube and $A_c$ is the cross-sectional area of the combustion chamber. Empirical evidence has shown that the ratio of the exhaust tube to combustion chamber cross-sectional areas determines if the fundamental mode of vibration is of the quarter-wave or Helmholtz resonance. The area ratio at which this transition occurs may be computed as follows.

$$\frac{A_e}{A_c} = \left(\frac{\pi}{2}\right)^2 \frac{l_e l_c}{(l_e + l_c)^2} \tag{4}$$

By determining whether a given pulse combustor design area ratio falls above or below that calculated with the above equation, it can be decided if the equation for Helmholtz (for values below) or Schmidt (for values above) should be used to compute the resonant frequency.

*Performance*

The established metric for characterizing the performance of an acoustically resonant pulse combustor is the Raleigh efficiency $\eta$ defined as follows [14].

$$\eta = \frac{R}{MC_p P\dot{Q}_t} \int_0^1 \dot{Q}'(\zeta)P'(\zeta)d\zeta \quad (5)$$

where $\eta$ is the Rayleigh's efficiency; $C_p$ is the specific heat at constant pressure; $R$ is the universal gas constant; $M$ is the mole averaged molecular weight; $P$ is the mean pressure; $P'(\zeta)$ is the perturbation of pressure about the mean; $P'(\zeta) = P - P(\zeta)$; $\dot{Q}$ is the mean energy release rate; $\dot{Q}'(\zeta)$ is the perturbation of the instantaneous energy release rate about the mean; $\dot{Q}'(\zeta) = \dot{Q} - \dot{Q}(\zeta)$; and $\zeta$ represents the normalized time.

The Rayleigh efficiency is a measure of the coupling between the instantaneous energy release and the resonant pressure wave which strongly correlates with the effectiveness of the device to perform in a pressure gain combustion system. However, in practice the parameters necessary to compute the Rayleigh efficiency are exceedingly difficult to measure experimentally. For this reason a single, easily measured parameter that is representative of the Rayleigh's efficiency is highly desired. As shown in Fig. 13, Keller et al. [14] has demonstrated that the second moment of the pressure probability distribution (Prms) exhibits the same trend as the measured Rayleigh efficiency. This indicates that the Prms as measured at the combustion chamber is indicative of the performance of the combustor. A useful metric as the Prms can be easily measured with inexpensive pressure traducers allowing for simple diagnostic setups for evaluating varied pulse combustor parameters and there relative influence on device performance.

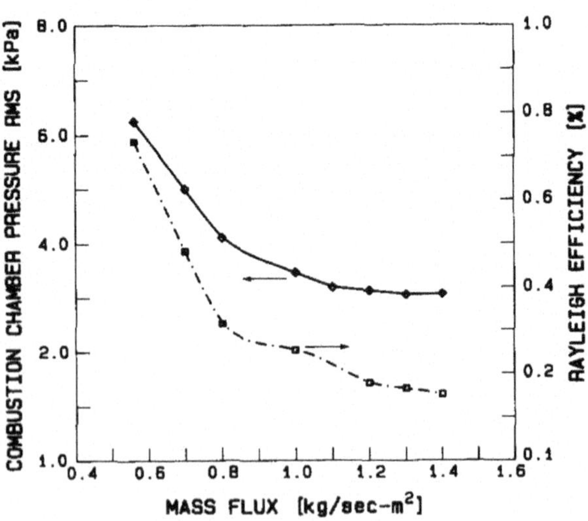


**Fig. 13** Comparison of Rayleigh's efficiency and the magnitude of the oscillating component of the combustion chamber pressure as a function of mass flux for an acoustically resonant pulse combustor [14]

Design for Pressure-gain Combustion Systems

In order for a pulse combustor to effectively replace a conventional gas turbine combustion chamber it should be capable of operating stably with conventional fuels across a wide range of powers. The operating frequency of an acoustically resonant pulse combustor is a direct function of the geometric design parameters (length, combustion chamber volume, exhaust section cross-sectional area) and, as such, a minimum for each of these physical dimensions exists below which fuel limitations also exist. On the fuel side, these limitation are a function of the laminar flame speed, ignition delay, and characteristic mixing times of the individual fuel. For instance, a hydrogen power pulse combustor is generally capable of operating at significantly higher frequencies than a Jet-A powered device. For this reason the overall size of the pressure gain system is, up to a point, dictated by the desired operating fuel.

In addition to operating frequency, the desired bypass ratio must be considered in design. As the gains in stagnation pressure are only realized in the fluid that passes directly through the pulse combustor the mass flow rate through the combustor inlet must be high enough such that a meaningful overall gain is achieved. The limitation preventing all of the working fluid from being directed through the pulse combustor is the need to reduce the gas temperature down to levels acceptable for the turbine inlet. The flow rate through the combustor inlet is a function of both the operating frequency and the inlet area, both of which scale with the overall length of the combustor.

For a practical system the pulse combustor must be robust enough to operate for an extended duration with a very low failure rate. This potentially eliminates most forms of passive mechanical valves for practical systems. Cyclic loads in a high temperature reaction zone with the requirement of being very low mass creates enormous difficulty in designing a practical passive mechanical valve that is capable of maintaining structural integrity for long periods of operation. For this reason many of the studies seeking to develop pulse combustor based pressure gain systems have been designed around valveless devices. Despite their apparent advantages valveless pulse combustors present inherent difficulties as discussed in Sect. 3.2 rendering practical design difficult. Rotary valves may provide a suitable mechanical inlet valve but few examples of resonant pulse combustors operating with actively driven rotating inlet valves exist in the literature. Further development of rotary valve technology may better enable the practical design of pulse combustor pressure gain combustion technology.

## *4.1 Unsteady Ejectors*

Conventional gas turbine technology has long been designed and optimized to operate with the steady flow present at the exit of a isobaric combustion chamber. This provides a challenge when designing an inherently unsteady pressure gain combustion device to be capable of operating with readily available turbomachinery designed for steady flow technology. For turbines designed with steady flow inlets the unsteady flow exiting a pulse combustor would obtain a large efficiency penalty in the form of decreased turbine performance. This penalty could be so large that any benefits obtained by the gain in stagnation pressure would not be realized as an overall gain in efficiency.

For this reason pulse combustor pressure gain combustion systems have harnessed an unsteady ejector (also known as a thrust augmenter) reduce the combustion driven velocity fluctuations. Experimental evidence has shown that a properly designed pulse combustor ejector system can reduce the mean flow velocity fluctuations to 4% of the magnitude present on the exit plane of the pulse combustor [6]. This level of fluctuation is low enough such that the decrease in turbine efficiency would be insignificant.

### 4.1.1 Principle of Operation

The unsteady ejector is a fluidic device that harnesses the energy in the high enthalpy flow exiting the pulse combustor exhaust pipe to efficiently mix the exhaust products with bypass flow. This reduces the temperature of the working fluid at the inlet of the turbine and smooths out velocity fluctuations generated by the periodic combustion process. This is accomplished by using specially designed geometry to enable the high enthalpy flow leaving the combustor to pull additional bypass air through the ejector, evening the flow velocity and temperature. The process occurs effectively in two phases [34], the cycle is initiated as flow is pushed from the exhaust tube of the pulse combustor and the jet of exhaust products forms a toroidal vortex which impacts with the inlet of the ejector. This accelerates the fluid mass inside of the ejector body and entrains additional bypass flow, thereby reducing the velocity of the initial jet. The inertia of the mixture of exhaust products and bypass air inside the ejector body isolates the flow leaving the ejector exit plane from the unsteady jets at the inlet providing the desired smoothing of the velocity field. These two phases are illustrated in Fig. 14.

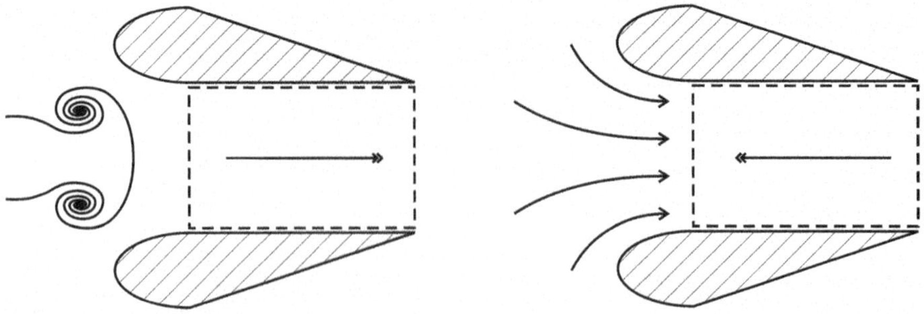

**Fig. 14** Unsteady ejector operation [34]

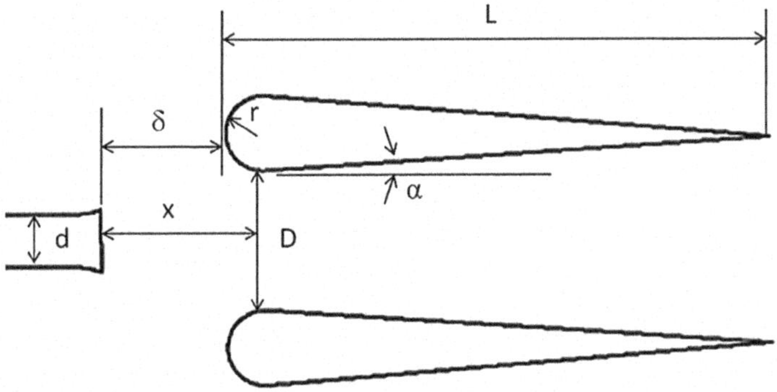


**Fig. 15** Key dimensions of common unsteady ejector design

### 4.1.2 Ejector Design for Pulse Combustor Based Pressure Gain Combustion System

The key design parameters of an unsteady ejector in pressure gain combustion applications is the ability to effectively smooth fluctuations in the flow velocity while maintaining the stagnation pressure gain provided by the unsteady combustor. A typical ejector geometry with key dimensions is shown in Fig. 15. As this technology is still developing there is not yet an established design methodology for selection of the ejector geometric parameters. Research efforts have yielded some general guidelines based on empirical results as summarized in the literature [24, 25, 33].

In effort to reduce the number of necessary components some efforts have been made to combine the unsteady ejector with turning vanes for coupling with a turbine. These are important considerations when attempting to move this technology from laboratory research to practical application.

## *4.2 Pulse Combustor Ejector Shroud Pressure Gain Combustion Systems*

Modern experiments that have been successful in evaluating the capability of pulse combustor based pressure gain combustion systems have largely been composed of three primary components. The acoustically resonant pulse combustor, an unsteady ejector to smooth velocity fluctuations, and a shroud to enclose these components and allow for adequate bypass flow to ensure practical turbine inlet temperatures. For this reason some researchers have designated these systems pulse-combustor, ejector, shroud or PCES.

The PCES represents the core set of components which together may replace a conventional isobaric combustor in a gas turbine engine. Although each of these components has undergone development separately it is ultimately important that they be purposefully designed to work together. This may yield an improved performance and ease of operation relative to systems that used off the shelf components.

Overview

The PCES system may provide a credible replacement for conventional steady flow combustors in gas turbine engines. Implementation of these systems would logically be realized in an annular arrangement as in current gas turbine designs and would likely, at least initially, result in an increase of overall mass due to the necessary inclusion of additional components. For this reason the technology would most likely see adaptation for ground based and naval power generation applications.

One challenge of designing such a system is the potential for acoustic interaction between the pulse combustor and geometry that encloses the bypass flow. This problem was highlighted in the study of Paxson [26] but was alleviate through the installation of an acoustic liner to dampen out resonant waves in the shroud.

The performance of the system is characterized by the measurement of the stagnation pressure ratio across the pulse combustor and unsteady ejector as a function of the total temperature ratio. In a well designed system the flow forward of the combustor should be fairly uniform and measurement of the stagnation pressure and temperature may be obtained at any position reasonably far from the wall. Aft of the combustor, if the system is performing well, the ejector should smooth velocity fluctuations allowing for the stagnation pressure measurement to be obtained from a single probe as in the case with the forward measurement. However, the total temperature distribution across the diameter of the shroud is unlikely to be uniform and instead requires a rake to capture an averaged value, this is demonstrated in the apparatus in Fig. 16.

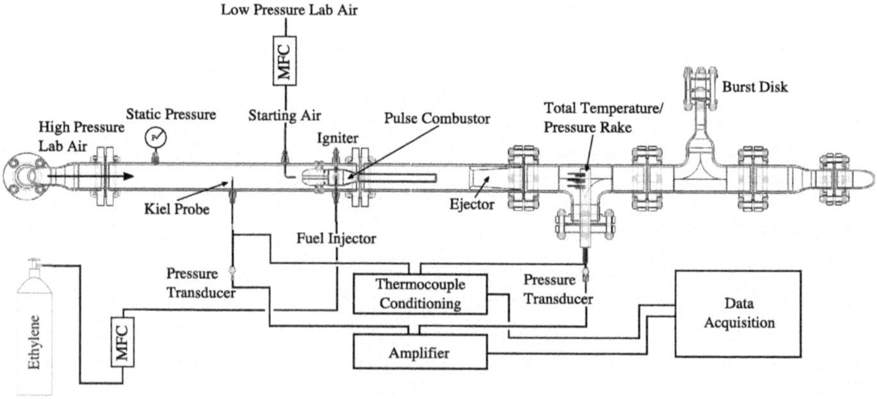


**Fig. 16** Pressure gain combustion system at King Abdullah University of Science and Technology (KAUST) Clean Combustion Research Center (CCRC) High Pressure Combustion Laboratory

Conclusion

Pulse combustor driven pressure gain combustion has great potential to bring a step increase to gas turbine efficiency in the near future. Further efforts to develop and understand resonant pulse combustors in the context of a pressure gain combustion system should allow for the design of effective experimental systems. If these systems can demonstrate that the performance and emissions characteristics can be preserved in typical gas turbine operating conditions, then adaptation of these devices into commercial systems will be possible.

## References


1. Akbari P, Nalim R, Mueller N (2006) A review of wave rotor technology and its applications. J Eng Gas Turbines Power 128:717
2. Akbari P, Nalim R (2009) Review of recent developments in wave rotor combustion technology. J Propul Power 25:833–844
3. Bellest FE (1986) Pulse combustion. Prog Energy Combust Sci 12:43–79
4. Dean AJ (2007) A review of PDE development for propulsion applications. In: 45th AIAA aerospace sciences meeting & exhibit
5. Gemens R, Richards G, Janus M (1995) Pressure-gain combustion for gas turbines. In: Advanced coal-power systems '95 review meeting. Morgantown, West Virginia
6. Heffer J, Miller R (2009) Performance of choked unsteady ejector-nozzles for use in pressure gain combustors. In: 47th AIAA aerospace sciences meeting including the new horizons forum and aerospace exposition
7. Hishida M, Fujiwara T, Wolanski P (2009) Fundamentals of rotating detonations. Shock Waves 19:1–10
8. Kailasanath K (2000) Review of propulsion applications of detonation waves. AIAA J 38:1698–1708
9. Kailasanath K (2009) Research on pulse detonation combustion systems? a status report. AIAA Paper
10. Kentfield JAC, Rehman M, Marzouk ES (1977) A simple pressure-gain combustor for gas turbines. J Eng Power 99:153
11. Kentfield JAC, Yerneni P (1987) Pulsating combustion applied to a small gas turbine. Int J Turbo Jet Engines 4:45–54
12. Kentfield JAC, Fernandes LCV (1990) Improvements to the performance of a prototype pulse, pressure- gain, gas turbine combustor. J Eng Gas Turbines Power 112:67–72
13. Kendfield JAC (1991) Performance implications of gas-turbine pressure-gain combustors. In: 27th joint propulsion conference
14. Keller JO, Bramlette TT, Dec JE, Westbrook CK (1989) Pulse combustion: the importance of characteristic times. Combust Flame 75:33–44
15. Lefebvre AH (2010) Gas turbine combustion. Taylor & Francis, United Kingdom
16. Lisanti JC, Roberts WL (2016) Design of an actively valved and acoustically resonant pulse combustor for pressure-gain combustion applications. In: 54th AIAA aerospace sciences meeting. San Diego, California
17. Lockwood RM (1962) Summary report on investigation of miniature valveless pulsejets. Hiller Aircraft Company
18. Lu F, Braun EM, Massa L, Wilson DR (2011) Rotating detonation wave propulsion: experimental challenges, modeling, and engine concepts. In: 47th AIAA/ASME/SAE/ASEE joint propulsion conference & exhibit, vol 30, pp 1125–1143

19. Meng X, de Jong W, Kudra T (2016) A state-of-the-art review of pulse combustion: principles, modeling, applications and R&D issues. Renew Sustain Energy Rev 55:73–114
20. Muller J (1971) Theoretical and practical aspects of the application of resonant combustion chambers in gas turbines. J Mech Eng Sci 13:137–150
21. Nicholls JA, Wilkinson HR, Morrison RB (1957) Intermittent detonation as a thrust-producing mechanism. J Jet Propul 27:534–541
22. Nikitin VF, Dushin VR, Phylippov YG, Legros JC (2009) Pulse detonation engines: technical approaches. Acta Astronaut 64:281–287
23. Pavri R, Moore, GD (2001) Gas turbine emissions and control. GE Energy Services
24. Paxson DE, Wilson J, Dougherty KT (2002) Unsteady ejector performance: an experimental investigation using a pulsejet driver. In: 38th AIAA/ASME/SAW/ASEE joint propulsion meeting and exhibit. Indianapolis, Indiana
25. Paxson DE, Wernet MP, Wentworth J (2004) An experimental investigation of unsteady thrust augmentation using a speaker-driven jet. In: 42nd aerospace sciences meeting and exhibit. Reno, Nevada
26. Paxson DE, Dougherty K (2005) Ejector enhanced pulsejet based pressure gain combustors: an old idea with a new twist. In: 41st AIAA/ASME/SAE/ASEE joint propulsion conference & exhibit. Tucson, Arizona
27. Paxson DE, Dougherty K (2008) Ejector enhanced pulsejet based pressure gain combustors: an old idea with a new twist. In: 46th aerospace sciences meeting and exhibit. Reno, Nevada
28. Porter CD (1958) Valveless-gas-turbine combustors with pressure gain. In: ASME 1958 gas turbine power conference and exhibit. 58-GTP-11:1–9
29. Pressure Gain Combustion Program Committee (2016) Resources. The American Institute of Aeronautics and Astronautics. https://info.aiaa.org/tac/pc/PGCPC/Resources/Resources.aspx. Accessed 20 Oct 2016
30. Rayleigh J (1976) Theory of sound. Dover Publications, United States of America
31. Reynst F (1961) Pulsating combustion: the collected works of FH Reynst. Pergamom Press, United Kingdom
32. Sawyer RT, Hawthorne WR (1994) Reflections on United Kingdom aircraft gas turbine history. J Eng Gas Turbines Power 116:495–510
33. Ward CM, Duffin M, Miller RJ, Heffer JH (2010) Unsteady ejectors; the effect of driver jet mark-space ratio. In: 48th AIAA aerospace sciences meeting including the new horizons forum and aerospace exposition. Orlando, Florida
34. Ward CM, Miller R (2012) Performance analysis of an ejector enhanced pressure gain combustion gas turbine. Aerospace Sciences Meetings, Nashville, Tennessee
35. Wolański P (2013) Detonative propulsion. Proc Combust Inst 34:125–158
36. Wood A (1945) Acoustics. Interscience Publishers Inc., New York

# Lean Blowout (LBO) Prediction Through Symbolic Time Series Analysis

**Swarnendu Sen, Soumalya Sarkar, Rajendra R. Chaudhari, Achintya Mukhopadhyay and Asok Ray**

**Abstract** Lean combustion is preferred in combustors for cleaner emission. But, the major problem in running the combustor in lean condition is the flame blow-out, due to lower burning velocity in lean condition. To sustain lean combustion, early prediction of lean blowout (LBO) is needed. This need calls for developing new strategies to predict LBO. This work is on development of strategies based on symbolic analysis of time series data obtained from the combustor. Optical signal (CH* chemiluminescence) based time series data has been captured from a laboratory scale dump combustor. Gaseous fuel has been used at different premixing regimes to run the combustor. The time series data obtained are partitioned to different cells to get symbol strings and subsequently generate a probabilistic finite state automata (PFSA). The constructed PFSA is called $D$-Markov machines. These machines have a deterministic algebraic structure and the states are represented by symbol blocks of length $D$ or less, where $D$ is a positive integer. The estimated state probability vector is found out on the basis of number of occurrence of each symbol over a time duration. A reference state vector is calculated at a condition that is sufficiently away from the state of LBO. For a given state, the deviation of the state vector from the reference vector can be computed and used as an anomaly measure for early detection of LBO. The problem of LBO prediction is posed in a pattern classification setting and the underlying algorithms have been tested on experimental data at different extents of fuel-air premixing and fuel/air ratio. It is found that, over a wide range of fuel-air premixing, performance of $D$-Markov machines, as LBO predictors, is better with $D > 1$ than those with $D = 1$.

---

S. Sen (✉) · R.R. Chaudhari · A. Mukhopadhyay
Department of Mechanical Engineering, Jadavpur University, Kolkata 700032, India

S. Sarkar · A. Ray
Department of Mechanical & Nuclear Engineering, Pennsylvania State University, University Park, State College, PA 16802, USA



A.K. Agarwal et al. (eds.), *Combustion for Power Generation and Transportation*, DOI 10.1007/978-981-10-3785-6_8

## 1 Introduction

To satisfy the stringent emission norms, lean combustion is popularly used to reduce the emission of oxides of nitrogen (NOx). The fuel lean situation in a combustor is susceptible to thermo-acoustic instabilities. This may lead to blowout of flames in combustor, which is known as lean blowout (LBO) problem. LBO is severely detrimental for combustor operation in land based as well as aviation gas turbine engines. Land-based gas turbines may suffer from productivity loss due to repetitive engine shutdown and re-ignition because of LBO. Loss of engine thrust may be experienced in aircraft engines due to LBO, which may lead even to a serious consequence. This leads to the necessity for a real-time LBO prediction and to take an appropriate measure to control it.

This LBO phenomenon depends on various parameters pertinent to the combustor configuration and operating conditions. To monitor this we require a complex as well as expensive measurement system.

Muruganandam et al. [14, 15, 17] and Muruganandam and Sietzman [16] characterized lean blowout phenomenon in swirl stabilized, dump plane, methane fueled laboratory scale gas turbine combustor using optical (OH chemiluminescence) signature of flame. They observed extinction and reignition events near blowout regime, which they called precursor events. They reported that the frequency of those precursor events were increased monotonically as the extinction limit was reached. During extinction event if the flame packets were unable to reignite the reactant, the flame finally blew off.

Zilwa et al. [29] studied blowout in dump combustors with and without swirl. The effects of mixture composition gradients and velocity oscillations on the blowoff dynamics of a bluff-body stabilized burner was investigated by Chaudhuri and Cetegen [3, 4]. They captured CH chemiluminescence signal through photomultiplier tubes for blowout characterization. Chaudhuri et al. [5] and Stohr et al. [25] used combined particle image velocimetry/planar laser induced fluorescence (PIV/PLIF)-based techniques to investigate blowout dynamics for premixed and partially premixed flames.

Nair et al. [18], Nair and Lieuwen [19] and Prakash et al. [20] used acoustic and optical sensors time series for early detection and control of LBO in laboratory-scale gas turbine combustors. Nair and Lieuwen [19] identified a number of parameters and various techniques through these parameters for LBO prediction. Yi and Gutmark [28] identified two indices, namely, normalized chemiluminescence root mean square and normalized cumulative duration of LBO precursor events, for LBO prediction in real time.

These above works related to LBO prediction demonstrated their techniques in premixed combustors. Chaudhari [1], on the other hand, showed that these techniques are not equally suitable for partially premixed combustors, which should be close to practical combustors. Chaudhari [1] and Chaudhari et al. [2] developed a strategy based on flame color to detect LBO. The authors used the ratio of red and blue intensities emitted by the flame as the metric for LBO detection. The intensity values were captured by an inexpensive CCD camera. The method was inexpensive in terms of both cost and computation time. But, real time detection could be a problem as capturing optical data from operational combustor is difficult.

This shortcoming motivated to develop a time series data (may be optical, pressure etc.) driven strategy for early detection of LBO, which could be used in real time for premixed as well as partially premixed combustors. Algorithms have been developed upon a data-driven symbolic-dynamics-based technique, called *D*-Markov machine [12, 21 to detect LBO. A considerable work has been carried out in this area by Chaudhari [1], Mukhopadhyay et al. [13], Sarkar et al. [22] and Sarkar et al. [23]. In the present paper, we describe the early prediction techniques of LBO based on a constructed generalized D-Markov machine. The efficacy of the technique is also described with the variation of the value of D.

## 2 Experimental

Based on the earlier works of Williams et al. [27], Meier et al. [11], Muranganandam et al. [17] and Yi and Gutmark [28], a swirl-stabilized dump combustor is designed as a laboratory-scale model of a generic gas turbine combustor. The combustor consists of a premixing section, the combustor and an exhaust chamber. All metal parts made of type SS316 stainless steel enable the device to withstand high temperatures. All three sections are aligned coaxially. A swirler is provided at the end of the premixing section to stabilize the reacting mixture. The combustor design with a swirling injection of reactants and a dump-plane with rear facing step produces a swirling flow field with large recirculation zones. The combustor section has an inner diameter of 60 mm and a length of 200 mm. This compares favorably with an actual gas turbine combustor, the "combustor can" of the Rolls-Royce Trent 60 DLE (a 58-MWe aeroderivative gas turbine), which is approximately 200 mm in diameter and 700 mm long [9]. An assembled view of the combustor is shown in Fig. 1.

The premixing section consists of a 375 mm long SS pipe of outer diameter 43 mm and inner diameter 23 mm. At one end of the pipe, a SS plate of 65 mm diameter and 15 mm thickness has been welded. An internally threaded 15 mm diameter hole tapped into the plate serves as the air inlet. Five 6 mm diameter holes are drilled on the wall of the pipe at intervals of 50 mm. The first hole is at a distance of 20 mm from the upstream end of the premixing tube where the plate is welded for air inlet. These holes act as alternative positions for fuel inlet. We name this fuel inlet ports as port 1, port 2,… port 5. Variation in the position of the fuel inlet alters the length available for the premixing of fuel and air (Characterized by the premixing length $L_{fuel}$). This allows one to investigate the effects of extent of fuel/air premixing on the combustion characteristics.

One slot for positioning the swirler has been provided at the downstream end of premixing tube just prior to dump plane. The swirler is provided to add swirl in the flow which increases the residence time as the flow takes helical path instead of

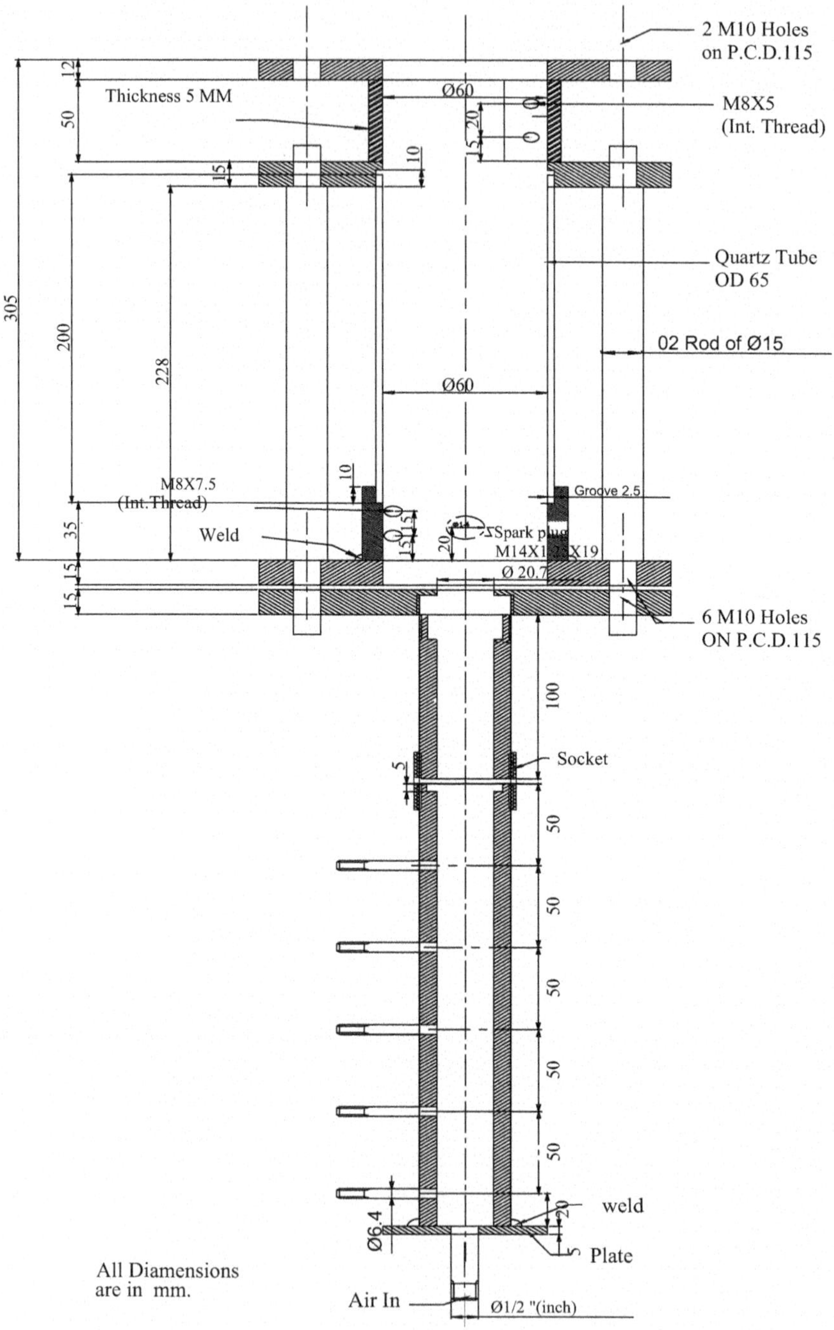


**Fig. 1** Assembled view of the combustor

straight path in combustor which enhances the combustion efficiency even in short length combustor. After swirler, the sudden expansion in the flow area (dump plane) slows down the flow velocity and generates re-circulation regions which ensure continuous heat-radicals feedback to incoming mixture. The combination of swirl and dump plane is a common method for modern gas turbine combustor. The same arrangement is used in the present study in the laboratory scale model.

The combustor is designed to provide optically accessible swirling flames. It consists of a 200 mm long quartz tube of inner diameter 60 mm and outer diameter 65 mm to allow optical access.

In the present study, Photo Multiplier Tube (PMT) (model 931A Hamamatsu) is employed to capture the heat release fluctuation of flame in the form of CH* Chemiluminescence emitting from flame as a measure of reaction rate and heat release rate. The PMT is held in a PMT holder where the interference filter, centered at 430 nm and with a full-width-half-maximum(FWHM) of 10 nm is kept in front of the PMT so that only CH* chemiluminiscence variation is captured on it. The output signal from PMT is electric current, proportional to the intensity of CH* radiation received on it which is later converted into voltage by the built-in amplifier which operate from a 15 V DC source.

Simultaneously a Sony cam recorder is used to capture the movie from which color images of flame are extracted at 30 frames per second in order to visualize LBO phenomenology and correlate the same with optical signal. Color images of the flame are also acquired using the digital DSLR camera (Nikon D3000) at suitable exposure to avoid pixel saturation.

For sampling the optical and pressure signal from combustor, the National Instrument hardware equipped with Labview software is used. The output signals (voltage) are fed to a NI-PXI 1050 chassis having built-in 08-AI channel signal conditioner module, subsequently the conditioned signal from sensors is digitalized and acquired to a personal computer using 16 bit M-series 6250 NIDAQ device. The signals are further processed on a computer by using NI software LABVIEW 7.1 and MATLAB 8.

For measuring the Air flow rate ($Q_{air}$) Aalborg made mass flow controller (MFC) is used (Range 0-500 LPM). The fuel (LPG, 60% propane, 40% butane) is metered ($Q_{fuel}$) by a second mass flow controller (Make: Aalborg, Range 0-10 LPM). The uncertainty in the flow of the all the MFC stated by manufacture is in the range of 1% of full range. The resolution of equivalence ratio can vary up to 0.001 in low range mass flow controller whereas the resolution achieved 0.01 in the case of high range MFC.

Figure 2 shows the photographs of the actual experimental setup of the swirl stabilized, premixed, dump plane combustor which simulates a laboratory scale model of a generic gas turbine combustor used in the current study.

The experiments are carried out for different air flow rates which give different Reynolds numbers. Five different ports provided on the premixing tube are used for injecting the fuel and thereby vary the premixing length ($L_{fuel}$) available for fuel/air mixing. The arrangement allows characterizing the flame dynamics in the regimes

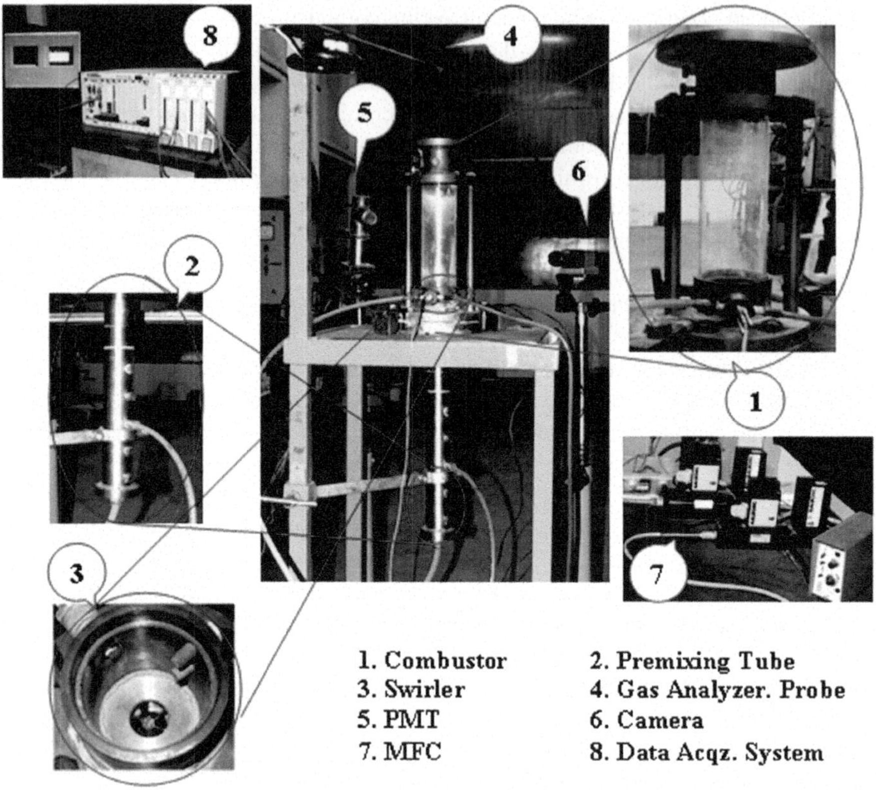


**Fig. 2** Photographs of the experimental set-up

of different degrees of premixing. The underline strategy is the higher premixing length ($L_{fuel}$) allows greater time for fuel/air mixing and thus improves air-fuel premixing.

## 3 Method of LBO Prediction

Symbolic time series analysis (STSA) [7] was built upon the concept of symbolic dynamics suggested by Lind and Marcus [10]. STSA is used as a tool to extract features for pattern classification. In this method a time series of signals can be represented as a sequence of symbols. Finally this develops a 'probabilistic finite state automata' (PFSA) [8, 26].

In STSA, the data range is partitioned into a number of non-overlapping zones, assigned with a symbol to each zone. The collection of symbols is known as the *alphabet*. The data falling in a zone is coded with the assigned symbol of that zone.

In this way, the time series can be written as a symbol sequence, known as *symbolic time series*. The data may be divided using different approaches. One approach develops the divisions in a way that the data-rich segment can be divided into smaller size partitions and larger size partitions for sparse data segment.

If the entire data is partitioned in a way to keep an equal number of data in each region that will maximize the information entropy about the data. This is called *maximum entropy partitioning*. Chaudhari [1] and Mukhopadhyay et al. [13] have adopted this type of partitioning. They sorted the time series data in ascending order. Starting from the first data in the sorted list, every consecutive data segment of length $\lfloor N/|\sum|\rfloor$ constitutes a distinct element of the partition where $N$ and $\Sigma$ represent the total number of data and number of symbols used and $\lfloor x \rfloor$ represents the greatest integer less than or equal to $x$. After partition generation, the time series is converted into a symbol sequence. The array containing numbers of data in all the partitions constitute the *state vector*.

A condition which is away from the blowout limit is taken as a nominal state for LBO prediction as combustion is stable at this condition. At the nominal state, each partition contains the same number of data due to maximum entropy partitioning. The same partitions are used for all subsequent operating conditions. The nominal state can be chosen as at equivalence ratio unity. When the equivalence ratio is reduced, we can obtain a new set of data. Every time series data is to be divided according to the same partitions decided for the nominal state. The state vector computed for this new data deviates from the nominal state vector. A suitable measure of deviation from the nominal state vector can be taken as the anomaly measure for the new data. This anomaly measure can be used to predict LBO.

For this, two methods of generating partitions are used. In one approach, the time series data is directly used. This has been referred to as *simple partitioning* (simple P). In the other approach, complex analytic functions are generated by Hilbert transform of the time series data. The algorithm and its efficacy have been discussed in details in Chaudhari [1], Mukhopadhyay et al. [13] and Sen et al. [24].

The work extended by Sarkar et al. [22, 23] in the context of dynamic data-driven application systems (DDDAS) [6] are delineated below:

(1) *State splitting and state merging*: The states can be splitted and merged on the basis of the information contents. In the first case, one state can be divided into more than one states, while the merger agglomerates more than one states into one [12].
(2) *Accommodation of a longer memory of chemiluminescence time series*: While Chaudhari [1], Mukhopadhyay et al. [13] and Sen et al. [24] used $D = 1$ for the constructed D-Markov machine, the new work redefined the machine with $D > 1$. In the first case, one state was associated only with the previous time. With $D > 1$, the machine can extract low-dimensional features with a longer time history.
(3) *Bias removal to achieve leaner operating conditions*: For removal of the bias, the time series has been transformed into a zero mean based data sequence.

The nonzero mean based prediction may result in richer operating conditions with associated penalties of emission.

(4) *Information-theoretic anomaly measure for LBO prediction*: To increase the sensitivity of the prediction, the anomaly measure is defined on the basis of Kullback-Leibler divergence [21].

5) *Pattern classification based on the features extracted from chemiluminescence time series*: The prediction of LBO is posed as a pattern classification problem based on different ranges of the equivalence ratio at several premixing levels.

The detailed algorithm may be found in Sarkar et al. [23].

## 4 Results and Discussion

In this section, the predictions of LBO based on the constructed D-Markov machine have been presented. The time series data are generated from the experimental set-up of swirl-stabilized dump combustor described in Sect. 2. A number of experiments have been conducted at airflow rates of 150, 175 and 200 lpm with liquefied petroleum gas (LPG) as fuel. Tests have been carried out at different equivalence ratios—starting at stoichiometric to LBO limit. Three different premixing lengths (*Lfuel* = 350 mm, 250 mm, and 150 mm) have been used—named as Port 1, Port 3, and Port 5, respectively. Reynolds numbers based on cold flow conditions have been up to 18, 700 at 200 lpm.

It is observed from the experiments that the flame for highest premixing length, when fuel is entering through port 1, is steady and conical in shape when the fuel-air ratio is stoichiometric or near stoichiometric as shown in Fig 3a. The flame

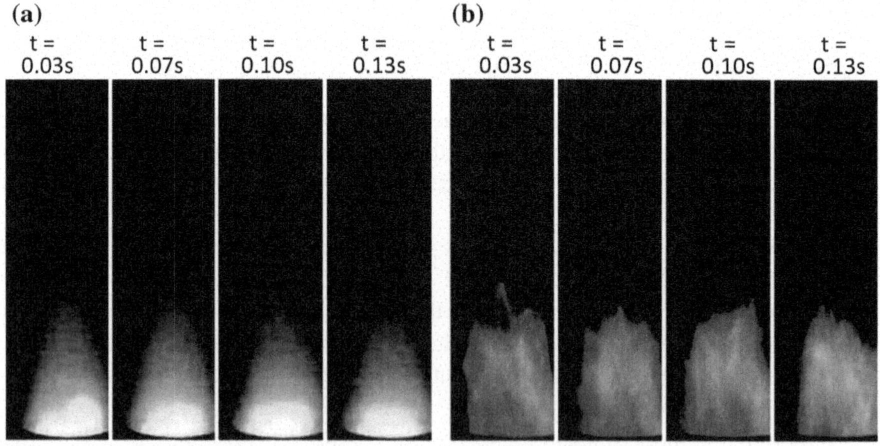


**Fig. 3** Still images of the flame at $\Delta t = 33$ ms intervals away from LBO for **a** $\Phi = 1.0$ and **b** $\Phi = 0.81$ $L_{fuel} = 35$ cm (port 1) [1]

is bright and has a long reddish tip. As the equivalence ratio is reduced, a significant change is observed in flame color. The flame color is shifted from reddish to bluish one, as observed in Fig 3b. When the equivalence ratio is further decreased and kept close to LBO, the flame shape is changed to columnar one due to reduced reaction rate. The reduced burning velocity also contributes to this shape. The flame shows an oscillation of random nature. Sometimes the flame is lifted from the dump plane and again attaches to it through reignition phenomenon. This dancing behavior of flame is observed for a few milliseconds prior to the occurrence of LBO.

The observations with the minimum premixing length (port-5) are significantly different. When the flame is observed at near stoichiometric condition, the flame structure is more or less similar to that of in case of port-1. But, as the equivalence ratio is reduced to near LBO, the flame does not show any lift off pattern as shown in Fig. 4. The flame is mostly attached to the dump plane. However, the flame intensity is significantly reduced. The reignition phenomenon before LBO is present here also but the flame is not oscillating like the earlier case. But, in this case, the presence of oscillatory precursor events is not so strong as that was in earlier cases. The prediction parameters used for premixed flames by earlier researchers were mainly based on these precursor events. As the number of such events decreases, in

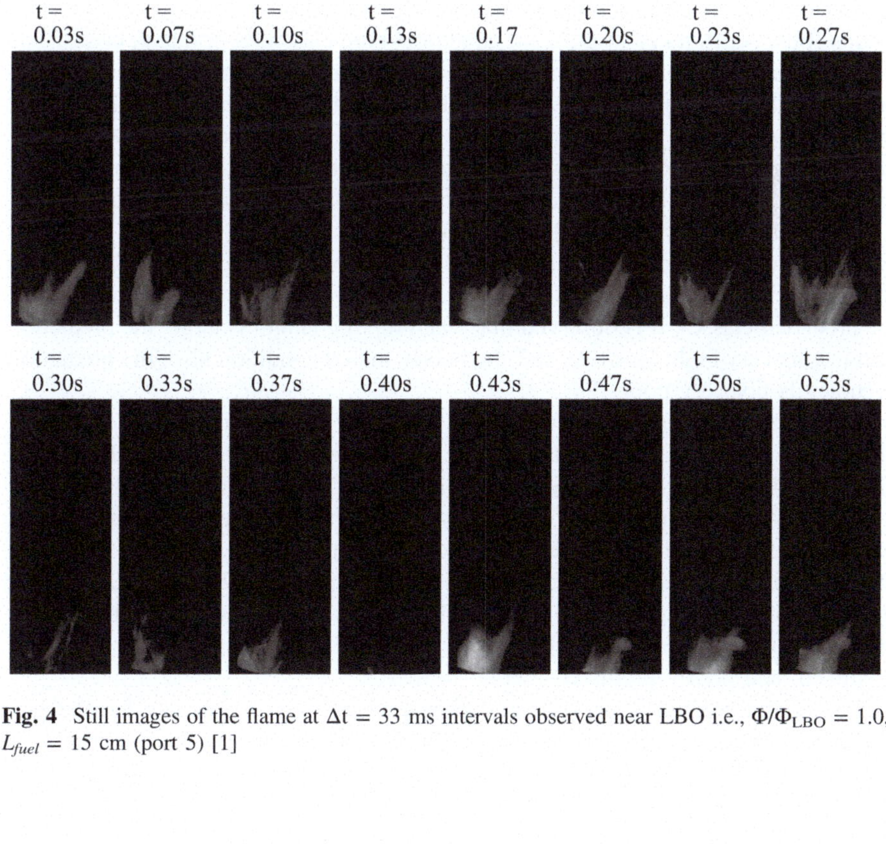


**Fig. 4** Still images of the flame at $\Delta t = 33$ ms intervals observed near LBO i.e., $\Phi/\Phi_{LBO} = 1.0$, $L_{fuel} = 15$ cm (port 5) [1]

case of departure from premixed flames, these prediction methods fail. This actually triggers the need of developing a method that can capture minute events near LBO.

Now the LBO measure is computed. This computation is done on the basis of the reference condition which is far from LBO (in present case—stoichiometric). The normalized LBO measure is assigned as zero to the reference condition. The normalization of LBO measures is done to compare them at different airflow rates.

Performances of the proposed LBO measures for D > 1 are shown in Figs. 5a–c. The figures show the performance for one port condition at different air flow rates (i.e., 150, 175 and 200 lpm). The three figures are drawn for three different port conditions—port-1, port-3, and port-5—premixed, well mixed and poorly mixed status. For premixed and well-mixed cases, it is seen that up to $\varphi/\varphi_{LBO} = 1.2$, the slope of the curve changes gradually with a very low gradient. As $\varphi/\varphi_{LBO}$ falls below 1.2 (i.e. the condition approaches LBO) the slope is increasing steeply to a normalized value of unity at LBO condition. This feature is a desirable condition for early prediction of LBO. The change in the gradient can annunciate for an approaching LBO condition.

The similar trend of gradient change is not observed in the case of poorly-mixed, shown in Fig. 4c. We can say that the normalized LBO measure for the poorly-mixed is not as sensitive as it is in the cases of well-premixed and partially-premixed conditions. The absence of strong precursor events at this condition may be attributed to this feature.

The method of classification is applied on to a number of test data. The test results (time series taken at different conditions) are classified in three classes. The *Nominal* class is defined when $\varphi/\varphi_{LBO}$ is greater than 1.2. When $\varphi/\varphi_{LBO}$ is less than or equal to 1.2 but more than 1.1, the class is called *Progressive LBO* (*PLBO*). When $\varphi/\varphi_{LBO}$ is less than or equal to 1.1 but more than 1, the class is *Impending LBO* (*ILBO*). For prediction purpose, an alarm can be generated when a data falls either in PLBO or ILBO class.

The results of this classification method are presented in Table 1 in form of a confusion matrix for different mixing levels. The rows of the confusion matrix are the actual classes (i.e., as found from the experimental result) and the columns are the predicted classes. It is observed that when mixing length is large, the prediction accuracy is very high. However, for the poorly mixed condition also, the prediction accuracy is not very poor.

The confusion matrix in Table 1 shows that the *D*-Markov machine is capable of classifying different LBO situations with reasonable accuracy. The method could also detect LBO even at low levels of premixing, i.e. in case of absence of precursor events. This is important as other methods reported in the literature for lean premixed flames can not work satisfactorily for poorly mixed cases [1]. This classification scheme, in general, predicts the proximity of LBO fairly accurately for a wide range of air-flow rate. The *D*-Markov machine parameters of the pattern classifier are now presented for LBO prediction for well-premixed, partially-premixed, and poorly premixed fuel-air flow.

To use the method in a practical combustor, time series data can be obtained during the test run. A near stoichiometry test data can be taken as the nominal case.

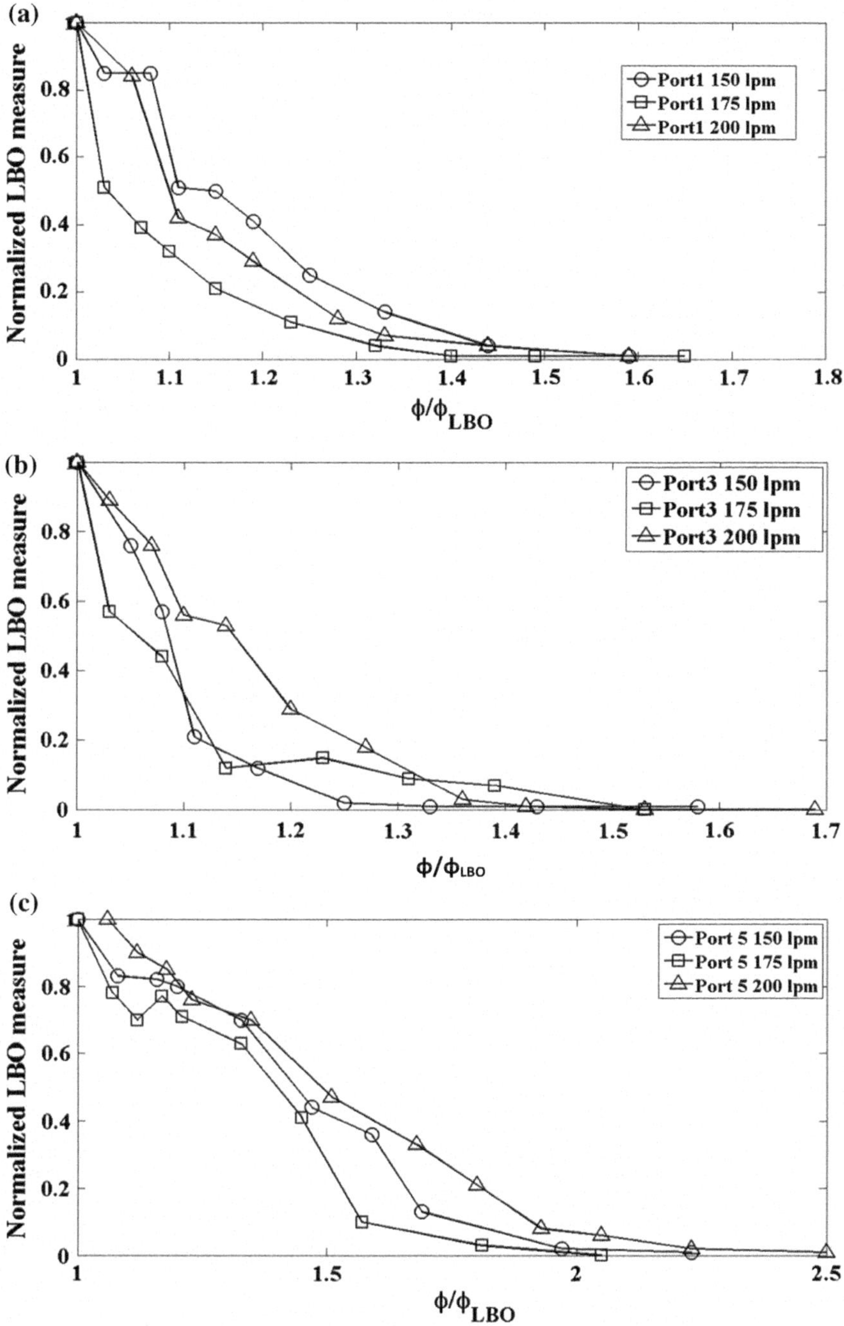


**Fig. 5** Dependence of Normalized LBO measure on $\varphi/\varphi_{LBO}$ for three airflow rates of 150, 175 and 200 lpm at **a** port-1, **b** port-3, and **c** port-5

**Table 1** Confusion matrix before lean blow out for different premixing (port 1, port 3 and port 5)

| | Predicted | | | | | | | | | |
|---|---|---|---|---|---|---|---|---|---|---|
| ACTUAL | | Port-1 | | | Port-3 | | | Port-5 | | |
| | | Nominal | PLBO | ILBO | Nominal | PLBO | ILBO | Nominal | PLBO | ILBO |
| | Nominal | 12 | 1 | 0 | 12 | 1 | 0 | 16 | 3 | 0 |
| | PLBO | 0 | 7 | 0 | 0 | 6 | 0 | 0 | 6 | 0 |
| | ILBO | 0 | 0 | 9 | 0 | 0 | 9 | 0 | 2 | 6 |

In actual combustor, if it is not possible to obtain optical data, the dynamic pressure data can also be used. Test can be carried out at different equivalence ratios to obtain the classes like PLBO, ILBO etc. Finally, a confusion matrix can be generated to check the confidence level of the prediction.

*D*-Markov machines with $D = 1$ has been reported by Chaudhari [1], Mukhopadhyay et al. [13] and Sen et al. [24]. It is found that the performance of the machines with D = 1 is similar or sometimes better than the other prediction methods. When, the D-Markov machine with D > 1 is devised, it is required to judge its performance with respect to that of the machines with D = 1. This comparison is done on the basis of two metrics.

(1) Total mis-prediction = (Actual ILBO, predicted PLBO) + (Actual ILBO, predicted Nominal) + (Actual PLBO, predicted Nominal); and
(2) Total false alarm = (Actual Nominal, predicted PLBO) + (Actual Nominal, predicted ILBO) + (Actual PLBO, predicted ILBO).

Table 2 shows the classification performance of the *D*-Markov machine with D = 1 and $D > 1$. It is observed that the performance is better in case of D > 1. It is an interesting finding of Table 2 that the performance of the machine with D > 1 turns increasing better relative to that of $D = 1$ when the premixing quality degrades. When the mixing level is high (port 1 and port 3), the performances of the machines with D = 1 and D > 1 are comparable. This is due to the presence of intense and clearly visible precursor events before the onset of LBO. When the mixing level is poor, visibility of precursor events is not very frequent. This physical phenomenon makes the prediction more difficult. The last row of Table 2 shows that the *D*-Markov machine with $D > 1$ performs well for all ports including Port 5 too, whereas the *D*-Markov machine with $D = 1$ fails to predict LBO as it

**Table 2** Comparison of classification performance for LBO prediction of D-Markov machines with D = 1 and D > 1

| Port number | Number of test cases | | Total misprediction | Total false alarm |
|---|---|---|---|---|
| Port 1 | 29 | D = 1 | 2 | 3 |
| | | D > 1 | 0 | 1 |
| Port 3 | 28 | D = 1 | 4 | 4 |
| | | D > 1 | 0 | 1 |
| Port 5 | 33 | D = 1 | 6 | 9 |
| | | D > 1 | 2 | 3 |

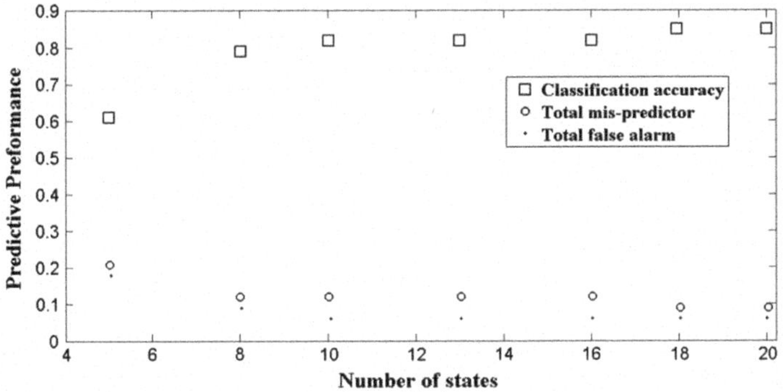


**Fig. 6** Predictive performance variation of the *D*-Markov machine on the number of states for alphabet size 5

yields unacceptable levels of total mispredictions and total false alarms. Thus, the development and validation of the *D*-Markov machine with $D > 1$ for LBO prediction is satisfactorily implemented.

It is observed that as the number of PFSA states $|Q|$ increases, the classification performance initially increases and then saturates from a value of number of states ($|Q| = 10$ in present case). Figure 6 shows this variation for a poorly mixed case (fuel inlet through port 5). The number of PFSA states $|Q|$ in the *D*-Markov machine is increased by state splitting and state merging for an alphabet size $|\Sigma| = 5$. Investigation of the fact, whether the phase space dimension of the combustion dynamics consistently converges in the vicinity of LBO, is a topic of future research.

The computation time will increase for a D-Markov machine with D > 1 in comparison with D = 1. But, the increment is not too high to be used for real time calculation.

## 5 Conclusions

The work addresses the development of prediction method of LBO through a symbolic time series analysis. The time series is first coded with symbols and then the state vectors are created. These vectors constitute a D-Markov machine. The machine has a deterministic structure and their states are represented by symbol blocks of length D (a positive integer) or less.

A Kullback-Leibler divergence based anomaly measure is constructed to predict the LBO successfully. This anomaly measure becomes increasingly sensitive to small changes in the equivalence ratio as the combustion process approaches LBO.

Over a wide range of fuel-air premixing and air flow rates, it is shown, that the *D*-Markov machine with $D > 1$ performs significantly better than the *D*-Markov machine with $D = 1$ for prediction of LBO. The results suggest the potential capability of $D$-Markov machines to precisely predict regions close to LBO in a laboratory-scale combustor. In this way, the operating condition in a combustor could be extended to a leaner equivalence ratio without significantly risking LBO.

## References

1. Chaudhari RR (2011) Investigation of thermoacoustic instabilities and lean blowout in a model gas turbine combustor. PhD thesis, Jadavpur University, India
2. Chaudhari RR, Sahu R, Ghosh S, Mukhopadhyay A, Sen S (2013) Flame color as a lean blowout predictor. Int J Spray Combust Dyn 5:49–66
3. Chaudhuri S, Cetegen BM (2008) Blowoff characteristics of bluff body stabilized conical premixed flames under upstream mixture gradients and velocity oscillations. Combust Flame 153:616–633
4. Chaudhuri S, Cetegen BM (2009) Blowoff characteristics of bluff body stabilized conical premixed flames in a duct with upstream mixture gradients and velocity oscillations. Combust Sci Technol 181:555–569
5. Chaudhuri S, Kostska S, Renfro MW, Cetegen BM (2010) Blowoff dynamics of bluff body stabilized turbulent premixed flames. Combust Flame 157:790–802
6. Darema F (2005) Dynamic data driven applications systems: new capabilities for application simulations and measurements. In: 5th international conference on computational science, Atlanta, GA, USA
7. Daw C, Kennel M, Finney C, Connolly F (1998) Observing and modeling nonlinear dynamics in an internal combustion engine. Phys Rev E 57(2811):2819
8. Dupont P, Denis F, Esposito Y (2005) Links between probabilistic automata and hidden markov models: probability distributions. Learn Models Ind Algorithms Pattern Recogn 38:1349–1371
9. Lieuwen TC, Yang V (2005) Combustion instabilities in gas turbine engines. American Institute of Aeronautics and Astronautics, Reston
10. Lind D, Marcus B (1995) Symbolic dynamics and coding. Cambridge University Press, Cambridge
11. Meier W, Weigand P, Duan XR, Giezendanner-Thobben R (2007) Detailed characterization of the dynamics of thermoacoustic pulsations in a lean premixed swirl flame. Combust Flame 150:2–26
12. Mukherjee K, Ray A (2014) State splitting and state merging in probabilistic finite state automata for signal representation and analysis. Sig Process 104:105–119
13. Mukhopadhyay A, Chaudhari RR, Paul T, Sen S, Ray A (2013) Prediction of lean blow-out in gas turbine combustors using symbolic time series analysis. J Propul Power 29:950–960
14. Muruganandam TM, Nair S, Neumeier Y, Lieuwen T, Seitzman J (2002) Optical and acoustic sensing of lean blowout precursors. In 38th AIAA/ASME/SAE/ASEE joint propulsion conference and exhibit, Indianapolis, Indiana
15. Muruganandam TM, Kim B, Olsen R, Patel M, Romig B, Seitzman JM (2003) Chemiluminescence based sensors for turbine engines. In: 39th AIAA/ASME/SAE/ASEE joint propulsion conference & exhibit, Huntsville, Alabama
16. Muruganandam TM, Seitzman JM (2005) Characterization of extinction events near blowout in swirl-dump combustors. In: 41st AIAA/ASME/SAE/ASEE joint propulsion conference and exhibit, Tucson, Arizona
17. Muruganandam TM, Nair S, Scarborough D, Neumeier Y, Jagoda J, Lieuwen T, Seitzman J, Zinn B (2005) Active control of lean blowout for turbine engine combustors. J Propul Power 21:807–814

18. Nair S, Muruganandam TM, Olsen R, Meyers A, Seitzman J, Zinn BT, Lieuwen T, Held T, Mongia H (2004) Lean blowout detection in a single nozzle swirl cup combustor. In: 42nd AIAA aerospace sciences meeting and exhibit, Reno, NV, USA
19. Nair S, Lieuwen T (2005) Acoustic detection of blowout in premixed flames. J Propul Power 21:32–39
20. Prakash S, Nair S, Muruganandam TM, Neumeier Y, Lieuwen T, Seitzman JM, Zinn BT (2005) Acoustic based rapid blowout mitigation in a swirl stabilized combustor. In: ASME turbo expo, Reno, USA
21. Ray A (2004) Symbolic dynamic analysis of complex systems for anomaly detection. Sig Process 84:1115–1130
22. Sarkar S, Ray A, Mukhopadhyay A, Chaudhari RR, Sen S (2014) Early detection of lean blow out (LBO) via generalized D-Markov machine construction. In: American control conference, Portland, Oregon, USA
23. Sarkar S, Ray A, Mukhopadhyay A, Sen S (2015) Dynamic data driven prediction of lean blowout in a swirl stabilized combustor. Int J Spray Combust Dyn 7:209–242
24. Sen S, Chaudhari RR, Mukhopadhyay A (2014) Lean blowout detection in partially premixed flames in a swirl-stabilized dump combustor. In: Agarwal K, Pandey A, Gupta AK, Aggarwal SK, Kushari A, (eds) Novel combustion concepts for sustainable energy development (A). Springer. ISBN: 978-81-322-2210-1
25. Stohr M, Boxx I, Carter C, Meier W (2011) Dynamics of lean blowout of a swirl stabilized flame in a gas turbine model combustor. Proc Combust Inst 33:2953–2960
26. Vidal E, Thollard F, de la Higuera C, Casacuberta F, Carrasco R (2005) Probabilistic finite-state machines—part I and part II. IEEE Trans Pattern Anal Mach Intell 27:1013–1039
27. Williams TC, Schefer RW, Oefelein JC, Shaddix CR (2007) Idealized gas turbine combustor for performance research and validation of large eddy simulations. Rev Sci Instr 78: 035114-1–035144-9
28. Yi T, Gutmark EJ (2007) Real-time prediction of incipient lean blowout in gas turbine combustors. AIAA J 45:1734–1739
29. Zilwa SD, Uhm JH, Whitelaw JH (2000) Combustion oscillations close to the lean flammability limit. Combust Sci Technol 160:231–258

# Theoretical Formulation for the Investigation of Acoustic and Entropy-Driven Combustion Instabilities in Gas Turbine Engines

**Balasubramanian Singaravelu, Sathesh Mariappan and Avijit Saha**

**Abstract** In gas turbine combustors, the unsteady flame is a source of acoustic and entropy waves, leading to combustion noise. When these fluctuations couple with the flame and establish a positive feedback mechanism, they grow in amplitude resulting in combustion instability. Combustion instability can be driven either by the acoustic waves or acceleration of entropy waves. Entropy-driven instability is one of the dominant cause of low frequency combustion instability in industrial gas turbines, where the flow exiting the combustor is accelerated by the turbine nozzle guide vanes. This chapter presents the theoretical framework to model the generation, convection, acceleration and reflection of acoustic and entropy waves in gas turbine combustors with variable area geometry. We also discuss in detail the procedure to solve the equations and present a comparison between acoustic-driven instability and entropy-driven instability.

## 1 Introduction

In order to decrease emissions and conserve fuel, it is imperative to design gas turbine combustors that can employ lean premixed combustion in an effective manner. Several investigations have shown that lean combustors can be unstable at their design operating conditions [10] due to the appearance of combustion instability. Unstable combustors give rise to severe structural vibrations and high heat fluxes on chamber walls causing irreparable damage to the system. The ability to predict combustion instability allow us to mitigate the problem in wide variety of systems such as gas turbines, ramjets and rocket engines. The study of combustion instability is a

B. Singaravelu (✉) · S. Mariappan
Indian Institute of Technology Kanpur, Kanpur 208016, India
e-mail: sbala.space@gmail.com

S. Mariappan
e-mail: sathesh@iitk.ac.in

A. Saha
Indian Institute of Technology Kharagpur, Kharagpur 721302, India
e-mail: avijitaero@iitkgp.ac.in



A.K. Agarwal et al. (eds.), *Combustion for Power Generation and Transportation*, DOI 10.1007/978-981-10-3785-6_9

challenging topic due to the nonlinear nature of the phenomena and the various couplings that are involved. In this study we present a theoretical formulation to model the unsteady interactions in gas turbine combustors. This framework can be used as a tool to understand and predict combustion instability.

The reacting chemistry of the flame and turbulence in the chamber gives rise to acoustic (mostly pressure, velocity and isentropic temperature fluctuations) and entropy (mostly non-isentropic temperature fluctuations) fluctuations inside a combustion chamber [24]. The pressure fluctuations generated by unsteady volume expansion are a source of noise, which contains broadband frequencies [28]. On the other hand, the non-isentropic temperature fluctuations are intrinsically silent and produce sound when they propagate through a non uniform steady state flow field. The noise generated by acoustic waves (isentropic pressure fluctuations) is called direct combustion noise and those generated by the acceleration of local “hot spots” or entropy waves (non-isentropic temperature fluctuations) is termed as entropy noise [18, 21]. The acceleration of vorticity waves is also a source of combustion noise but its contribution is less when compared to entropy noise. Both entropy and vorticity noise together is termed as indirect combustion noise. When a feedback cycle is established between the unsteady heat release rate and the acoustic/entropy waves, the low-amplitude broadband fluctuations amplify into high-amplitude discrete tones. This discrete frequency combustion-driven oscillation is called *combustion instability*. Similar to combustion noise, it can be driven by either acoustic or entropy waves. The former is termed as acoustic-driven instability, while the latter is called as entropy-driven instability [20].

The area of acoustic-driven instability has been extensively investigated in a number of earlier investigations [4, 14, 15, 26]. Therefore, in this study we limit ourselves to the topic of entropy-driven instability. The subject of entropy-driven instability is more relevant because, in real combustors entropy waves are accelerated through variable area sections such as turbine guide vanes and nozzles. In the process, they reflect back as acoustic waves affecting the flame dynamics. Due to the relatively low convective speed of the entropy waves (in comparison to the acoustic speed), the resonant frequencies of entropy-driven instability are expected to be well below the fundamental acoustic modes of the chamber. Keller et al. [9] pointed out in their studies that the frequency of the pressure oscillations in industrial combustion chambers can be below those of the fundamental acoustic modes. These modes strongly depend on the flow speed, unlike the case of acoustic-driven instability. This observation can be used as a criterion for distinguishing entropy-driven instability from acoustic-driven instability [20]. Experimental investigations by [2, 5, 7, 9, 12, 21] also suggest that acceleration of entropy waves significantly affects the observed resonant frequency in combustors.

Muthukrishnan et al. [21] asserted that in the presence of strong pressure gradient, entropy noise was the prime contributor to the core engine noise. Macquisten et al. [12] observed combustion instability with a dramatic change in frequency and mode shape, when a choked nozzle is inserted at the exit of a combustor. They also observed that the stability of the system is reduced by increasing the inlet Mach number or equivalence ratio. Hield and Brear [7] observed striking difference in the spectra between a duct with open and choked exit. They attributed their observation to the interaction of entropy disturbances with the choked nozzle. Bake and Fischer [2] observed a linear relation between temperature fluctuations and sound pressure level in a duct terminated with a convergent-divergent nozzle. Keller et al. [9] observed a similar phenomena when secondary air is injected at the location where temperature and pressure oscillations are in phase. The injection of secondary air amplified the entropy waves and drove the system to instability. Eckstein et al. [5] also observed the generation of entropy waves in self-excited flames. But they found that dispersion of entropy waves is high in the passage from the flame to the nozzle. They concluded that even though entropy waves reflect back as pressure waves, they are too weak to affect the thermoacoustic modes of the combustion chambers. They attributed the observed low frequency oscillations to the unsteady heat release due to equivalence ratio fluctuations. We believe that the difference in the conclusion drawn by [5] and others [2, 7] can be attributed to the difference in the aspect ratio of the combustors used in their investigations. Large eddy simulations by [19] show that the mechanism for low-frequency combustion instability involves both acoustic and entropy waves. A recent paper by [18] gives an excellent review on the status of the progress made in understanding entropy noise. Furthermore, our group has recently observed the transition from entropy-driven noise to entropy-driven instability in experiments with laboratory scale swirl stabilized gaseous fuel combustor [20].

It is clear from the above investigations that entropy waves should be properly taken into consideration to determine the stability of unsteady interactions in combustors. From theoretical point of view, it is essential to develop a framework that can take into account, the generation, convection, dissipation, dispersion and reflection of entropy waves to gain a better understanding of their role in low-frequency instability observed in real combustors.

Theoretical approach for modeling the convection of entropy waves through variable geometry cross sections was given by [13] in their seminal work. They solved the equations using compact nozzle assumption for a supercritical nozzle, sub critical nozzle and nozzle with a shock. They noted that the difference between the use of compact and finite length nozzle is a shift in the phase of the response. Their one dimensional flow model with non-uniform steady state flow, suggested that a 2% fluctuation in upstream temperature is sufficient to cause a significant pressure fluctuation of 0.01 atmosphere, which translates to 154 dB in air.

In general, there are two approaches followed in the theoretical analysis. One is the time-domain and the other is frequency-domain approach. They can be further sub-classified based on the degree of geometrical and mathematical complexity. The time-domain approach allows us to gain deeper insight by understanding the route to instability. But this approach is computationally more expensive than the frequency-domain method if we wish to study a large batch of geometric models and operating conditions. The frequency-domain approach is a good first estimate of the periodical effects generated by the system. In this study we use the frequency-domain approach in the quasi-one dimensional (along the axial direction) framework and compute the stability of combustors for a range of geometrical configurations and operating conditions. The frequency domain approach uses the linearized form of the governing equations. There are several challenges involved in reproducing the real physics in the linearized model. Three important challenges and the available solutions are discussed below.

The first challenge in the linearized approach is modeling the heat release rate: both the steady and unsteady state components. The flame is assumed to be confined in a small region and the steady state heat release is modeled using an analytical distribution function for the static temperature [22]. The steady state heat release is then proportional to the spatial gradient of the total temperature along the direction of the flow. The unsteady heat release rate is modeled using the $\eta - \tau$ model proposed by [3]. Polifke et al. [23] used this approach to show that entropy waves can interfere in both constructive and destructive manner, leading to either enhancement or reduction of the stability of the system.

The second challenge in the linearized approach is introducing the mean flow effects. Nicoud and Wieczorek [22, 31] independently introduced the effects of mean flow in the calculation of thermoacoustic instabilities. Nicoud and Wieczorek [22] showed that even for Mach number as low as 0.05, ignoring mean flow effects gives rise to large deviations in the resonant frequency and its stability. Yu et al. [31] using linearized Euler equations and natural acoustic boundary conditions observed a decrease in resonant frequencies and shifts in the locations of the nodes for various combustor configurations.

The third challenge is to incorporate the effects of dissipation and dispersion of entropy waves. Sattelmayer [25] introduced spatial dispersion in the propagation of entropy waves due to equivalence ratio fluctuations. They argued that if dispersion is ignored, the generation of instabilities is highly over predicted. According to [25], entropy waves deteriorate the system stability only if dispersion is not considered in the analysis. This claim has been refuted in the works of [6], who asserted that entropy waves do play a role in the stability of the combustor even when dissipation and dispersion are taken into account. Recent numerical investigations by [16] also suggest that dissipation of entropy waves is negligible and significant strength for the entropy waves remain at the combustor exit. Morgans and Li [17] noted that shear dispersion has the potential to strongly affect thermoacoustic resonant modes.

This study takes into account the following: 1. unsteady heat release due to pressure and velocity fluctuations, 2. the mean flow effects and 3. dissipation of acoustic and entropy waves in computing the resonant frequencies and their growth rates for the chosen combustor configuration. However, the dispersion of entropy waves is not modeled in the present investigation. The outline of this study is as follows: In Sect. 2, we present the governing equations and the models used for various phenomena. In Sect. 3, we present the numerical schemes and methods used to solve the governing equations. In Sect. 4, we discuss the procedure to distinguish between acoustic-driven and entropy-driven instability. Finally, in Sect. 5, we present the concluding remarks of this chapter.

# 2 Theory

In this section we discuss the equations governing the generation and convection of acoustic and entropy waves in a system with heat addition and variable area cross section. First, we describe the geometry of the system. Next, we present the governing equations. Then, we linearize and non-dimensionalise the equations. After that, we discretize the equations using central difference scheme. Finally, we pose the model as a nonlinear eigenvalue problem and solve the same.

## *2.1 Geometry of the System*

The geometry of our gas turbine combustor is a straight duct terminated by a converging nozzle at the exit (Fig. 1). The combustor is described by its length $L$, the location of the center of the flame $x_f$, the thickness of the flame $\delta_f$ and location of the nozzle that is fitted at $x_N$ downstream of the flame. The axially varying cross section area for the nozzle, $x_N < x \leq L$ is given by $A(x)$. This framework does not impose compact nozzle assumption. The parameters known at the inlet are temperature $T_i$, Mach number $M_i$ (or velocity $u_i$) and inlet area $A_i$. At the exit, pressure $p_e$ and area $A_e$ are known. The maximum temperature $T_h$ reached by the hot gases is also given.

## *2.2 Steady State Values of the Flow Variables*

For the given geometry and boundary conditions, we first need to compute the steady state values of the flow variables pressure $p$, density $\rho$, temperature $T$, and Mach number $M$ along the axis of the combustor. The fluctuations are measured with the steady state values as reference. For this purpose we split the geometry into three portions. They are as follows: portion 1 is before the flame ($0 < x < x_f$), portion 2 is after the flame ($x_f < x < x_N$) and portion 3 is the nozzle portion ($x_N < x < L$).

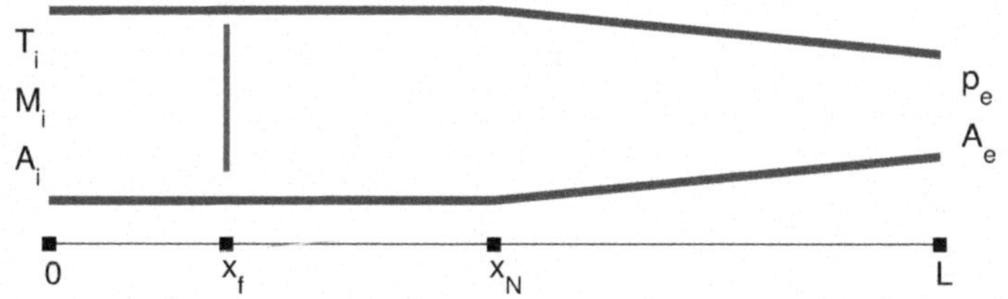


**Fig. 1** Geometry of our variable area gas turbine combustor. The system is of length $L$ with the center of the flame at $x_f$ and a variable area nozzle beginning at $x_N$. The conditions imposed at the inlet and exit are also shown. The flame has a thickness of $\delta_f << L$ along the axial direction

Using $T_i$ and $T_h$, the static temperature profile across the flame is computed using the following relations given by:

$$T(x) = \frac{T_h + T_i}{2} + \frac{T_h - T_i}{2} \tanh\left(3\frac{x - x_f}{\delta_f}\right) \tag{1}$$

The steady state heat release rate $\dot{Q}$ per unit volume is obtained from the temperature profile as follows:

$$\dot{Q} = \rho u C_p \frac{dT_0(x)}{dx} \tag{2}$$

where $C_p$ is the specific heat capacity and $T_0$ is the total temperature along the axis. The total temperature is computed from the static temperature and the local flow velocity [1]. It is further assumed that the pressure remains constant across the flame. Figure 2 shows the spatial distribution of the steady state values of temperature, heat release rate, Mach number and pressure for a chosen configuration. In portion 1, the steady state values are the same as the inlet conditions. When the flow passes through the flame, there is a rise in temperature but the pressure remains the same.

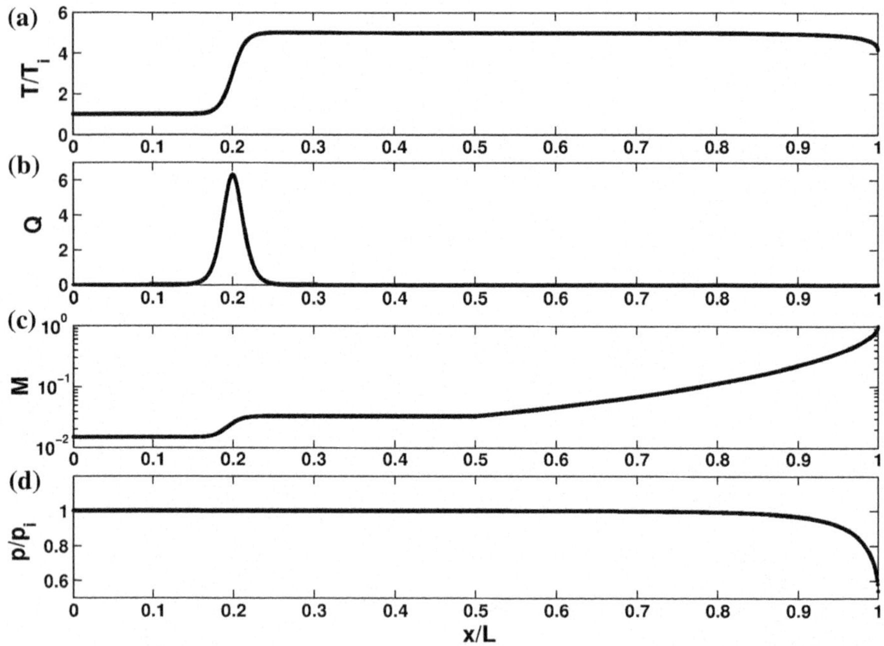


**Fig. 2** Steady state values of the flow variables as a function of $x$ along the axis of the combustor: (*a*) Temperature $T/T_i$, (*b*) steady state heat release rate $\dot{Q}$, (*c*) Mach number $M$ and (*d*) Pressure $p/p_i$. The geometric parameters chosen for this study are $L = 1$ m, $x_f/L = 0.2$, $\delta_f/L = 0.05$, $x_N/L = 0.9$ and $A_e = 0.0579A_i$. The operating conditions are $M_i = 0.015$, $T_{0i} = 300K$, $T_{0e}/T_{0i} = 5$

We assume that there is no pressure drop across the flame. To keep $p = \rho RT$ an invariant ($R$ being the characteristic gas constant), the density decreases with rise in temperature. For subsonic flow, the Mach number $M$ (or velocity $u$) increases with $T$ to keep the mass flow rate $\dot{m}$ a constant. The variation of Mach number $M$ in the nozzle portion is also computed by holding $\dot{m}$ a constant. The mass flow rate $\dot{m}$ is given by

$$\dot{m} = \frac{A(x)p_0}{C_p T_0}\frac{\gamma M}{\sqrt{\gamma - 1}}\left(1 + \frac{\gamma - 1}{2}M^2\right)^{-(\gamma+1)/2(\gamma-1)}, \tag{3}$$

where $A(x)$ is the area variation in the nozzle portion. Using Mach number $M$, the variation in other variables can be found using the isentropic relations [1].

## 2.3 Governing Equations

The equations that govern the flow through a variable cross section area combustor with unsteady heat release are as follows:

$$\left[\frac{\partial \tilde{\rho}}{\partial \tilde{t}} + \tilde{u}\frac{\partial \tilde{\rho}}{\partial \tilde{x}}\right] + \frac{\tilde{\rho}}{\tilde{A}}\frac{\partial \tilde{u}\tilde{A}}{\partial \tilde{x}} = 0, \tag{4}$$

$$\tilde{\rho}\left[\frac{\partial \tilde{u}}{\partial \tilde{t}} + \tilde{u}\frac{\partial \tilde{u}}{\partial \tilde{x}}\right] = -\frac{\partial \tilde{p}}{\partial \tilde{x}}, \tag{5}$$

$$\left[\frac{\partial \tilde{p}}{\partial \tilde{t}} + \tilde{u}\frac{\partial \tilde{p}}{\partial \tilde{x}}\right] + \frac{\gamma \tilde{p}}{\tilde{A}}\frac{\partial (\tilde{u}\tilde{A})}{\partial \tilde{x}} = (\gamma - 1)\tilde{\dot{q}}_v, \tag{6}$$

$$\tilde{\rho}\left[\frac{\partial \tilde{s}}{\partial \tilde{t}} + \tilde{u}\frac{\partial \tilde{s}}{\partial \tilde{x}}\right] = R\tilde{\dot{q}}_v. \tag{7}$$

Equations (4) and (5) correspond to the continuity and momentum equations respectively. While Eqs. (6) and (7) represent the energy equation in pressure and entropy form. In Eqs. (6) and (7), $\tilde{\dot{q}}_v$ represents the volumetric heat release rate. In addition to the above equations we also need the following two state equations:

$$\tilde{p} = \tilde{\rho}R\tilde{T} \tag{8}$$

$$\tilde{s} - \tilde{s}_i = C_p \ln\frac{\tilde{T}}{\tilde{T}_i} - R\ln\frac{\tilde{p}}{\tilde{p}_i} \tag{9}$$

to close the model. In this study we take the reference entropy at the inlet $\tilde{s}_i$ to be 0. The *tilde* above a quantity (example $\tilde{p}$) denotes that the quantity is dimensional. In Sect. 2.5 we non-dimensionalise the equations and the tildes are dropped for the non-dimensional quantities.

## 2.4 Linearized Governing Equations

We linearize the governing Eqs. (4)–(7) around the steady state by splitting the variables into their mean and fluctuating components. The variables $p$, $u$, $s$, $\rho$ and $\dot{q}_v$ are split as follows:

$$\tilde{p} \rightarrow \bar{p} + \tilde{p}', \quad \tilde{u} \rightarrow \bar{u} + \tilde{u}', \quad \tilde{s} \rightarrow \bar{s} + \tilde{s}', \quad \tilde{\rho} \rightarrow \bar{\rho} + \tilde{\rho}', \quad \tilde{\dot{q}}_v \rightarrow \bar{\dot{q}}_v + \tilde{\dot{q}}'_v$$

We use a *bar* (for example $\bar{p}$) above the variable name to denote the steady state component. The fluctuations are denoted by a prime (for example $\tilde{s}'$). The linearized state equations are given by:

$$\frac{\tilde{p}'}{\tilde{p}} = \frac{\tilde{\rho}'}{\tilde{\rho}} + \frac{\tilde{T}'}{\tilde{T}} \tag{10}$$

$$\frac{\tilde{p}'}{\gamma\tilde{p}} - \frac{\tilde{\rho}'}{\tilde{\rho}} = \frac{\tilde{s}'}{c_p} \tag{11}$$

After linearizing the governing Eqs. (4)–(7), the resulting equations are simplified further using the linearized state Eqs. (10) and (11). These operations result in the following equations:

$$\frac{\partial \tilde{p}'}{\partial \tilde{t}} + \tilde{u}\frac{\partial \tilde{p}'}{\partial \tilde{x}} + \gamma\tilde{p}\frac{\partial \tilde{u}'}{\partial \tilde{x}} + \frac{\gamma\tilde{p}\tilde{u}'}{A}\frac{dA}{d\tilde{x}} + \frac{\gamma\tilde{p}'}{A}\frac{\partial(uA)}{\partial x} - \tilde{u}'\frac{d\tilde{p}}{d\tilde{x}} = (\gamma - 1)\dot{q}'_v \tag{12}$$

$$\frac{\partial \tilde{u}'}{\partial \tilde{t}} + \tilde{u}\frac{\partial \tilde{u}'}{\partial \tilde{x}} + \left[\frac{\tilde{u}'}{\tilde{u}} + \frac{\tilde{p}'}{\gamma\tilde{p}} - \frac{\tilde{s}'}{c_p}\right]\tilde{u}\frac{d\tilde{u}}{d\tilde{x}} = -\frac{1}{\tilde{\rho}}\frac{\partial \tilde{p}'}{\partial \tilde{x}} \tag{13}$$

$$\tilde{p}\left[\frac{\partial \tilde{s}'}{\partial \tilde{t}} + \tilde{u}\frac{\partial \tilde{s}'}{\partial \tilde{x}}\right] + \left[\frac{\tilde{p}'}{\tilde{p}} + \frac{\tilde{u}'}{\tilde{u}}\right]\frac{d\tilde{s}}{d\tilde{x}} = R\tilde{\dot{q}}'_v \tag{14}$$

## 2.5 Linearized Non-dimensional Equations

The linearized equations (12)–(14) are non-dimensionalised using the following scales:

$$\begin{aligned} M_i &= \tfrac{u_i}{c_i}; \quad t = \tfrac{c_i}{L}\tilde{t}; \quad x = \tfrac{\tilde{x}}{L}; \quad u' = \tfrac{\tilde{u}'}{u_i}; \quad p' = \tfrac{\tilde{p}'}{\gamma M_i p_i}; \\ s' &= \tfrac{\tilde{s}'}{c_p}; \quad u = \tfrac{\tilde{u}}{u_i}; \quad p = \tfrac{\tilde{p}}{p_i}; \quad A = \tfrac{\tilde{A}}{A_i}; \quad \dot{q}'_v = \tfrac{L}{p_i c_i}\tilde{\dot{q}}'_v \end{aligned} \tag{15}$$

The linearized non-dimensional equations are

$$\frac{\partial p'}{\partial t} + uM_i\frac{\partial p'}{\partial x} + p\frac{\partial u'}{\partial x} + \frac{pu'}{A}\frac{dA}{dx} + \frac{\gamma M_i p'}{A}\frac{\partial(uA)}{\partial x} + \frac{u'}{\gamma}\frac{dp}{dx} + \zeta_p^d p' = \frac{K}{M_i}\dot{q}'_v \tag{16}$$

$$\frac{\partial u'}{\partial t} + uM_i\frac{\partial u'}{\partial x} + \left[\frac{u'}{u} + \frac{M_i p'}{p} - s'\right] uM_i\frac{du}{dx} + \frac{1}{\rho}\frac{\partial p'}{\partial x} = 0 \tag{17}$$

$$\frac{\partial s'}{\partial t} + uM\frac{\partial s'}{\partial x} + \left[\gamma M_i\frac{p'}{p} + \frac{u'}{u}\right] uM_i\frac{ds}{dx} + \zeta_s^d s' = \frac{K}{p}\dot{q}_v' \tag{18}$$

where $\zeta_p^d$ and $\zeta_s^d$ are coefficients which control the frequency dependent damping and $K = (\gamma - 1)/\gamma$. The unsteady heat release term $\dot{q}_v'$ is written using the $\eta - \tau$ model proposed by [3] as follows:

$$\dot{q}_v' = \dot{Q}\left[\eta_p p'(t - \tau_p) + \eta_u u'(t - \tau_u)\right] \tag{19}$$

where $\eta_p$ and $\eta_u$ are the interaction index for pressure and velocity respectively. In this study we set $\eta_p = 0$ for all test cases. This is equivalent to the assumption that unsteady heat release is only caused by unsteady velocity fluctuations, which is true for gas turbine configurations [32]. For sake of completeness we do include $\eta_p$ in the remainder of the derivation. The parameters $\tau_p$ and $\tau_u$ are the time lags for pressure and velocity coupling respectively.

## 3 Numerical Scheme

There are several techniques available for solving the governing equations presented in the previous section. In this study we use frequency domain approach to capture the periodical effect arising from instability. Frequency domain approach allows us to compute the resonant frequency and growth/decay rate of the oscillations. In this section, we first transform our non-dimensional linearized equations to frequency-domain. Next, we discretize the equations using appropriate schemes. Then, we briefly discuss the boundary conditions. After that, we rewrite the equations in the form of a nonlinear eigenvalue problem.

### *3.1 Time-Domain to Frequency-Domain Transformation*

Assuming harmonic oscillations, the fluctuations $\{p', u', s'\}$ are written as

$$p' = \Re[\hat{p}e^{i\omega t}]; \quad u' = \Re[\hat{u}e^{i\omega t}]; \quad s' = \Re[\hat{s}e^{i\omega t}] \tag{20}$$

where $\omega$ is a complex number of the form $\omega_r + i\omega_i$. The $\Re(\omega) = \omega_r$, corresponds to the resonant frequency and $-\Im(\omega) = -\omega_i$ corresponds to the growth rate of the oscillations. Substituting the above relations in Eqs. (16)–(18) gives the following set of equations:

$$i\omega\hat{p} + uM_i\frac{d\hat{p}}{dx} + p\frac{d\hat{u}}{dx} + \frac{p}{A}\frac{dA}{dx}\hat{u} + \frac{\gamma M_i}{A}\frac{\partial(uA)}{\partial x}\hat{p} + \frac{1}{\gamma}\frac{dp}{dx}\hat{u} \tag{21}$$

$$+\zeta_p\Re(\omega)\hat{p} = \frac{(\gamma-1)}{\gamma M_i}\dot{Q}\left[\eta_p\hat{p}e^{-i\omega\tau_p} + \eta_u\hat{u}e^{-i\omega\tau_u}\right] \tag{22}$$

$$i\omega\hat{u} + uM_i\frac{d\hat{u}}{dx} + \left[\frac{\hat{u}}{u} + \frac{M_i\hat{p}}{p} - \hat{s}\right]uM_i\frac{du}{dx} + \frac{1}{\rho}\frac{d\hat{p}}{dx} = 0$$

$$i\omega\hat{s} + uM_i\frac{d\hat{s}}{dx} + \left[\gamma M_i\frac{\hat{p}}{p} + \frac{\hat{u}}{u}\right]uM_i\frac{ds}{dx} + \zeta_s\Re(\omega)\hat{s} = \tag{23}$$

$$\frac{\gamma-1}{\gamma p}\dot{Q}\left[\eta_p\hat{p}e^{-i\omega\tau_p} + \eta_u\hat{u}e^{-i\omega\tau_u}\right]$$

The frequency dependent damping terms $\zeta_p^d$ and $\zeta_s^d$ are written as $\zeta_p\Re(\omega)$ and $\zeta_s\Re(\omega)$ respectively. Initially the system is solved without damping i.e. by setting $\zeta_p = \zeta_s = 0$. Then for each computed $\omega$, $\zeta$ times $\Re(\omega)$ is introduced as frequency dependent damping to compute the correct growth rate for that particular $\omega$.

We then group the variables in Eqs. (21)–(23) to simplify the representation in the following form:

$$C'_\alpha\frac{d\hat{p}}{dx} + C'_\beta\frac{d\hat{u}}{dx} + C'_\gamma\hat{p} + C'_\delta\hat{u} = -i\omega\hat{p} + C'_\mu\hat{p}e^{-i\omega\tau_p} + C'_\epsilon\hat{u}e^{-i\omega\tau_u} \tag{24}$$

$$C'_\zeta\frac{d\hat{p}}{dx} + C'_\alpha\frac{d\hat{u}}{dx} + C'_\kappa\hat{p} + C'_\lambda\hat{u} + C'_\xi\hat{s} = -i\omega\hat{u} \tag{25}$$

$$C'_\phi\hat{p} + C'_\psi\hat{u} + C'_\alpha\frac{d\hat{s}}{dx} + C'_\nu\hat{s} = -i\omega\hat{s} + C'_\chi\hat{p}e^{-i\omega\tau_p} + C'_\iota\hat{u}e^{-i\omega\tau_u} \tag{26}$$

The expressions for the coefficients $\{C'_\alpha, \ldots\}$ after discretizing the governing equations are presented in the following subsection.

## 3.2 Discretization of the Equations

Finite difference scheme is used to discretize the linearized non-dimensional governing equations. Gradient of fluctuating pressure and velocity are approximated using 2nd order central difference scheme in order to account for the fact that acoustic wave can travel both upstream and downstream. However for gradient of entropy fluctuation 2nd order backward finite difference scheme is used because entropy fluctuations are only convected downstream. Furthermore, second order schemes do not produce dissipation.

For the given system of length $L$, we generate two sets of grids in one dimensional space. One is the normal grid which spans from 0 to $L$ with $N$ grid points. Another is a staggered grid starting from $h$ to $L - h$ with $N - 1$ grid points, where $h = 1/(2N - 2)$ (Fig. 3). The scalar quantities (for example $\hat{p}$, $\hat{s}$) are discretized in the normal grid and the vector quantities (for example $\hat{u}$) are discretized in the staggered grid. The reason for choosing two different sets of grids is to achieve point to point convergence. A brief discussion on the necessity of staggered grids for this model can be found [29].

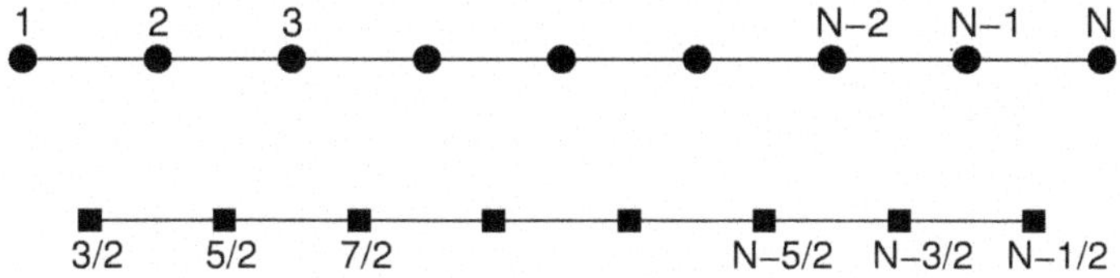


**Fig. 3** Staggered grid based discretization. The scalar quantities (for example $\hat{p}$, $\hat{s}$) are discretized using the normal grid $(1, 2, \ldots, N)$ and vector quantities (for example $\hat{u}$) are discretized using the staggered grid $(3/2, 5/2, \ldots, N-1/2)$

Discretizing equations (21)–(23) results in the following three equations:

$$C'_\alpha \frac{\hat{p}_{n+1} - \hat{p}_{n-1}}{2h} + C'_\beta \frac{\hat{u}_{n+1/2} - \hat{u}_{n-1/2}}{h} + C'_\gamma \hat{p}_m + C'_\delta \frac{\hat{u}_{n+1/2} + \hat{u}_{n-1/2}}{2} \tag{27}$$

$$= -i\omega \hat{p}_m + C'_\mu \hat{p}_n e^{-i\omega\tau_p} + C'_\epsilon \frac{\hat{u}_{n+1/2} + \hat{u}_{n-1/2}}{2} e^{-i\omega\tau_u} \tag{28}$$

$$C'_\zeta \frac{\hat{p}_n - \hat{p}_{n-1}}{h} + C'_\alpha \frac{\hat{u}_{n+1/2} - \hat{u}_{n-3/2}}{2h} + C'_\kappa \frac{\hat{p}_n + \hat{p}_{n-1}}{2} + C'_\lambda \hat{u}_{n-1/2}$$

$$+ C'_\xi \frac{\hat{s}_n + \hat{s}_{n-1}}{2} = -i\omega \hat{u}_{n-1/2} \tag{29}$$

$$C'_\phi \hat{p}_m + C'_\psi \frac{\hat{u}_{n+1/2} + \hat{u}_{n-1/2}}{2} + C'_\alpha \frac{3\hat{s}_n - 4\hat{s}_{n-1} + \hat{s}_{n-2}}{2h} + C'_\nu \hat{s}_m$$

$$= -i\omega \hat{s}_m + C'_\chi \hat{p}_m e^{-i\omega\tau_p} + C'_\iota \frac{\hat{u}_{n+1/2} + \hat{u}_{n-1/2}}{2} e^{-i\omega\tau_u}$$

Absorbing the $h$s into the coefficients $\{C_\alpha, \ldots\}$ and rearranging the terms gives us

$$-C_\alpha \hat{p}_{n-1} + C_\gamma \hat{p}_n + C_\alpha \hat{p}_{n+1} + (C_\delta - C_\beta)\hat{u}_{n-1/2} + (C_\delta + C_\beta)\hat{u}_{n+1/2} \tag{30}$$

$$= -i\omega \hat{p}_n + C_\mu \hat{p}_n e^{-i\omega\tau_p} + C_\epsilon \left(\hat{u}_{n+1/2} + \hat{u}_{n-1/2}\right) e^{-i\omega\tau_u} \tag{31}$$

$$\left(C_\kappa - C_\zeta\right) \hat{p}_{n-1} + (C_\zeta + C_\kappa)\hat{p}_n - C_\alpha \hat{u}_{n-3/2} + C_\lambda \hat{u}_{n-1/2} + C_\alpha \hat{u}_{n+1/2}$$

$$+ C_\xi \hat{s}_n + C_\xi \hat{s}_{n-1} = -i\omega \hat{u}_{n-1/2} \tag{32}$$

$$C_\phi \hat{p}_n + C_\psi \left(\hat{u}_{n+1/2} + \hat{u}_{n-1/2}\right) + 3C_\alpha \hat{s}_n - 4C_\alpha \hat{s}_{n-1} + C_\alpha \hat{s}_{n-2}$$

$$= -i\omega \hat{s}_n + C_\chi \hat{p}_n e^{-i\omega\tau_p} + C_\iota \left(\hat{u}_{n+1/2} + \hat{u}_{n-1/2}\right) e^{-i\omega\tau_u}$$

where $n$ corresponds to a particular grid point and the coefficients are given as follows:

$$\left.\begin{aligned}
C_\alpha &= \frac{uM_i}{2h}, & C_\phi &= \frac{\gamma u M_i^2}{p}\frac{ds}{dx},\\
C_\beta &= \frac{p}{h}, & C_\psi &= \frac{M_i}{2}\frac{ds}{dx},\\
C_\gamma &= \frac{\gamma M_i}{A}\frac{\partial(uA)}{\partial x} + \zeta_p \Re(\omega), & C_\mu &= \frac{\gamma-1}{\gamma M_i}\dot{Q}\eta_p,\\
C_\delta &= \frac{1}{2}\left(\frac{p}{A}\frac{dA}{dx} + \frac{1}{\gamma}\frac{dp}{dx}\right), & C_\epsilon &= \frac{\gamma-1}{2\gamma M_i}\dot{Q}\eta_u,\\
C_\zeta &= \frac{1}{\rho h}, & C_\chi &= \frac{\gamma-1}{\gamma p}\dot{Q}\eta_p,\\
C_\kappa &= \frac{1}{2}\left(\frac{uM_i^2}{p}\frac{du}{dx}\right), & C_\iota &= \frac{\gamma-1}{2\gamma p}\dot{Q}\eta_u,\\
C_\lambda &= M_i\frac{du}{dx}, & C_\nu &= \zeta_s \Re(\omega).\\
C_\xi &= -\frac{uM_i}{2}\frac{du}{dx}, & &
\end{aligned}\right\} \tag{33}$$

## 3.3 Boundary Conditions

The following boundary conditions are applied to the discretized equations obtained in the previous section. Since we have three unknowns $\{\hat{p}, \hat{u}, \hat{s}\}$ we introduce three boundary conditions. They are

$$\hat{p}_1 = -Z_i \hat{u}_1,\ \hat{p}_N = Z_e \hat{u}_N,\ \&\ \hat{s}_1 = 0 \tag{34}$$

where $Z_i$ & $Z_e$ are non-dimensional inlet and exit impedances respectively. $\hat{s}_1 = 0$ implies that entropy fluctuation at the inlet is zero. Since $\hat{u}$ is defined only in the staggered grid, $\hat{u}_1$ & $\hat{u}_N$ are found using Lagrange interpolating polynomial, which is given by

$$P(x) = \sum_{j=1}^{n} P_j(x) \tag{35}$$

where

$$P_j(x) = y_j \prod_{\substack{k=1\\ k\neq j}}^{n} \frac{x - x_k}{x_j - x_k} \tag{36}$$

Since we are performing 2nd order finite difference approximation, we use Lagrange polynomial of order 2 in order to find $\hat{u}_1$ & $\hat{u}_N$ respectively. Thus the expressions of $\hat{u}_1$ & $\hat{u}_N$ are

$$\hat{u}_1 = \frac{15}{8}\hat{u}_{3/2} - \frac{5}{4}\hat{u}_{5/2} + \frac{3}{8}\hat{u}_{7/2} \tag{37}$$

$$\hat{u}_N = \frac{15}{8}\hat{u}_{N-1/2} - \frac{5}{4}\hat{u}_{N-3/2} + \frac{3}{8}\hat{u}_{N-5/2} \tag{38}$$

Substituting expressions of $\hat{u}_1$ & $\hat{u}_N$ and rewriting the impedance boundary conditions gives the following two equations

$$\hat{p}_1 + Z_i\left[\frac{15}{8}\hat{u}_{3/2} - \frac{5}{4}\hat{u}_{5/2} + \frac{3}{8}\hat{u}_{7/2}\right] = 0 \tag{39}$$

$$\hat{p}_N - Z_e\left[\frac{15}{8}\hat{u}_{N-1/2} - \frac{5}{4}\hat{u}_{N-3/2} + \frac{3}{8}\hat{u}_{N-5/2}\right] = 0 \tag{40}$$

## 3.4 *Matrix Representation of the Nonlinear Eigenvalue Problem*

Equations (30)–(32) are written for all the grid points and finally represented using the matrix notation given below.

$$\mathbf{M1p}\hat{P} + \mathbf{M1u}\hat{U} + \mathbf{M1s}\hat{S} = i\omega\mathbf{M1r}\hat{P} + e^{-i\omega\tau_p}\mathbf{M1q}\hat{P} + e^{-i\omega\tau_u}\mathbf{M4q}\hat{U} \tag{41}$$

$$\mathbf{M2p}\hat{P} + \mathbf{M2u}\hat{U} + \mathbf{M2s}\hat{S} = i\omega\mathbf{M2r}\hat{U} \tag{42}$$

$$\begin{aligned}\mathbf{M3p}\hat{P} + \mathbf{M3u}\hat{U} + \mathbf{M3s}\hat{S} &= i\omega\mathbf{M3r}\hat{S} + e^{-i\omega\tau_p}\mathbf{M2q}\hat{P} \\ &\quad + e^{-i\omega\tau_p}\mathbf{M3q}\hat{U} + e^{-i\omega\tau_u}\mathbf{M5q}\hat{U}\end{aligned} \tag{43}$$

The matrices **M1p, M1u**, … are matrices whose entries are functions of the coefficients $\{C_\alpha, C_\beta, \ldots\}$. The above equations are combined to give

$$\begin{bmatrix}\mathbf{M1p} & \mathbf{M1u} & \mathbf{M1s}\\ \mathbf{M2p} & \mathbf{M2u} & \mathbf{M2s}\\ \mathbf{M3p} & \mathbf{M3u} & \mathbf{M3s}\end{bmatrix}\begin{bmatrix}\hat{P}\\ \hat{U}\\ \hat{S}\end{bmatrix} = i\omega\begin{bmatrix}\mathbf{M1r} & 0 & 0\\ 0 & \mathbf{M2r} & 0\\ 0 & 0 & \mathbf{M3r}\end{bmatrix}\begin{bmatrix}\hat{P}\\ \hat{U}\\ \hat{S}\end{bmatrix}$$
$$+ e^{-i\omega\tau_p}\begin{bmatrix}\mathbf{M1q} & 0 & 0\\ 0 & 0 & 0\\ \mathbf{M2q} & \mathbf{M3q} & 0\end{bmatrix}\begin{bmatrix}\hat{P}\\ \hat{U}\\ \hat{S}\end{bmatrix} + e^{-i\omega\tau_u}\begin{bmatrix}0 & \mathbf{M4q} & 0\\ 0 & 0 & 0\\ 0 & \mathbf{M5q} & 0\end{bmatrix}\begin{bmatrix}\hat{P}\\ \hat{U}\\ \hat{S}\end{bmatrix} \tag{44}$$

The sub-matrices in Eq. (44) are combined to form a single block matrix equation given by

$$\mathbf{ML}\chi = i\omega\mathbf{MR}\chi + e^{-i\omega\tau_p}\mathbf{MR1}\chi + e^{-i\omega\tau_u}\mathbf{MR2}\chi \tag{45}$$

where the column vector $\chi = \{\hat{P}, \hat{U}, \hat{S}\}$. The above equation is multiplied by $\mathbf{MR}^{-1}$ to yield:

$$\mathbf{A}\chi = i\omega\mathbf{I}\chi + e^{-i\omega\tau_p}\mathbf{B}\chi + e^{-i\omega\tau_u}\mathbf{C}\chi \tag{46}$$

$$\mathbf{A}\chi = \lambda\mathbf{I}\chi + e^{-\lambda\tau_p}\mathbf{B}\chi + e^{-\lambda\tau_u}\mathbf{C}\chi \tag{47}$$

where $\mathbf{A} = \mathbf{MR}^{-1}\mathbf{ML}, \mathbf{B} = \mathbf{MR}^{-1}\mathbf{MR1}, \mathbf{C} = \mathbf{MR}^{-1}\mathbf{MR2}$ and $\lambda = i\omega$. Equation (46) corresponds to a nonlinear eigenvalue problem.

### *3.5 Procedure to Find the Characteristic Roots and the Eigenvectors*

Equation (46) is highly difficult to solve for large $N$ ($N >> 2$). Wieczorek [29] used Arnoldi iteration to solve a similar equation. This method requires the knowledge of the frequency spectrum in which the eigenvalues lie. In this study we use the spectral discretization method developed by [30] to find the characteristic roots of time-delayed systems.

Since the eigenvalues and eigenvectors are computed after the transformation $\lambda = i\omega$, $\Re(\lambda)$ gives the growth rate of the oscillations and $\Im(\lambda)$ gives the resonant frequency of the oscillations. If $\Re(\lambda) < 0$, the system is stable and if $\Re(\lambda) > 0$, the system is unstable. The eigenvector corresponding to each characteristic root $\lambda$ gives the mode shapes of the oscillations. The eigenvectors are computed using the singular value decomposition (SVD) of the matrix

$$\mathbf{MM} = \lambda\mathbf{I} - \mathbf{A} + e^{-\lambda\tau_p}\mathbf{B} + e^{-\lambda\tau_u}\mathbf{C} \tag{48}$$

The SVD of $\mathbf{MM}$ gives the eigenvector for the corresponding eigenvalue $\lambda$. The eigenvector is completed by including the boundary conditions given by Eq. (34). The obtained eigenvectors are then normalized using $\hat{p}(x = 0) = \hat{p}_1$.

### *3.6 Procedure to Distinguish Between Acoustic-Driven Instability and Entropy-Driven Instability*

In Sect. 3.4 we demonstrated that the instability of the system can be computed from the characteristic roots of the Eq. (46). If the real part of the root is greater than zero, the system is unstable. To establish whether a given mode is acoustic or entropy, we use both the locus of the eigenvalues in the spectrum and their mode shapes. We start with a configuration in which only acoustic modes are present and track the locus of the modes in the spectrum by varying one parameter at a time. This method is explained briefly in our earlier study [27]. Another supplementary technique, is to track the mode shapes and phases of $\hat{p}$, $\hat{u}$ and $\hat{s}$ to determine whether the modes are acoustic. If a modes ceases to show the characteristics of the acoustic mode, then it should be either an entropy mode or spurious mode. We dismiss a mode as spurious mode if it does not show smooth variation in the spectrum for small change in parameters.

## 4 Results

In this section, we solve the nonlinear eigenvalue problem posed in Sect. 3.4 and distinguish between acoustic-driven and entropy-driven instability. First, we present the geometric configuration and the operating conditions, explored in this study. Next, we define two case studies to distinguish the two types of instability. Then, we fix all the parameters and vary the time lag $\tau_u$ to show the transition from a stable regime to an unstable regime. Using the locus of the eigenvalues of the modes in the spectrum and the associated mode shapes, we distinguish between acoustic-driven instability and entropy-driven instability.

### *4.1 System Chosen for Study*

To compute the stability of a system we need to set its geometry and operating conditions. We perform two case studies and the parameters chosen for both are given in Table 1. A large fraction of the parameters are fixed for both the cases. The significant difference between the two case studies is the choice of exit area $A_e$. Moreover, the interaction index $\eta_u$ and the range for time lag $\tau_u$ are chosen different for both the cases.

In Table 1, all the parameters are listed in dimensional form except for impedances ($Z_i$, $Z_e$) and damping ($\zeta_p^d$, $\zeta_s^d$). The dimensional parameters are non-dimensionalised using the scales given in Eq. (15). Since the exit area $A_e$ is chosen to be same as the inlet area $A_i$ in case #1, we expect to observe acoustic-driven instability. In case #2, the exit area is chosen such that the Mach number at the exit just becomes unity ($M_e = 1$). Hence, we explore the possibility of entropy-driven instability. For all the configurations, the number of grid points chosen in the normal grid is 200.

**Table 1** Geometrical configuration and operating parameters used for the case studies

| Parameter | Symbol | Case #1 | Case #2 |
|---|---|---|---|
| Length of the system | $L$ | $1\,m$ | $1\,m$ |
| Location of the center of the flame | $x_f$ | $0.2\,m$ | $0.2\,m$ |
| Thickness of the flame | $\delta_f$ | $0.05\,m$ | $0.05\,m$ |
| Location of the beginning of the nozzle | $x_N$ | $0.9\,m$ | $0.9\,m$ |
| Inlet area | $A_i$ | $10^{-2}\,m^2$ | $10^{-2}\,m^2$ |
| Exit area | $A_e$ | $A_i$ | $0.0579A_i$ |
| Inlet impedance | $Z_i$ | $\infty$ | $\infty$ |
| Exit impedance | $Z_e$ | 0 | 0 |
| Acoustic damping | $\zeta_p^d$ | 0.02 | 0.02 |
| Entropy damping | $\zeta_s^d$ | 0.01 | 0.01 |
| Inlet Mach number | $M_i$ | 0.015 | 0.015 |
| Total inlet temperature | $T_{0i}$ | 300 K | 300 K |
| Total temperature rise | $T_{0f}/T_{0i}$ | 1–5 | 1–5 |
| Exit pressure | $p_e$ | 101325 *Pa* | 101325 *Pa* |
| Interaction index for *p* | $\eta_p$ | 0 | 0 |
| Interaction index for *u* | $\eta_u$ | 0–0.6 | 0–0.4 |
| Time lag | $\tau_u$ | 0–0.14 | 0–0.4 |
| Number of grid points | $N$ | 200 | 200 |

## 4.2 *Case Study Showing Acoustic-Driven Instability*

The objective of investigating case #1 is to show the transition from stable to acoustic-driven instability. First, for the chosen system configuration we distinguish between the acoustic and entropy modes in the resonant frequencies. Then, holding all parameters fixed, we vary the time lag $\tau_u$ to show the transition of the system from stable regime to unstable regime.

We know from previous investigations (Sect. 1) that acoustic-driven instability is more prone in a combustor with area ratio, $A_e/A_i = 1$. Therefore, in this case study, we choose the inlet area $A_i$ to be equal to the exit area $A_e$ with no area variation from the inlet to the exit. The unsteady interaction index is fixed as $\eta_u = 0.6$ and a range of time lag explored is $0 < \tau_u < 0.14$.

In both the studies to distinguish the acoustic modes from the entropy modes, we start with a configuration for which the acoustic modes are known from analytical expressions available in the literature. In this case, we choose the configuration of a constant cross section area straight duct in the absence of heat source. Then, we add the complexities in the following manner for case #1 and track the variation in the acoustic modes:

- The total temperature ratio ($T_{0f}/T_{0i}$) across the heat source is increased from 1 to 5 in steps of 0.1, keeping $\eta_u = \tau_u = 0$.
- Then the total temperature ratio is fixed at 5 and $\eta_u$ is increased from 0 to 0.6 in steps of 0.01, keeping $\tau_u = 0$.
- Finally, the total temperature ratio is fixed at 5, $\eta_u$ is set as 0.4 and the time lag $\tau_u$ is varied from 0 to 0.14 in steps of 0.05.

In a similar manner, investigations for case #2 is also performed in three steps. The range chosen for the parameters $T_{0f}/T_{0i}$, $\eta_u$ & $\tau_u$ are shown in Table 1.

### 4.2.1 Distinguishing Between Acoustic and Entropy Modes

For a straight duct with acoustically closed-open boundary condition ($Z_i = \infty$, $Z_e = 0$), in the absence of heat release, the frequency of the acoustic modes are given by the equation

$$f_n^a = (2n - 1)\frac{1 - M_i^2}{4} \tag{49}$$

where $f_n^a$ is the non-dimensional acoustic resonant frequency and $n$ corresponds to the $n$th harmonic [11]. Similar expression was given for entropy modes by [8] as follows.

$$f_n^s = (2n - 1)\frac{M_H}{2}\frac{1}{1 - x_f/L} \tag{50}$$

where $f_n^s$ is the non-dimensional entropy mode and $M_H$ corresponds to the Mach number of the hot gases downstream of the compact flame.

For $M_i = 0.015$ and the parameters tabulated for case #1, we calculated $M_H = 0.264$. The first five acoustic and entropy modes computed using Eqs. (49) and (50) are given in Table 2 for reference.

Note that the acoustic and entropy modes are computed for different scenarios. In case of acoustic modes, the associated resonant frequencies are calculated for the inlet temperature $T_i$, while the entropy modes are calculated for the temperature of the hot gases $T_h$. For this reason we do not draw any quantitative inference from

**Table 2** The first five non-dimensional resonant frequencies for acoustic and entropy modes, associated with $M_i = 0.015$ and $M_H = 0.264$

| Mode number | $f_n^a$ | $f_n^s$ |
|---|---|---|
| 1 | 0.249 | 0.165 |
| 2 | 0.749 | 0.494 |
| 3 | 1.249 | 0.824 |
| 4 | 1.749 | 1.154 |
| 5 | 2.249 | 1.484 |

Table 2. One qualitative inference, which is worthy to note is that the frequency of the entropy modes $f_n^s$ are lesser than the corresponding acoustic modes $f_n^a$. This is true in general for all the test cases. The entropy modes have a lower resonant frequency when compared with the acoustic modes. This arises due to the lower propagation speeds of the former in comparison to the latter. Expressions (49) and (50) are valid only for simple configurations.

In an earlier study [27], we used the following approach to assert whether the given mode is acoustic or entropy. We started from the simple configuration for which the acoustic modes are known from the above analytical expression. Next, we varied one parameter at a time to track the locus of the eigenvalues on the spectrum. We assume for infinitesimal variation in the system parameters there is only a infinitesimal variation in the eigenvalues. For example we observe from Fig. 4, for small variation in $\tau_u$, the eigenvalues show a smooth variation in the spectrum. In this study, we recall the important characteristics of the locus of the modes and in addition provide an alternate argument in terms of the mode shape and phase of $\hat{p}$, to justify that the unstable mode is the acoustic mode.

In Fig. 4, the blue-green branch corresponds to case where the temperature rise $(T_{0f}/T_{0i})$ is increased from 1 to 5 in steps of 0.1. When $T_{0f}/T_{0i} = 1$, we exactly recover the acoustic modes given by Eq. (50). Note that $\Im(\lambda)/2\pi$ indicates the resonant oscillating frequency, which can be compared to $f_n^a$ in Table 2. The first three acoustic modes are $\lambda/2\pi = -0.002 + 0.249i$, $-0.007 + 0.749i$ and $-0.012 + 1.249i$. For $T_{0f}/T_{0i} = 1$ no entropy modes are present. When $T_{0f}/T_{0i}$ is increased, the resonant frequencies of the acoustic mode increase as shown in Fig. 4. This is because

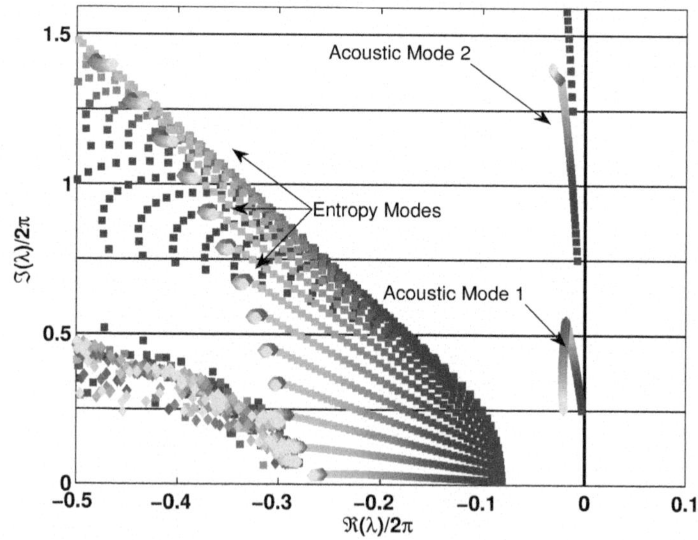


**Fig. 4** Locus of eigenvalues for case #1, showing the acoustic modes and entropy modes for variation in total temperature (*blue-green* branch) and interaction index $\eta_u$ (*red-yellow* branch)

with rise in temperature of the medium, the speed of sound increases and the resonant frequency is proportional to this quantity. For $T_{0f}/T_{0i} = 5$, the first two acoustic modes are $\lambda/2\pi = -0.018 + 0.546i$ and $-0.024 + 1.353i$. The third acoustic mode lies outside the region of interest in the spectrum (highly damped). Therefore we do not discuss it further.

Next, we fix the temperature rise and increase the interaction index $\eta_u$ from 0 to 0.6 in steps of 0.01. The locus of the eigenvalues is given by the red-yellow branch in Fig. 4. In this scenario, the resonant frequency of the first acoustic mode decreases with increase in $\eta_u$. For $\eta_u = 0.6$, the eigenmode drops to $\lambda/2\pi = -0.021 + 0.252i$. On the other hand, the second acoustic mode increases at a relatively slow pace and settles at $\lambda/2\pi = -0.032 + 1.394i$ for $\eta_u = 0.5$. We do not discuss the entropy modes in detail because they decay much faster than the acoustic modes and do not contribute significantly to the unsteady interactions in this case.

### 4.2.2 Transition from Stable to Unstable Regime

In Fig. 5 we observe that for $\tau_u = 0$, the real part of all the modes are in the left half of the spectrum (i.e. $\Re(\lambda) < 0$). Some spurious modes (blue shaded region) are present in the range $-0.5 < \Re(\lambda)/2\pi < -0.25$ and $0 < \Im(\lambda)/2\pi < 0.5$. These modes do not show any well defined variation with parameters and their mode shapes are found to be irregular. As $\tau_u$ is increased from 0, most of the modes shift towards the right i.e. $\Re(\lambda)$ increases. In particular the second acoustic mode that originates at $\lambda/2\pi = -0.032 + 1.394i$ becomes unstable ($\lambda/2\pi = 0.001 + 1.39i$) for $\tau_u = 0.09$. This implies that the system is unstable for this configuration. For higher values of $\tau_u$, namely $0.09 < \tau_u < 0.14$, the mode keeps translating to the right of the spectrum. Therefore the system remains unstable for increase in time lag. We also infer from Fig. 5 that the first acoustic mode $\lambda/2\pi = -0.021 + 0.252i$ shifts to the left for increase in $\tau_u$, which implies that this mode decays faster for increasing time lag. All the entropy modes remain damped in comparison to the acoustic modes. Therefore, the instability obtained in this case is an acoustic-driven instability.

### 4.2.3 Mode Shape and Phase of $\hat{p}$, $\hat{u}$ and $\hat{s}$

Fig. 6 shows the difference between the mode shape and phase of $\{\hat{p}, \hat{u}, \hat{s}\}$ for a stable and unstable system. The time lag associated with the stable and unstable system is $\tau_u = 0$ and $\tau_u = 0.14$ respectively. The corresponding eigenvalues are $\lambda/2\pi = -0.018 + 1.394i$ and $\lambda/2\pi = 0.039 + 1.369i$.

In case of $\hat{p}$, the mode shape remains the same for both the stable and the unstable configuration (Fig. 6*a*, *b*). There is an anti-node at $x = 0$ and two nodes: one close to $x = x_f/L$ and another at $x = 1$. This is reflected as a sudden phase change of $\hat{p}$ close to $x = x_f$. For the stable case the phase change is $\pi$. While, for the unstable case the phase change is close to $\pi/2$.

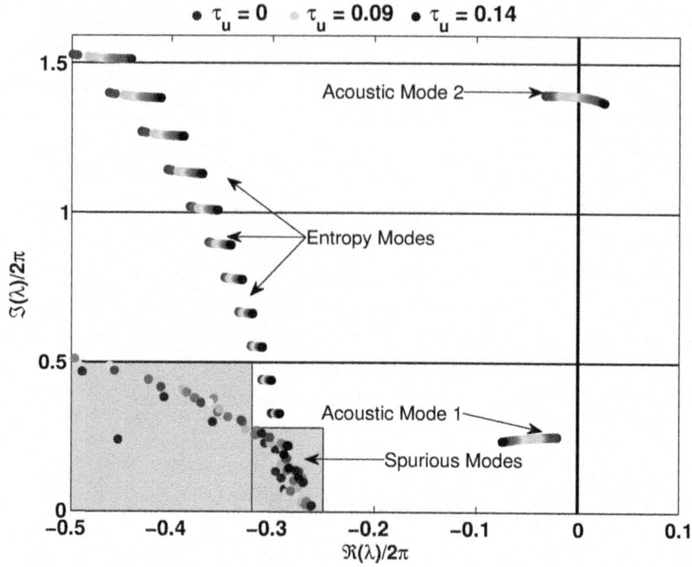


**Fig. 5** Locus of eigenvalues for case #1, showing the transition from stable region ($\mathfrak{R}(\lambda) < 0$) to unstable region ($\mathfrak{R}(\lambda) > 0$) as time lag parameter $\tau_u$ is increased from 0 to 0.14 holding all other parameters constant. A number of spurious modes (*blue* shaded region) can be observed in the *bottom left* corner

Also for the case of $\hat{u}$, the mode shapes qualitatively remain the same for both stable and unstable configurations (Fig. 6*c*, *d*). In both the cases there is a sharp increase in $|\hat{u}|$ at $x = x_f$ due to the presence of heat source. Further, the mode shapes of $\hat{p}$ and $\hat{u}$, indicate that acoustically driven instability occurs in case # 1.

Since the entropy fluctuation at the inlet is zero, the mode shape of $\hat{s}$ (Fig. 6*e*) stays at zero for $0 \leq x < x_f/L$. At $x = x_f/L$ both the stable and unstable configuration show a sharp rise in $\hat{s}$ due to the presence of the flame. After which, the entropy wave is convected downstream, which is indicated by a monotonic negative variation in the phase of $\hat{s}$ (Fig. 6*f*).

## *4.3 Case Study Showing Entropy-Driven Instability*

In case study #2, we show for variation in system parameters, the transition of the system to entropy-driven instability. In Sect. 4.2.1, we presented the procedure to distinguish acoustic and entropy modes, which is followed again in this section. As before, we vary time lag $\tau_u$ to show the transition from a stable regime to unstable regime.

In Sect. 1, we recalled the various experimental and numerical investigations that studied entropy-driven instability. It is evident from these studies that entropy-driven

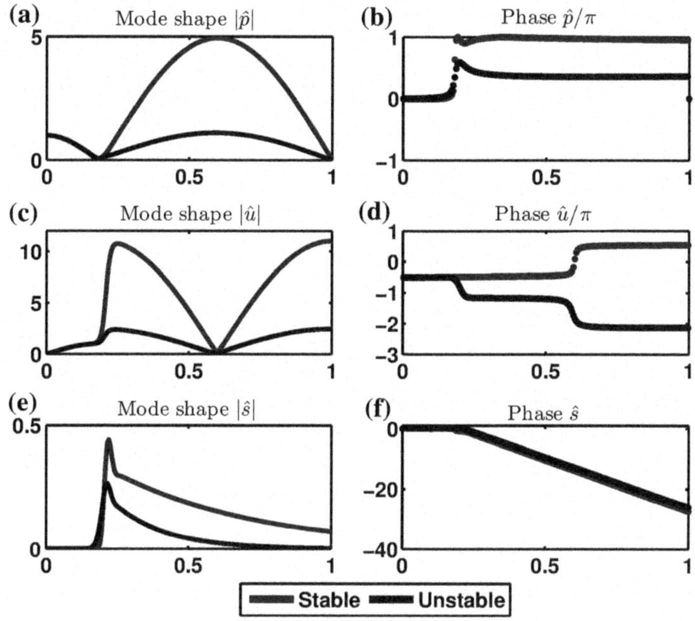


**Fig. 6** Figure showing the mode shape $(a, c, e)$ and phase $(b, d, f)$ of $\hat{p}$, $\hat{u}$ and $\hat{s}$ for case study #1 showing acoustic-driven instability. The stable configuration corresponds to $\tau_u = 0$ and $\lambda/2\pi = -0.0183 + 1.394i$, while the unstable configuration corresponds to $\tau_u = 0.14$ and $\lambda/2\pi = 0.039 + 1.369i$

instability is more prone in combustors with strong flow acceleration at the exit. Therefore, we choose a geometry similar to one in Sect. 4.2, but a nozzle fitted at the exit. The exit area of the nozzle $A_e$ is chosen such that at the chosen operating condition, the Mach number at the exit is unity. Note that the nozzle exist in the region $0.9L \leq x \leq L$. The unsteady interaction index is set as $\eta_u = 0.4$ and time lag is uniformly increased in the range $0 < \tau_u < 0.4$.

### 4.3.1 Distinguishing Between Acoustic Modes and Entropy Modes

For case #2, we observe that the ratio of the length of the nozzle to the length of the combustor is 0.1. Hence, the nozzle is compact compared to the length. If the exit of the compact nozzle is choked, one can approximate the boundary at the location $x_N$ as acoustically closed. In this situation, the resonant frequencies of the acoustic modes inside the straight section can be approximated as follows [11]:

$$f_n^a = (2n - 1)\frac{1 - M_i^2}{2} \tag{51}$$

**Table 3** The first five non-dimensional acoustic modes for $M_i = 0.015$ in case of an acoustically closed-closed boundary condition

| Mode number | $f_n^a$ |
|---|---|
| 1 | 0.499 |
| 2 | 1.499 |
| 3 | 2.499 |
| 4 | 3.499 |
| 5 | 4.499 |

For $M_i = 0.015$, the first five modes computed using Eq. (51) are given in Table 3.

Similar to Sect. 4.2.1, we start with a simple configuration for which the acoustic modes are given by the Eq. (51). Then, for variation in parameters we identify the locus of the acoustic modes and the entropy modes. Unlike case#1, in this case study, both the acoustic and entropy modes are equally dominant and hence the plots are cluttered. Therefore, we simply mark the acoustic branch for convenience. Figure 7 shows both the acoustic and entropy branch for variation in time lag $\tau_u$.

### 4.3.2 Transition from Stable to Unstable Regime

Figure 7 shows that for $\tau_u = 0$ all the modes are in the left half of the spectrum ($\Re(\lambda) < 0$). This implies that the modes are stable because they have a negative growth rate. Compared to Fig. 5, all the modes here have relatively similar growth rates. This suggests that both the acoustic and entropy modes are equally important in the study of a choked system. When time lag $\tau_u$ is increased, initially all the modes shift towards the right becoming less stable. After reaching a critical value of $\tau_u$, several modes cross the zero line ($\Re(\lambda) = 0$) and become unstable. If the time lag $\tau_u$ is increased further, low frequency modes continue to remain unstable while the high frequency modes reverse direction and their growth rates decrease. For all the values of the time lag explored in this case study, the acoustic mode remains stable. For $\tau_u = 0$, the acoustic modes has eigenvalue $\lambda/2\pi = -0.115 + 1.475i$. As $\tau_u$ is increased, the mode translates to the right half of the spectrum. After reaching a critical value of time lag (here $\tau_u = 0.18$), the mode reverses direction and decays faster for further increase in $\tau_u$. Therefore the instability observed here is not an acoustic-driven instability.

In addition to the above discussed acoustic mode, several entropy modes are observed in the spectrum, shown in Fig. 7. All these modes shift towards the right half of the spectrum for increase in time lag $\tau_u$. For $\tau_u = 0.19$, one of the entropy modes $\lambda/2\pi = 0.74i$ crosses the zero line and becomes unstable. After that for increasing value of $\tau_u$ several entropy modes cross to the unstable region of the spectrum. Around $0.25 < \tau_u < 0.28$, some of the entropy modes reverse direction and move towards the stable regime. But the low frequency entropy modes continue to grow unstable. The same trend continues up to $\tau_u = 0.4$. At $\tau_u = 0.4$, three modes $\lambda/2\pi = 0.007 + 0.597i, 0.013 + 0.498i$ and $0.007 + 0.407i$ are unstable. Since these

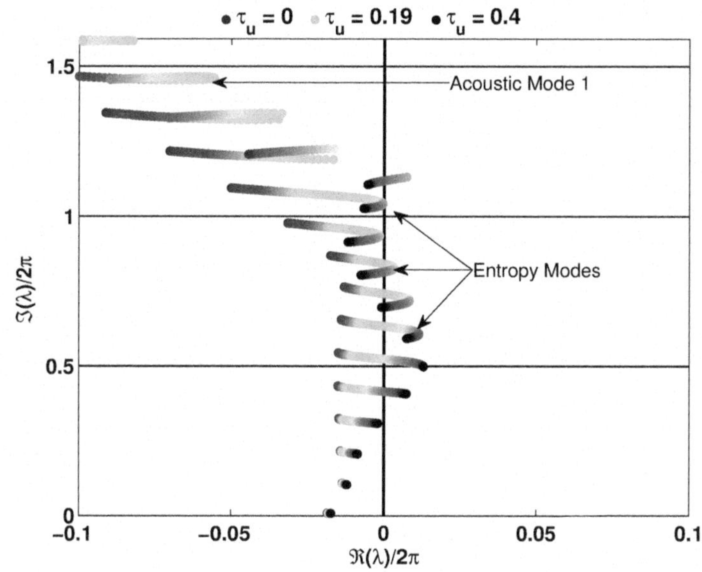


**Fig. 7** Locus of eigenvalues for entropy-driven instability (case # 2). The plot shows transition of eigenvalues from stable region ($\Re(\lambda) < 0$) to unstable region ($\Re(\lambda) > 0$) as time lag $\tau_u$ is increased from 0 to 0.4 holding all other parameters constant. For the chosen configuration both the acoustic and the entropy modes are equally dominant

modes arise from the entropy fluctuations the instability in this case is an entropy-driven instability.

### 4.3.3 Mode Shape and Phase of $\hat{p}$, $\hat{u}$ and $\hat{s}$

Figure 8 shows the difference between the mode shape and phase of $\{\hat{p}, \hat{u}, \hat{s}\}$ for a stable and unstable configuration. The corresponding time lags and eigenvalues are $\tau_u = 0$, $\lambda/2\pi = -0.0148 + 0.544i$ and $\tau_u = 0.4$, $\lambda/2\pi = 0.014 + 0.498i$ respectively.

The mode shape of $\hat{p}$ is significantly different between the stable and the unstable configuration. Both the stable and the unstable configuration have an anti-node at $x = 0$ and node at $x = 1$. The node at $x = 1$ is due to the impedance boundary condition enforced at this end. Note that there is a pressure anti-node at the location of the nozzle inlet ($x_N/L = 0.9$), indicating the effects of choked nozzle. Also the stable system has a node in the middle which appears to be absent in the unstable configuration. The occurrence of this node is clearly captured in the phase of $\hat{p}$, where there is a phase change of $\pi$ at the location of the node. On the other hand, the unstable configuration shows no phase change in $\hat{p}$.

Similar difference is observed in the mode shape of $\hat{u}$. Here both the stable and unstable configuration has a node at $x = 0$. The phase plot of $\hat{u}$ shows two phase

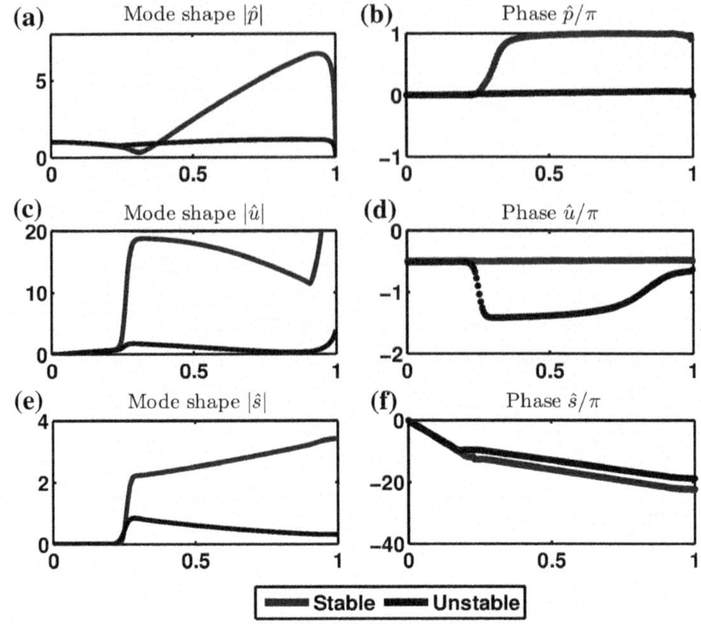


**Fig. 8** Figure showing the mode shape $(a, c, e)$ and phase $(b, d, f)$ of $\hat{p}$, $\hat{u}$ and $\hat{s}$ for case study #2 showing entropy-driven instability. The stable configuration corresponds to $\tau_u = 0$ and $\lambda/2\pi = -0.0148 + 0.544i$. The unstable configuration corresponds to $\tau_u = 0.4$ and $\lambda/2\pi = 0.014 + 0.498i$

changes for the unstable configuration whereas no phase change is observed for the stable configuration.

In the mode shape of $\hat{s}$ (Fig. 8*e* and *f*), since at the inlet the value is chosen to be 0, it remains 0 till $x = x_f/L$. At $x = x_f/L$ both the stable and unstable configurations show a sharp rise in $\hat{s}$ due to the presence of the flame. After which the entropy wave is accelerated.

### 4.4 Discussion

Figure 9 shows a comparison between the mode shape and phase of $\hat{p}$, $\hat{u}$ and $\hat{s}$ obtained for acoustic-driven instability and entropy-driven instability. In case of acoustic-driven instability, the mode shape and phase of $\hat{p}$ indicates clearly the presence of the second acoustic mode (Fig. 9*a* and *b*). Whereas for entropy-driven instability, one cannot associate the obtained mode shape with any of the known acoustic mode shapes. The corresponding phase variation (Fig. 9*b*) appears to be constant during entropy-driven instability. For $\hat{u}$, acoustic-driven instability again clearly shows the modes shape of the second acoustic mode. Entropy-driven instability shows one phase change of $\pi$ at the location of the flame ($x_f$). The mode shape of $\hat{s}$ is signif-

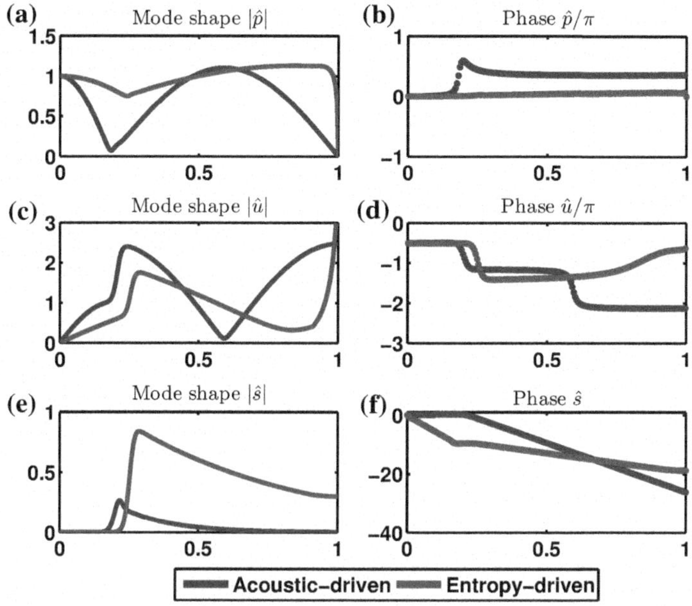


**Fig. 9** Comparison of the mode shapes obtained for acoustic-driven instability and entropy-driven instability. This plot shows the mode shape $(a, c, e)$ and phase $(b, d, f)$ of $\hat{p}$, $\hat{u}$ and $\hat{s}$ for the unstable configuration corresponding to both case#1 and case #2. For acoustic-driven instability, the eigenvalue is $0.039 + 1.369i$ and for the entropy-driven instability the eigenvalue is $0.014 + 0.498i$

icantly different between acoustic-driven instability and entropy-driven instability. In particular, the relative dominance of the entropy fluctuations $\hat{s}$ in comparison to the fluctuations in pressure $\hat{p}$ and velocity $\hat{u}$ is more in the mode associated with the entropy-driven instability than the acoustic-driven instability. This is the crucial difference portraying the role of entropy waves in entropy-driven instability.

In Sects. 4.2 and 4.3, we discussed the transition of the acoustic and entropy modes of a system from stable to unstable regime in the spectrum. We showed that entropy-driven instability is the dominant form of instability in combustors with choked exit. This is in agreement with a number of previous investigations detailed in Sect. 1. Here we present the comparison between some of the salient features observed in both the case studies. Most of these comparisons are only qualitative in nature, unless specified.

In case of acoustic-driven instability only the second acoustic mode becomes unstable for variation in time lag. Remaining modes are always stable. However, in the case of entropy-driven instability for certain time lags several entropy modes are unstable. Some modes also reverse direction and become stable after a critical value of time lag. The same is not observed in the case of acoustic-driven instability. Acoustic-driven instability is observed for relatively small values of time lag when compared to entropy-driven instability. In this study acoustic instability occurred at $\tau_u = 0.09$ and entropy-driven instability occurred for $\tau_u = 0.19$. Probable reason for this could be the relatively longer time taken by the entropy fluctuations to convect to the exit of the system, where they get reflected as acoustic waves.

## 5 Conclusion

This chapter presented a framework to model the generation, convection, acceleration and reflection of acoustic and entropy waves in a gas turbine combustor. We have considered the effects of the unsteady heat release from the flame, mean flow and dissipation of entropy and acoustic waves in our model. The framework is valid for variable geometry configurations. Quasi one dimensional linearized governing equations are used. Frequency domain approach is implemented. In the end, the equations are posed as a nonlinear eigenvalue problem and solved using spectral discretization method. The computed values of the resonant frequencies agree with the analytical expressions available for simple configurations. We discussed two techniques, based on the resonant frequency and mode shape to distinguish acoustic modes from entropy modes in the computed spectrum. We computed the spectrum for two case studies. Case study #1 considers a constant cross section area geometry. We observed the unstable modes have the characteristics of acoustic waves. The modes associated with entropy waves are observed to be highly damped and hence never become unstable. Therefore, acoustic-driven instability occurs in geometry, where there is negligible flow acceleration.

On the other hand, the exit area of the combustor configuration is choked in case study #2. In the spectrum, apart from the acoustic waves, entropy waves are also present. From their growth rates, both the modes are found to be equally dominant. In the present study, we observed that the system become unstable, when one of the entropy mode achieved positive growth rate. Hence, in case study #2, entropy-driven instability occurs.

Mode shapes and phases of pressure, velocity and entropy fluctuations are significantly different between acoustic-driven instability and entropy-driven instability. It is evident from these findings that entropy waves should be properly taken into account when modeling unsteady interactions in gas turbine combustors with choked exit.

**Acknowledgements** The authors are grateful to Prof. T. Poinsot, Research Director, Institut de Mecanique des Fluides de Toulouse, CNRS, France for his insightful suggestions and providing recent references, which formed the foundation of the reported work. The first and third authors would like to thank Indian Institute of Technology Kanpur for providing the financial support through the Institute Post-Doctoral and SURGE Fellowships respectively, which made this work possible.

## References


1. Anderson JD (2010) Fundamentals of aerodynamics. Tata McGraw-Hill Education
2. Bake F, Fischer A (2009) Experimental investigation of the entropy noise mechanism in aero-engines. Int J Aeroacoustics 8
3. Crocco L, Cheng S (1956) Theory of combustion instability in liquid propellant rocket motors, vol 8. Cambridge Univ Press
4. Dowling A, Morgans A (2005) Feedback control of combustion oscillations. Ann Rev Fluid Mech
5. Eckstein J, Freitag E, Hirsch C, Sattelmayer T (2006) Experimental study on the role of entropy waves in low-frequency oscillations in a rql combustor. J Eng Gas Turbines Power 128(2): 264–270
6. Goh C, Morgans A (2013) The influence of entropy waves on thermoacoustic stability of a model combustor. Combust Sci Technol 185(2):249–268
7. Hield P, Brear M (2008) Comparison of open and choked premixed combustor exits during thermoacoustic limit cycle. AIAA 46(2):517–526
8. Keller J (1995) Thermoacoustic oscillations in combustion chambers of gas turbines. AIAA 55:2280–2287
9. Keller JJ, Egli W, Hellat J (1985) Thermally induced low-frequency oscillations. Zeitschrift für angewandte Mathematik und Physik ZAMP 36(2):250–274
10. Lieuwen T, Zinn BT (1998) The role of equivalence ratio oscillations in driving combustion instabilities in low nox gas turbines. In: Symposium (International) on combustion, vol 27. Elsevier, pp 1809–1816
11. Lieuwen TC (2012) Unsteady combustor physics. Cambridge University Press
12. Macquisten MA, Dowling AP (1993) Low-frequency combustion oscillations in a model after-burner. Combust Flame 94(3):253–264
13. Marble F, Candel S (1977) Acoustic distrubance from gas non-uniformities convected through a nozzle. J Sound Vibr 55(2):225–243
14. Matalon M (2007) Intrinsic flame instabilities in premixed and nonpremixed combustion. Ann Rev Fluid Mech 39:163–191
15. McManus KR, Poinsot T, Candel SM (1993) A review of active control of combustion instabilities. Prog Energ Combust Sci 19(1):1–29
16. Morgans A, Goh C, Dahan J (2013) The dissipation and shear dispersion of entropy waves in combustor thermoacoustics. J Fluid Mech 733:R2–1– R2–11
17. Morgans A, Li J (2015) The effect ofentropy noise on combustion instbility in the presence of advective shear dispersion. In: The 22nd international congress on sound and vibration
18. Morgans AS, Duran I (2016) Entropy noise: a review of theory, progress and challenges. Int J Spray Combust Dyn 1756827716651791
19. Motheau E, Nicoud F, Poinsot T (2014) Mixed acoustic-entropy combustion instabilities in gas turbines. J Fluid Mech 749
20. Murugesan M, Singaravelu B, Kushwaha A, Mariappan S (2016) Onset of entropy-driven combustion instability in turbulent combustors. J Fluid Mech (under review)
21. Muthukrishnan M, Strahle WC, Neale DH (1978) Separation of hydrodynamic, entropy and combustion noise in a gas turbine combustor. AIAA 16:320–327
22. Nicoud F, Wieczorek K (2009) About the zero mach number assumption in the calculation of thermoacoustic instabilities. Int J Spray Combust Dyn 1:67–112
23. Polifke W, Paschereit C, Döbbeling K (2001) Constructive and destructive interference of acoustic and entropy waves in a premixed combustor with a choked exit. Int J Acoust Vib 6:135–146
24. Putnam A, Faulkner L (1983) An overview of combustion noise. J Energy 7(6):458–469
25. Sattelmayer T (2003) Influence of the combustor aerodynamics on combustion instabilities from equivalence ratio fluctuations. pp V002T02A003–V002T02A003
26. Schadow KC, Gutmark E (1992) Combustion instability related to vortex shedding in dump combustors and their passive control. Prog Energy Combust Sci 18(2):117–132

27. Singaravelu B, Mariappan S, Saha A (2016) Role of entropy waves in the stability of unsteady motions in combustors. In: Proceedings of Asian congree of Gas Turbines (in press)
28. Strahle WC (1978) Combustion noise. Prog Energy Combust Sci 4(3):157–176
29. Wieczorek K (2010) Numerical study of Mach number effects in combustion instability. Ph.D. thesis, Université Montpellier II
30. Wu Z, Michiels W (2012) Reliably computing all characteristic roots of delay differential equations in a given right half plane using a spectral method. J Comput Appl Math 236:2499–2514
31. Yu YC, JCS, Sankaran V, Anderson W E (2010) Effects of mean flow, entropy waves, and boundary conditions on longitudinal combustion instability. Combust Sci Technol 182(7): 739–776
32. Zinn BT, Lieuwen TC (2006) Combustion instabilities: basic concepts—Combustion Instabilities in Gas Turbine Engines: operational experience, fundamental mechanisms, and modeling. AIAA, USA

# Chronology of Dynamic Stall Characteristics Through Simultaneous PIV and Surface Pressure Measurements

**Deepakkumar M. Sharma and Kamal Poddar**

**Abstract** Wind tunnel experiments were carried out at low speed aerodynamics lab at IIT Kanpur to investigate dynamic stall characteristics for a flow past an oscillating NACA 0015 airfoil at various reduced frequencies $\kappa$. The NACA airfoil model was designed and developed to incorporate simultaneous surface pressure measurement at the mid-span of the airfoil model along the chord and 2D PIV measurements of the flow-field downstream near the Trailing Edge (TE) vicinity of the oscillating airfoil model subjected to the higher reduced frequency $\kappa$ up to 0.5. The main objective of this work was to critically assess the effect of $\kappa$ in the unsteady domain from fully develop to partially develop dynamic stall regimes. For a given constant Reynolds number Re of $2 \times 10^5$, the instantaneous 2D PIV images were captured for an oscillating airfoil at varied $\kappa$. Case studies at $\kappa = 0.01$ and 0.10 falling under fully developed dynamic stall regimes are presented and critically assessed to provide qualitative information and flow visualization of the flow field to trace various dynamic stall events which mainly include the formation, growth and shedding of Dynamic Stall Vortex (DSV) from the TE.

**Keywords** Dynamic stall · PIV · Dynamic stall vortex · Leading edge vortex

## 1 Introduction

Dynamic stall phenomena are the results of an airfoil (wing) undergoing ramp or oscillatory motion having a maximum angle of attack greater than the static stall angle. The unsteady flows induced by the dynamic stall phenomena are characterized by massive separation and formation of large-scale vortical structures. The fundamental understanding of the dynamic stall onset is important to the rotary as

D.M. Sharma (✉)
TSI Instruments India Private Limited, Bangalore, India

K. Poddar
Indian Institute of Technology, Kanpur, India



A.K. Agarwal et al. (eds.), *Combustion for Power Generation and Transportation*, DOI 10.1007/978-981-10-3785-6_10

well as the fixed wing aircraft configurations [1–4]. Recent interests in exploiting the dynamic vortex lift on super-maneuverable aircraft and other applications such as wind turbine rotors, compressor blades, etc., require a thorough understanding of the dynamic stall process before utilizing the potential of these energetic flows. It is evident, however, that the knowledge of the aerodynamics for these applications is far from complete, especially the phenomena associated with the boundary layer behavior in the context of dynamic stall. This has prompted extensive experimental and computational research in all aspects of the low Reynolds number regimes, which includes, wind tunnel testing of the airfoils, developing more efficient computational schemes, numerical and analytical modeling of the laminar separation bubble (LSB) and dynamic stall vortex (DSV) [5, 6]. Thus, understanding the physics involved in the behavior of the dynamic stall characteristics in this Reynolds number regime would contribute to the various aerodynamic applications.

## 2 Experimental Methodology

Wind tunnel experiments were conducted on an oscillating, $\alpha = 10° + 15° \sin(\omega t)$ NACA-0015 airfoil model, having the chord length of 0.31 m spanning the test-section width of 0.305 m [7]. The geometric blockage is within 10% for the maximum angle of attack of 25°. The measured free-stream turbulence level in the 2D wind tunnel test section is within 0.15%. The designed airfoil model, as shown Fig. 1, incorporates simultaneous surface pressure measurement at the mid-span and PIV measurements of the flow-field downstream.

Other oscillating parameters which may affect the hysteresis behavior such as airfoil geometry, the amplitude of oscillation, the mean angle of incidence are kept unchanged.

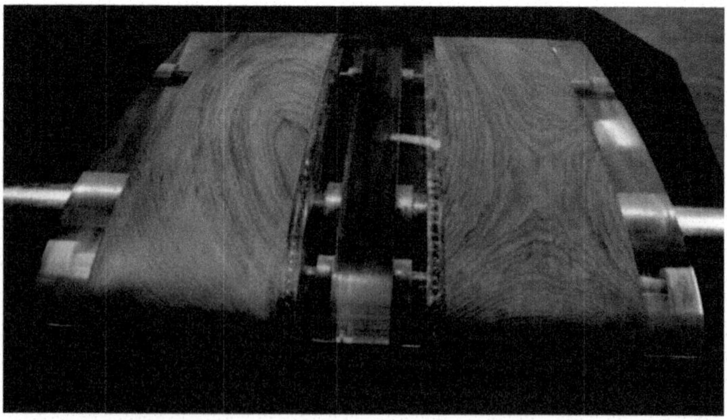

**Fig. 1** NACA0015 airfoil model for simultaneous Surface pressure and PIV measurements

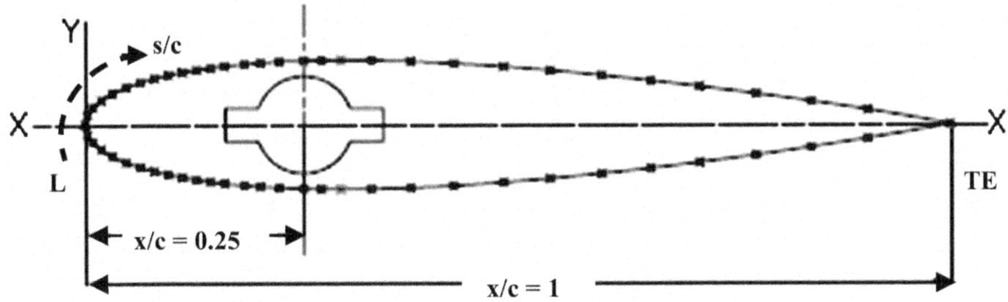


**Fig. 2** Pressure port locations opted for NACA0015 airfoil model

The surface pressure measurements at the mid-span along the chord length (c) are done using two piezo-resistive 32-port ESP pressure scanners from PSI, USA having 20 kHz multiplexing frequency and 0.07% of full scale accuracy. Uncertainty in the calculation of the surface pressure coefficient is about 0.1%. Schematic sketch of pressure port locations on the NACA airfoil model are shown in Fig. 2. The surface pressure tapings (#60) are unevenly distributed on the upper and lower surfaces such that the pressure taps were more clouded towards the LE as compared to the TE. This helps in precisely capturing the surface flow phenomena and boundary layer characteristics emerging from the LE [8, 9].

PIV measurements were carried out using the TSI's 2D PIV system. Figure 3 displays a schematic sketch for 2D-PIV acquisition for a field of View of 550 × 550 mm using a 4MP CCD camera. Flow was seeded with a fog generator. TSI

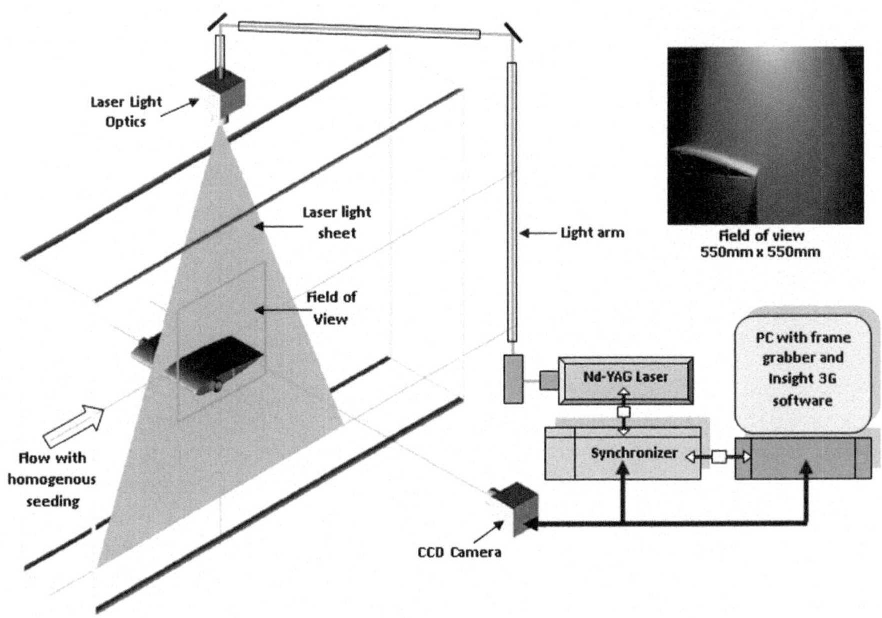


**Fig. 3** Measurement and instrumentation chain for 2D-PIV measurement

Insight 3G software was used for image data acquisition, processing and analysis. TecPlot software was used for the velocity vector field visualizations and plotting.

PIV is a non-intrusive, flow-field measurement technique, by which a velocity field is measured based on sequentially acquired images of illuminated flow-field within a known interval of time. Seeding particles are introduced into the fluid medium and these particles follow the fluid motion. Particles are illuminated by a thin sheet of light produced by a laser, and successive images of the illuminated seeding particles are recorded with a camera, from which average in-plane particle displacements are found. Velocity, vorticity, and streamline fields are then readily computed from the particle displacement information [10–13]. An error associated with estimation of the displacement of the particles in pixel unit is 0.1 pixels [14] because the Fast Fourier transform (FFT) was applied for correlation analysis. The inherent uncertainty is in the spatial location of the peak in the correlation function. This is widely accepted to be about 1/10 of a pixel, so by looking at the displacements in the PIV images (we can process with no calibration, and the units will be in pixels) we can determine the uncertainty. Displacements of about 10 pixels give an uncertainty of about 1%.

## 3 Results and Discussions

*PIV and Surface Pressure Analysis:* Comparative pressure and PIV analysis are carried out for the following selected cases [15]. PIV analyses are well supported by the time series of the coefficient of pressure $C_P$ distribution and the sectional aerodynamic characteristics drawn from the integration of phase-averaged surface $C_P$ distribution. Number of instantaneous frames of PIV images has been acquired for every oscillatory cycle of the motion. Certain instantaneous frames at varied angle of attack, covering a distance of $\approx 1.5c$ downstream from the TE, during the pitch-up and pitch-down motion are considered here for detailed analysis for the following cases.

### *3.1 Case-Study-1; κ = 0.01:* (*Figs. 4 and 5*)

Figure 4a and b represents the flow-field at pre-stall angles $\alpha = 12.23°$ and $14.74°$ for $\kappa = 0.01$. They show no sign for the inception of DSV during the pitch-up motion. The center of pressure is in the vicinity of quarter-chord line. Figure 4b indicates the $C_n$-surge point where flow reversal appears and the boundary layer begins to thicken at the rear portion of the airfoil followed by the moment stall. Center of pressure shows incremental jump and move downwards at the aft of quarter-chord line. Figure 4c at $\alpha = 17.08°$ shows the growth of wake in the downstream TE vicinity because of the development of TEV. The primary LEV (DSV) moves on the upper surface of the airfoil model towards the TE followed by the lift stall where the DSV completely sheds-off the TE. Figure 4d at $\alpha = 19.09°$ follows the lift stall where the large DSV is clearly visible downstream at the aft of TE enlarging the wake. It also triggers the movement of secondary vortex likes structure on the upper surface of the airfoil moving towards the TE. Further, the progressive DSV and TEV completely leaves the airfoil TE and move downstream. Although potentially both DSV and TEV tend to move down the wake in the downstream region, the rotation sense of both DSV and TEV are opposite. DSV

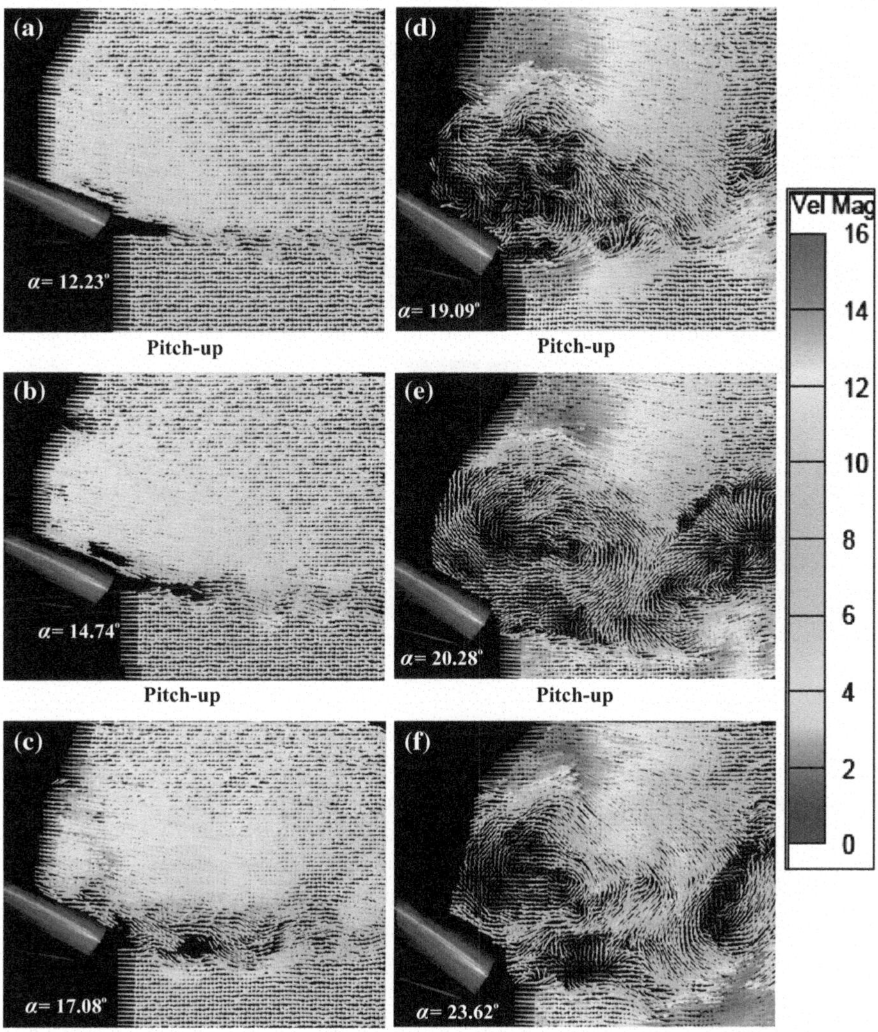


**Fig. 4** **a–f** PIV images at instantaneous angle of attack for $\kappa = 0.01$ at $Re = 2 \times 10^5$. **g–l** PIV images at instantaneous angle of attack for $\kappa = 0.01$ at $Re = 2 \times 10^5$

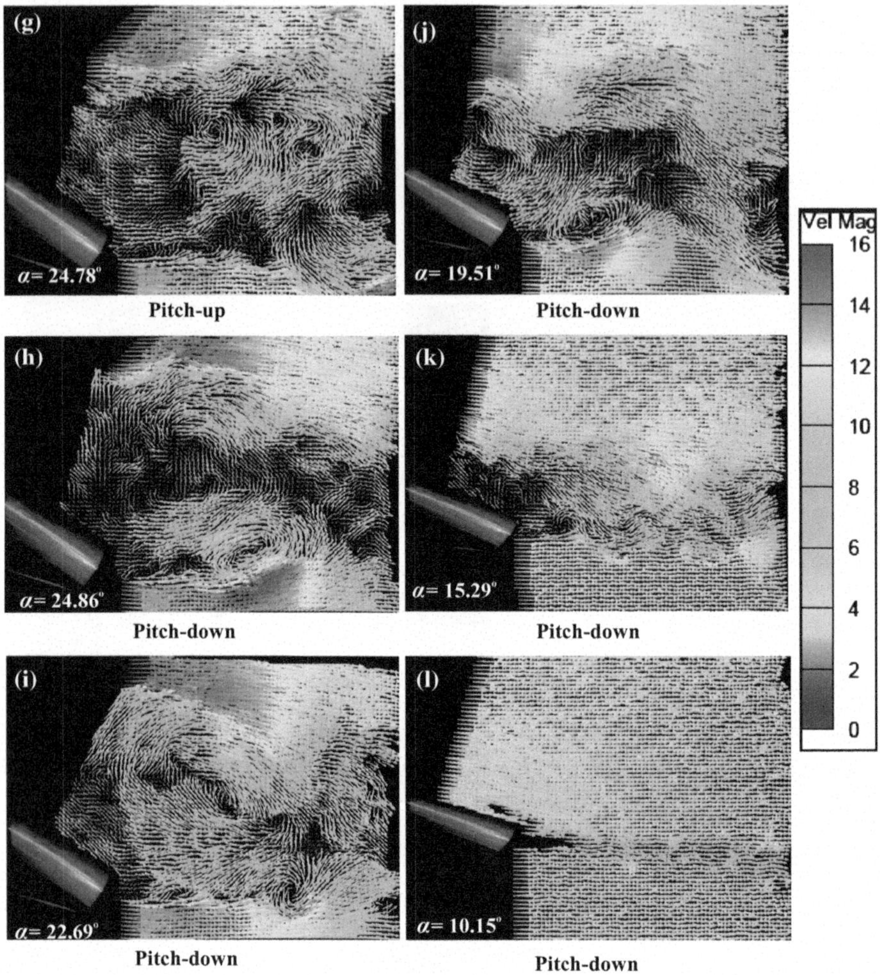


**Fig. 4** (continued)

shows clockwise rotation and TEV shows anticlockwise rotation, both advances towards downstream to form large wake. Even secondary vortices which followed the DSV, completely sheds-off the TE at point 'e' as represented in Fig. 4e at $\alpha = 20.28°$. The flow gets fully separated and at this location, the center of pressure gets paused ($X_{cp} \approx 0.38$) with no further incremental changes in its position. Fully separated flow in shown Fig. 4f at $\alpha = 23.62°$ where the DSV and TEV collides and creates mushroom-like flow structure in the downstream wake region.

The convective velocity of TEV seems to be lesser than the convective velocity of DSV. TEV also referred as starting vortex develops and progresses prior to the inception and growth of DSV. At an instant due to the differences in the strength

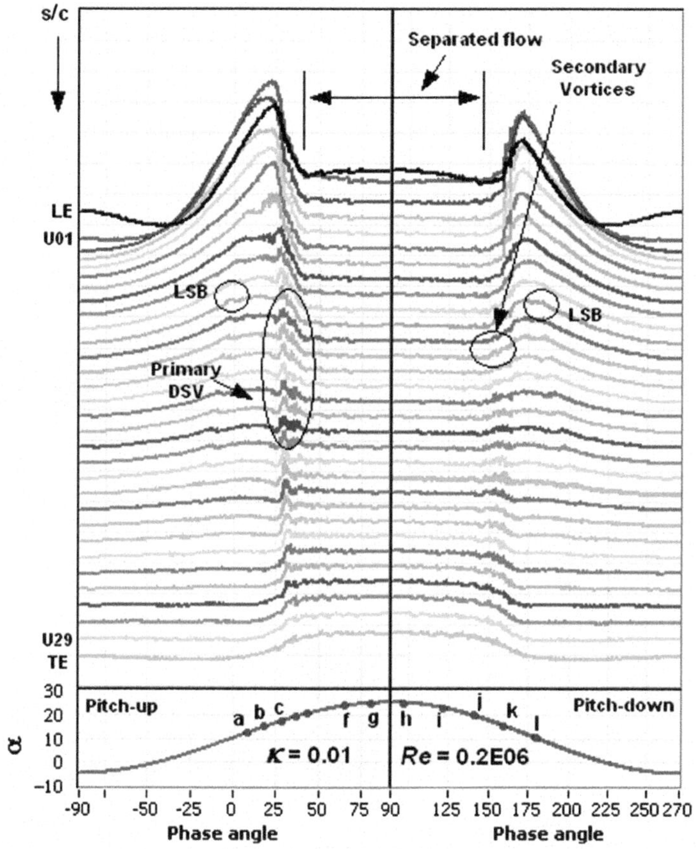


**Fig. 5** Time series $C_P$ distribution for $\kappa = 0.01$ at $Re = 2 \times 10^5$

and convective speed of DSV and TEV, causes the generation of mushroom-like flow structure. This structure gets perturbed further as the airfoil advances the maximum incidence at $\alpha = 24.78°$ shown in Fig. 4g.

Figure 4h at $\alpha = 24.86°$ indicates the start of pitch-down motion with large separated flows. As shown in Fig. 4i at $\alpha = 22.69°$, the flow still continues to be fully separated. For Fig. 4j at $\alpha = 19.51°$, the inception of secondary vortices occurs and provide low order fluctuations on the upper surface of the airfoil model. Figure 4k at $\alpha = 15.29°$ shows the initialization of the progressive reattachment process followed by the complete reattachment. Figure 4l at $\alpha = 10.15°$ shows the complete attached flow with thin wake in the downstream. The entire chronology of dynamic stall events are well captured in Fig. 4a–l. All the time line events from a–l are well placed on the time series $C_P$ distribution plot as shown in Fig. 5. At the downstream of the flow, the progressive motion within the wake is oscillatory in nature.

### 3.2 Case-Study-2; κ = 0.1: (Figs. 6 and 7)

For higher reduced frequency within the fully developed dynamic stall regime at $\kappa = 0.10$, hysteresis loop being large and PIV images captured are well within this loop except of the first image which is in the pre-stall region. Figure 6a shows a completely attached flow during the pitch-up motion at $\alpha = 16.18°$. Figure 6b at

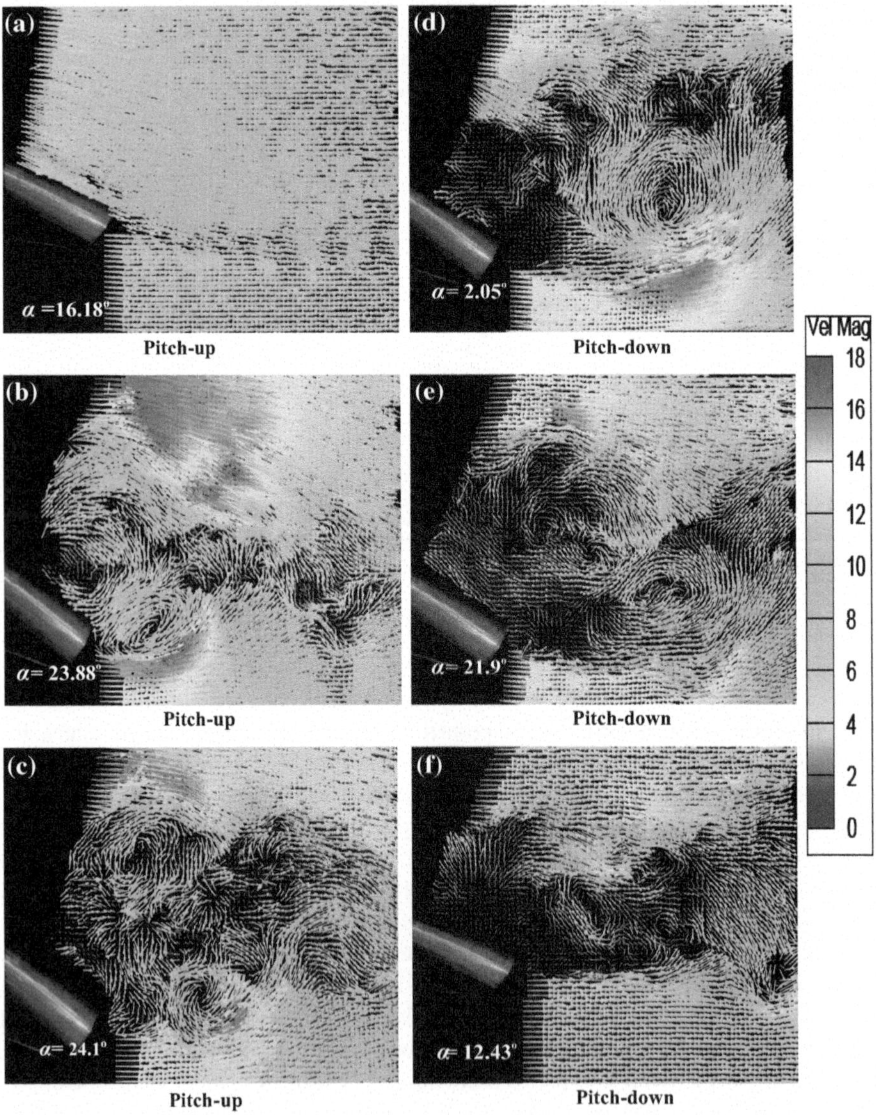


**Fig. 6 a–f** PIV images at instantaneous angle of attack for $\kappa = 0.10$ at $Re = 2 \times 10^5$

$\alpha = 23.8^{\circ}$ shows progression of DSV and TEV from TE towards the downstream wake. Further rise in the angle of attack at $\alpha = 24.1^{\circ}$ depicts some larger growth in wake as DSV and TEV separates away from each other as shown in Fig. 6c.

The airfoil reaches the maximum incidence and advances further for the pitch-down motion. Figures 6d and e shows the fully separated flows at $\alpha = 22.05^{\circ}$ and $\alpha = 21.9^{\circ}$ respectively. Further at point ‘f’ of Fig. 7, the entrainment energy gets introduced from the upper free-stream flow to initiate the reattachment process. As a result, the wake size reduces which is shown in Fig. 6f at $\alpha = 12.44^{\circ}$. Further rise in pitch-down motion leads to the re-introduction of LSB, which collides with the secondary vortices from the upper surface, to track the transition path and to form complete reattached flow. Due to significant rise in hysteresis, comparative adverse changes are also observed in the center of pressure during the pitch-down motion prior to reattachment.

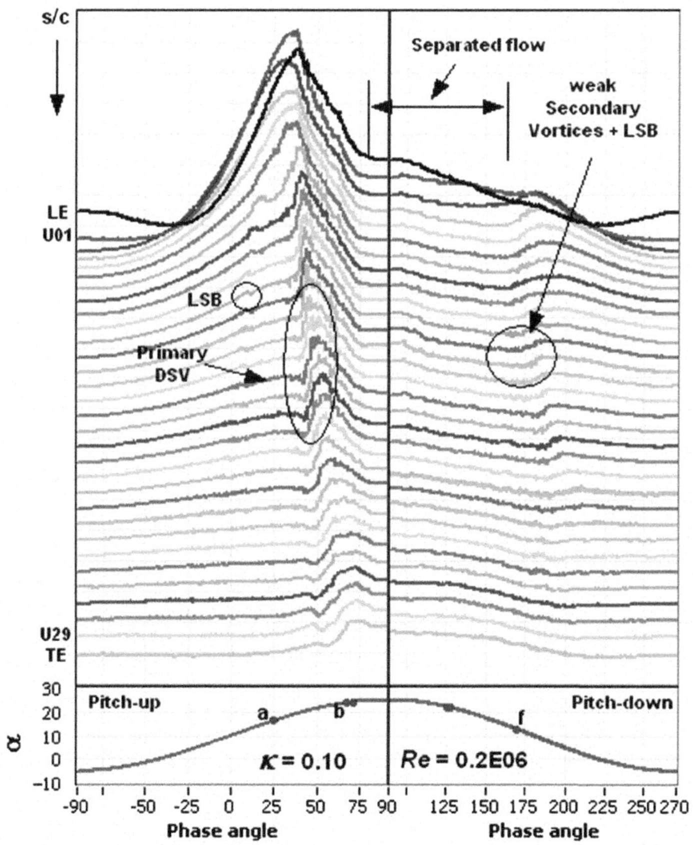


**Fig. 7** Time series $C_P$ distribution for $\kappa = 0.01$ at $Re = 2 \times 10^5$

# 4 Conclusion

PIV results demonstrate qualitative flow visualization of various dynamic stall chronology at varied reduced frequencies for a given *Re*. Distinct flow structures of DSV, TEV, mushroom-like flow structures, flow separation, flow reversal, wakes etc. are clearly depicted for these PIV images. It helped in understanding the dynamic stall phenomena to a greater extent along with the sectional aerodynamic characteristics derived from the surface pressure distribution measured simultaneously along with the PIV image capturing.

For a given constant $Re = 2 \times 10^5$, it has been observed that the change in reduced frequency from 0.01 to 0.10, causes negligible change in the onset of LSB. Increase in reduced frequency only attempts to make the LSB weaker.

Reflection issues on the upper surface of airfoil due to laser light sheet persist to a certain extent which fails to resolve the near-surface information more distinctly as compared to the surrounding flow-field.

Surface pressure and PIV measurements help in resolving the local and flow-field characteristics surrounding the oscillating airfoil but only limited to the mid-span (sectional domain) where the flow has a closer resemblance to 2D configurations.

**Acknowledgements** We convey our sincere thanks to AR&DB (Aeronautical Research & Development Board), INDIA for funding the project and acknowledge the support for faculty members, Technical and administrative staff at the Department of Aerospace Engineering (IIT Kanpur) and National Wind Tunnel Facility (IIT Kanpur). The in time technical support provided from TSI Inc. USA is also appreciated.

# References

1. Carr LW, McAlister KW, McCroskey WJ (1977) Analysis of the development of dynamic stall based on oscillating airfoil experiments. NASA TN D-8382
2. McCroskey WJ (1981) The phenomenon of dynamic stall. NASA TM 81264
3. Shih C, Lourenco L, Van Dommelen L, Krothapalli A (1992) Unsteady flow past an airfoil pitching at a constant rate. AIAA J 30(5):1153–1161
4. Shih C, Lourenco LM, Krothapalli A (1995) Investigation of flow at leading and trailing edges of pitching-up airfoil. AIAA J 33(8):1369–1376
5. Muti Lin JC, Pauley LL (1996) Low-Reynolds number separation on an airfoil. AIAA J 34(8):1570–1577
6. Lee T, Gerontakos P (2004) Investigation of flow ver an oscillating airfoil. J Fluid Mech 512:313–341
7. Abbott IH (1958) Theory of wing sections. Dover Publications, New York
8. Holm R, Gustavsson J (1999) A PIV study of separated flow around a 2D airfoil at high angles of attack in a low speed wind tunnel. FFA TN 1999-52
9. Berton E, Favier D, Maresca C, Benyahia A (2002) Flow field visualizations around oscillating airfoils. LABM Laboratory, UMSR, Marseille, France
10. Tinar E, Cetiner O (2006) Acceleration data correlated with PIV images for self-induced vibrations of an airfoil. Exp Fluids 41(2):201–212

11. Ferreira CS, Kuik G (2009) Visualization by PIV of dynamic stall on a vertical axis wind turbine. Exp Fluids 46:97–108
12. Adrian RJ (1991) Particle-imaging techniques for experimental fluid-mechanics. Annu Rev Fluid Mech 23:261–304
13. Raffel M, Willert CE, Kompenhans J (1998) Particle image velocimetry—a practical guide. Springer, Berlin
14. Westerweel J (1997) Fundamentals of digital particle image velocimetry. Meas Sci Technol 8:379–1392
15. Sharma DM (2010) Experimental investigations of dynamic stall for an oscillating airfoil. PhD Thesis, IIT Kanpur

# Part IV
# Internal Combustion Engines

# Lasers and Optical Diagnostics for Next Generation IC Engine Development: Ushering New Era of Engine Development

**Avinash Kumar Agarwal and Akhilendra Pratap Singh**

**Abstract** Researchers are trying to improve engine efficiencies and emission characteristics in an extremely challenging environment. This demands complying with ever-increasingly tightening emission norms and fuel economy norms globally, which are the prime forces driving the automotive industry globally. Optical diagnostic techniques have emerged as a valuable tool in development of engine powertrain and emission control technologies in last one decade. This chapter deals with variety of optical diagnostic techniques such as particle image velocimetry (PIV), phase Doppler interferometry (PDI) and endoscopy, applied to internal combustion (IC) engines. Laser based measurements in real time in the engine combustion chamber have ushered a new era of research for finding answers to some of the most intriguing in-cylinder processes, which were only speculated until recently. These optical diagnostic techniques have enabled engine researchers to sharpen their modeling tools using experimental data from realistic geometry engines firing under varying loads. Fuel-air mixing and combustion are mainly affected by in-cylinder air-flow and fuel spray characteristics. Air-flow structures developed inside the engine combustion chamber significantly influence the fuel-air mixing process. For validate this phenomenon, researchers investigated in-cylinder air-flow characteristics of motored engine using time-resolved tomographic particle imaging velocimetry (TPIV). Fuel spray characteristics could be determined using PDI, which provides information about droplet size distribution and 3D-velocity distributions. To gain visual access to the combustion process in a production engine's combustion chamber at high loads, another optical diagnostic technique "engine endoscopy" can be used, which can enable spatial combustion visualization, soot distribution as well as temperature distribution.

**Keywords** In-cylinder flow-field · Fuel sprays · Phase doppler interferometry · Engine endoscopy · Flame-temperature distribution · Soot distribution

A.K. Agarwal (✉) · A.P. Singh
Indian Institute of Technology Kanpur, Kanpur, India



A.K. Agarwal et al. (eds.), *Combustion for Power Generation and Transportation*, DOI 10.1007/978-981-10-3785-6_11

# 1 Introduction to Optical Diagnostics

Internal combustion (IC) engines are an important technology, affecting human life for more than a century. Reliability and efficiency improvement in IC engine were the two domains of improvement until the 20th century however in the 21st century the focus has shifted to exhaust gas emissions control and fuel consumption reduction. Increasing concerns about the environmental degradation and human health due to the harmful gases emitted from IC engines have imposed relatively stricter emission regulations for automotive manufactures. In 1970, Clean Air Act (1970) of the EPA started the new paradigm for control of emission levels, which became more stringent in successive norms. Numerous techniques such as turbocharging, exhaust gas recirculation (EGR), high speed direct injection (HSDI), gasoline direct injection (GDI), low temperature combustion (LTC) and after-treatment devices have been developed and implemented in IC engines in order to optimize the combustion and reduce emission. However, these techniques have their own limitations. Many researchers focused on in-cylinder conditions to improve combustion by controlling the fuel injection parameters and in-cylinder flows. They looked at injection parameters such as fuel injection pressure (FIP) and start of injection (SoI) timings along with injector nozzle configuration, which directly influence the spray characteristics such as spray tip penetration, spray cone angle, droplet size distribution, etc. These spray characteristics play an important role in fuel-air mixing and affect engine combustion, performance and emissions characteristics.

Therefore detailed investigation of in-cylinder characteristics becomes important for further improvement of IC engine performance and emissions reduction. Over the years, different optical diagnostics techniques have been developed to visualize and understand the in-cylinder phenomenon of IC engines in which laser based techniques such as particle image velocimetry (PIV) and phase Doppler interferometry (PDI) are the most important ones. These methods are well-known for their high temporal and spatial resolution. In most of these methods, the measurements are based on light and molecule interactions. This interaction generates a signal, which depends on the molecule's state and their quantity. Generated signal is acquired and processed to obtain specific in-cylinder information. Main difficulty with these methods in IC engines is to gain optical access. In the early years of research, probes were used to gain insights into the combustion chamber. However probes themselves may affect/ alter the actual physical condition in the measurement volume, leading to unreliable results. Use of existing ports, small optical windows, endoscopes and transparent cylinder liners has made these optical techniques possible for IC engine diagnostics. In IC engines, in-cylinder processes are mainly controlled by fluid flows such as air-flow, fuel-air mixing, flow of combustion products, etc. Recent upgrades in computing power have established Computational Fluid Dynamics (CFD) as a key design tool in fluid mechanics. Most CFD models are close approximation to a given fluid flow, but they generally require experimental validation because these flows cannot be simulated without appropriate boundary conditions. Combustion in IC engines is far from fully understood and there are complete lack of understanding about formation of pollutants and oxidation within the engine combustion chamber, which needs to be better understood. Considering the pollutant formation inside the combustion chamber, the fuel-air mixing is of paramount importance for emission reduction. Optimized fuel-air mixing is fundamentally dependent on the charge air motion inside the engine cylinder. Therefore the focus is set on understanding the connections between spray process and the formation of pollutants such as soot particles and unburned hydrocarbons using optical diagnostics. It also becomes important to study and understand the air-flow behavior inside the combustion chamber of a modern diesel engine. Engines modified for optical studies, allowing optical access to the combustion chamber have been used in different investigations, and these have been presented in this chapter. This chapter is focused on understanding different in-cylinder processes such as spray diagnostics, air-flow characteristics, fuel-air mixing, combustion and emission formation.

## 2 Air-Flow Characterization

In-cylinder air-flow of an engine is mainly governed by the intake manifold, valve design and piston geometry. Air induced in the intake stroke of an engine cycle interacts with the piston and cylinder walls to form complex 3-D flow structures. These flow structures affect the fuel-air mixing, fuel distribution and heat transfer from the engine cylinder, which directly influences the engine performance. These in-cylinder flow structures can be controlled by several parameters such as inlet manifold design, cylinder head and piston crown geometry. Different geometries such as helical intake port, directed intake port, flat-crown piston and concave crown piston have been explored to improve the air flow. Engine speed and piston position also strongly affect in-cylinder flow characteristics and it is reported that large flow structures formed in the intake stroke distort and breaks into smaller structures in later stages of the compression stroke, which enhances turbulence at the time of fuel injection [27]. Figure 1 shows that average velocity of inlet air is significantly higher at higher engine speeds, however during the compression stroke, the difference between average air velocities at different engine speeds reduced.

Swirl and tumble are commonly known flow structures generated inside the cylinder. These flow structures are vortices with axis of rotation being parallel and perpendicular to the cylinder axis respectively. Swirl is believed to survive the compression stroke with increasing angular velocity, while tumble is known to break into smaller flow structures of high turbulence intensity near TDC [34]. Combination of these flow patterns produces complex and highly turbulent flow structures. These flow structures lead to more uniform fuel distribution, resulting in leaner engine operation. However, this can lead to misfire in case of SI engine because bulk motion causes charge to redistribute away from the spark plug. For CI

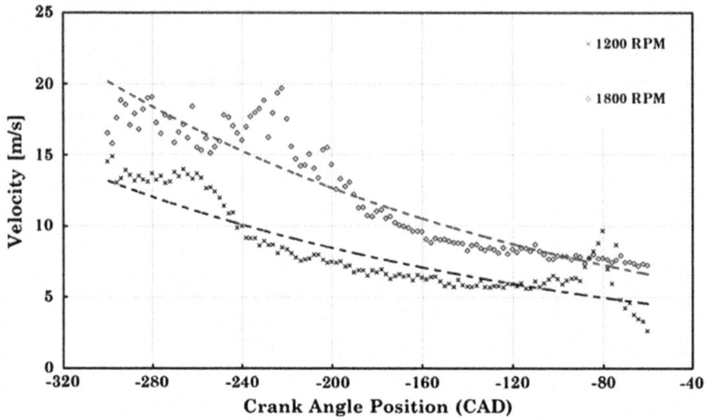


**Fig. 1** Average velocity variation at 1200 and 1800 rpm during intake and compression stroke [27]

engines, turbulence is beneficial to a certain extent because current engine technology requires different level of turbulence according to their operating conditions. Implementation of different engine modifications and techniques to enhance the air motion result in decreased volumetric efficiency. Therefore it becomes important to quantify the in-cylinder flows.

## 2.1 Air-Flow Characterization Techniques

Different methods have been used by researchers to measure and visualize the in-cylinder air-flow characteristics. Hot wire anemometry (HWA), laser Doppler anemometry (LDA) and PIV are widely known methods for flow measurements. HWA is an old method used for studying laminar, transitional and turbulent flows. This is an intrusive velocity measurement technique, which provides flow information at a single point in the flow-field. Subramanian et al. [86] used HWA in a high compression diesel engine to obtain quantitative flow information. They analyzed the effects of engine speed and compression ratio on mean velocity and turbulence intensity. They used DISA 55M10 CTA bridge in combination with a DISA 55A75 high temperature hot-wire probe for velocity measurement at four positions inside the piston cavity. Figure 2 shows the positions of probes inside the engine combustion chamber.

LDV is another non-intrusive velocity measurement technique, which gives velocity information at a single point of interest. In this technique, two coherent laser beams are made to intersect at the point of measurement to form a fringe pattern. The tracer particles in the seeded flow pass through the fringe pattern and scatter light. The scattered light is received by a photomultiplier. The frequency of the scattered light depends on the particle velocity. By comparing changes in the frequency of the incident and the scattered light, the particle velocity is calculated.

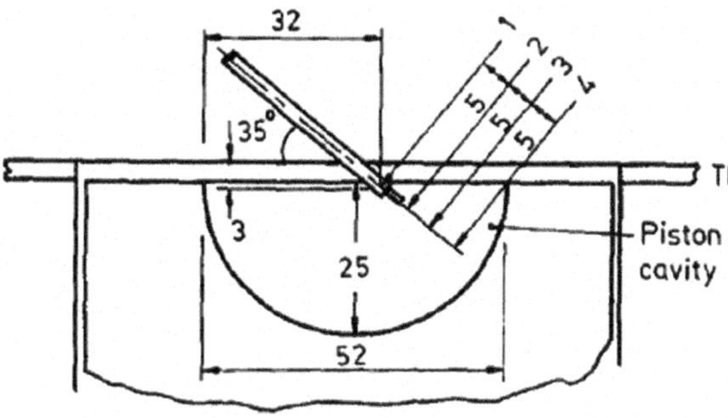


**Fig. 2** HWA probe positions for in-cylinder velocity measurement [88]

Chan [14] applied LDV for 2-component and 3-component velocity measurements in a motored SI engine. For 3-component velocity measurement, all six laser beams were guided into the cylinder volume through a semi-circular window incorporated in the cylinder head. The axial, radial and tangential velocity components were measured at different crank angle positions (Fig. 3).

They also described the difficulties in applying LDV in IC engines. In spite of advantages such as non-intrusive to the flow, directional sensitivity and high temporal resolution, LDV technique deals with only a small area in the flow-field. Thus volumetric flow visualization of large area becomes difficult for LDV technique, which is a severe limitation. PIV is a non-intrusive volumetric flow visualization method, which resolved difficulties experienced in conventional

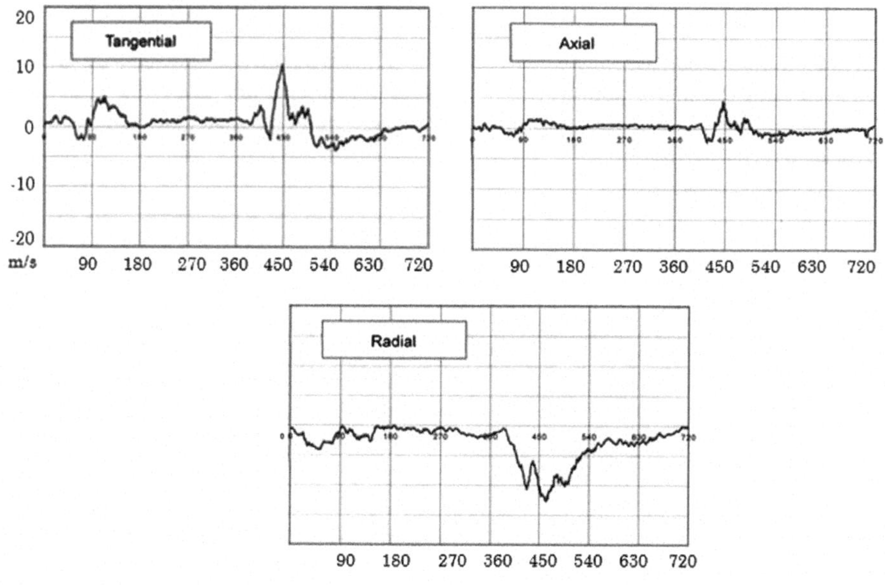


**Fig. 3** Three component velocity data (Tangential, axial and radial) using LDV [14]

techniques and emerged as an effective flow measurement technique. Tomographic particle imaging velocimetry (TPIV) can be used to investigate the in-cylinder flow characteristics within a volume.

## 2.2 Particle Imaging Velocimetry (PIV)

PIV is an optical, indirect flow visualization and measurement technique, in which illuminated tracer particles are captured on a photographic film or digital camera and captured images are further processed to obtain instantaneous velocity field. In this visualization technique, flow-field is seeded with small tracer particles, which are believed to faithfully follow the flow. The area of interest in the flow-field is illuminated by a double pulsed, high intensity laser sheet at two instant with small time lag between them. Light scattered by the tracer particles is captured by cameras, usually placed at 90° to the illumination plane. Triggering of these cameras is synchronized with laser pulses using a external synchronizer. For the flow-field acquisition, cameras can be operated either in single frame mode (Flow-field for consecutive illuminations is captured on a single frame) or double frame mode (separate frames are used for each illumination). Acquired recordings are post-processed by dividing each image into small interrogation regions, which are auto-correlated or cross-correlated to obtain average displacement vectors of tracer particles in photographic coordinates. These camera scale displacements are converted into global coordinates by using a calibration process. Further, with known time separation between consecutive laser illuminations, velocity vectors in the area of interest are calculated. Depending on the experimental requirements and desired data from the flow-field, the nature and fundamental requirements of PIV setup changes. These requirements include seeder material, laser pulse energy, pulse separation, laser thickness and camera settings. A general schematic of PIV setup is shown in Fig. 4.

## 2.3 PIV Configurations

Different arrangements, such as 2-D PIV, time-resolved PIV, stereo PIV, holographic PIV and TPIV are developed with the passage of time for 2-component and 3-component velocity measurements either in a plane or volume. It is possible to perform TPIV measurements using 2 camera projections [55]. Various researchers implemented PIV for flow analysis in diverse experiments [7, 8, 81, 83, 96, 98]. Wernert et al. [95] applied the PIV for flow-field analysis around an aero foil. Kawaguchi et al. [45] studied the spatial velocity distribution of drag reducing surfactant in a 2-D channel the PIV technique. Reeves et al. [69] applied PIV in a four-valve, optical engine to study large-scale flow structures. Depending on the experimental setup, there are numerous variations in PIV arrangement.

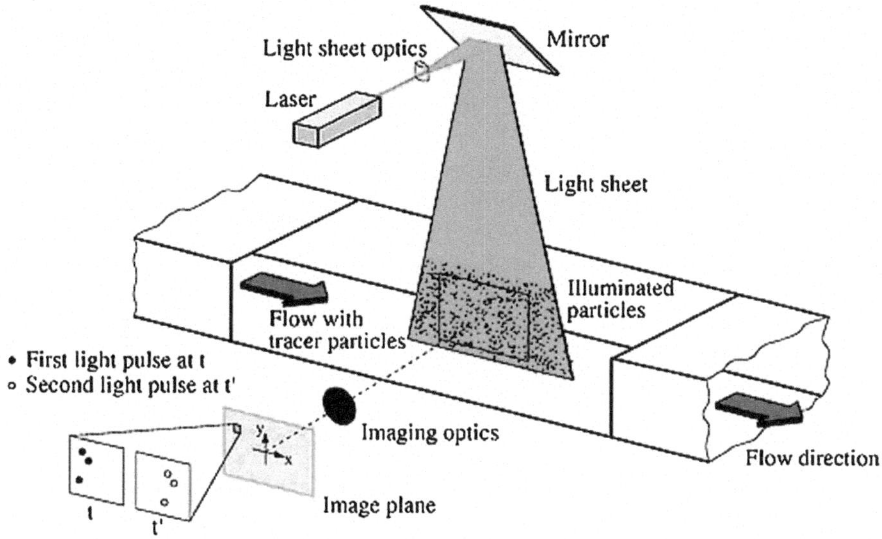


**Fig. 4** Typical PIV setup for flow velocity measurements [37]

2-D PIV (2D-2C) is the most common PIV arrangement used for planar flow measurements and visualization. In this technique, the illuminated 2-D flow-field is captured on two consecutive images. These images are further auto-correlated or cross-correlated to obtain the flow velocity information. This technique gives 2-components of the planar velocity vectors. Time-resolved PIV is similar to 2-D PIV, except that instead of a single pair of images, it takes a number of image pairs, thus providing temporal development of the flow-field. The limitation of this technique is the acquisition speed of the camera, i.e. the time delay between the successive image pair capture. With rapid development in camera technology, the temporal resolution of PIV technique is expected to increase. Stereoscopic PIV (2D-3C) removes the inability of 2-D PIV to measure the 3rd, out-of-plane component of the flow velocity. It uses 2 cameras with different viewing directions to capture the illuminated flow-field. The three-components of velocity vectors are calculated by using images from both cameras. For calculation of 3rd velocity component, both these cameras are required to focus on the same plane in the acquisition flow-field.

Holographic PIV (3D-3C) allows 3-D measurement of all three velocity components. In this technique, the positions of tracer particles are recorded on the holographic plate by making the scattered light of seeders to interfere with the reference laser beam. This generates a hologram, which can be used to reconstruct the flow-field by illuminating it with the original reference beam. This reconstructed flow-field is interrogated using 3-D cross-correlation to obtain velocity field. Despite advantages 3D-3C velocity measurement, issues such as distortions in reconstructing the flow-field, speckle noise, coherent imaging degrades the holographic PIV results. TPIV is another 3D-3C velocity measurement technique. Velocity measurement principle used in TPIV is same as 2D PIV, except the fact that the flow-field projections are captured by several cameras. These projections are further used to reconstruct the 3-D flow-field. This 3-D flow-field is analyzed by 3-D cross-correlation, in order to calculate the 3-D three component velocity field. Foundation of TPIV is based on algorithm used to reconstruct the 3-D field. Generally TPIV uses the "Multiplicative Algebraic Reconstruction Technique" (MART) to reconstruct three dimensional flow-field. TPIV technique with 2 cameras has been implemented in this study for in-cylinder flow analysis in a CI engine.

## 2.4 Application of PIV

In the early years of PIV development for IC engines, it was difficult to gain the in-cylinder access while keeping the modified engine geometry and working close to the actual engine. Different techniques such as endoscopy, small optical windows, engine head ports and valves were used to visualize the engine flows. Sweetland and Reitz [89] applied the planar PIV technique to measure gas velocity and turbulence inside the piston bowl of a motored diesel engine. In this study, in-cylinder optical access was gained through the fuel injector port and the exhaust valve. The seeder particles of $TiO_2$ were illuminated using a Nd:YAG laser, which was guided through the injector port. The flow images were captured on photographic film for flow visualization. To validate PIV measurements, the acquired turbulent jet data were compared with the measurements made using conventional techniques and the simulation results using KIVA-3 code. Though the measurement was affected by the out-of-plane velocity component, it was inferred that the high and low estimates of turbulence intensity can be obtained from these limited PIV measurements. Flows inside a specific geometry can also be visualized by using the endoscopic technique. Nishiyama et al. [61] applied endoscopic stereo PIV to study the exhaust gas flow in the exhaust manifold of an SI engine. It is believed that the impingement of exhaust flow on the oxygen sensor and the catalyst affected the engine performance. A special twin-endoscope and optical window arrangement was developed to investigate the exhaust gas flow. The developed endoscopic stereo PIV system was tested on a steady flow rig and the results were compared with HWA, LDV and 2-D PIV measurements.

In the early days of development of PIV, photographic films were used for image acquisition and auto-correlation was used for image evaluation, which resulted in directional ambiguity. Reeves et al. [69] applied auto-correlation based 2-D PIV for in-cylinder flow measurements and visualization in a SI engine. They derived the velocity vector maps in vertical and horizontal planes for different crank angle positions. Their study discussed the difficulties experienced in implementing PIV in the engines. The flow images of olive oil droplets, illuminated by Nd:YAG laser were acquired on a photographic film. In-cylinder flow visualization revealed breakdown of large-scale flow patterns into homogeneous micro-turbulence, due to distortion and shear as the cycle progressed from the intake to the compression stroke. To remove the directional ambiguity of the velocity data, different techniques were used. Stucky et al. [85] used two-color planar PIV to remove this kind of ambiguity in the velocity data from the engines. This study analyzed implementation of planar PIV in IC engines and reported the advantages of cross-correlation over auto-correlation technique using computer simulations. They also investigated the importance of cross-correlation over auto-correlation by analyzing the effects of seed density on signal-to-noise ratio of auto and cross-correlation techniques. They found significantly higher signal-to-noise ratio for cross-correlation compared to auto-correlation (Fig. 5).

Developments in digital photography and computational power changed the method of image acquisition from conventional photographic film to digital cameras. Reeves et al. [69] showed the application of 2-D digital PIV in an engine, which allowed acquisition of thousands of flow visualization sequences and velocities at different crank angle positions. Their results showed that swirl center drifted and flow patterns changed significantly with crank angle. The observed flow structures were further studied to understand the cyclic variability. Though the recorded digital PIV data was of reduced accuracy and lower spatial resolution, it provided the quantification of major flow structures inside the engine cylinder.

Significant effects of in-cylinder flows on engine performance and emissions have led researchers to improve the engine flow patterns by changing the engine design. The in-cylinder flow-field is mostly affected by the intake manifold geometry, valve and piston shapes. Some of the flow-field affecting designs are helical inlet, directed inlet, shrouded valve, masked valve, deep and shallow bowl piston, flat crown and concave piston. Intake valves being the only energy source for in-cylinder flows, analysis of flows near the intake valve during its opening and closing is of considerable importance. Lee and Farrell [52] investigated the two dimensional flow-field near the intake valves using 2-D PIV, and studied the unsteady effects of the cam-speed and valve lift on flows near intake valve. The flow measurements were obtained at valve half open (during opening), fully open intake valve, valve half open (during closing) and after full valve closure. Their

**Fig. 5** Effect of seed density on signal-to-noise ratio [87]

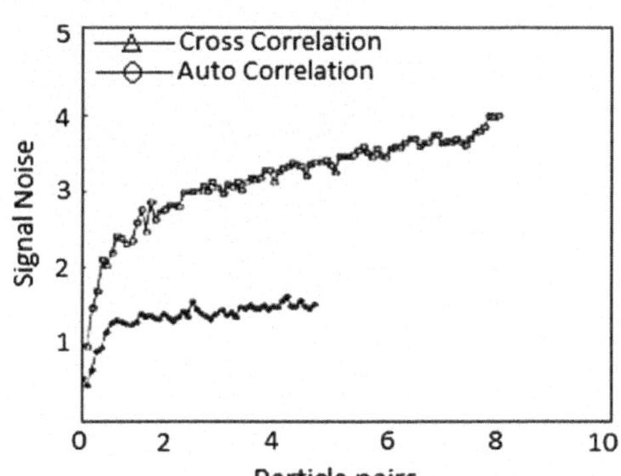

results showed that the intake jet velocity increased with increasing cam angle. The comparison of dynamic and static valve flow at same valve lift and pressure drop across the intake valve showed that the static valve flow has higher velocity than the dynamic valve. The flow study also showed the cyclic variations in flow patterns and these variations were reduced with high intake jet velocity. Valentoni et al. [91] extended this research to obtain better 3-D flow patterns around the valve. They analyzed the flow in a certain valve area. Flow patterns from these measurement planes showed the development of wall jet type of flow on the backside of the valve. This study also showed that in case of two intake valves, flow from one of the intake valves dominated the initial flow. Figure 6 shows the velocity map of fully open inlet valve at 500 rpm.

Huang et al. [40] analyzed the inception, establishment, development and destruction of in-cylinder flow structures of a two-valve, motored engine with a flow deflection valve in the intake manifold. The flow structures were quantified by swirl and tumble ratio using PIV measurement. The flow data were correlated with the engine performance to show the effect flow has on the engine combustion. The deflection valve was found to generate high tumble and swirl structures inside the cylinder, while without inlet stream deflection, no swirl and tumble effects were observed. The valve deflection was also noticed to produce positive effects on the engine power and torque. Similar research was performed by Perini et al. [67] using a CFD model. The effects of variable throttle orientation in the intake manifold to generate swirl motion were obtained using KIVA code and the same were validated using PIV measurements. Justham et al. [44] studied the inter-relationship of intake manifold design and in-cylinder flow-fields by implementing digital PIV setup on a motored SI engine. They reported that the relationship between flow-field intake runner and cylinder was stronger during the intake stroke however it died out in compression stroke due to breakdown of large-scale motions. Large-scale bulk motions were affected by cyclic variations in the intake flow. Figure 7 shows that

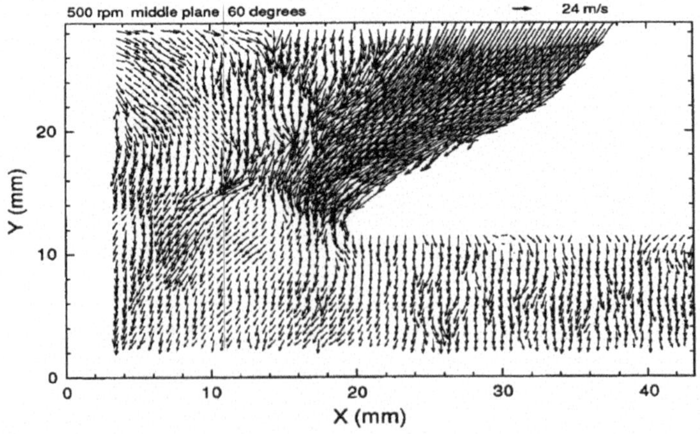


**Fig. 6** Velocity map in the center plane of inlet valve at 60° aTDC [52]

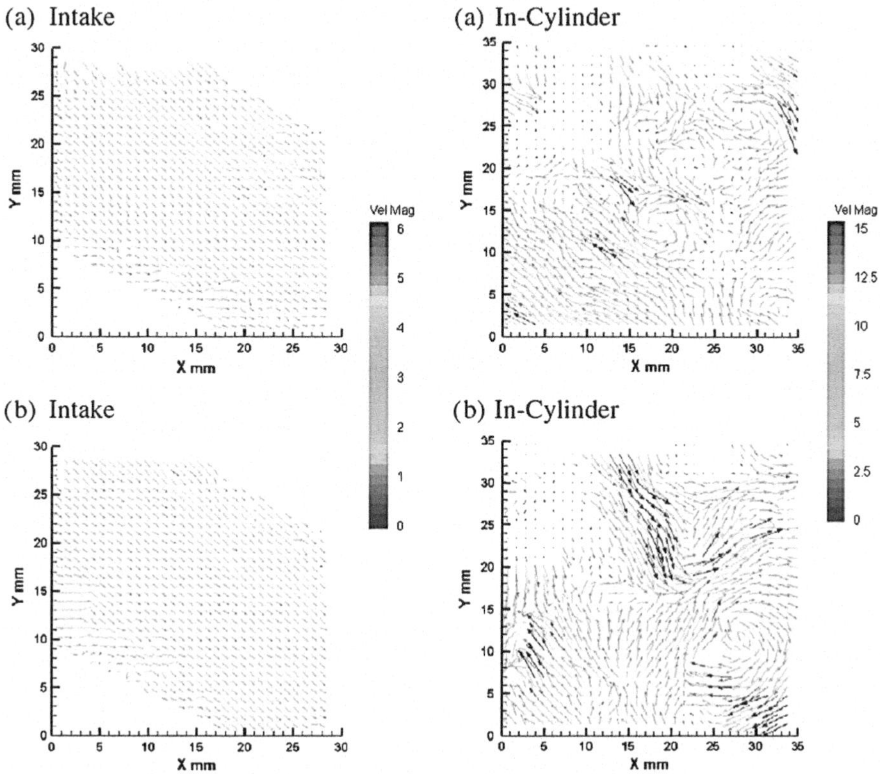


**Fig. 7** Comparison of intake runner flow structure with in-cylinder flows [44]

despite of similar intake runner flows, in-cylinder flow is considerably different, leading to cyclic variations.

Piston geometry also has a significant impact on the flow patterns in the compression stroke. PIV enables complete visualization of these flow patterns and their effects on engine combustion. Deslandes et al. [21] studied the in-cylinder aerodynamics of a diesel engine using planar PIV in a transparent single cylinder engine for two different piston configurations; bowl-in-piston and flat piston in the intake and compression strokes. In the intake stroke, the intake jet flow and cylinder walls produce large flow structures, which hit the piston to form small structures. Their results showed that the bowl-in-piston has centered swirl however it remained unsymmetrical. It was also observed that the experimentally obtained squish intensity was relatively weaker than the theoretical expectations.

It is also possible to study the fuel spray-air flow interaction using PIV. Sasaki et al. [76] studied the flow-field near non-evaporating diesel spray using PIV. They reported temporal variation of air flow around pattern the fuel spray, using PIV auto-correlation technique. It was observed that the fuel-air interference leads to a complicated spray surface. Major finding of this research was that the air flow ahead of spray tip has very low velocity compared to droplet velocity and the fuel-air interaction mainly happens in the mid-spray region and at the spray tip. Researchers observed that for detailed analysis of fuel-air interaction, detailed information about 3-D flow-field is required. Therefore different flow visualization techniques have been implemented to fully develop the complex 3-D flow patterns inside the engine cylinder. Implementation of planar PIV in different in-cylinder planes has emerged as an extensively used method for understanding 3-D flow structures. Dannemann et al. [18] implemented the planar PIV in different in-cylinder vertical planes to visualize 3-D flow-field evolution (Fig. 8). They analyzed the flow by studying the turbulent kinetic energy (TKE) and vorticity at 1500 rpm engine speed. Their results showed that the maximum TKE and its spatial variance decreased with increasing crank angle. Observations of in-cylinder flow-field suggest the existence of 3-D flow-fields, which call for the use of 3-D measurement techniques such as holographic and TPIV to completely analyze the in-cylinder flows.

Choi et al. [16] validated STAR-CD CFD code model by analyzing the in-cylinder flows using planar PIV measurements of a single cylinder diesel engine. The flow-field was visualized in the compression stroke at 600 rpm to show the streamlines, generation and the disappearance of tumble and squish. The study reported the difference of ~26% between experimental PIV and CFD results. For CFD code, both standard Reynolds k-e and RNG k-e turbulence model were used and it was shown that the standard Reynolds k-e model is more suitable for low engine speeds. In a similar study, Das and Chmiel [19] validated KIVA-3V code for in-cylinder flows in a GDI engine using the PIV technique. Experiments to study the flow-field were performed by PIV, while the computational study was done using RNG K-e turbulence model in KIVA-3V multi-dimensional code. Three planes of flow-field were analyzed by PIV; center plane, valve plane and swirl plane at 90°, 180°, 270° CAD aTDC. Results showed that the intake generated tumble motion weakened with a downward motion of the piston during the intake stroke and this tumble motion was enhanced with upward motion of the piston in the compression stroke. The discrepancies observed in the results of KIVA-3V code and PIV experiments were attributed to uneven valve seat surface.

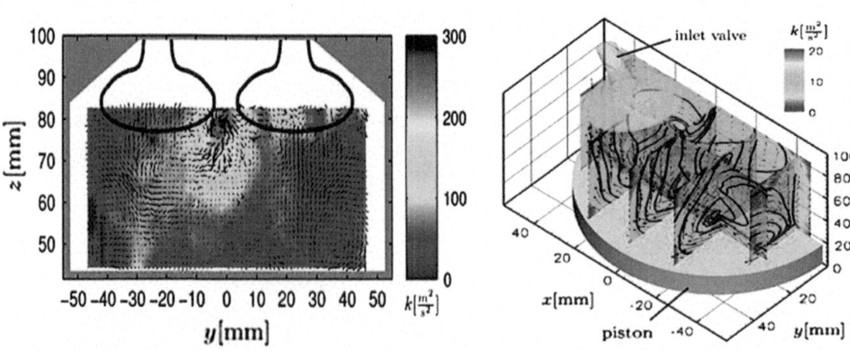


**Fig. 8** 2-D and 3-D TKE at 80° and 240° aTDC respectively [18]

Despite visualizing the flow evolution in different planes inside the engine cylinder, complete 3-D flow behavior cannot be fully understood till the volumetric flow-field is captured. Measurement techniques such as holographic PIV and TPIV enable complete visualization of the 3-D flow behavior. Baum et al. [10, 11] were one of the first researchers, who applied TPIV in IC engines. They investigated the 3-D flow-field of direct injection spark ignition (DISI) engines using TPIV. They used the principle of conservation of mass to address the uncertainty in 3-D velocity measurement and found the precision of 9% in the measured velocity data. This precision of velocity measurement is of the same order as volumetric flow measurement in a generic configuration indicating applicability of TPIV in IC engines. The flow-field analysis of the intake and compression stroke showed tumble formation and the velocity reduction from intake to compression (Fig. 9). The fluctuations in out-of-plane velocity component were analyzed by comparing the 2-D and 3-D TKE. Their results of 3-D-TKE at different crank angle positions showed the turbulence dissipation due to flow-cylinder wall interaction. Obtained flow-field was further analyzed by the Zentgraf et al. [97]. They studied the in-cylinder turbulence by analyzing the Reynolds-stress gradient tensor and anisotropic invariant tensor with homogeneous and isotropic turbulence as a reference. Because of the good agreement of RSGT data from TPIV and PIV, the authors were able to investigate the flow-field for larger field of view using planar PIV.

During the compression stroke near TDC, flow was affected by upward piston motion, leading to inhomogeneous turbulence. They reported that apart from crank angle position and engine speed, the flows also varied with their spatial position inside the cylinder. During the intake stroke, flow recirculation zone was approximated by isotropic turbulence levels, while intake-jet was approximated as anisotropic contraction and flow of the stagnation zone approximated as anisotropic expansion. These TPIV flow-field results were also used by Baum et al. [9] to validate the large eddy simulations developed for in-cylinder flows. Recently Singh et al. [80] applied TPIV for in-cylinder flow characteristics of an engine operated at different engine speeds, intake air temperatures and intake valve deactivation. The photograph of the experimental setup is shown in Fig. 10.

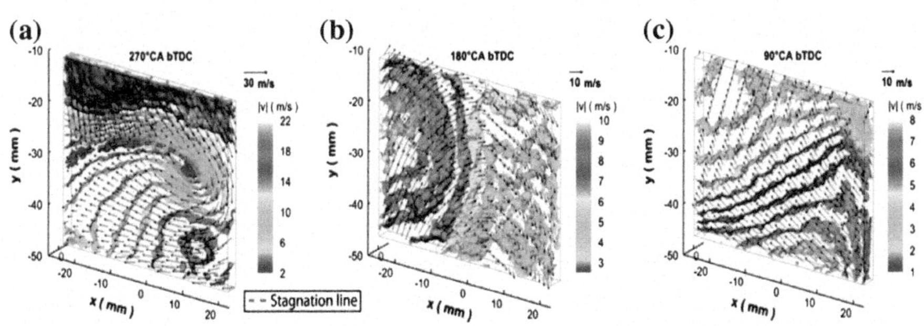


**Fig. 9** Velocity iso-surfaces at 270°, 180° and 90° bTDC at 800 rpm [10, 11]

**Fig. 10** Experimental setup for TPIV [80]

They reported that high velocities generated in the first half of the intake stroke started to dissipate in later stages of the intake stroke. A large part of flow energy of the intake stroke dissipated during the compression stroke. Therefore in the compression stroke, the flow became more homogeneous except locations near to the cylinder head. A higher rate of energy dissipation and velocity fluctuations were present in the intake stroke compared to the compression stroke. Higher piston speed and volumetric efficiency at higher engine speeds led to increased intake jet velocity and turbulence.

In another study by Singh et al. [81], flow evolution was investigated at different crank angle positions. They used TPIV and analyzed the flow behavior at different locations inside the combustion chamber. They concluded that flow behavior and different components of velocity were significantly affected by in-cylinder location. In similar study, Singh et al. [81] also investigated the effect of different intake port configurations, namely swirl port open and tangential port open configurations. They reported that swirl port open configuration provided highest average air velocity and vorticity, followed by both port open configuration. Tangential port configuration showed the highest rate of energy dissipation, which resulted in minimum absolute air velocity (Fig. 11).

## 3 Spray Characterization

Understanding fuel spray characteristics is very important for combustion in IC engines because it is an interface between fuel injection and combustion. It plays a very important role in fuel distribution, which directly affects fuel-air mixing. The main aim of spray investigation is to control the fuel injection system, and to deliver

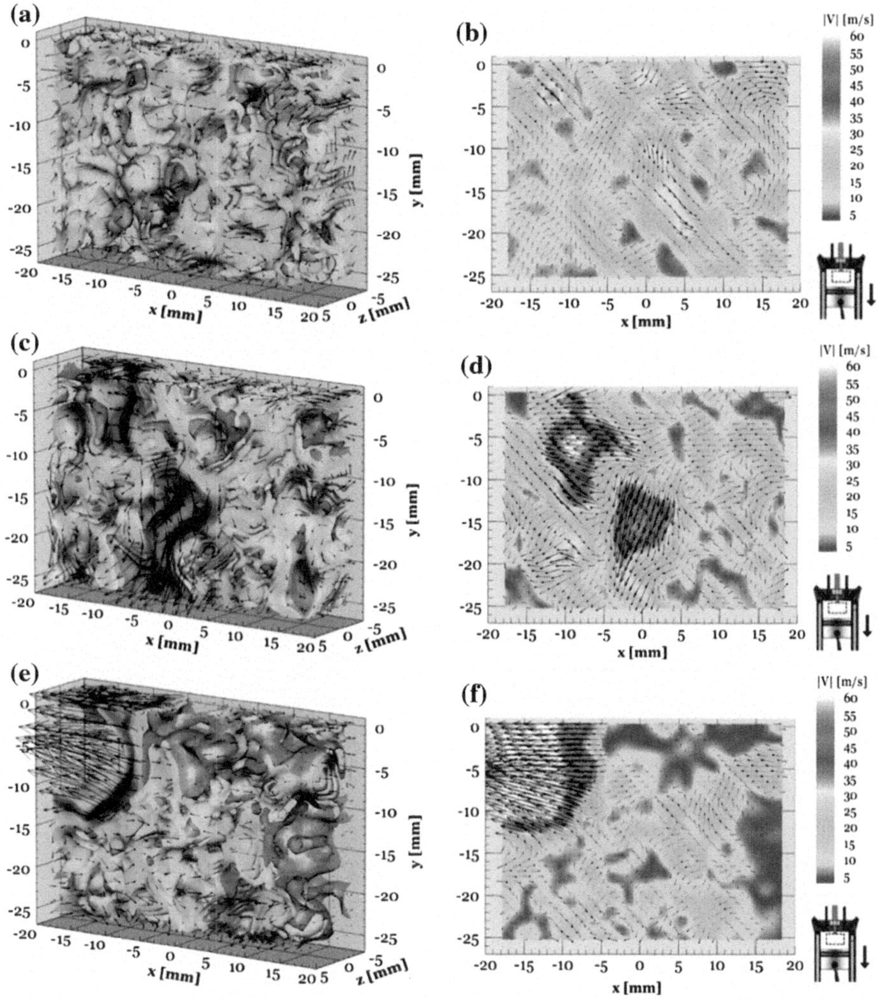


**Fig. 11** Ensemble velocity profile at mid-intake stroke at 1500 rpm using different port configuration as **a** and **b** both ports open; **c** and **d** swirl port open; **e** and **f** tangential port open [80]

precise fuel quantity for obtaining optimal engine performance and lower engine exhaust emissions. Fuel-air mixing and auto-ignition of the charge is highly dependent on spray formation process starting from start of injection (SoI) to full spray development. Spray development after fuel injection and fuel cut-off is very critical for formation of different pollutant species during combustion. The spray formed due to pressure difference between the spray holes of the injector nozzle and ambient. After leaving the nozzle, spray jet becomes turbulent and partially mix with the soundings air. It flows away from the nozzle and the spray velocity

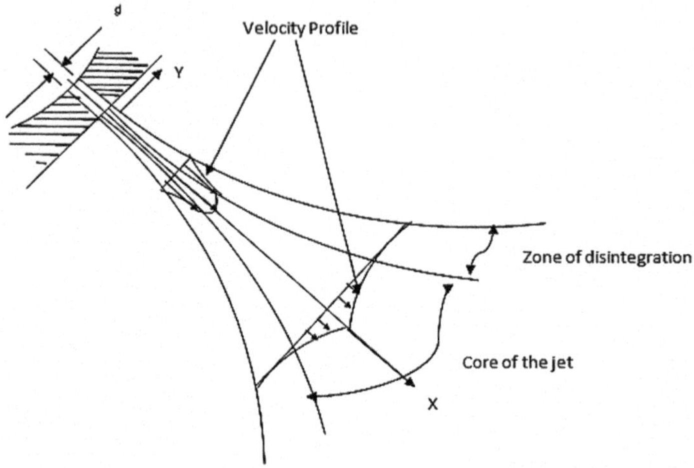


**Fig. 12** Formation of fuel spray from a single hole [33]

increases in X-direction. At the same time, it also spreads in Y-direction, which reduces the jet velocity in X-direction due to momentum conservation. In the X-direction, jet velocity further decreases due to frictional drag forces [35, 36]. Hence the fuel spray structure can be controlled by optimizing various microscopic and macroscopic parameters and spray atomization characteristics. In an engine, these characteristics can easily be determined using optical visualization technique such as endoscope however it needs high intensity lighting arrangement. Endoscopes are especially designed for low light applications. They are coupled with an optical fiber to illuminate the chamber during spray imaging. Using image analysis methods, these characteristics can be determined and compared for different operating conditions (Fig. 12).

## *3.1 Macroscopic Spray Characterization*

Macroscopic spray characteristics consist of several geometrical spray parameters, which control the shape of the spray and help in optimizing the spray development process. In spray images, geometrical parameter can be measured by image analysis and can be optimized by varying fuel injection parameters. These parameters are as follows (Fig. 13).

- Spray breakup length
- Spray tip penetration
- Spray cone angle
- Spray area

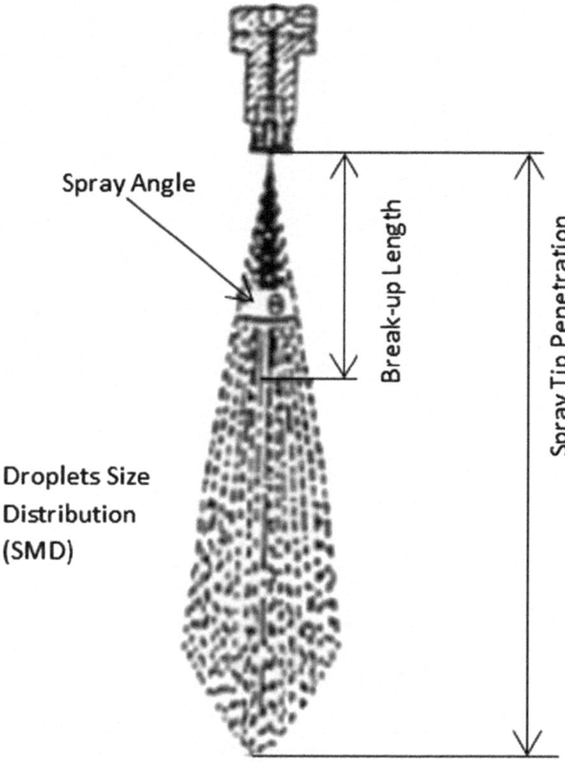


**Fig. 13** Spray parameters [39]

### Spray Breakup Length

It is defined as the length of the spray from the nozzle tip to point of discontinuity, where spray changes from dense packed zone of bulk liquid to finely atomized droplets. These droplets move in surrounding air so they have some relative velocity. This develops into non-uniform pressure wave, which further breaks the droplets into smaller droplets due to deformation. The detachment of liquid core to larger droplet is primary breakup and its further breaks up into smaller droplets is called secondary breakup [43]. Therefore, droplet density is the highest at the spray breakup point. These small droplets of liquid fuel reflect the illuminating light, while capturing the images. Hence higher density of droplets reflects higher illumination light rays. Therefore in spray images, breakup length can be measured as a distance from the nozzle tip to the maximum intensity point along the middle line of the spray (Fig. 14).

### Spray Tip Penetration

Spray tip penetration (S) is defined as the maximum length from the nozzle tip to spray tip at a given time. In spray images, this length can easily be measured and

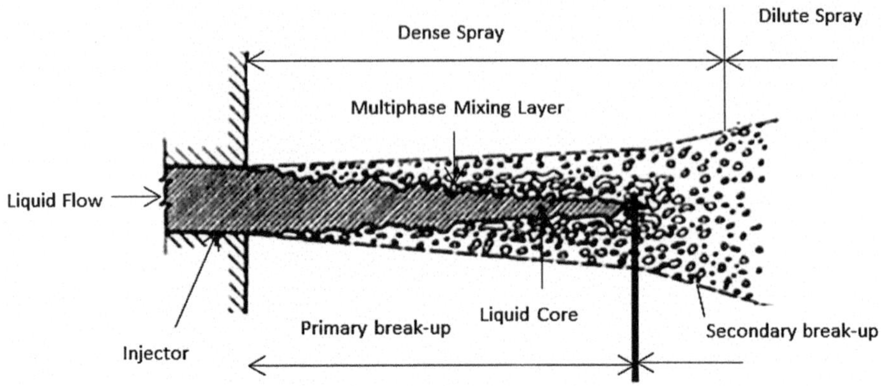


**Fig. 14** Flow pattern of atomized spray [26]

multiplied by camera magnification factor. Spray length should be optimized for efficient combustion and pollutant formation. If spray length is too short, it results in poor combustion. If it is too long, it can hit the combustion chamber walls, which leads to fuel wastage and encourages soot formation. Spray length is significantly affected by FIP. Dent [20] derived the equation, which is based on jet gas theory and states that spray tip penetration increases with fuel injection pressure.

$$S = 3.07\left(\frac{\Delta p}{\rho_g}\right)^{1/4}(td_n)^{1/2}\left(\frac{294}{T_g}\right)^{1/4} \quad (1)$$

Here, $\Delta p$ is the pressure difference between the injection and ambient conditions, $\rho_g$ denotes the ambient gas density, t represents the transmit time, $d_n$ is the projection diameter and $(295/\ T_g)^{1/4}$ represents additional factor imposed on the relationship for accurate results.

Hiroyasu and Arai [39] measured spray penetration length with the help of photographic technique and correlated their data as mentioned below;

$$t < t_{break} \qquad S = 3.07\left(\frac{\Delta p}{\rho_g}\right)^{1/4}(td_n)^{1/2}\left(\frac{294}{T_g}\right)^{1/4} \quad (2)$$

$$t > t_{break} \qquad S = 2.95\left(\frac{\Delta p}{\rho_g}\right)^{1/4}(d_n t)^{1/2} \quad (3)$$

where $t_{break} = \frac{29\rho_l d_n}{(\rho_g \Delta p)^{1/2}}$.

## Spray Cone Angle

Lefebvre et al. [53] suggested the definition of Spray cone angle as the angle between lines drawn from the nozzle tip to the outer periphery of spray at the downstream of the nozzle. It can be calculated from Eq. (4)

$$\tan\frac{\theta}{2} = \frac{1}{A}4\pi\left(\frac{\rho_g}{\rho_l}\right)^{1/2}\frac{\sqrt{3}}{6} \tag{4}$$

Spray angle can be determined from spray images using linear regression method. Firstly spray contours are plotted on the basis of spray images and linear regression method is used for finding the upper and lower limiting lines, where intensity is rapidly reduced. Only point between injector and 60% of the total length is considered for applying linear regression (Fig. 15). The angle between these two lines is considered as spray cone angle [66].

Above mentioned parameters affect the overall spray behavior, which in-turn influences vaporization of fuel droplets and mixing of fuel and air inside the chamber. The emissions can be minimized and the efficiency and power output from an engine can be enhanced considerably, if spray can be optimized. Spray visualization can be done in a spray chamber as well as in an engine cylinder using a high-speed camera synchronized with a high speed laser and white light source for obtaining transient liquid droplet evolution for a complete injection. Spray penetration length and cone angle in a cold non-reacting environment using a common-rail direct injection (CRDI) system was first studied by Postrioti and Grimaldi [68]. Both experimental and numerical analysis of microscopic and macroscopic spray parameters using a common rail in a diesel engine was performed by Lee and Park [50]. They used particle motion analysis system to determine the spray tip penetration and spray cone angle, while the phase Doppler particle analyzer (PDPA) was used to study droplet's Sauter mean diameter (SMD) and axial velocity. They varied the axial distance between injector tip and measurement point and reported that as the axial distance increased from 10 to 30 mm, SMD and axial velocity of droplets decreased rapidly. On further moving downwards to 40 mm, these parameters almost remained constant. The reason was that as the spray formation zone was separated from the spray atomization zone, their rates became equal.

Many researchers have reported that all fuels exhibit longer spray penetration, wider cone angle and better fuel atomization at higher FIPs. Park et al. [62] studied

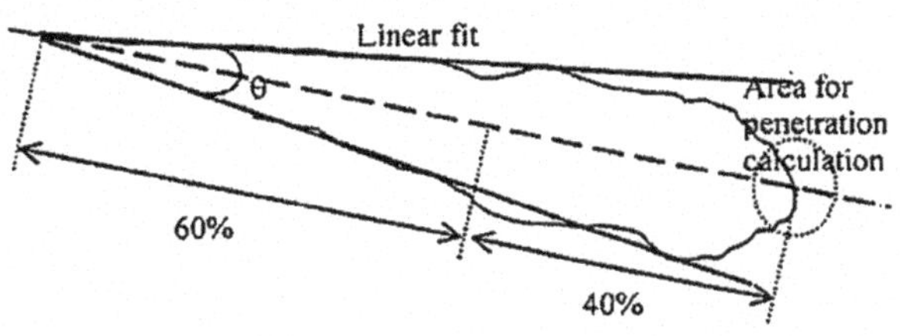


**Fig. 15** Measurement of spray cone angle [66]

the effect of FIP on macroscopic spray parameters using a CRDI system. Increasing FIP resulted in longer spray penetration for both mineral diesel as well as biodiesel. The results were also supported by Suh et al. [89]. Wang et al. [94] investigated the effect of FIP on a spray cone angle and spray penetration length for mineral diesel. They reported that spray penetration length increased with increasing FIP and spray cone angle decreased. The effect of fuel properties on the spray characteristics and atomization was first studied by Lefebvre [53]. Abiola et al. [1] studied these parameters using mineral diesel and biodiesel in evaporating and non-evaporating environments. They reported that at very high FIPs, mineral diesel showed relatively shorter spray penetration compared to biodiesel. This was attributed to the higher volatility of mineral diesel, which resulted in higher evaporation compared to biodiesel, irrespective of FIP.

Biodiesel showed poor atomization characteristics compared to mineral diesel. Spray tip penetration and SMD were marginally higher for biodiesel, while spray dispersion and spray cone angle were relatively lower. Higher liquid length of biodiesel spray was attributed to its higher boiling range and higher heat of vaporization [174]. He et al. [38] analyzed the shape of biodiesel spray compared to mineral diesel. They reported that the spray tip penetration and spray cone angle of biodiesel were larger than baseline mineral diesel because of higher density, viscosity, surface tension and bulk modulus of biodiesel. Kim et al. [47] used two test fuels (biodiesel and DME) to study the spray characteristics in a constant volume spray chamber (CVCS). Suh et al. [89] compared fuel spray behavior on the basis of injection rate, spray penetration length, spray droplet-size distribution and axial velocity distribution for mineral diesel and biodiesel. Ejim et al. [24] and Abiola et al. [1] suggested that biodiesel had inferior volatility characteristics compared to mineral diesel, which resulted in inferior spray atomization. Therefore in order to improve fuel injection and atomization characteristics of low volatility fuels, they are blended with fuels having higher volatility.

Suh et al. [89] reported that blending ratio of biodiesel had an insignificant effect on their spray characteristics. Development of spray in case of mineral diesel was more advanced than biodiesel because of lower shear resistance offered by mineral diesel in the injector nozzle. Properties such as viscosity, bulk modulus, density and surface tension played an important role in providing shear resistance. Spray penetration length was also observed to be the same with or without considering the blending ratio. Gao et al. [29] investigated the reasons for reduction in spray cone angle with increasing biodiesel blending ratio. They reported that variation in spray cone angle showed local maxima, in which spray cone angle first increased with increasing biodiesel blending ratio and then decreased with further increase in biodiesel blending ratio. They explained this behavior as diffusion of smaller droplets, which resulted in narrower spray cone angle. Park et al. [64] concluded that the size of spray droplets decreased with increasing biodiesel blending ratio. It was found that gasoline blended with mineral diesel showed relatively smaller droplet sizes than baseline mineral diesel hence it can be used for improving the spray characteristics.

Researchers also investigated fuel spray characteristics using experimental and numerical techniques to correlate their findings. They concluded that fuel atomization and spray characteristics were two important aspects, which need to be improved in order to enhance in-cylinder fuel-air mixture homogeneity. As the spray droplet velocity distribution and the ambient air density increased, aerodynamic and viscous effects became more dominant and the spray breakup process became more complex. Small droplets offered larger surface area, enabling higher heat transfer and energy interactions compared to larger droplets for the same amount and volume of fuel injected.

## 3.2 *Microscopic Spray Characterization*

Microscopic analysis of spray indicates the spray droplet-size distribution, three velocity components, number density, time of arrival of droplets, and volume flux measurements. Laser Doppler velocimetry (LDV) or laser Doppler anemometry (LDA) techniques are used for microscopic spray analysis. These techniques measure the velocity of spray droplets however they do not give information about the spray droplet-size distribution. Therefore, these are substituted by phase Doppler interferrometry (PDI). PDI technique, formerly known as phase Doppler particle analysis (PDPA) was developed by Dr. William Bachalo and Mike Houser in 1984. The principle of LDV is same as that of PDI. Phase Doppler interferometry is used to understand the intricate spray behavior in various fields such as in the turbulence generated by different flow regimes, high pressure injectors used in rockets, unsteady spray characterization in gasoline engines and compression ignition engines. PDI is a non-intrusive technique, which simultaneously measures the size and velocity of individual spherical spray droplets in inconsistent shapes, sizes and mass distributed flow environment. This technique is based on the principle of light scattering interferometry, which takes into account the wavelength of light. Since the intensity of light does not play an important role in measurement, thereby distorted light intensity has negligible effect on the performance of droplet-size determination. Parameters such as wavelength of the laser, focal length of the transmitter and receiver lenses, and detector separation are taken into consideration. Since the signals produced by the system are sinusoidal in nature, they are easily perceived even in presence of noise using Fast Fourier Transform (FFT). The schematic of basic PDI system is shown in the Fig. 16.

In PDI system, the laser is bifurcated into two beams of equal intensities using Bragg's cell and a beam splitter. The beams are then concentrated using a transmitter lens. Bragg's cell is used to shift the frequency of beams so as to provide exactness to the droplet direction. The laser, beam splitter and transmitter lens constitute a complete transmitter module. Two diodes pumped solid state lasers (DPSS) are kept inside the transmitter module, which have wavelengths of 532 nm

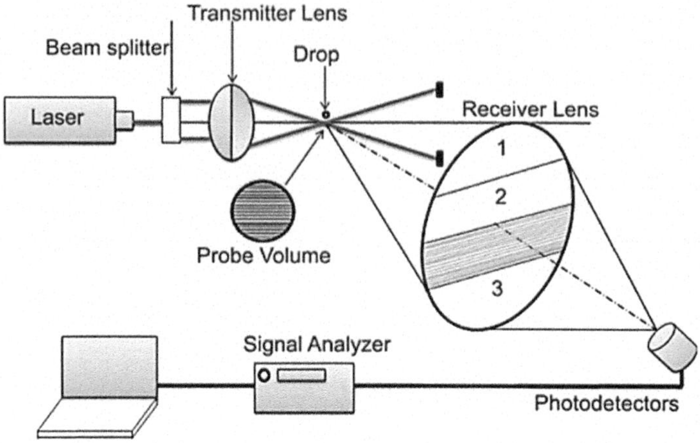


**Fig. 16** Schematic of PDI system [54]

(green beam) and 471 nm (blue beam). Three photo-detectors, photomultiplier tubes and receiver lenses constitute a complete receiver module.

Spray droplets, while passing through the intersection point of the laser beams scatter light (refraction/ reflection), which is received by the photo-detectors of the receiver module through the receiver lenses. A single aperture is used to focus the refracted light from the receiver lens to the photo-detectors. A spherical spray droplet at the intersection of two laser beams (with enlarged view) is shown in Fig. 17.

As a result, an interference pattern is formed around the droplet. This interference pattern produces signals of sinusoidal intensity distribution, which when projected onto a plane gives hyperbolic signals. Once the separation between the detectors (s) is known, the spatial wavelength (Λ) and phase shift (ϕ) is calculated, which is helpful in finding the droplet size.

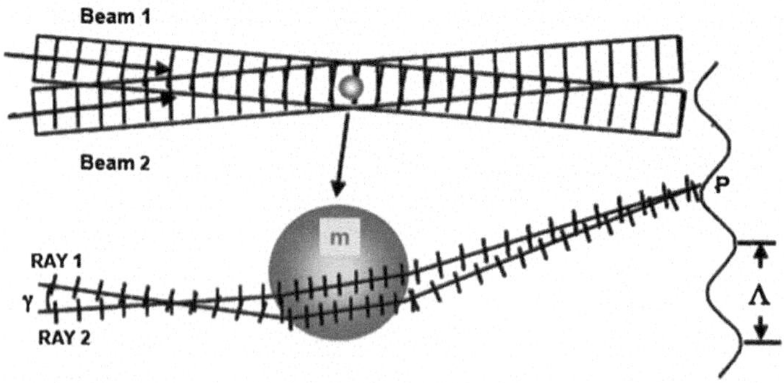


**Fig. 17** Phase shift by two beams of light to reach a common point [54]

### 3.3 Application of Spray Parameters in IC Engines

The homogeneity in fuel-air mixing and combustion process inside the engine cylinder, reduction in emissions, and enhancement in the engine performance are some of the compelling reasons to study the spray formation, spray breakup and atomization in an engine. Thermal efficiency and power output of an engine can be increased by optimizing spray parameters. Physical and thermal properties of test fuels such as kinematic viscosity, bulk modulus of compressibility, volatility, surface tension, rate of vaporization, boiling temperature, etc. and ambient gas density play a prominent role in spray formation and fuel-air mixture formation. Apart from these properties, cavitation inside the injector nozzle and turbulence at the exit of the injector nozzle tip are also very important, which need to be looked for complete study of spray behavior. To investigate this, apart from an experimental perspective, theoretical and numerical approach has been adopted by a large number of researchers [47, 50, 62, 74]. Most of the experimental investigations are carried out in reacting or non-reacting environment in a specially built spray chamber. This CVSC is designed and developed in such a manner that it should create actual engine combustion chamber conditions except temperature and turbulence. Once the spray is analyzed for a particular injector nozzle type, the design of the nozzle geometry can be ameliorated. Hence the basic spray behavior is determined by the injector nozzle design. The study of fuel sprays behavior for diesel, vis-à-vis vegetable oils, biodiesel and blends in simulated compression ignition engine conditions needs to be thoroughly investigated. The fuel is introduced by an injector nozzle by creating a pressure difference between the ambient and the fuel injector. Generally, the ambient pressure and the FIP inside the combustion chamber were kept in the range of 1–60 bar and from 150 to 2000 bar respectively. This range depends on the size of the engine and the fuel injection system used. Combustion in a CI engine is highly dependent on the quality of spray and atomization during the fuel injection. A good spray should ensure that the injected fuel jet enters into the combustion chamber at sufficiently high velocity and consequently atomizes into small droplets to enable rapid vaporization and mixing with ambient air and form combustible mixture. When mineral diesel is replaced by biodiesel, it is essential to optimize the fuel injection system for the new test fuel. Different straight vegetable oils (SVOs), biodiesels and their blends with mineral diesel have significantly different physical properties, depending on the chemical structure of the biodiesel feedstock. Because of higher viscosity, surface tension and density, biodiesel and their blends exhibit inferior spray characteristics hence poor air-fuel mixing. Liquid fuel distribution visualization in an engine cylinder or a constant volume spray chamber provides useful information on the evolution of spray, spray penetration, spray cone angle, spray-wall interaction and spray area. These parameters depend on fuel droplet dispersion, axial and radial velocity distributions, number density, and volume flux measurements. Study of these factors in a gasoline or diesel engine combustion chamber makes the task extremely chaotic and challenging. Therefore so far no research has been reported in the open literature, which investigates these parameters inside a firing engine. All reported research is carried out in a CVSC only.

For measuring droplet size and spray velocities, Kim and Lee [46] and Koh et al. [49] used image processing methods. However, this method does not provide complete information because it was very difficult to study coalescing droplets. Schunemann et al. [79]; Park and Lee [65] and Lee and Reitz [51] used PDPA technique for studying spray atomization. Most researchers investigated different alternative fuels and adopted various visualization and quantitative methods to study spray behavior. Kim et al. [47] used two fuels, namely biodiesel and DME to study the sprays in a CVSC and varied the FIP and ambient pressure. The injection system used in this study consisted of a common rail and a high pressure pump. They also varied the axial and radial distance of the injector tip from the central axis of the measurement point and studied the spray characteristics at different locations. They reported that with increasing axial distance in the downstream direction, SMD of biodiesel droplets increased, however SMD of DME remained almost constant. The reason stated was that DME, due to its lower viscosity and density, atomizes very early. Hence the spray formation and atomization zones occur in the very early stage for DME and this leads to constant SMD droplets throughout the measurement volume. The spray images obtained by the high speed camera were converted to cell-images using image processing technique. This technique is called 'threshold method'. To differentiate between the dark (black) and bright (white) spray pixels, the values of light intensity were selected from 0 to 255. Here 0 indicates that the pixel has a brightness level of 0 grayscale or a darkness level of 256 gray scales and 255 indicates vice versa. The spray area is defined by the sum of the total number of pixels selected. They reported that because of adequate atomization in the extreme region of the spray, light intensity levels were lower. However in the central region of the spray, light intensity was higher.

Ghurri et al. [30] used image posterization processing method instead of the threshold method. According to them, posterization method gives more detailed information about the spray because it divides a complete image in layers of various gray level scales. Segmentation of image processing technique is generally employed to separate the captured images and the spray background. The threshold method of image processing transfigures gray images (as captured by the camera) into shades of pure black and pure white. In order to differentiate between black and white images, a threshold value is selected such that if the image is brighter than this value, it will be referred to as a white image otherwise it will be referred to as a black image. This lengthy procedure of segmentation has to be applied to each image. This problem was rectified by posterization image processing method, in which a threshold value was applied to a single image and on the basis of that value, other images were interrogated. Gao et al. [29] simulated the biodiesel sprays using Star-CD software. The simulations obtained were in good agreement with the experimental data. Galle et al. [28] used shadowgraphy technique to capture the spray images with the help of a high speed camera. Sakaguchi et al. [74] measured the droplet sizes of diesel spray with the help of laser-2-focus ($L_2F$) velocimetry technique, in which there were two focal points of the probe and the distance between them was 36 μm. To increase the data rate and to improve the accuracy, multiple scattering of the particles are circumvented. This $L_2F$ technique was first introduced by Schodl [78]. In order to calculate the flow measurement and to increase the signal-to-noise ratio in the impeller of a centrifugal compressor, $L_2F$ technique was used. The work of Schodl [78] was further advanced by Chaves et al. [15]. They used $L_2F$ technique in engines for measuring the diesel spray droplets, in which, the distance between the two focal points was 60 μm. The diameter of the probe focus was 10 μm. Kuti et al. (2013) studied the effect of Mie-scattering in reacting spray environment and the effect of OH Chemiluminescence on the flame lift-off length.

An important contribution to the field of spray structure was made by Park et al. [62]. They studied the effect of fuel temperature and back pressure on the spray behavior in a diesel engine. The fuel used in this study was diesel and soybean oil methyl ester (SOME). The experimental results using PDPA were compared with numerical simulation results obtained using KIVA-3 code. The parameters tested were local and overall SMD at a variable distance from the nozzle tip. The spray breakup models used for the primary and secondary breakup were Kelvin-Helmholtz (KH) and Rayleigh-Taylor (RT) models. A computational model for studying the spray behavior and injector characteristics for different fuels (diesel and biodiesel) was developed by Som et al. [82]. Parameters such as cavitation and turbulence at the nozzle outlet were studied using simulations. In order to ensure that the efficiency of the fuel injected is the maximum, they investigated various injection pressures and back pressure conditions in a diesel engine. CONVERGE software based on Kelvin Helmholtz-Aerodynamics Cavitation Turbulence (KH-ACT) model was used to completely understand the spray behavior and spray characterization inside the engine combustion chamber. They compared the results of this model with KH-RT model and reported that spray penetration of biodiesel was longer than diesel using KH-ACT model. Due to turbulence and cavitation, apart from lower fuel viscosity and surface tension, atomization of biodiesel spray was faster, which resulted in a smaller spray droplets-size distribution and smaller penetration length compared to diesel. Since KH-RT model does not take into consideration the effect of turbulence and cavity, the results showed the opposite trend. Park et al. [63] also investigated the effect of multiple fuel injection strategy in reacting and non-reacting environment in a diesel engine and used spray visualization and PDPA techniques. Elkotb [25] developed a relationship for determination of SMD of a single phase flow using fuel properties like viscosity, surface tension, pressure difference and density. Ejim et al. [24] verified that this relationship can be applied to dense diesel and biodiesel sprays. This relationship is given as:

$$SMD = 6156\nu^{0.385}\sigma^{0.737}\delta_f^{0.737}\delta_a^{0.06}\Delta P^{-0.54}$$

Here $\nu$ and $\sigma$ are viscosity and surface tension, and $\Delta P$ is the difference in fuel injection pressure and ambient pressure. It is to be noted that SMD, $\nu$, $\sigma$, $\delta$ and $\Delta P$ have units of μm, $m^2/s$, N/m, $kg/m^3$ and bar respectively.

# 4 Combustion Diagnostic Techniques

Strict emission legislations and fuel economy compliances are very important for the automotive industry. They can be controlled by analysis of different combustion and spray parameters, which can be used to optimize engine design and fuel injection parameters. Combustion and sprays can be diagnosed by several techniques. There are several classical techniques such as optical detectors (Photodiodes, Photo multiplier tubes), and mechanical sampling methods, which are of no interest to the engine researchers, hence these are not covered in this chapter. Ion current probe method, spectroscopy, optical sensors, optical particle sizing methods and optical visualization technique are some of the advanced techniques, which are currently used for combustion and spray diagnostics by the automotive industry and researchers. These techniques are briefly discussed in following sub-sections.

## *4.1 Ion Current Probe Method*

It is based on electrical characteristics of combustion reaction. Docquier and Candel [23] suggested that the main source of ions in flames is chemi-ionization process, which occurs, when the energy of reactants together with reaction energy is large enough to ionize the product. $CHO^+$ and $H_3O^+$ are the positive ions in the combustion flames, which are mainly responsible for ion current. The chemical reactions of generating ions are as follows:

$$CH + O \leftrightarrow CHO^+ + e^-$$
$$CHO^+ + H_2O \leftrightarrow H_3O^+ + CO$$
$$H_3O^+ + e^- \leftrightarrow H_2O + H$$

These positive ions $CHO^+$ and $H_3O^+$ are mainly generated from CH, which originates from fuel. Ions are also generated by thermal reactions of NO, as shown by Zeldovich mechanism. This method is mainly used in SI engines. The spark plug is used as an Ion current probe. Tanaka et al. [92] used this method for combustion

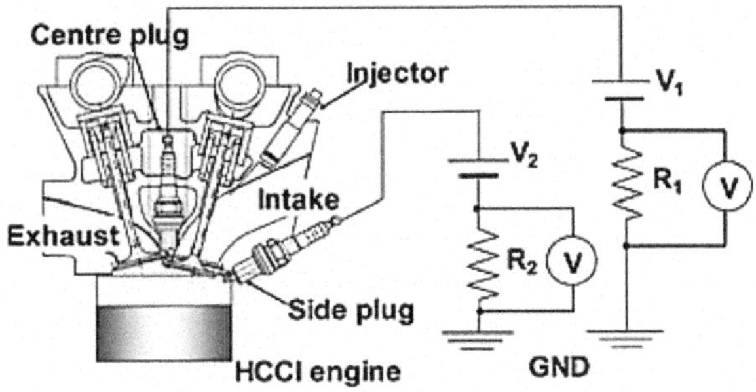


**Fig. 18** Schematic of the ion current detection circuit [92]

analysis of HCCI in SI engine. They used two spark plugs for detecting the Ion current; one was mounted in the center and the other was mounted near the wall. Their experimental setup and current detection circuit are shown in Fig. 18.

They detected ion current and pressure signal w.r.t. crank angle and found one peak on HCCI mode and more than one peak in normal SI mode. It clearly suggested absence of $NO_x$ emissions in HCCI mode. They also found that starting timing of current signal is dependent on gas temperature, while the peak of current signal is influenced by the gas composition. Length of the peak was the maximum for stoichiometric mixture, which suggested that large number of ions were generated in stoichiometric mixture. This technique is good for SI engines, because spark plug works as a current probe and no extra arrangement is required, however in CI engines, it is very difficult to access probes in different locations of the combustion chamber.

## 4.2 Spectroscopy

It is used to determine combustion temperature and species emitted in combustion process. It measures the intensity of light spectrum emitted by combustion gases. Wavelengths of this spectrum lie in infrared, visible and ultraviolet regions. Combustion temperature can be determined according to Planks or Wine's Law and Kirchhoff law. Radiance is measured at a particular wavelength and temperature can be calculated according to Wien's law, which states that the maximum flux is attained at lower wavelengths upon increasing the temperature (Web 2009).

$$L_\lambda^b(T) = \frac{c_1}{\left\{\pi\lambda^5\left(e^{\frac{c_2}{\lambda T}} - 1\right)\right\}}$$

$\lambda$ = wavelength; $c_1$, $c_2$ = radiation constants; $L_\lambda^b(T)$ = spectral radiance and T = Temperature.

## 4.3 Optical Sensors

Various types of optical sensors have been developed for determining emissions, presence of flames and temperature. These sensors are usually installed in the inlet and exhaust pipes. Gaseous emissions ($NO_x$, CO, etc.) are normally determined by semi-conductor sensors, which are made of metal oxides such as $TiO_2$, $ZrO_2$, and $SnO_2$ [23]. Working principle of $TiO_2$ and $SnO_2$ sensors is based on the change of semi-conductor resistance with partial pressure of oxygen, while in $ZrO_2$ sensors, the presence of species can be determined from the change in current and voltage. Apart from this, there are several other sensors like resistive temperature detectors, pressure sensors and flow sensors, etc., which are used for determining exhaust temperature and chamber pressure. Main problem with the optical sensor is that they each sensor has limited applications. Hence different types of sensors are required for different applications.

## 4.4 Fraunhofer Diffraction Method

This technique is typically used for spray analysis. When a beam of monochromatic light falls upon a large diameter droplet/ particle ($d \gg \lambda$), light is scattered and it forms Fraunhofer diffraction patterns [36]. Particle diameter is determined from the angular distribution of scattered light in the diffraction pattern. This technique is good for detecting large diameter droplets/ particles. Lafebvre et al. (1989) found that there are inaccuracies, while measuring the particle less than 5 μm diameter. The major drawback of this technique is in dense sprays due to multiple scattering. It also does not work well for non-spherical particles (Fig. 19).

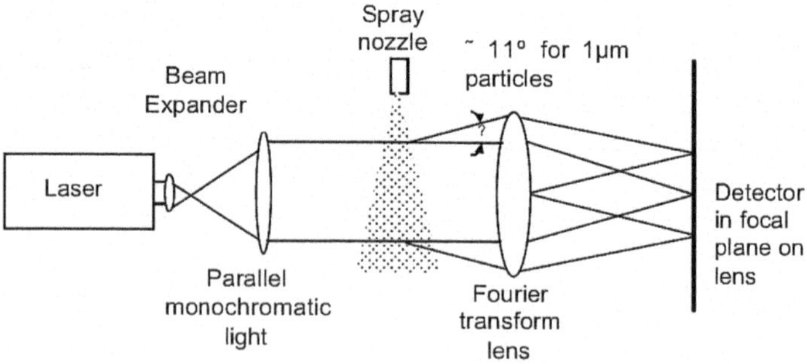


**Fig. 19** Schematic of Fraunhofer diffraction technique employed in Malvern particle arrangement [77]

## 4.5 Laser Sheet Droplet Sizing

This technique is a combination of Mie scattering and LIF for calculating SMD of dense sprays. Laser sheet is used to measure the image cross-section of sprays and illuminate the droplets/ particles. A part of the energy is absorbed by these droplets/ particles, which excite and emit fluorescence. Remaining part of incident light is scattered and experienced by elastic light scattering. Intensity of scattered light signal from a droplet of diameter $D_d$ is calculated using Lorenz Mie theory of elastic scattering;

$$S_{Mie} = C_{Mie}(D_d)^2$$

The droplet's molecules emit fluorescence signal, which is $S_{LIF.}$

$$S_{LIF} = C_{LIF}(D_d)^3$$

Ratio of florescence and Mie scattering signal from a large number of droplets is proportional to

$$\frac{S_{LIF}}{S_{Mie}} = \frac{C_{LIF} D_d^3}{C_{Mie} D_d^2} \propto \frac{D_d^3}{D_d^2} \; or \; SMD$$

Hence SMD can be calculated by taking ratio of these two signals and its map in image plane can be generated by a ratio of two images and divided by the calibration constant. This technique works well in dense sprays, but it is limited to measuring only SMD. It also works with maximum diameter up to 20 μm [36].

## 4.6 Applications of Optical Diagnostics in Engines

Following combustion features can be determined using combustion visualization techniques:

- Combustion temperature,
- Start of combustion and ignition delay,
- Flame propagation and flame velocity in SI engines.

### 4.6.1 Combustion Temperature

Combustion in IC engines generates heat, and releases hot exhaust gases through the exhaust valve. Performance and emissions from an engine are directly related to in-cylinder temperature. In diesel engines, $NO_x$ (which include NO and $NO_2$) is a major pollutant and its formation mainly depends on the peak combustion temperature. NO, which is a toxic and colorless gas, is formed during oxidation of atmospheric nitrogen at high temperature (Zeldovich mechanism) in post-flame combustion processes [4]. $NO_2$ is formed during a reaction of NO with $HOO^-$ radicals in combustion zone and it is converted back into NO by reacting with atomic oxygen. This $NO_2$ decomposition reaction freezes, when high temperature exhaust gas rapidly mixes with ambient air. Therefore, $NO_2$/ NO concentration ratio was higher at lower engine loads due to higher probability of high temperature gas coming in contact with cold ambient air [36]. $NO_x$ concentration is significantly lower in SI engines as compared to CI engines therefore it can be controlled by using catalytic devices. However in diesel engines, it is controlled by using EGR. Agarwal et al. [4] suggested that $NO_x$ emission reduced in CI engine on increasing EGR, primarily due to reduction in peak combustion temperature in the cylinder. Hence the combustion temperature is very important combustion characteristics for controlling $NO_x$ emissions. Optical visualization techniques such as engine endoscopy are very useful for finding the combustion temperature using combustion imaging and optical pyrometery. Normally 'ThermoVision' software with endoscope is used for finding the temperature during in-cylinder combustion however colour temperature methods are also helpful for comparing the combustion temperature distributions for different fuels.

### 4.6.2 Ignition Delay and Combustion Duration

Fuel after being injected from the nozzle undergoes physical and chemical processes before the ignition takes place. There is a delay between the start of injection and start of combustion, which is called 'ignition delay'. Physical delay part of ignition delay includes atomization, evaporation and mixing with air, while the chemical delay part of the ignition delay is due to pre-combustion reactions in the fuel-air mixture, which leads to auto-ignition. Mtui and Hill [58] suggested that ignition delay can be measured experimentally as time between start of injection and 10% mass burn fraction. It can also be found from the combustion images. Start of combustion can be easily found by visualizing the combustion chamber during combustion.

### 4.6.3 Flame Propagation and Flame Velocity in SI Engines

In SI engines, flame propagation is very important phase of combustion process because it directly affects the knocking in the engine. Higher flame propagation velocity reduces knocking combustion due to shorter flame propagation duration. Spicher et al. [84] visualized combustion process and reported that the flame propagates in unburned mixture in a spherical wave form with a wrinkled surface. Experiments were performed in a DISI engine with homogenous and stratified charge using an endoscope (Fig. 20).

Koch et al. [48] developed the phenomenological model in a DISI engine for calculating the flame front velocity with the help of optical experiments, which

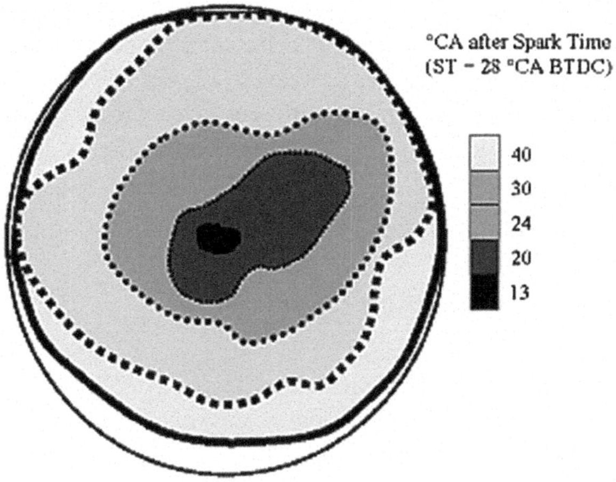


**Fig. 20** Spherical flame-front in DISI engine with homogenous mode [86]

were performed on a single cylinder research engine. They captured the combustion images and analyzed flame propagation velocity, flame front area and suggested that air-fuel ratio changed with flame front propagation. Rhodes and Keck [70] as well as Metghalchi and Keck [56] suggested that the laminar flame speed mainly depends on the air-fuel ratio, temperature, pressure and mass burn fraction in a SI engine. In a diesel engine during pre-combustion phase, motion of gases affects the

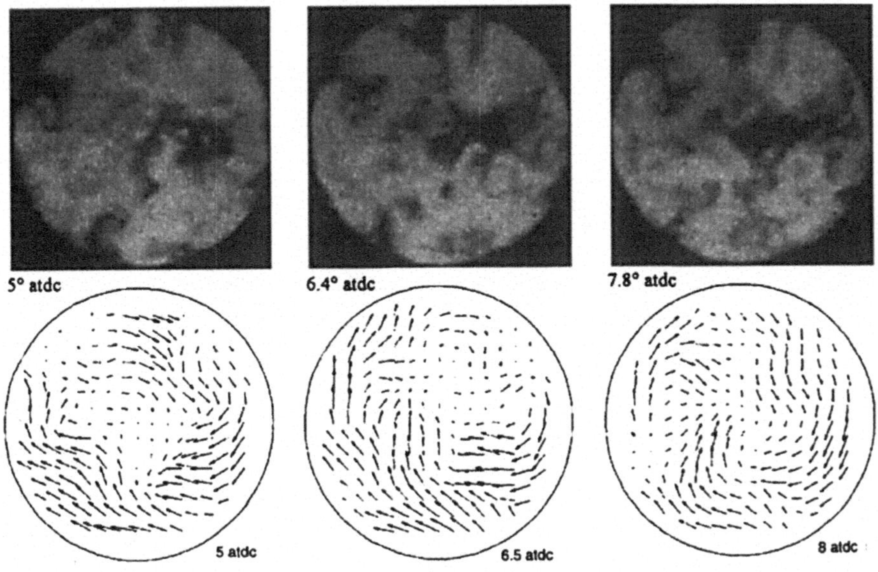


**Fig. 21** Combustion images with velocity vector plot [90]

post-combustion processes. Sun et al. [90] measured the flow-field of gases using high speed combustion photography and cellular cross-correlation technique and analyzed its effect on the combustion processes (Fig. 21).

They plotted the velocity vector for different swirl ratios and found that with increasing swirl, flame velocity also increased and forced vortex dominated the flow in the combustion chamber.

# 5 Combustion Visualization

In IC engines, most emission and performance tests have been done passively and such investigations do not provide vital spatial and time resolved information about the formation of pollutants during combustion. It is possible only with direct visualization of combustion and sprays in the engine cylinder, which can be done by using sophisticated optical visualization techniques. Hence combustion and spray diagnostics in engines are very important because it is one of the most effective ways to understand and resolve the engine emission formation in the engine combustion chamber. In-cylinder combustion visualization can be carried out by using a high-speed CCD camera and associated visualization equipment in a customized engine. From these images, various combustion parameters and spray characteristics can be analyzed with the help of image processing software. However optical access to the engine combustion chamber is the biggest challenge in applying optical diagnostic techniques to engines. There are different ways of gaining optical access to the engine cylinder and it differs according to the requirements of the measurement and available financial resources. Two main types of optical access arrangements can be made namely; full optical access such as in an optical research engine; and limited optical access such as in engine endoscopy of a production grade engine.

## *5.1 Optical Research Engine*

An optical research engine allows full optical access to the engine combustion chamber. This special engine is used for studying the fuel-air mixing, in-cylinder flows, combustion, and flame propagation through the optical liner, which is made of high quality quartz and a transparent piston. Early efforts on gaining optical access were by using a transparent acrylic resin, which was assembled on the extended part of the cylinder. A quartz window was installed on top of the extended piston, which enclosed a mirror inclined at 45° to observe the motion of in-cylinder flows. Bowditch [13] modified a production engine for optical access through the piston crown and developed it as a new tool for combustion research. After that, several researchers and automotive companies improved this type of engines and implemented this technique using a transparent window in strategic locations such as cylinder, cylinder head and piston for investigations in engine firing conditions. Fully transparent cylinder was also implemented in some engines for non-firing studies. Transparent windows were made of fused silica or high temperature quartz, which is capable to withstand high temperature and high pressure prevailing during combustion. This type of engine is fired for very short periods only because the visibility of the optical window decreases rapidly due to soot deposition. Therefore, regular cleaning is required after each firing experiment. Optical engines cannot be tested at higher loads and speeds [36]. Allen et al. [6] used new design of the optical research engine, developed by Lotus Engineering, UK. Fuel spray and combustion investigations were done in this engine by employing laser diagnostics. They tested the engine up to 5000 rpm and 60 bar peak in-cylinder pressure. Han et al. [32] used optically accessible direct injection diesel engine for soot and temperature analysis using laser diagnostics. The main advantage of this engine was that entire combustion event could be captured for one full combustion cycle using suitable high-speed charge coupled device (CCD) camera. Hence this type of engine was very helpful in analyzing various combustion characteristics, which were not understood well using conventional techniques. Huang et al. [41] used a transparent cylinder, which was made of Plexiglas and a section of aluminum cylinder. A chrome metal liner was inserted between the Plexiglas cylinder and the crankcase. An elongated hollow cylindrical aluminum frame was used instead of the original piston. The top end of the cylindrical frame was screwed to the piston head. Richter et al. [72] used standard Bowditch layout, in which a quartz cylinder of 80 mm height was mounted to access the combustion chamber (Fig. 22a).

Singh et al. [80, 81] used an optical research engine (AVL, 5408) for PIV experiments in motored conditions. The engine had full optical access through the optical liner made of quartz and a transparent piston, in the Engine Research Laboratory of IIT Kanpur, India. Lotus single cylinder optical research engine (SCORE) was designed for optical diagnostics as well. Its cylinder liner was made from fused silica and a sapphire window was inserted in the piston crown in order to provide full optical access to the combustion chamber (Fig. 22c). Optical diagnostics using full optical access are very useful for real time data analysis however optical engines have several limitations. Optical research engines suffer from several restrictions such as balancing difficulty of a single cylinder, which is due to the added mass of the elongated piston. This type of engine can only be operated from low-to-medium loads and speeds because the limited structural strength of the optical window, which is a limiting factor due to its large size. Quartz window is very expensive and a specific window is required for a particular engine according to its geometrical configuration. Hence this technique can't be used in a production grade engine. It can only be used for research in a highly customized research engine test cell environment, which is prohibitively expensive. The optical engine can be operated for short durations only due to low heat transfer rates through the optical components and lack of cooling of the optical liner. The optical walls must be cleaned and combustion products must be removed frequently. This is quite tedious task generally.

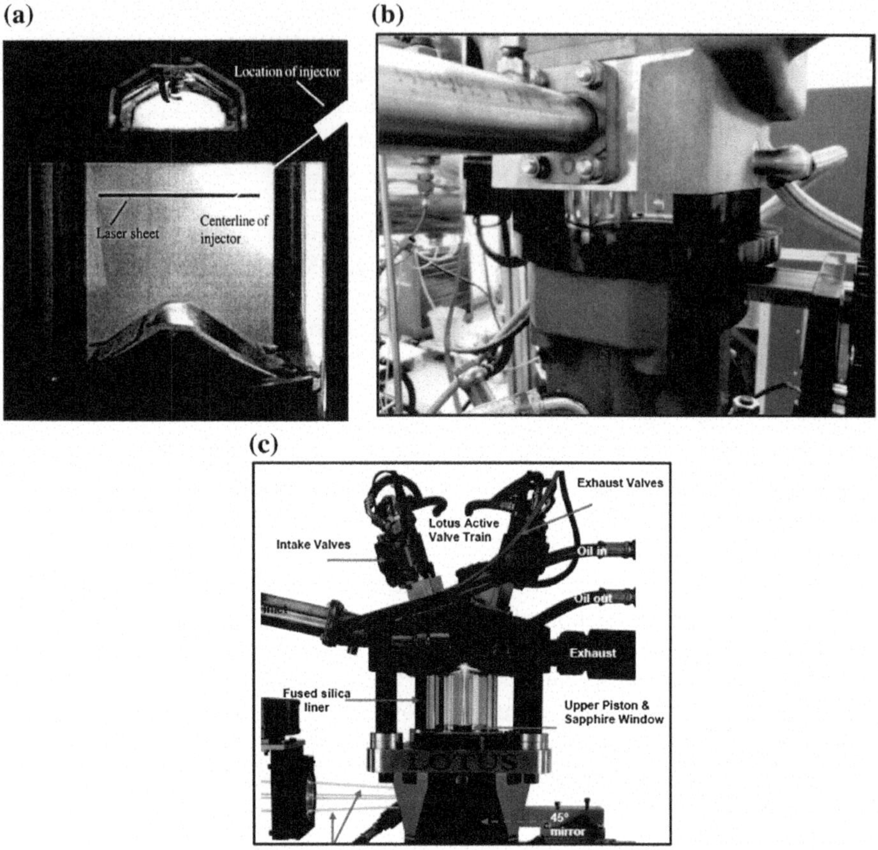


**Fig. 22** **a** Optical access with 80 mm quartz liner [72]; **b** AVL single cylinder optical research engine [80]; **c** Lotus single cylinder optical research engine [12]

## 5.2 *Engine Endoscopy*

Due to the above mentioned limitations of optical engines, several optical diagnostics techniques use optical access using an endoscope, which is relatively easier and endoscopic technique can be applied to any production grade engine with some modifications in the engine hardware. Endoscopy is an optical visualization technique used for several medical and industrial applications. Initially this technique was used by doctors in medical surgeries for capturing the view of interior parts of the human body. It was used for the first time in 1806 by Philippbonzzini in Mainz using light conductor for examining the cavity in a human body. After many improvements in this technique, Georg Wolf (1873–1938) introduced rigid endoscopes in 1906 and flexible endoscope in 1911. At the time of World War-II, several modern technologies emerged in medical science. In 1945, Karl Storz, who is known as "father of endoscopes" introduced a new endoscope, which was used with external light source in the human body. After that, this concept was widely used in industry for investigating problems which were associated with the interior part of machines in locations, where there is limited space for optical access. After that, it was designed for automotive applications for visual inspection of hollow spaces and areas, which are challenging for optical examinations such as engines, cylinder heads and intake manifolds of the engine. It can be used in myriads of applications in SI and CI engines, both for combustion diagnostics, spray formation, mixture formation etc.

### 5.2.1 Principle of Endoscopy

The endoscope is a large, thin tube, which consists of achromatic lens or rod lens combinations used for transferring image from the viewing end to the ocular (eye-piece). The image is transferred through image lenses and field lenses, which are set in the guide tube in alternate sequence. These pairs of lenses form an image reversal system, which repeats the reversal of image and transfers the image straight through the endoscope (Fig. 23).

The image of an object is first projected at plane B1 (Fig. 23) and from there, it is transferred through reversal lenses (L1/ L2) and projected on B2 plane. Image on plane B2 is a mirror image of plane B1. In the same way, it is transferred to Plane B3, where image is similar to image plane B1. C mount adapter is attached to the ocular, which consists of focus adjustment arrangement [22].

### 5.2.2 Methodology of Engine Endoscopy

Normally for engine applications, two types of endoscopy can be used: (i) combustion chamber endoscopy and (ii) general cavity endoscope. Combustion chamber endoscopy is used for high temperature and high pressure applications such as visualization of in-cylinder combustion process. This type of endoscope is

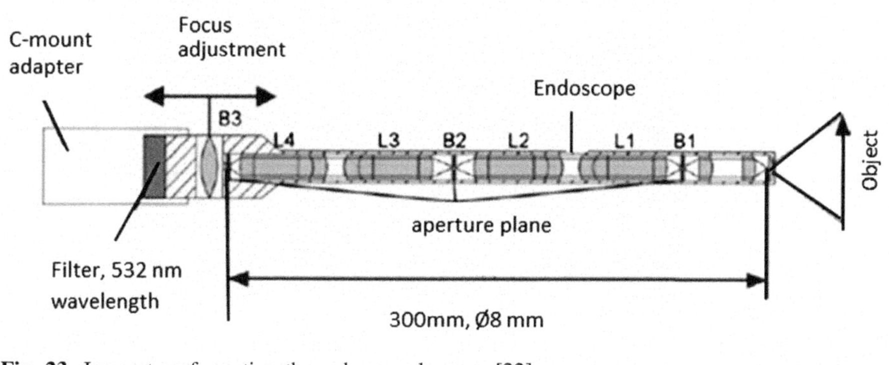


**Fig. 23** Image transformation through an endoscope [22]

coupled with a cooling arrangement in order to protect it from high temperature and high pressure combustion gases. Normally combustion is captured in the absence of light because sufficient light is emitted by the soot formed in the combustion chamber however fuel spray needs to be illuminated by an external light source. Dry and dust-free air is circulated through the endoscope assembly at ~6 bar pressure to keep it cooler and to avoid its damage during the experiment. General cavity endoscopy is used in low temperature, low pressure conditions, where cooling is not required, e.g., visualization of mixture formation process in SI engines. Optical fiber is coupled to the lens system in the endoscope for transmission of light, which is used for illuminating the area of capturing the images.

Engine endoscopes gain access into the engine combustion chamber through the cylinder head. For combustion and spray diagnostics in an IC engine, a hole is made in the cylinder head such that its endoscopic view can capture the image of the illuminated area of interest and it covers a maximum possible view of the combustion and fuel spray. Schematic of the engine combustion chamber endoscopic access and its components are shown in Fig. 24.

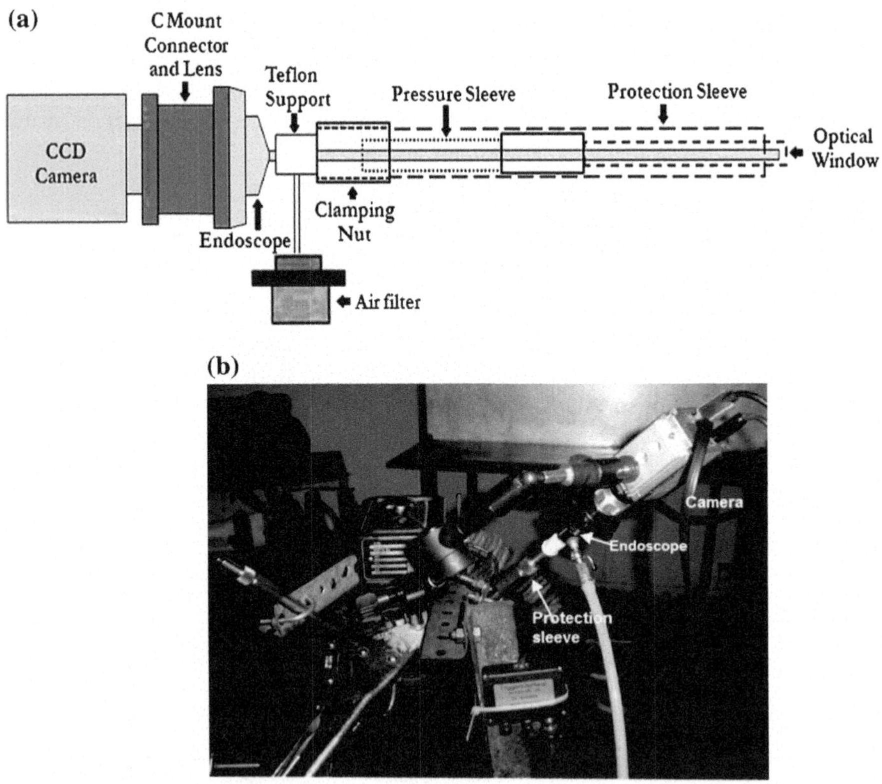


**Fig. 24** Components of the engine endoscopy and its mounting on the engine: **a** schematic of the endoscopic access system, **b** common arrangement of engine endoscope with camera [2]

For a long time, endoscopic technique has been used to investigate the combustion process in CI engines. Different combustion and spray parameters have been determined in diesel engines using this technique. Further, it is combined with lasers to provide access to complex engine geometries and used for characterization of combustion and mixture formation in indirect and direct injection spark ignition engines. It was used for modifications of combustion chamber shapes for optimizing the fuel utilization and reducing soot formation. Recently it has been used in several advanced applications in automotive industry, which were otherwise very difficult for conventional techniques such as determining piston temperature, studying the pilot injection combustion of diesel and the effect of intake charge condition on the in-cylinder combustion temperature and soot distributions. Several researchers have used endoscopy in different areas of engine research in SI and CI engines.

### 5.2.3 Application of Engine Endoscopy in Engines

Several researchers have used the endoscopic technique to diagnose the effect of different engine parameters on combustion properties. Hampson and Reitz [31] investigated the advantages of double injection over single injection on combustion characteristics such as combustion temperature and soot distribution using endoscopic visualization technique. They captured combustion images in modern heavy-duty diesel engines equipped with common rail electronically-controlled, high-pressure fuel injection system, which could inject the fuel four times in one engine cycle. Combustion temperature and soot distribution were analyzed using a two-colour pyrometry technique. They also developed a multi-dimensional simulation model for verifying the data and got a good agreement between experimental and simulated results. They explained the reason for reducing soot with split injection and advancing the injection by endoscopic images in conjugation with modelling (Fig. 25).

A combustion phenomenon was also observed by Spicher et al. [86] in a DISI engine using endoscopy. They analyzed combustion characteristics such as flame propagation speed, and flame radiation and compared them in homogeneous and stratified combustion modes. Optical spark plug and endoscope were inserted in place of one outlet valve of a 4-valve engine. Flame fronts of combustion were captured at a difference of 0.2° CA and plotted in contours (Fig. 26).

Several studies were also conducted for comparing combustion characteristics of different alternative fuels with baseline diesel with the help of an endoscope. The endoscopy proved to be very helpful in such study because different real time characteristics were analyzed by changing different parameters of the engine and fuel injection system. Miers et al. [57] conducted a similar study with Sun-diesel, which is an alternative biodiesel derived from wood-chips. The endoscope was installed in Mercedez A-class engine, which was equipped with CRDI system. They analyzed soot concentration factor from combustion images for both fuels at different SoI timings.

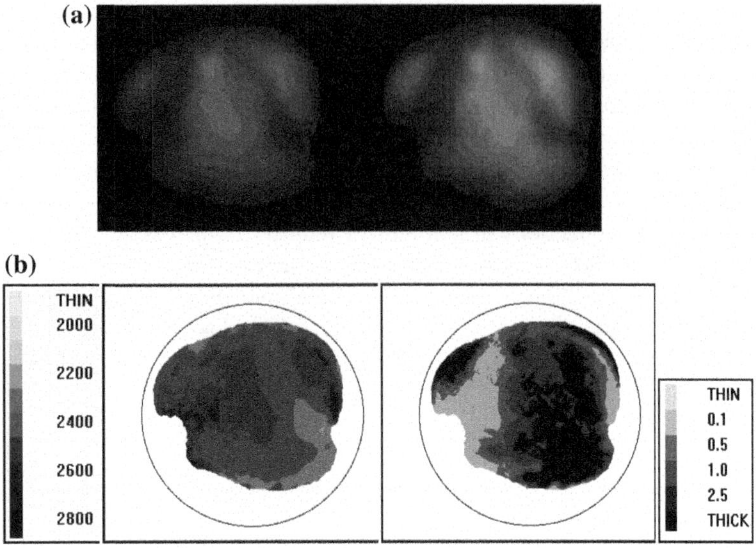


**Fig. 25** **a** Combustion image captured by endoscope, **b** Temperature and soot distribution obtained from engine endoscopy [31]

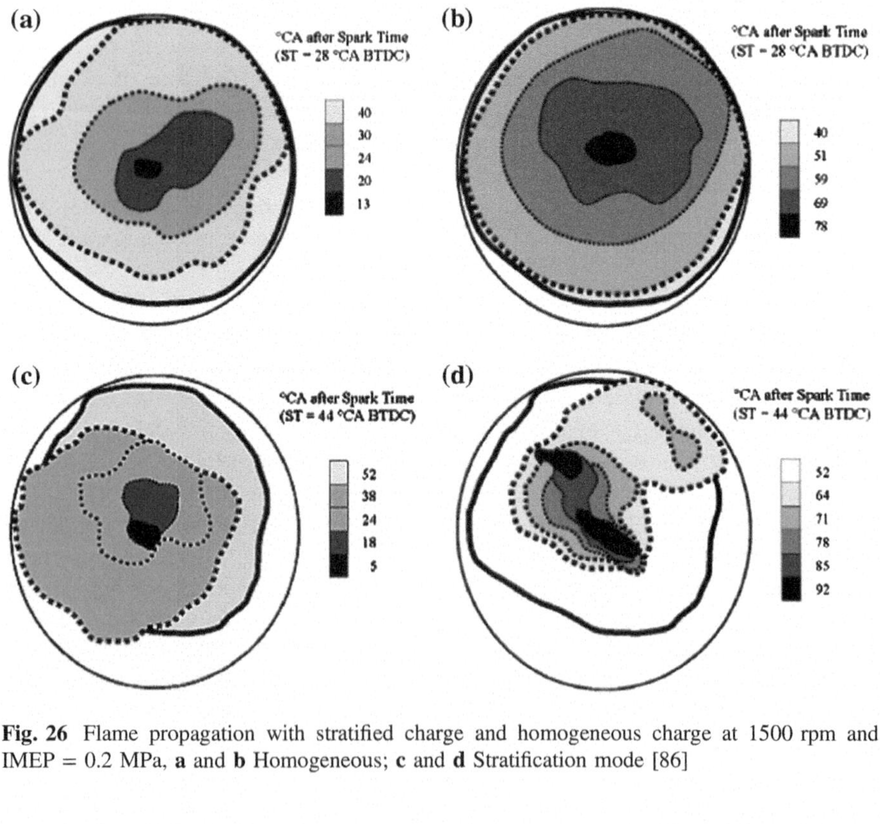


**Fig. 26** Flame propagation with stratified charge and homogeneous charge at 1500 rpm and IMEP = 0.2 MPa, **a** and **b** Homogeneous; **c** and **d** Stratification mode [86]

They found that the peak of soot concentration factor becomes higher at advanced SoI timings (Fig. 27). However in late combustion cycle, soot concentration factor is higher at retarded SoI timings due to decreased turbulence and cylinder pressure, which reduces the soot oxidation rate in late combustion cycles.

In Toyota Motors, Sakata et al. [75] developed a Toyota Reflex Burn (TRB) system to optimize the fuel-air mixing in diesel engines. At that time, gaseous and particulate emissions from small diesel engines created a huge environmental furor. They modified the combustion chamber shape, which generated complex air motion with high turbulence. Endoscope with a camera was used to visualize the mixture formation process, spray and flame behavior to achieve optimum impingement intensity. TBR system generated complex air motion from impingement with the reflex edge in the combustion chamber cavity, which created highly turbulent motion of air. They found that deflection of fuel spray away from the reflex edges reduced wall-wetting, which decreased particulate emissions (Fig. 28).

Nikolic et al. [60] used endoscope in a single cylinder rapid compression machine (RCM), which was used for simulating diesel combustion. They studied the effect of $CO_2$ concentration, which is used for diluting the intake air in the EGR to control $NO_x$ formation. Soot distribution and ignition properties were analyzed

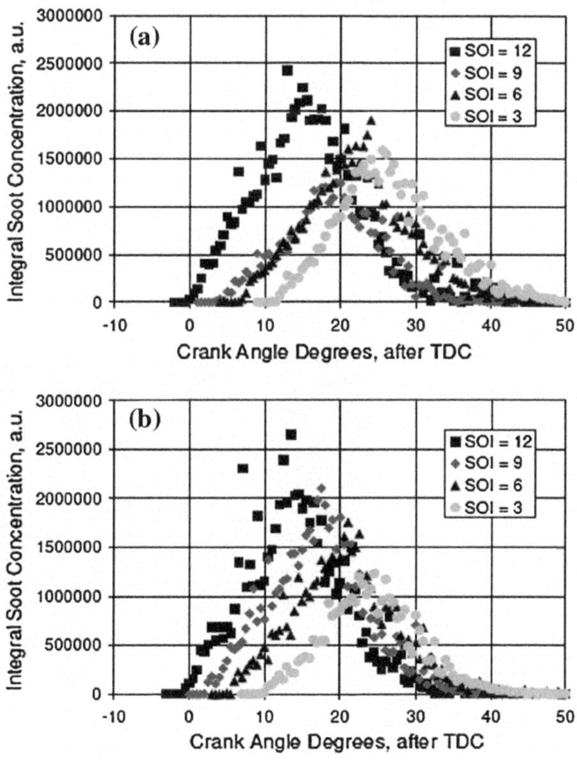


**Fig. 27** Soot concentration factor for four different SoI timings: (a) Diesel, (b) Sun-diesel [57]

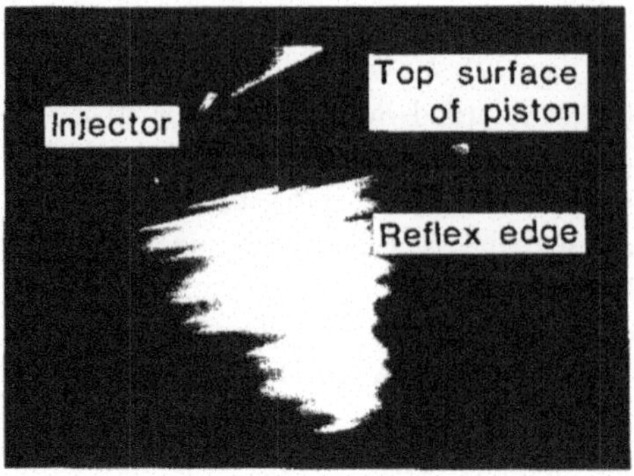


**Fig. 28** Combustion cavity captured by endoscopy [75]

with the help of combustion imaging. They captured the combustion images at 0, 4.3, 9.5, 14.3% $CO_2$ concentrations and compared the combustion and ignition parameters. They found that flame temperature and maximum soot concentration decreased with increasing $CO_2$ concentration in the intake gas, which made ignition delay longer (Fig. 29).

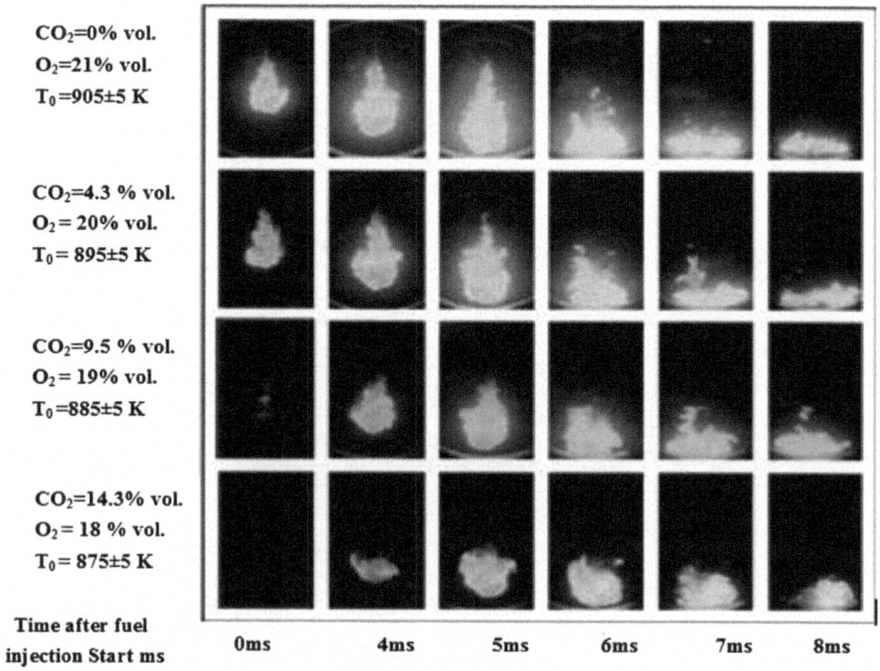


**Fig. 29** Diesel spray and combustion images at different conditions [60]

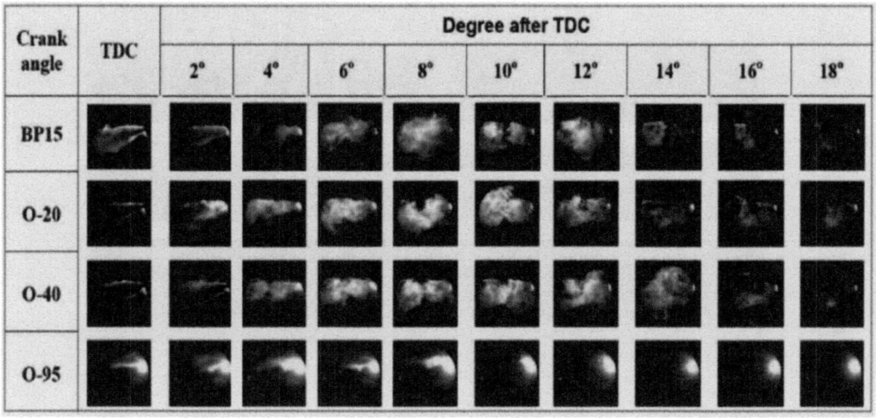


**Fig. 30** Spray and combustion images of a DI diesel engine fueled by diglyme-diesel blends [5]

Endoscopic technique is very useful in determining spray characteristics in the engine because this technique provides direct visual access to the combustion chamber and focuses on the nozzle tip. Several studies were carried out for spray visualization, but they were limited to CVSC, but not in a real firing engine, because there were no arrangements for direct access to the belly of the engine which were capable of withstanding high temperature and pressure conditions during combustion. Using endoscopes with lighting arrangement proved to be very helpful for spray analysis in the engine. General cavity endoscopes are used for low light applications. Light is passed through an optical fiber, which is coupled with the endoscope. Alam et al. [5] captured the images of sprays in diesel engine for different blends of dyglyme blended with diesel, namely 20% w/w dyglyme blended with BP15 ("O-20"), 40% w/w dyglyme blended with BP15 ("O-40") and 95% w/w dyglyme blended with BP15 ("O-95"). They compared the start of injection timings and spray penetration length for these fuels and found that O-20 showed earliest SoI and O-95 showed shortest spray penetration length (Fig. 30).

This technique was used for investigating the effect of different parameters on combustion and ignition sites at different engine speeds [71, 73]. Werlberger and Cartellieri [96] used endoscopic high-speed combustion photography in a HSDI engine to investigate the effect of pilot injection on fuel-air mixing and combustion. They also observed the effect of wall temperature on combustion characteristics. Wallner et al. (2007) applied endoscopic technique in a hydrogen fueled DI engine to investigate the effects of fuel injection parameters on fuel-air mixing and combustion characteristics. They found the endoscopic technique to be an excellent tool for in-cylinder combustion diagnostics. Agarwal et al. [2] carried out engine endoscopy in a production grade CI engine fueled by mineral diesel and biodiesel blends and investigated the radiation flame temperature and soot distribution (Fig. 31). They reported that combustion visualization using endoscopy is not effective for highly oxygenated fuels because the intensity and color of the

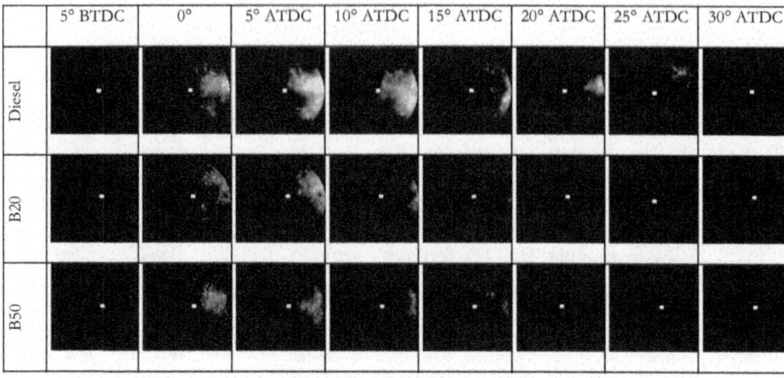


(a) Spatial and time resolved combustion endoscopy images for different fuels at various positions in an engine cycle

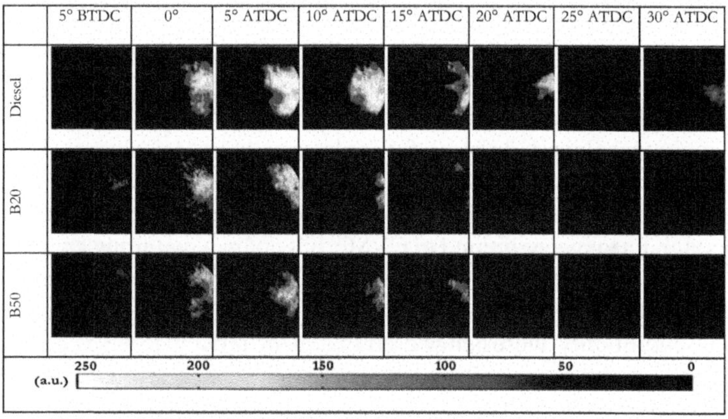


(b) Contours of R intensity for different fuels at various positions in an engine cycle

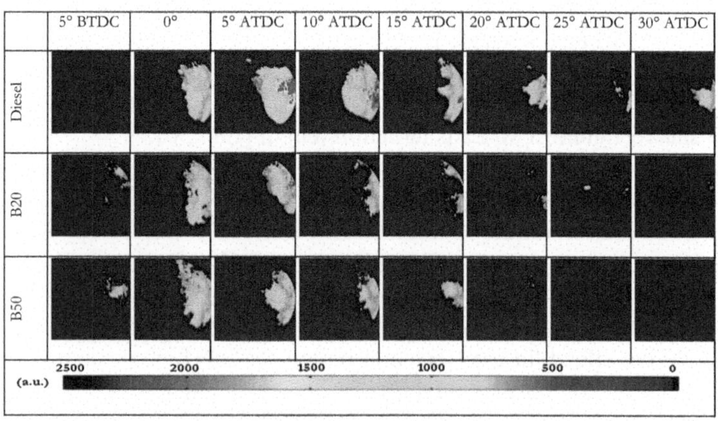


(c) Spatial and time resolved in-cylinder flame temperature distribution for different fuels at various positions in anengine cycle

**Fig. 31** Results of engine endoscopy using mineral diesel and biodiesel blends [2]

combustion images was slightly lower due to lower emission of radiations from unburnt soot particles. They found that higher biodiesel blends led to lower flame temperatures due to lower soot radiations, which indicated higher bulk temperatures. This also justified the higher $NO_x$ formation during biodiesel combustion.

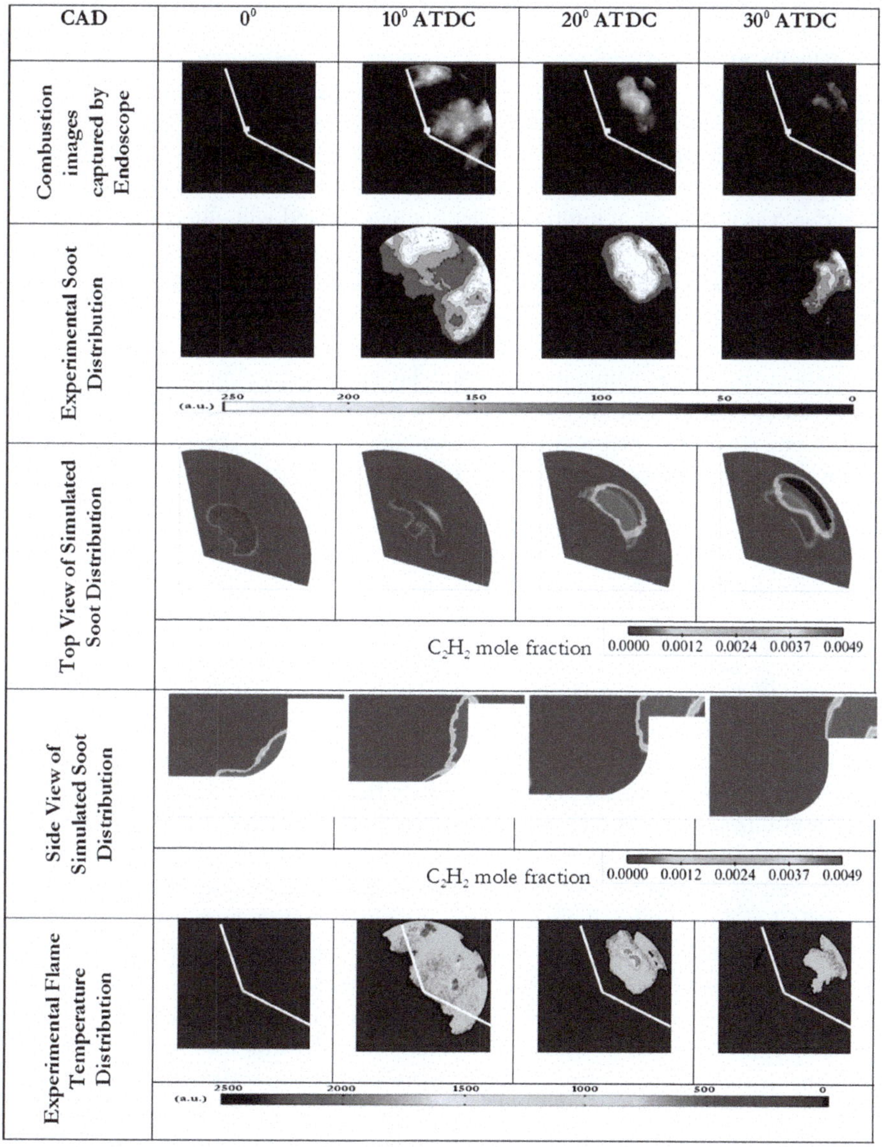


**Fig. 32** Comparison of experimental and simulation results of diesel combustion image sequence, spatial soot contours and spatial flame temperature contours at no load [3]

In several studies related to flow pattern analysis, endoscopic technique replaced conventional optical windows due to ease of operation and accessibility. Nauwerck et al. [59] used endoscopic optics to investigate the in-cylinder flow structures during the gas exchange process inside a small two-stroke engine. They successfully implemented this system for PIV under a variety of operating conditions and speeds up to 6000 rpm. Agarwal et al. [3] also compared the engine endoscopy results with simulation results and reported a good agreement between them (Fig. 32).

Schänzlin et al. [77] developed and implemented endoscopic technique for a spray-guided DISI gasoline engine. They examined the in-cylinder behavior to provide optical information on the liquid phase distribution in the combustion chamber at different times after the SoI. They performed numerical analysis for liquid phase distribution and found a good qualitative similarity among them. This technique was also used in multi-cylinder engines fuelled with alternative fuels like alcohols and biofuels to investigate the air-fuel mixing, flow pattern inside the combustion chamber and complex combustion characteristics. Several researchers used this technique for comparing the combustion characteristics of different fuels such as sun-diesel, diesel, diesohol and mineral diesel [57, 95]. Mtui and Hill [58] used endoscopy for investigating ignition delay and combustion duration in a CNG fueled diesel engine. Recently endoscopy was used for various advanced engine applications such as piston temperature evaluation, spray diagnostics, charge flow pattern, flame propagation and mixture formation in SI engines [42].

# 6 Conclusions

The primary objective of this chapter is to provide an overview of fundamentals and applications of important optical diagnostic techniques for IC engine applications. These techniques are capable of measurements of in-cylinder air-flow characteristics; macroscopic and microscopic spray analysis and combustion diagnostics. A common feature of most optical diagnostics described here is their capability of being non-intrusive and in situ measurements with high spatial and temporal resolution. Hence, these techniques are particularly suited for turbulent flow conditions encountered in IC engines. For each technique, basic theory and the principle were included. For in-cylinder flow analysis, mainly PIV has been used by a large number of researchers. Several new capabilities of advanced configurations of PIV technique such as velocity component analysis, spatial variations of velocity inside the cylinder make this technique even more versatile. Further improvements in combustion process can be achieved by enhancing the fuel-air mixing process by providing a suitable combustion environment and by optimization of parameters that critically influence the combustion. Engine endoscopy is an important optical technique for analysis of engine combustion and spray formation process. It can be used in a wide variety of engine investigations ranging from fuel-air mixing in SI engines to combustion visualization in CI engines. Endoscopy can be also used to optimize the engine design e.g. shape of the combustion chamber, piston bowl, etc. in addition to analyzing the effect of designs of different components on engine performance, combustion and emission characteristics. Engine endoscopy is widely used for determining spatial flame temperature distribution and spatial soot distribution. Spray visualization in the engine combustion chamber is another important application of endoscopy. Endoscopy offers several advantages in engine experiments due to limited space used for optical access. The applicability of this technique in any production grade engine is another major advantage of endoscopy.

In summary, optical diagnostic techniques have contributed significantly to the knowledge of in-cylinder flow characteristics, fuel-air mixture formation and combustion in IC engines. With ever more stringent emission legislations, use of advanced optical techniques will continue to grow and provide greater understanding of complex air-flows, mixing and combustion phenomena in IC engines.

## References


1. Abiola K, Zhu J, Nishida K, Wang X, Huang Z (2013) Characterization of spray and combustion processes of biodiesel fuel injected by diesel engine common rail system. Fuel 104:838–846
2. Agarwal AK, Agarwal A, Singh AP (2015) Time resolved in-situ biodiesel combustion visualization using engine endoscopy. Measurement 69:236–249
3. Agarwal AK, Singh AP, Agarwal A, Jeon J, Lee CS, Park S (2016) Spatial combustion analysis of biodiesel fueled engine using combustion chamber endoscopy and modeling. Renew Energy 98:292–303
4. Agarwal AK, Singh SK, Sinha S, Shukla MK (2004) Effect of EGR on the exhaust gas temperature and exhaust opacity in compression ignition engines. Sadhana 29(3):275–284
5. Alam M, Song KH, Boehman A (2005) Spray and combustion visualization of a direct injection diesel engine operated with diglyme fuel blends. Int J Engine Res 7:503–521
6. Allen J, Law D, Pitcher G, Williams P (2000) A new optical access research engine for advanced fuel spray and combustion analysis using laser based diagnostics. In: Monograph of 33rd international symposium on automotive technology and automation (ISATA), Dublin, Ireland
7. Ashwood AC, VandenHogen SJ, Rodarte MA, Kopplin CR, Rodríguez DJ, Hurlburt ET, Shedd TA (2015) A multiphase, micro-scale PIV measurement technique for liquid film velocity measurements in annular two-phase flow. Int J Multiph Flow 68:27–39
8. Ayati AA, Kolaas J, Jensen A, Johnson GW (2015) Combined simultaneous two-phase PIV and interface elevation measurements in stratified gas/liquid pipe flow. Int J Multiph Flow 74:45–58
9. Baum E, Peterson B, Böhm B, Dreizler A (2014) On the validation of LES applied to internal combustion engine flows: Part 1: Comprehensive experimental database. Flow Turbul Combust 92:269–297
10. Baum E, Peterson B, Surmann C, Michaelis D, Böhm B, Dreizler A (2012) Tomographic PIV measurements in an IC engine. In: 16th international symposium on applications of laser techniques to fluid mechanics, Lisbon, Portugal, 09–12 July 2012
11. Baum E, Peterson B, Surmann C, Michaelis D, Bohmc B, Dreizler A (2013) Investigation of the 3D flow field in an IC engine using tomographic PIV. Proc Combust Inst 34:2903–2910

12. Binjuwair SA (2013) Characterization of flow structures inside an engine cylinder under steady state condition. Doctoral thesis, Department of Aeronautical and Automotive Engineering, Loughborough University
13. Bowditch FW (1961) A new tool for combustion research: a quartz piston engine. SAE Paper 610002
14. Chan VSS (2000) Velocity measurement inside a motored internal combustion engine using three-component laser Doppler anemometry. Opt Laser Technol 32:557–566
15. Chaves H, Kirmse C, Obermeier F (2001) Velocity measurements of dense diesel sprays in pressurized air. In: Proceedings of Spray
16. Choi K, Choi BC, Park H, Joo KJ, Lee JH (2003) In-cylinder flow analysis of a single cylinder DI diesel engine using PIV and CFD. SAE Paper 2003-01-1846
17. Course on IC Engine About Spectroscopy in Optical Engineering Laboratory, University of Warwick. http://www.eng.warwick.ac.uk/oel/courses/engine/ic029.htm
18. Dannemann J, Klaas M, Schröder W (2010) Three dimensional flow field within a four valve combustion engine measured by particle-image velocimetry. In: ISFV14—14th international symposium on flow visualization, EXCO Daegu, Korea, 21–24 June 2010
19. Das S, Chmiel DM (2003) Computational and experimental study of in-cylinder flow in a direct injection gasoline (DIG) engine. SAE Paper 2003-01-3083
20. Dent JC (1971) Basis for the comparison of various experimental methods for studying spray penetration. SAE Paper 710571
21. Deslandes W, Dupont A, Georges XB, Boree CJ (2003) PIV measurements of internal aerodynamic of diesel combustion chamber. SAE Paper 2003-01-3083
22. Dierksheide U, Meyer P, Hovestadt T, Hentschel W (2002) Endoscopic 2D particle image velocimetry (PIV) flow field measurements in IC engines. Exp Fluids 33:794–800
23. Docquier N, Candel S (2008) Combustion control and sensors: a review. Prog Energy Combust Sci 28:17–150
24. Ejim C, Fleck B, Amirfazli A (2007) Analytical study for atomization of biodiesels and their blends in a typical injector: surface tension and viscosity effects. Fuel 86:1534–1544
25. Elkotb MM (1982) Fuel atomization for spray modeling. Prog Energy Combust Sci 61–91
26. Faeth GM, Hsiang LP, Wu PK (1995) Structures and breakup properties of sprays. Int J Multiph Flow 21:9–127
27. Gadekar S, Singh AP, Agarwal AK (2016a) Tomographic PIV evaluation of in-cylinder flow evolution and effect of engine speed. SAE Paper 2016-01-0638
28. Galle J, Demuynck J, Vancoillie J, Verhelst S (2012) Spray parameter comparison between diesel and vegetable oils for non-evaporating conditions. SAE Paper 2012-01-0461
29. Gao Y, Deng J, Li C, Dang F, Liao Z, Wu Z et al (2009) Experimental study of the spray characteristics of biodiesel based on inedible oil. Biotechnol Adv 27:616–624
30. Ghurri A, Jae-duk K, Kyu-Keun S, Jae-Youn J, Hyung GK (2011) Qualitative and quantitative analysis of spray characteristics of diesel and biodiesel blend on common-rail injection system. J Mech Sci Technol 25:885–893
31. Hampson GJ, Reitz RD (1998) Two-color imaging of in-cylinder soot concentration and temperature in a heavy-duty DI diesel engine with comparison to multidimensional modeling for single and split injections. SAE Paper 980524
32. Han YT, Ki-bum K, Lee G (2008) The investigation of soot and temperature distributions in a visualized direct injection diesel engine using laser diagnostics. Meas Sci Technol 19:115402–115412
33. Handbook: Henein NA, Patterson DJ (1972) Emissions from combustion engine and their control. Ann Arbor Science Publishers, Ann arbor, MI
34. Handbook: Heywood JB (1988) Internal combustion engine. McGraw-Hill Book Company, New York
35. Handbook: Majewski WA, Khair MK (2006) Diesel emission and their control. SAE International, pp 367–377
36. Handbook: Pundir BP (2010) IC engines: combustion and emissions. Narosa Publishing House, New Delhi

37. Handbook: Raffel M, Willert CE, Wereley S, Kompenhans J (2007) Particle image velocimetry, a practical guide. Springer, Berlin, Heidelberg
38. He C, Ge Y, Tan J, Han X (2008) Spray properties of alternative fuels: a comparative analysis of biodiesel and diesel. Int J Energy Res 32:1329–1338
39. Hiroyasu H, Arai M (1990) Structures of fuels sprays in diesel engines. SAE Paper 1990-90-0475
40. Huang RF, Huang CW, Chang SB, Yang HS, Lina TW, Hsub WY (2005) Topological flow evolutions in cylinder of a motored engine during intake and compression strokes. J Fluids Struct 20:105–127
41. Huang Z, Lu H, Jiang D, Zeng K, Liu B, Zhang J, Wang X (2004) Combustion behaviors of a compression-ignition engine fuelled with diesel/methanol blends under various fuel delivery advance angles. Bioresour Technol 95(3):331–341
42. Husberg T, Gjirja S, Denbratt I (2005) Piston temperature measurement by use of thermographic phosphors and thermocouples in a heavy-duty diesel engine run under partly premixed conditions. SAE Paper 2005-01-1646
43. Julien L (2006) Characteristics of diesel sprays at high temperatures and pressures. PhD thesis, University of Brighton, UK
44. Justham T, Jarvis S, Clarke A, Garner CP, Hargrave GK, Halliwell NA (2006) simultaneous study of intake and in-cylinder IC engine flow fields to provide an insight into intake induced cyclic variations. J Phys: Conf Ser 45:146–153
45. Kawaguchi Y, Segawa T, Feng Z, Li P (2002) Experimental study on drag-reducing channel flow with surfactant additives—spatial structure of turbulence investigated by PIV system. Int J Heat Fluid Flow 23:700–709
46. Kim YD, Lee SY (2002) Application of Hough transform to image processing of heavily overlapped particles with spherical shapes. Atomization Sprays 12:451–461
47. Kim HJ, Park SH, Lee CS (2010) A study on the macroscopic spray behavior and atomization characteristics of biodiesel and dimethyl ether sprays under increased ambient pressure. Fuel Process Technol 91:354–363
48. Koch T, Schänzlin K, Boulouchos K (2002) Characterization and phenomenological modeling of mixture formation and combustion in a direct injection spark ignition engine. SAE Paper 2002-01-1138
49. Koh KU, Kim JY, Lee SY (2001) Determination of in-focus criteria and depth field in image processing of spray particles. Atomization Sprays 11:317–333
50. Lee CS, Park SW (2002) An experimental and numerical study on fuel atomization characteristics of high-pressure diesel injection sprays. Fuel 81:2417–2423
51. Lee CS, Reitz RD (2004) Investigation of spray characteristics from a low-pressure common rail injector for use in a homogeneous charge compression ignition engine. Meas Sci Technol 15:509–519
52. Lee J, Farrell PV' (1993) Intake valve flow measurements of an IC engine using particle image velocimetry. SAE 930480
53. Lefebvre AH (1989) Atomization and sprays, vol 12. Hemisphere
54. Manual: Operation of the phase doppler interferometer (PDI) for spray drop size and velocity measurement. Artium Technologies
55. Manual: Tomographic PIV, volumetric flow field imaging. Lavision, Germany
56. Metghalchi M, Keck JC (1982) Burning velocities of mixtures of air with methanol, isooctane, and indolene at high pressure and temperature. Combust Flame 48(1982):191–210
57. Miers SA, Ng H, Ciatti SA (2005) Emissions, performance, and in-cylinder combustion analysis in a light-duty diesel engine operating on a fischer-tropsch, biomass-to-liquid fuel. SAE Paper 2005-01-3670
58. Mtui PL, Hill PG (1996) Ignition delay and combustion duration with natural gas fueling of diesel engines. SAE Paper 961933
59. Nauwerck, A, Gindele J, Spicher U, Rosskamp H (2000) Investigation of the transient in-cylinder flow inside a two stroke engine with particle-image-velocimetry. SAE Paper 2000-01-0902

60. Nikolic D, Iida N (2007) Effects of intake $CO_2$ concentrations on fuel spray flame temperatures and soot formations. J Automobile Eng Proc IMech E 221:1567
61. Nishiyama A, Jeong H, Ikeda Y, Sawada R (2012) Application of endoscopic stereo PIV to 3D exhaust gas flow: measurements in a practical SI engine. In: 16th International symposium on applications of laser techniques to fluid mechanics, Lisbon, Portugal
62. Park SH, Kim HJ, Suh HK, Lee CS (2009) Experimental and numerical analysis of spray-atomization characteristics of biodiesel fuel in various fuel and ambient temperatures conditions. Int J Heat Fluid Flow 30:960–970
63. Park SH, Yoon SH, Lee CS (2011) Effects of multiple-injection strategies on overall spray behavior, combustion, and emissions reduction characteristics of biodiesel fuel. Appl Energy 88:88–98
64. Park SH, Youn IM, Lim Y, Lee CS (2013) Influence of the mixture of gasoline and diesel fuels on droplet atomization, combustion, and exhaust emission characteristics in a compression ignition engine. Fuel Process Technol 106:392–401
65. Park SW, Lee CS (2003) Macroscopic structure and atomization characteristics of high-speed diesel spray. Int J Automot Technol 4:157–164
66. Pastor JV, Garcia M, Molina S (2000) Analysis methodology of diesel spray and flame by means of in-cylinder endoscopic imaging. IEEE Savoy Place, London, vol 80
67. Perini F, Miles PC, Reitz RD (2014) A comprehensive modeling study of in-cylinder fluid flows in a high-swirl, light-duty optical diesel engine. Comput Fluids 105:113–124
68. Postrioti L; Grimald C, (2000) Experimental comparison between conventional and bio-derived fuels sprays from a common rail injection system, SAE 2000-01-1252
69. Reeves M, Garner CP, Dent JC, Halliwell NA (1996) Particle image velocimetry analysis of ic engine in-cylinder flows. Opt Lasers Eng 25:415–432
70. Rhodes DB, Keck JC, Burnign L (1985) Speed measurements of indolene-air-diluent mixtures at high pressures and temperature. SAE Paper 850047
71. Ricart LM, Xin J, Bower GR, Reitz RD (1997) In-cylinder measurement and modelling of liquid fuel spray penetration in a heavy-duty diesel engine. SAE Paper 971591
72. Richter M, Axelsson B, Aldén M, Josefsson G, Carlsson LO, Dahlberg M, Nisbet J, Simonsen H (1999) Investigation of the fuel distribution and the in-cylinder flow field in a stratified charge engine using laser techniques and comparison with CFD modeling. SAE Paper 1999-01-3540
73. Ryan TW, Shahed SM (1994) Injection pressure and intake air density effects on ignition and combustion in a 4-valve diesel engine. SAE Paper 941919
74. Sakaguchi D, Amida OL, Ueki H, Ishida M (2008) Measurement of droplet size distribution in core region of high-speed spray by micro-probe L2F. J Therm Sci 17:90–96
75. Sakata I, Ishisaka K, Yanagihara H, Sami H (1990) Development of TOYOTA reflex burn (TRB) system in DI diesel. SAE Paper 900658
76. Sasaki S, Akagawa H, Tsujimura K (1998) A study on surrounding air flow induced by diesel sprays. SAE Paper 980805
77. Schänzlin K, Koch T, Tzannis A, Boulouchos K (2001) Characterization of mixture formation in a direct injected spark ignition engine. SAE Paper 2001-01-1909
78. Schodl R (1974) $L_2F$ a laser-dual-beam method for flow measurements in turbomachines. ASME Paper No.74, GT-157
79. Schunemann E, Fedrow S, Leipertz A (1998) Droplet size and velocity measurements for the characterization of a DI-diesel spray impinging on a flat wall. SAE Paper 982345
80. Singh AP, Gadekar S, Agarwal AK (2016b) In-cylinder air-flow characteristics using tomographic PIV at different engine speeds, intake air temperatures and intake valve deactivation in a single cylinder optical research engine. SAE Paper 2016-28-0001
81. Singh AP, Gupta A, Agarwal AK (2015) Tomographic particle image velocimetry for flow analysis in a single cylinder optical engine. SAE Int J Mater Manuf 8(2):472–481
82. Som S, Longman DE, Ramírez AI, Aggarwal SK (2010) A comparison of injector flow and spray characteristics of biodiesel with petrodiesel. Fuel 89:4014–4024

83. Son SY, Kihm KD, Han JC (2002) PIV flow measurements for heat transfer characterization in two-pass square channels with smooth and 90 ribbed walls. Int J Heat Mass Transf 45:4809–4822
84. Spicher U, Kölmel A, Kubach H, Töpfer G (2000) Combustion in spark ignition engines with direct injection. SAE Paper 2000-01-0649
85. Stucky MJ, Nino E, Gajdeczko B, Felton PG (1994) Two-color particle image velocimetry technique for an internal combustion engine. Exp Thermal Fluid Sci 8:305–314
86. Subramaniyam S, Ganesan V, Rao SP, Sampath S (1990) Turbulent flow inside the cylinder of a diesel engine—an experimental investigation using hot wire anemometer. Exp Fluids 9:167–174
87. Suh HK, Roh HG, Lee CS (2008) Spray and combustion characteristics of biodiesel/diesel blended fuel in a direct injection common-rail diesel engine. J Eng Gas Turbines Power 130:032807
88. Sun JH, Yates A, Winterbone DE (1996) Measurement of flow field in diesel engine combustion chamber after combustion by cross correlation method. Exp Fluids 20:335–345
89. Sweetland P, Reitz RD (1994) Particle image velocimetry measurements in the piston bowl of a DI diesel engine. SAE Paper 940283
90. Tanaka T, Narahara K, Tabata M, Yoshiyama S, Tomita E (2005) Ion current measurement in a homogeneous charge compression ignition engine. Int J Engine Res 6:453–463
91. Valentino G, Kaufman D, Farrell P (1993) Intake valve flow measurements using PIV. SAE Paper 932700
92. Wang X, Huang Z, Kuti OA, Zhang W, Nishida K (2010) Experimental and analytical study on biodiesel and diesel spray characteristics under ultra-high injection pressure. Int J Heat Fluid Flow 31:659–666
93. Wattanavichien K, Azetsu, A (2004) Studies of visualized diesohol combustion phenomena in IDI engine. In: The international symposium on diagnostics and modeling of combustion in internal combustion engines 2004, vol 6, pp 423–430, 2004-08-02
94. Werlberger P, Cartellieri W (1987) Fuel injection and combustion phenomena in a high speed DI diesel engine observed by means of endoscopic high speed photography. SAE Paper 870097
95. Wernert P, Koerber G, Wietrich F, Raffel M, Kompenhans J (1997) Demonstration by PIV of the non-reproducibility of the flow field around an airfoil pitching under deep dynamic stall conditions and consequences. Aerosp Sci Technol 2:125–135
96. Xu D, Chen J (2013) Accurate estimate of turbulent dissipation rate using PIV data. Exp Thermal Fluid Sci 44:662–672
97. Zentgraf F, Baum E, Böhm B, Dreizler A, Peterson B (2014) Analysis of the turbulent in-cylinder flow in an IC engine using tomographic and planar PIV measurements. In: 17th international symposium on applications of laser techniques to fluid mechanics, Lisbon, Portugal, 07–10 July 2014
98. Zhang X, Su G, Yu J, Yao Z, He F (2015) PIV measurement and simulation of turbulent thermal free convection over a small heat source in a large enclosed cavity. Build Environ 90:105–113

# Exergy Destruction Study on Small DI Diesel Engine

**Veena Chaudhary and R.P. Gakkhar**

**Abstract** In this study, energy and exergy analyses have been carried out using the experimental data for a small naturally aspirated, direct injection, single cylinder, 4 stroke diesel engine. The exergy and performance parameters were computed from the measured data. It has been found that significant exergy destruction occurs at low loads. Exergy destruction percentage decreases as the load increases at higher speeds. For low speed and low load conditions the exergy destruction is higher as compared to that at high speed and higher load conditions.

**Keywords** Exergy · Exergy destruction · Second law · Diesel engine · Entropy generation

## Nomenclature

| | |
|---|---|
| A/F | Air fuel ratio |
| BSFC | Brake specific fuel consumption |
| Bth | Brake thermal efficiency |
| BP | Brake power |
| $c_{p,e}$ | Specific heat of exhaust gas |
| $c_{p,w}$ | Specific heat of cooling water |
| $\bar{h}$ | Specific enthalpy |
| $h_{in}$ | Input enthalpy |
| $h_{out}$ | Outlet enthalpy |
| LHV | Lower heating value |
| $\dot{m}_a$ | Mass flow rate of air |
| $\dot{m}_{eg}$ | Exhaust gas flow rate |
| $\dot{m}_f$ | Mass flow rate of fuel |

V. Chaudhary · R.P. Gakkhar (✉)
Mechanical & Industrial Engineering Department, IIT Roorkee, Roorkee 247667, Uttarakhand, India
e-mail: gakhrfme@iitr.ac.in

V. Chaudhary
e-mail: aim12dme@iitr.ac.in



A.K. Agarwal et al. (eds.), *Combustion for Power Generation and Transportation*, DOI 10.1007/978-981-10-3785-6_12

| | |
|---|---|
| $\dot{m}_w$ | Cooling water flow rate |
| N | Engine speed |
| $P_e$ | Exhaust gas pressure |
| $P$ | Products |
| $P_0$ | Ambient pressure |
| $\dot{Q}_{cw}$ | Heat carried away by cooling water |
| $\dot{Q}_{eg}$ | Heat carried away by exhaust gas |
| $\dot{Q}_{in}$ | Heat input |
| R | Universal Gas Constant |
| $r$ | Reactants |
| $S_{gen}$ | Entropy generation |
| $T$ | Torque |
| $T_{cw}$ | Cooling water temperature |
| $T_{ex}$ | Exhaust gas temperature |
| $T_0$ | Ambient temperature |
| $T_{wi}$ | Cooling water inlet temperature |
| $T_{wo}$ | Cooling water outlet temperature |
| $\dot{\zeta}_{destruction}$ | Exergy destruction |
| $\dot{\zeta}_{exhaust}$ | Exhaust exergy rate |
| $\dot{\zeta}_{in}$ | Exergy (fuel) input rate |
| $\dot{\zeta}_{Q}$ | Exergy rate accompanying with heat transfer |
| $\dot{\zeta}_{w}$ | Exergy rate accompanying with work transfer |
| $\eta_{II}$ | Second law efficiency |

## 1 Introduction

Diesel engines are widely used in transportation and power sectors because of their higher efficiency and ruggedness. However, they are also a major source of particulate matter and unburned hydrocarbon emissions.

There have been growing concerns about effect of emissions on global climate change and increasing awareness that world's energy resources will not last long. These concerns have prompted the scientific community to direct their efforts to (i) explore the feasibility of using environment friendly renewable energy sources like bio-diesel and (ii) have a closer look at energy conversion processes and systems for evolving better processes or developing improved systems for better utilization of fuels.

There have been several studies [1–3] concerning (ii) above. These studies involve the engine analysis based on combined first and second laws of thermodynamics. Analysis based on the first law of thermodynamics is quite useful in assessing the thermal performance of the system as well as the efficient conversion of supplied heat to useful shaft work. However, this alone is inadequate as it does not differentiate energies of different grades. It concerns only with the quantity of energy. It neither deals with the quality of energy nor its degradation during the process. Second law and its deductions on the other hand are useful in the prediction of best theoretical performance of thermal systems. They are also helpful in the quantitative evaluation of factors that prevent the attainment of best theoretical performance level [4]. Analysis based on the combined first and second laws of thermodynamics is commonly known as exergy analysis or availability analysis. The method of exergy analysis is quite suitable for achieving more efficient resource use as it aids in the determination of locations, types and true magnitudes of irreversibilities. This information is quite helpful for the design of thermal systems, for directing the efforts to reduce the sources of inefficiency in the existing systems and to evaluate system economics [4–6].

In thermodynamics, availability or exergy is defined as the maximum theoretical work which can be obtained as the system comes to equilibrium with its reference environment. Unlike energy, exergy is not a conserved quantity. It gets destroyed due to the presence of irreversibilities. Exergy analysis involves the concept of exergy transfer associated with heat, work and mass transfers and exergy destruction due to irreversibilities. The magnitude of exergy destruction is a quantitative measure of irreversibilities present in the system. Thus, the knowledge of the magnitude of exergy destruction and identification of the component in which maximum exergy destruction occurs can be helpful in the better design of the equipment and selection of appropriate processes for minimizing losses.

Reviews of investigations based on second law of thermodynamics for the study of internal combustion engines have been presented by Caton [1], Rakopoulos and Giakoumis [2] and Jelic and Ninic [3]. Caton [1] presented a review of studies carried over a period from 1950–2000 on both diesel and spark ignition engines. Representative results for both SI and CI engines have been discussed to high-light the information that can be obtained by using second law analysis. Instantaneous values for the exergy and overall values of energy and exergy have been described.

An extensive and comprehensive review is by Rakopoulos and Giakoumis [2]. Energy and exergy balance equations were applied to simulation models for engine cylinder and various subsystems. Second law balances for both steady and transient operations of the engine have been discussed. It also covers the second law applications to low heat rejection engines and engines using alternative fuels.

Rakopoulos et al. [7] studied the effect of operating conditions on the exergy transfer and destruction for a reduced heat rejection engine. Energy and exergy balances were applied to models for engine processes occurring inside the cylinder. From the measured data exergy parameters were computed. Results from this study indicate that the exergy destruction decreases and the exhaust gas exergy increases with increasing fuel-air ratio. Both exergy destruction and exhaust gas exergy increase with engine speed and slightly decrease with increase in injection duration. Also the exergy destruction decreases with increase in load.

The objective of this study is to assess the exergy destruction in an operating small DI diesel engine at various loads and speeds to ascertain optimum conditions for its best performance.

## 2 Experimental Setup

The experiments were performed on a 4-stroke single cylinder, direct injection diesel engine. Table 1 shows the specifications of the test engine in the present experimental work. The engine load was varied using AVL Dycon 201A. The intake airflow rate was measured by using the air-box method. The engine speed was measured using a magnetic speed pickup. AVL Ditest Gas 1000 light five gas analyzer was used to measure the carbon monoxide (CO), unburned hydrocarbon (HC), and nitrogen oxides ($NO_x$) emissions. Smoke density was measured using AVL 480 smoke meter. All measuring instruments were calibrated properly before being used. All the experiments were performed under steady-state conditions. Each test run was repeated twice randomly. The experimental setup layout is given in Fig. 1. The

**Table 1** Technical Specifications of test DI Diesel engine

| Engine | Greaves Cotton |
|---|---|
| Model | GL 600 WII |
| Number of cylinders | 1 |
| Bore (mm) | 92 |
| Stroke (mm) | 92 |
| Displacement (cc) | 611 |
| Maximum power (kW) at rated RPM | 8.1 |
| Rated rpm | 3000 |
| Compression ratio | 18 |

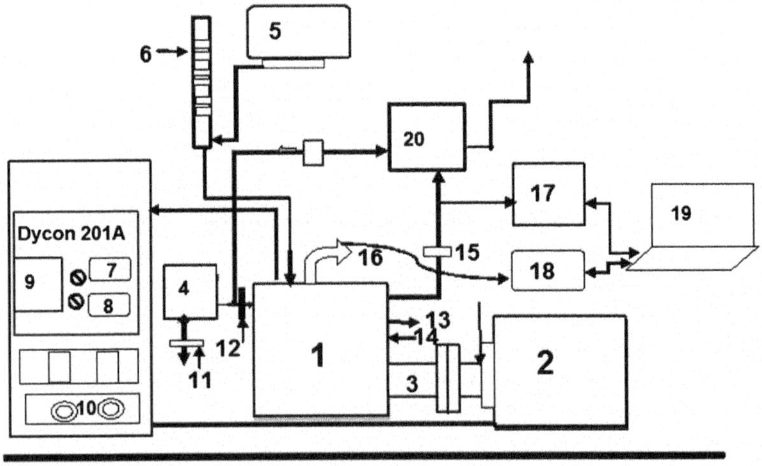


**Fig. 1** Schematic diagram of experimental setup. *1* DI Diesel Engine. *2* Eddy current dynamometer. *3* Universal coupling. *4* Air box. *5* Fuel tank. *6* Fuel consumption measurement. *7* Speed indicator. *8* Torque indicator. *9* Operating mode of dycon panel. *10* Throttle control. *11* U tube manometer of air box. *12* Intake air temperature. *13* Water outlet. *14* Water inlet. *15* Exhaust gas temperature. *16* Pressure transducer. *17* Exhaust gas analyzer. *18* DPM 800. *19* Personal computer. *20* Exhaust gas surge tank

steady state test data was used to perform energy and exergy balances and to compute performance and exergy related parameters. The reaction equation was determined using exhaust emission measurements.

## 3 Reaction Equation

Energy and exergy analyses of the test engine for each test can be performed after determining the reaction equation for which following simplifying assumptions were made:

(i) There is no water vapor in the combustion air.
(ii) The air contains oxygen and nitrogen only. All components except oxygen are grouped to-gather with nitrogen. Thus on molar basis air contains 21% oxygen and 79% nitrogen.
(iii) Nitrogen is considered as inert. It does not take part in combustion reactions to

Produce oxides of nitrogen. The chemical formula for the diesel was taken as mentioned in Pundir [8]. The molar basis reaction equation can be written as

$$\begin{gathered} aC_{14.6}H_{24.9} + b\,(O_2 + 3.76N_2) \rightarrow cCO_2 + dCO + eO_2 + fN_2 + gNO + hC_{14.6}H_{24.9} + iH_2O \\ \textit{Carbon balance:} \quad 14.6 * a = c + d + h\,(14.6); \\ \textit{Hydrogen balance:} \quad a * 24.9 = 2i + 24.9 * h \\ \textit{Oxygen balance:} \quad 2 * b = 2c + d + 2e + g + i \\ \textit{Nitrogen balance:} \quad 2 * b * 3.76 = 2f + g \end{gathered} \tag{1}$$

Using the concentrations of exhaust gases as indicated by the exhaust gas analyzer, the unknown coefficients as mentioned in the above equation were calculated to determine the loss of energy in the exhaust.

## 4 Energy Analysis

The engine is considered as an open system is shown in Fig. 2. Air and fuel enter the control volume and the combustion products exit the control volume. In another stream, cooling water enters and exits the control volume. Lubricating oil circulates within the control volume. It does not cross the control surface.

Following assumptions were made:

- The engine operates at steady state.
- The combustion air and the exhaust gas each forms an ideal gas mixture.

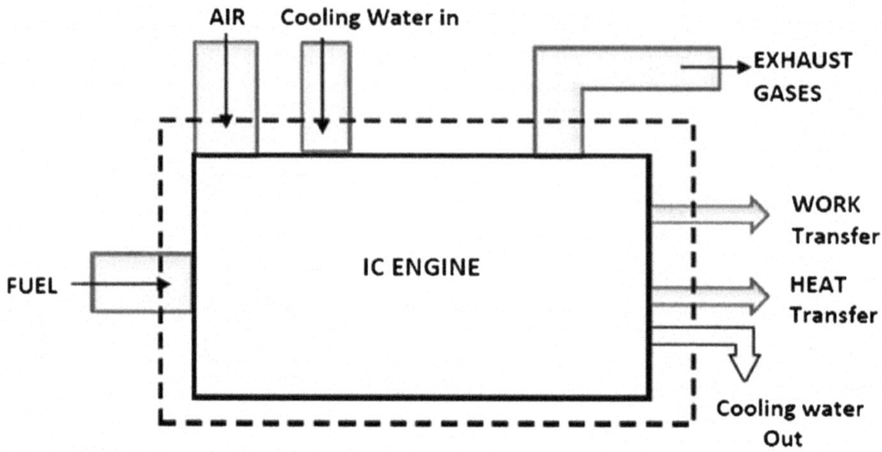


**Fig. 2** Engine as control volume

- Potential and kinetic energy effects of the incoming and outgoing fluid streams are ignored.

The governing equations of mass, energy and exergy conservation for a steady flow are:

$$\sum \dot{m}_{in} = \sum \dot{m}_{out} \tag{2}$$

where $\dot{m}$ mass flow rate and the subscript 'in' stands for inlet and 'out' stands for outlet.

The general equation of energy balance based on First law of thermodynamics

$$\dot{Q}_{in} + \sum \dot{m}_{in} h_{in} = \dot{W} + \sum \dot{m}_{out} h_{out} \tag{3}$$

Rate of energy entering the engine or the fuel power can be calculated by;

$$\dot{Q}_{in} = \dot{m}_f \times LHV; \quad \mathbf{kW} \tag{4}$$

The available shaft power or the brake power can be written as,

$$BP = \frac{2\Pi NT}{60,000}, \ \mathbf{kW} \tag{5}$$

where; T is torque in Nm, N is revolution per minute.

Brake specific fuel consumption is given by;

$$\mathbf{BSFC} = (\dot{\mathbf{m}}_\mathbf{f} / 3600 \times \mathbf{BP}) \quad \mathbf{kg/kWh} \tag{6}$$

Brake thermal efficiency is given by:

$$\eta_{bth} = \frac{BP}{\dot{Q}_{in}} \tag{7}$$

Total heat energy carried away by the cooling water is given by

$$\dot{Q}_{cw} = \dot{m}_{cw} \times c_w \times (T_{wo} - T_{wi}),\ kW \tag{8}$$

w here; $\dot{m}_{cw}$ is the mass flow rate of cooling water in kg/sec passing through the engine jacket,

Heat carried away by exhaust gasses is given by the following equation

$$\dot{Q}_{eg} = \sum_{p} n\,\bar{h}\,(T_{ex}) - \sum_{r} n\,\bar{h}\,(T_0) \tag{9}$$

here n represents the molar ratio of each exhaust components per unit of fuel based on the balance of Eq. 1. $\bar{h}\,(T_{ex})$ and $\bar{h}\,(T_0)$ represent the molar specific enthalpy values for exhaust components at the exhaust temperature and at the ambient temperature respectively.

The amount of the unaccounted losses determined by performing an energy balance is given by;

$$\dot{Q}_{unaccounted} = \left[\dot{Q}_{in} - (BP + \dot{Q}_{cw} + \dot{Q}_{eg})\right],\ \mathrm{kW} \tag{10}$$

## 5 The Exergy Analysis

Similar to energy analysis, same assumptions were made for exergy analysis; the whole engine was considered to be an open system as shown in Fig. 2 operating at steady state. The exergy balance under steady state conditions, for the control volume can be stated as;

$$-\dot{\zeta}_Q - \dot{\zeta}_W + \dot{\zeta}_{in} - \dot{\zeta}_{eg} - \dot{\zeta}_{destroyed} = 0 \tag{11}$$

In an IC engines, the exergy input is in the form of chemical exergy of the fuel. This input exergy is distributed in the following:

I. Useful work output or shaft work exergy ($\dot{\zeta}_W$)
II. Exergy transferred rate to cooling water ($\dot{\zeta}_Q$);
III. Exergy transferred rate to the exhaust gases ($\dot{\zeta}_{eg}$);
IV. Exergy destructions ($\dot{\zeta}_{destroyed}$) due to friction, radiation, heat transfer to surroundings etc.

Total exergy of fuel is the sum of thermo-mechanical and chemical exergies. The thermo-mechanical exergy component of fuel is zero as it enters the control volume at environment conditions. Similarly the chemical exergy component of the exit stream is zero as the reaction components are stable species at the dead state. Following relationship is used for computing the chemical exergy of liquid hydrocarbons having the chemical formula of $C_xH_y$ (x = 14.4, y = 24.9) proposed by Kotas [9];

$$\zeta_{in} = \dot{Q}_{in}\left[1.04224 + 0.011925\frac{y}{x} - \frac{0.42}{x}\right] \tag{12}$$

$$\text{Shaft work exergy } \zeta_W = \text{brake power of the engine} = \text{BP} \tag{13}$$

Exergy transfer accompanying cooling water is given by the following equation:

$$\zeta_Q = \sum\left(1 - \frac{T_0}{T_{cw}}\right)\dot{Q}_{CW} \tag{14}$$

where; $T_0$ is reference state temperature and $T_{cw}$ is the cooling water temperature.

The exergy of the exhaust gases can be calculated from the following Eq. (15).

$$\begin{aligned}\zeta_{exhaust} &= m_{eg}[(h_{e,0} - h_0) + T_0(S_{e,0} - S_0)]\\ &= \dot{Q}_{eg} + \dot{m}_{eg}T_0\left[c_{p,e}\ln\left(\frac{T_0}{T_{ex}}\right) - R\left(\frac{p_0}{P_{e,0}}\right)\right]\\ &= \dot{Q}_{eg} + \dot{m}_{eg}T_0\left[c_{p,e}\ln\left(\frac{T_0}{T_{ex}}\right)\right] \quad as\ p_0 \cong P_{e,0}\end{aligned} \tag{15}$$

where $\dot{m}_{eg}$; mass flow rate of exhaust gases

$$= \dot{m}_f(1 + A/F); \tag{16}$$

$$\text{Heat energy of exhaust gases} = \dot{m}_f\, c_{p,e}\,(1 + A/F)(T_{ex} - T_0) \tag{17}$$

$$\text{Exergy Destroyed; } \zeta_{destruction} = \zeta_{in} - (\zeta_W + \zeta_{cw} + \zeta_{eg}) \tag{18}$$

## 6 Second Law or Exergetic Efficiency

The second law efficiency measures how efficiently fuel exergy input is converted into output.

$$\eta_{II} = \frac{exergy\ associated\ with\ shaft\ work}{exergy\ input} \tag{19}$$

Entropy generation rate can be expressed as;

$$\dot{s}_{gen} = \left[\frac{\dot{\zeta}_{destruction}}{T_0}\right] \quad (20)$$

## 7 Results and Discussion

Energy and exergy analyses of the single cylinder diesel engine have been carried out for various loads and engine speeds.

The variation of BSFC for different loads is shown in Fig. 3. The BSFC decreases with increase in load. It is observed that most economical operating point, at all speeds is in the vicinity of full load. The BSFC curve rises sharply at low loads. At low load the power out-put is less while the friction power is almost constant and as a result brake specific fuel consumption is higher. With increase in load, power out-put increases consequently BSFC decreases.

Figure 4 shows the variation of thermal efficiency with load for various speeds. As expected, the thermal efficiency increases with increasing load at constant speed and also increases with increasing speed at constant load. Maximum thermal efficiency is observed to be in the mid range of load and speed. It is of the order of 28%.

Figure 5 shows the variation of exhaust gas temperature with engine load at different speeds. It is observed that exhaust temperature increases with increase in load as well as with increase in speed. However, the effect of speed is larger. These are the consequences of the variations in the relative magnitudes of heat transfer to the cylinder-wall and to the exhaust valve and port. With increase in load the mass and temperature of the exhaust gases during blow down pulse increase. This results

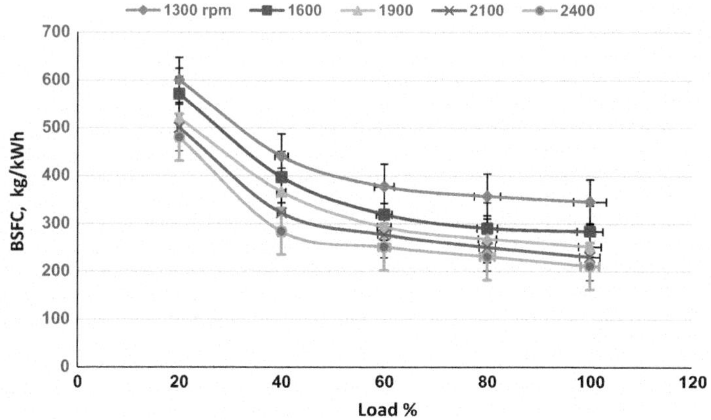


**Fig. 3** Variation of BSFC versus load at various speeds

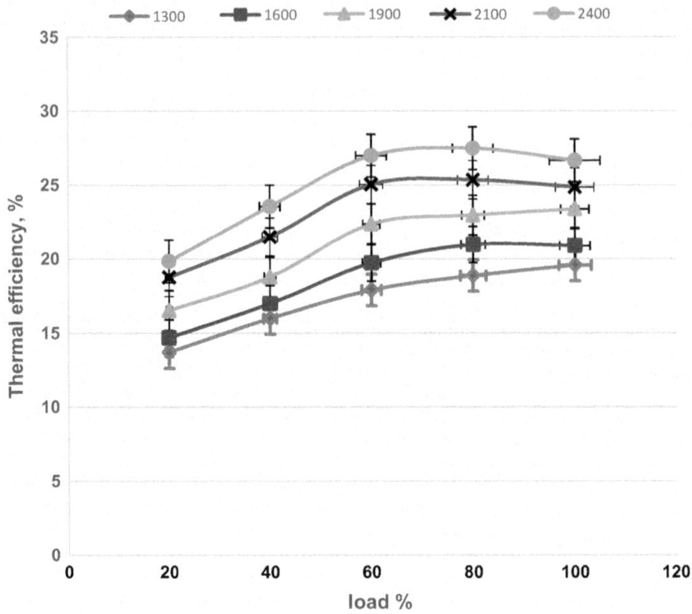


**Fig. 4** Variation of thermal efficiency versus load at various speeds

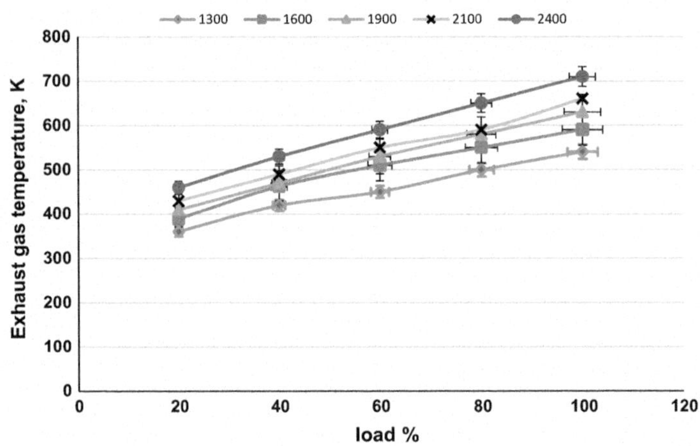


**Fig. 5** Variation of exhaust gas temperature versus load at various speeds

in higher mass averaged temperature of the exhaust. With increase in speed the time available for heat transfer to walls of combustion chamber and cylinder is decreased. Thus the burned gases will have higher temperature throughout the expansion and exhaust processes.

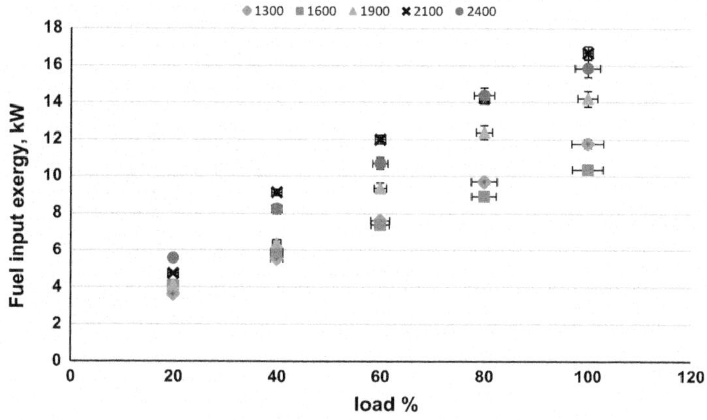


**Fig. 6** Variation of fuel input exergy versus load at different speeds

Figure 6 shows the variation of fuel input exergy with load at different speeds. Fuel input-exergy increases with increase in load as well as speed. Fuel input-exergy depends on the mass flow rate of the fuel. With increase in load as well as speed the mass of fuel injected per unit time also increases consequently the input-exergy increases.

Variation of exergy associated with exhaust gases is shown in Fig. 7. Exhaust gas exergy increases with increase in load and also with increase in speed at constant load. Increase in load as well as increase in speed result in higher exhaust gas temperature as explained earlier. The increase in temperature with load and speed causes the exhaust exergy to increase.

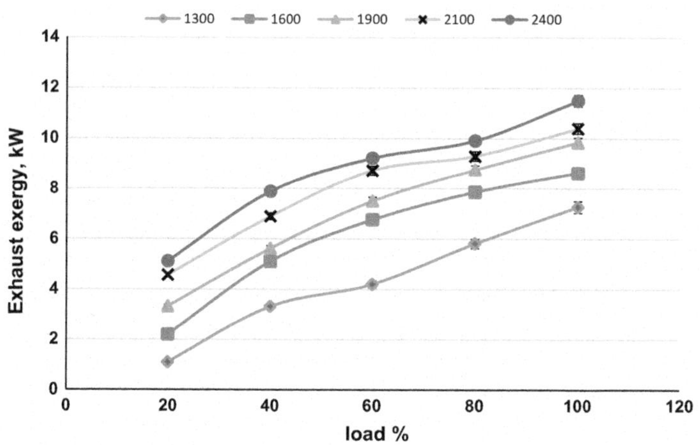


**Fig. 7** Variation of exhaust exergy versus load at different speeds

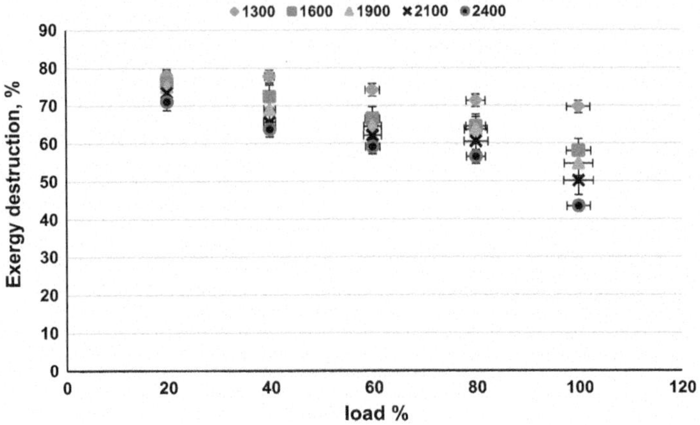


**Fig. 8** Variation of Exergy destruction % versus load at different speeds

Figure 8 shows the variation of exergy destruction as a percentage of fuel input-exergy, with load at different speeds. It is observed that higher percentage of exergy destruction occurs at low load. As load increases the percentage exergy destruction decreases. Also the percentage exergy destruction is lower at higher speeds. Unlike energy, exergy is not conserved. It gets destroyed during irreversible processes. Combustion is one of the irreversible processes in which significant exergy destruction occurs. It has been reported [7, 10] that exergy destruction due to combustion process decreases as the temperature of burned gases increases. As increase in load results in increase in the burned gas temperature, the exergy destruction, therefore, decreases with increase in load. Also, increase in speed results in higher peak temperature during compression process and consequently higher temperature during combustion. This causes the decrease in exergy destruction due to increase in speed.

Figure 9 shows the variation of second law efficiency with load at different speeds. It is observed that variation trend is similar to that for thermal efficiency (Fig. 4). However peak value of second law efficiency is obtained at full load and high speed. This is because exergy destruction is lower at these operating conditions.

Figures 10 and 11 show the exergy distribution as a percentage of fuel input—exergy at 2400 rev/min for 20% and 100% load respectively. Considerable exergy destruction occurs at low load. Exergy carried by cooling water is quite small. However, exergy carried by exhaust gases is of the same order as that corresponding to the brake power. Similar observations are made from Figs. 12 and 13 for exergy distribution for low speed operation of the engine. However, in this case, the exergy carried by the exhaust gases at low load is more that for the brake power.

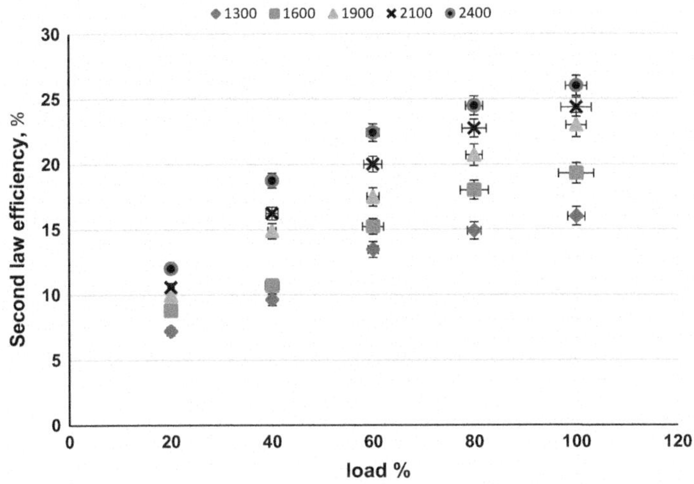


**Fig. 9** Variation of second law efficiency versus load at different speeds

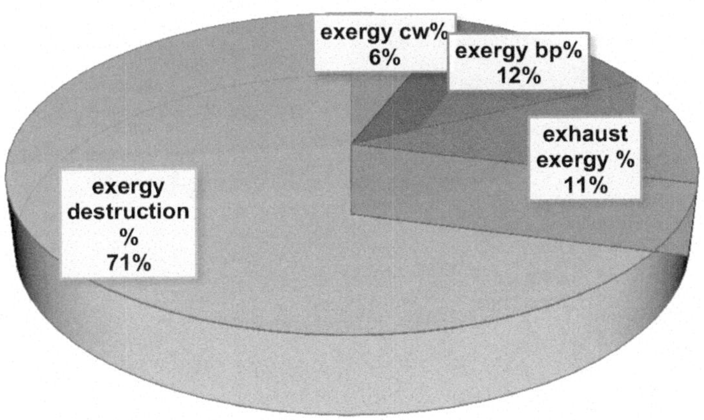


**Fig. 10** Exergy distribution for 20% at 2400 rev/min

Thus, from these figures, is clear that low load and low speed operation of the engine is not desirable because of considerable exergy destruction and exhaust exergy loss.

Considering above results, it is apparent that considerable fraction of exergy is destroyed because of irreversiblities. Combustion is one of the main causes of significant exergy destruction. However, exergy destruction on account of combustion can be reduced by increasing the temperature. Temperature can be increased by preheating the charge, by supercharging/turbocharging or by using thermal barrier coatings. Each of these methods have their problems. Another

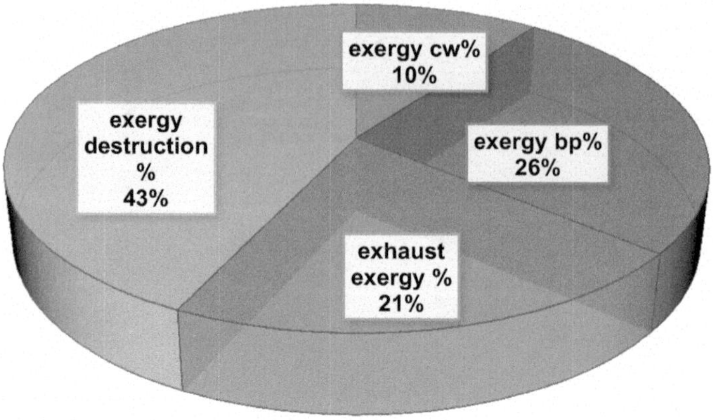


**Fig. 11** Exergy distribution for full load at 2400 rev/min. Where; Exhaust exergy% = exhaust exergy percentage with respect to input exergy; exergy destruction% = exergy destroyed percentage with respect to input exergy; exergy cw% = exergy carried away by cooling water percentage with respect to input exergy; exergy bp% = exergy correspond to brake power percentage with respect to input exergy

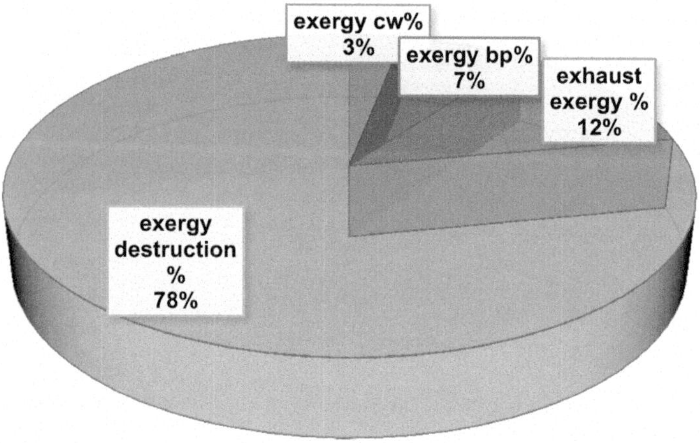


**Fig. 12** Exergy distribution for 20% load at 1300 rev/min

measure which needs to be further investigated is use of porous media combustion technology in which combustion takes place in porous media where super-adiabatic temperatures can be reached because of heat feed-back mechanism [11].

Further, it is observed that exergy carried by exhaust gases is of the order of the brake power. Therefore, it would always be desirable to recover this exergy by suitable heat recovery systems.

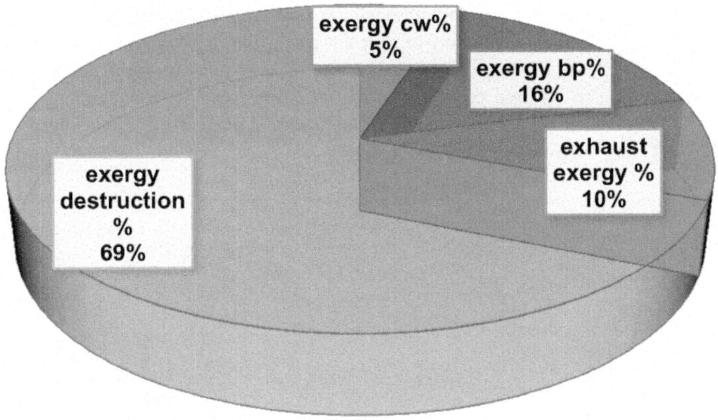


**Fig. 13** Exergy distribution for 100% load 1300 rev/min

# 8 Conclusions

First and second law analyses have been carried on a small DI diesel engine. It is concluded from the study that the best operating point for the engine is at full load and high speed. At this operating conditions the exergy destruction is less resulting in most effective utilization of fuel.

Exergy carried by exhaust is of the same order as that for brake power and needs to be recovered by appropriate heat recovery systems.

Exergy destruction decreases with increase in speed and load.

Second law efficiency increases with increase in load and speed.

# References

1. Caton J (2000) A review of investigations using the second law of thermodynamics to study internal-combustion engines. SAE paper no. 2000-01-1081. Society of Automotive Engineers, Inc, Warrendale, PA
2. Rakopoulos CD, Giakoumis EG (2000) Second-law analyses applied to internal combustion engines operation. Prog Energy Combust Sci 32:2–47
3. Jelic Maro & Ninic Neven (2008) Analysis of internal combustion engine thermodynamic using the second law of thermodynamics. Strojarstvo 50(2):85–94
4. Moran MJ, Shapiro HN (2010) Fundamentals of engineering thermodynamics, 6th edn. Wiley, New York
5. Bejan A (2006) Advanced engineering thermodynamics, 3rd edn. Wiley, New York
6. Bejan A (1995) Entropy generation minimization: the method of thermodynamic optimization of finite-size systems and finite-time processes, 1st edn. CRC Press
7. Rakopoulos CD, Andritsakis EC, Kyritsis DK (1993) Availability accumulation and destruction in a DI diesel engine with special reference to the limited cooled case. Heat Recovery Syst CHP 13:261–276

8. Pundir BP (2010) IC engines combustion and emissions. Narosa Publishing House
9. Kotas TJ (1985) The exergy method of thermal plant analysis. Butterworths, London
10. Caton JA (2000) On the destruction of availability (exergy) due to combustion processes-with specific application to internal combustion engines. Energy 25:1097–1117
11. Howell JR, Hall MJ, Ellzey JL (1996) Combustion of hydrocarbon fuels within porous inert media. Prog Energy Combust Sci 22:121–145

# Impact of Fuel Premixing Ratio and Injection Timing on Reactivity Controlled Compression Ignition Engine

**Mohit Raj Saxena and Rakesh Kumar Maurya**

**Abstract** Stringent emission mandates and harmful effect of engine emissions on human health as well as environment has driven the research into alternative combustion strategy for internal combustion engines. Low temperature combustion (LTC) concept is a promising approach to achieve higher fuel conversion efficiency along with lower $NO_x$ and particulate emissions. Among all the LTC strategies, reactivity controlled compression ignition (RCCI) strategy has potential to mitigate the limitations of conventional diesel combustion (CDC) engines. This study provides an overview of different LTC concepts and demonstrates the benefits of the RCCI combustion strategy over other LTC strategies. In present study, the impact of the fuel premixing ratio and fuel injection timing of high reactivity fuel on the combustion and emission characteristics of RCCI engine is investigated. Proportion of fuel premixing and injection events affect the combustion and emission characteristics of RCCI engine. Study shows that the RCCI engine has a lower operating range as compared to CDC strategy, which can be extended by using optimized premixing ratio and fuel injection events. Results show that RCCI combustion strategy has lower $NO_x$ and soot emissions as compared to CDC strategy while maintaining higher thermal efficiency. However, RCCI combustion has a higher HC and CO emissions.

## 1 Introduction

Crisis of fossil fuels depletion and degradation of the environmental conditions encourage the automotive industries/researchers to develop an efficient engine with lower emissions and higher thermal efficiency. Conventional compression ignition (CI) engine has higher thermal efficiency and power output as compared to

M.R. Saxena (✉) · R.K. Maurya
Department of Mechanical Engineering, Indian Institute of Technology Ropar, Rupnagar 140001, India
e-mail: mohitraj@iitrpr.ac.in

R.K. Maurya
e-mail: rakesh.maurya@iitrpr.ac.in



A.K. Agarwal et al. (eds.), *Combustion for Power Generation and Transportation*, DOI 10.1007/978-981-10-3785-6_13

spark-ignition (SI) engine while producing higher soot and nitrogen oxides ($NO_x$) emissions. Automotive industries are currently using post-treatment devices to meet the stringent emission mandates. However, these post-treatment devices (like a lean $NO_x$ trap, selective catalytic reduction, etc.) have a penalty on the fuel economy of the engine and these devices are also very expensive. However, to minimize the cost of post-treatment devices and to improve the fuel economy of the engine along with meeting the emission legislation limits, researchers focused toward the investigation of alternative combustion strategies. In an effort to get better thermal efficiency and lower emissions, several alternative combustion strategies are proposed by researchers. Proposed advance combustion strategies mainly premixed low temperature combustion (LTC) strategies, have higher or similar thermal efficiencies as in conventional diesel combustion (CDC) engines while having very low nitrogen oxides ($NO_x$) and soot emissions in comparison to CDC engines. In LTC strategies, lower combustion temperature leads to the reduction of $NO_x$ formation in the combustion chamber. In addition, premixed charge reduces the fuel rich zones in the combustion chamber, which leads to reduction in the soot formation. Among all the LTC strategies, homogeneous charge compression ignition (HCCI) and partially premixed charge compression ignition strategies are the most investigated combustion modes [1–5]. The HCCI combustion strategy has potential to achieve similar or higher thermal efficiency as compared to the CDC engine while having ultra-low $NO_x$ and soot emissions [6]. In HCCI combustion mode, fuel is injected into the intake manifold (for gasoline like fuels) or directly into the cylinder during an early stage of compression stroke (for diesel like fuels), which leads to the formation of a homogeneous charge before auto-ignition starts. This premixed homogeneous charge auto-ignite at several locations in the combustion chamber simultaneously, which leads to higher heat release rate and maximum pressure rise rate (MPRR). At higher load conditions, higher MPRR leads to produce excessive noise and vibrations. In HCCI combustion, there is no direct control on the ignition timing and combustion phasing similar to spark timing and fuel injection timing in the SI and CI engines respectively. In HCCI combustion, ignition timing is kinetically controlled and it depends upon the fuel/air charge composition, temperature and pressure inside the engine cylinder. Operating range of HCCI engine is limited by the high pressure rise rate and achieved HCCI operating range is lower than the CDC engine. Moreover, HCCI combustion engine has a higher carbon monoxide (CO) and unburned hydrocarbon (HC) emissions as compared to the CDC engine due to its lower in-cylinder combustion temperature [7]. These challenges limit the commercial application of HCCI combustion strategy in automotive engines.

To overcome the technical challenges of HCCI combustion strategy (such as high MPRR), partially premixed combustion (PPC) has been investigated. Studies [8, 9] show that PPC strategy has a potential for ignition timing control as well as has lower $NO_x$ and soot emissions. PPC strategy is considered as an intermediary combustion mode between the HCCI and CDC strategy. In PPC combustion fuel is injected during the compression stroke in the cycle before the top-dead-center position. Thus, fuel injection timing is used for varying the fuel stratification in the cylinder which controls the combustion phasing of PPC engine. This is the advantages of PPC over the HCCI combustion strategy. In addition to this, the duration between the end-of-injection and start-of-injection (SOI) controls the degree of air-fuel mixing. However, it has a lower degree of air-fuel mixing as compared to HCCI combustion. Therefore, it has better control on the ignition delay and thus heat release rate (HRR). Required ignition delay can be obtained by using enhanced charge motion and excessive exhaust gas recirculation (EGR). Excessive EGR leads to decrease the combustion temperature, which results in lower $NO_x$ emissions. However, the major technical challenge with diesel PPC is the requirement of excessive EGR to achieve lower $NO_x$ and soot emissions at higher engine load conditions. Excessive EGR leads to reduce the combustion efficiency and it also increases the HC and CO emissions [10]. A study reported that gasoline PPC strategy has a higher thermal efficiency as compared to the CDC, while maintaining lower $NO_x$ and soot emissions below the 0.30 g /kWh and 0.30 FSN respectively [11]. Kalghatgi et al. [12] also conducted an experimental investigation to compare the gasoline and diesel fuel for various SOI timing, engine operating load and EGR rates for PPC combustion engine. They observed that gasoline has lower smoke emissions as compared to diesel fuel for all the engine-operating conditions due to longer ignition delay. Moreover, they suggested that two injections (pilot and main injections) could be used to attain the PPC combustion at higher load conditions. Similar results are also reported in another recent study [13].

Generally, light-duty diesel engines operated in LTC mode are termed as premixed charge compression ignition (PCCI). In PCCI combustion strategy, fuel is injected via either early or late direct injection [14]. PCCI combustion strategy is not completely homogeneous like HCCI combustion. In PCCI engine, spontaneous ignition occurs at several locations as occurs in the compression ignition engine. However, PCCI strategy is related to varying levels of the stratification of fuel in the cylinder [15]. PCCI combustion has longer ignition delay period as compared to CDC strategy [16] and required ignition delay could be achieved through an excessive EGR. Higher EGR leads to increase the ignition delay, which results into better premixing of air/fuel ratio and hence less local fuel rich zones are present in the spray. Lower combustion temperature reduces the formation of $NO_x$ and lower local equivalence ratio reduces the formation of soot particle [16]. However, the PCCI combustion strategy is limited by combustion stability and combustion efficiency at lower load condition. Moreover, PCCI combustion has higher CO and HC emissions at lower load condition [11].

Figure 1 shows the different LTC strategies proposed in the past few decades to develop advanced combustion engines. Figure 1 shows that conventional combustion strategies (SI and CI) are converging towards the moderate fuel reactivity (between gasoline and diesel). In addition to this, all the premixed LTC strategies have a higher fuel conversion efficiency while having lower $NO_x$ and soot emissions. Higher fuel conversion efficiency in premixed LTC combustion strategies attributed to lower heat transfer loss and reduced combustion duration.

In the past few years, researchers mostly focused on the investigation of the reactivity controlled compression ignition (RCCI) combustion strategy due to its better ignition control and higher fuel conversion efficiency along with simultaneous

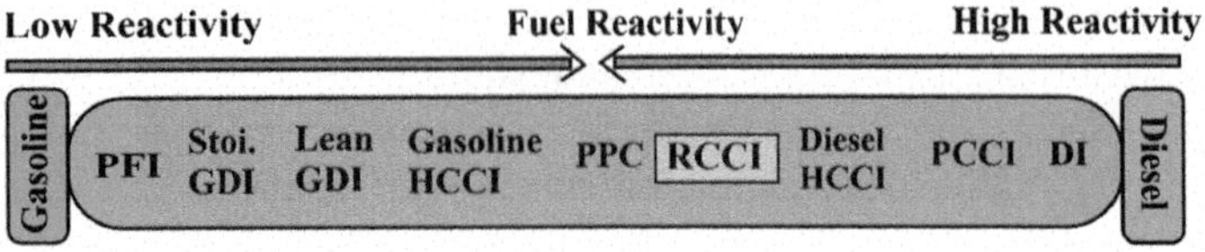


**Fig. 1** Proposed combustion strategies based on fuel reactivity (adapted from [17])

reduction in $NO_x$ and soot to very low level. Results indicated that RCCI combustion strategy has higher combustion phasing control potential among other LTC strategies [18]. RCCI combustion is a dual fuel combustion strategy, which operates through in-cylinder blending of two fuels of different reactivity. In RCCI combustion mode, multiple injections can be used to control the fuel reactivity inside the cylinder. The varying level of fuel reactivity, controls the combustion phasing and the combustion duration [19]. The RCCI combustion mode has better ignition control while retaining the advantages of the HCCI combustion strategy. Present study is mainly focused on the effect of premixing ratio and injection timing of high reactivity fuels on the performance and emission characteristics of RCCI engine.

RCCI strategy is a premixed combustion strategy and reactivity of the charge is controlled by injection of two fuels of different reactivity. Typically low reactivity fuel (gasoline like fuels) is injected by using the port-injection system while the high reactivity fuel (diesel like fuels) is injected using direct injection system. Figure 2 represents the schematic diagram of the dual fuel injection system for RCCI engine.

**Fig. 2** Schematic diagram of dual fuel injection system for RCCI combustion mode

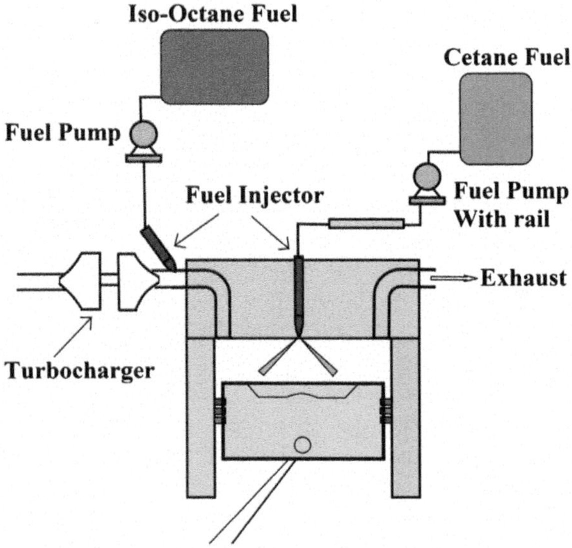

Percentage of low reactivity fuel and injection events of high reactivity fuel can be used to control the combustion phasing and to run the engine over a wide operating range [20]. Inagaki et al. [21] investigated the dual fuel premixed compression ignition (PCI) strategy. They observed that by varying fuel reactivity, combustion can be controlled in dual fuel PCI strategy. Injection of two fuels of different reactivity can easily control the local equivalence ratio and fuel reactivity stratification. Figures 3 and 4 indicates the two staged HRR in dual fuel PCI combustion. Low temperature heat release (first peak) is mainly due to high reactivity fuel. Additionally two staged ignition also leads to increase the combustion duration. It can be depicted from the Fig. 3 that lower HRR obtained with dual fuel PCI combustion, as compared to HCCI combustion while $NO_x$ emissions are found comparable. Inagaki et al. [21] observed that engine could be operated within the range of 2–12 bar IMEP by changing the ratio of low and high reactivity fuels. High reactivity fuel reacts with the low-temperature reactions and ignites the premixed charge of low reactivity fuel and air. Furthermore, the ratio of the two fuels of different reactivity helped to control combustion phasing.

In general, the combustion phasing depends upon the local equivalence ratio, while the local equivalence ratio depends upon the injection timing of directly injected high reactivity fuel. Figure 4 demonstrates the effect of injection timing on the HRR of dual fuel PCI combustion strategy. Figure 4 shows that the HRR is delayed by advancing the injection timing of diesel fuel. Advanced injection timings of diesel enhances the premixing of fuel, which leads to the formation of fuel-air charge with the lower local equivalence ratio resulting into delayed HRR.

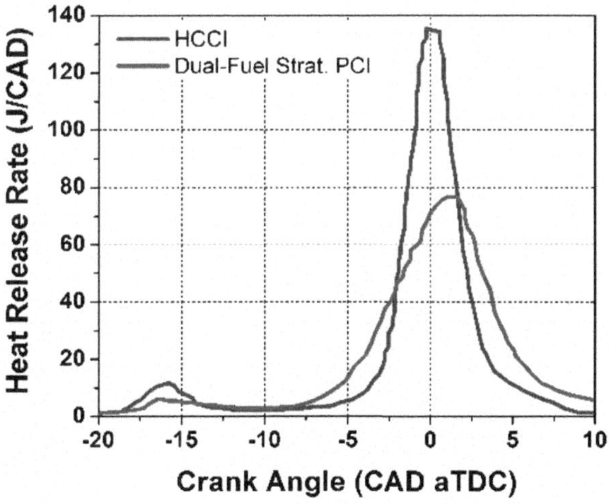


**Fig. 3** Comparison of heat release rate for HCCI and dual fuel PCI strategy at equivalence ratio = 0.35. (adapted from [21])

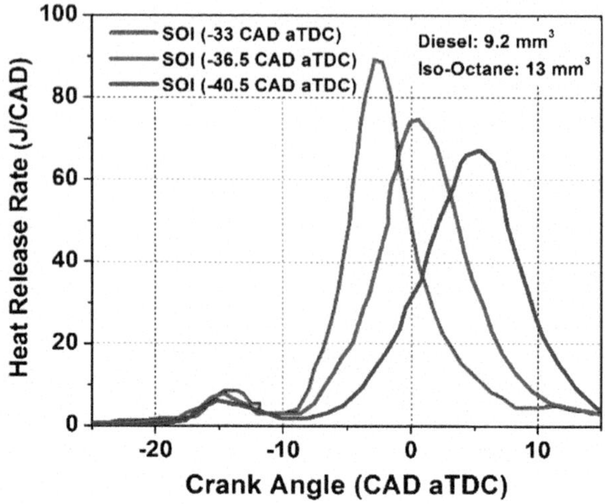


**Fig. 4** Effect of injection timing on the HRR of dual fuel PCI combustion mode at equivalence ratio = 0.35. (adapted from [21])

## 2 Combustion Characteristics of RCCI Engine

In this section, first operating range of the RCCI engine with conventional as well as bio-fuels is discussed. After discussion of the RCCI operating range, effect of the fuel premixing ratio and SOI timing along with the effect of multiple injections on combustion characteristics of RCCI engine is discussed.

An automotive engine needs to run over a wide range of operating conditions, from idle to higher engine load conditions. RCCI engine has superior performance at medium engine operating load as compared to HCCI engine and has a better combustion control [22]. Still, the RCCI engine operating range is lower as compared to CDC strategy (Fig. 5). The upper operating load range of RCCI engine is limited by the MPRR and combustion noise while the lower operating load range is limited by higher CO and HC emissions. It is possible that at lower load conditions,

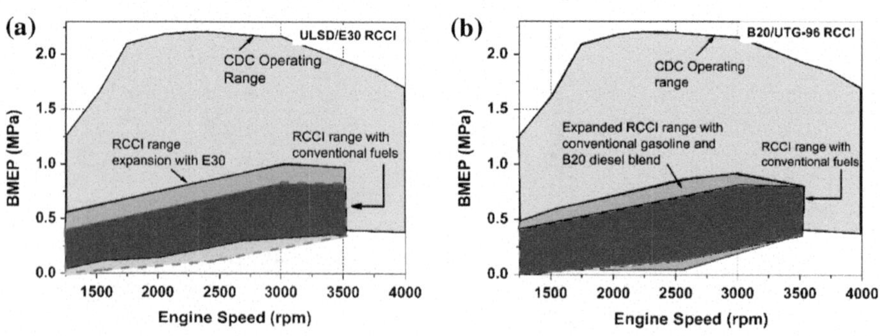


**Fig. 5** Comparison of performance maps for RCCI combustion with different fuels (adapted from [23])

CDC can be used because of similar efficiency and $NO_x$ emissions while having lower HC and CO emissions as compared to RCCI engine.

Figure 5 shows the operating range extension of RCCI engine using ethanol and bio-fuel blends. It can be observed from Fig. 5a that RCCI engine has a higher operating range with ethanol blend as compared to conventional fuels. It could be due to the combined effect of higher octane number and charge cooling effect of ethanol [23]. Figure 5b indicates that bio-fuel RCCI engine has a higher operating range as compared to the RCCI engine with conventional fuels. The premixing ratio of fuels can be increased by using bio-fuels without sacrificing the stability of the RCCI engine [23]. Load range expansion is necessary to operate the RCCI engine in the wide range of driving cycle map while maintaining higher efficiency with lower $NO_x$ and soot emissions. By optimizing the premixing ratio and fuel injection events, the engine operating range can be increased. A study showed that double injection strategy can be used for low to medium engine load while single injection strategy could be used for higher operating load conditions [24]. RCCI engine has a capability to operate from 6 bar to 24 bar BMEP while overcoming the $NO_x$-soot trade-off characteristic of the CDC engine and having a better fuel economy as compared to the CDC engine [24]. Another study revealed that separate direct injectors for low and high reactivity fuels can also be used for higher load operation of RCCI engine [25]. Moreover, by the development of a co-axial dual-fuel injector, clean RCCI combustion could be achieved at the higher engine load [25].

Figure 6a shows the variation of indicated thermal efficiency with gross indicated mean effective pressure (IMEP). Figure 6a reveals that the RCCI combustion has more than 55% indicated thermal efficiency at few load conditions. Thermal efficiency is a function of engine operating load. A study indicates that the RCCI combustion strategy has a higher thermal efficiency as compared to CDC strategy [26]. Results of another study already demonstrated that RCCI engine has 60% gross indicated thermal efficiency at 6.5 bar IMEP by optimizing the injection characteristics [27]. Another study showed that RCCI engine can be operated efficiently at full load condition, i.e. 23 bar gross IMEP while achieving 46.6% thermal efficiency and MPRR of 15 bar per crank angle degree [28]. Figure 6b shows the variations of combustion efficiency with BMEP. Combustion efficiency is also a function of the engine load. Combustion efficiency increases with engine operating load due to higher combustion temperature at higher engine load. Figure 6b also indicates that RCCI engine has a lower combustion efficiency as compared to CDC engine.

A study demonstrated that LTC strategies have a higher gross indicated thermal efficiency while having a coefficient of variations of IMEP in the acceptable range [29]. In RCCI combustion, since there is a higher gradient of fuel reactivity (between high reactivity and low reactivity fuels), combustion duration is longer and lower MPRR as compared to HCCI and PCCI combustion strategy [29]. In addition to this, RCCI combustion has a lower combustion efficiency as compared to PCCI and HCCI combustion.

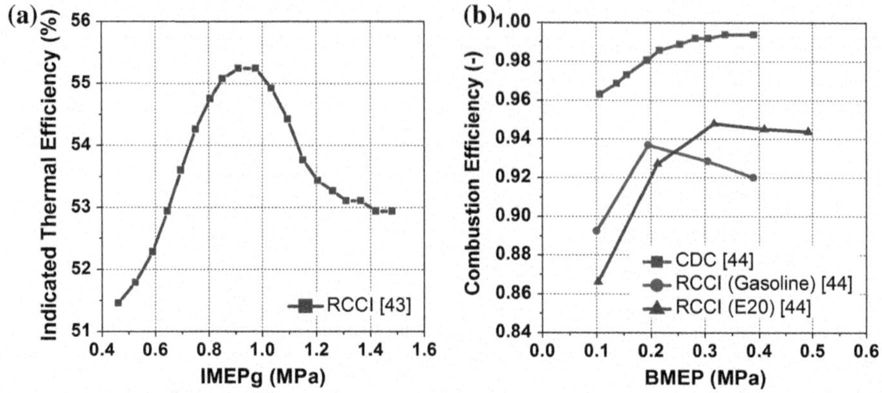


**Fig. 6** Variations of **a** indicated thermal efficiency with $IMEP_g$ and **b** combustion efficiency with BMEP in RCCI Engine

## 2.1 *Effect of Fuel Premixing Ratio and Start of Injection on Combustion Characteristics*

Combustion phasing and ignition characteristics of the RCCI engine are typically controlled by premixing ratio of the fuels and the start of injection (SOI) timings of high reactivity fuel. Retarded high reactivity fuel injection timing or lower premixing ratio of fuels lead to increase the inhomogeneity of charge in the combustion chamber because of lower mixing time. Higher inhomogeneity of charge results into the formation of higher $NO_x$. An excessive advanced SOI timing results into lower combustion progression. Figure 7 shows the effect of premixing ratio and SOI timings of high reactivity fuel on the CA50 (crank angle position for 50% heat release) in the RCCI engine for two different fuel injection quantity (40 mg/cycle and 50 mg/cycle). Figure 7 shows that for any fixed proportion of gasoline with retard SOI timing, an advanced CA50 position observed. A possible reason for this observation is that, retard injection timing increases the charge stratification, which leads to increase the local equivalence ratio and the reactivity of charge. Higher reactivity and higher local equivalence ratio of charge leads to shorten the ignition delay and combustion duration, which results in an advanced CA50 position [30]. It can also be depicted from Fig. 7 that both injected fuel masses in the cycle follow the similar trend.

The effect of premixing ratio and SOI timings of high reactivity fuel on the IMEP of the RCCI engine for two different fuel quantity (40 mg/cycle and 50 mg/cycle) are presented in the Fig. 8. It can be noticed from Fig. 8 that higher IMEP obtained for 50 mg/cycle injected fuel as compared to 40 mg/cycle fuel because of higher quantity of fuel burn in the combustion chamber. It can also be observed from the Fig. 8 that for each proportion of gasoline, higher IMEP obtained

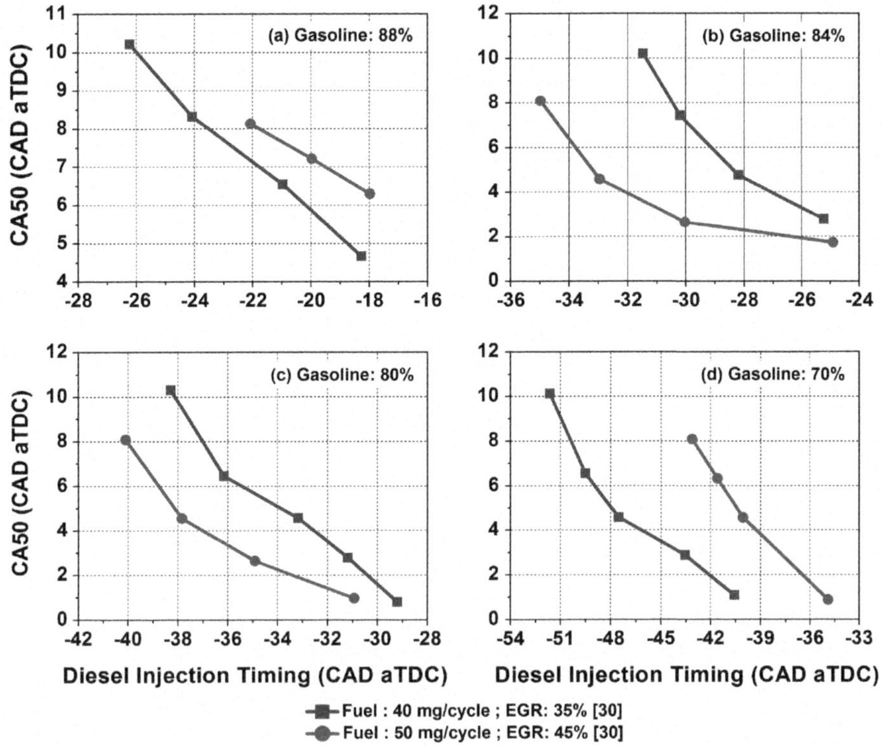


**Fig. 7** Effect of premixing ratio and injection timing on $CA_{50}$ in RCCI (adapted from [30])

in a certain range of SOI timings of diesel fuel. Excessive advance SOI timing of diesel leads to the leaner distribution of diesel fuel in the combustion chamber during start of combustion, which results in slower combustion rate. Slower combustion rate at the start of combustion may lead to decrease the IMEP. As the SOI timing of diesel is slightly retarded, diesel fuel ignites the premixed charge (homogeneous charge) in the combustion chamber, which leads to increase the IMEP. Homogeneous distribution of charge in the combustion chamber leads to higher IMEP [31].

With further retard in the SOI timings, diesel fuel leads to increase the inhomogeneity of air-fuel mixture in the combustion chamber because of lower mixing time. Figure 8 also reveals that maximum IMEP is obtained with 84 and 80% gasoline proportion in total fuel. Wang et al. reported that further increase in the gasoline proportion (above 88% gasoline) leads to unstable combustion because of lower injected quantity of diesel (high reactivity) fuel [30]. However, both injected fuel quantity follows a similar trend in IMEP with SOI timings of high reactivity fuel (Fig. 8).

Figure 9 shows the effect of premixing ratio and SOI timings on the MPRR in the RCCI engine for two different fuel injection quantity (40 mg/cycle and

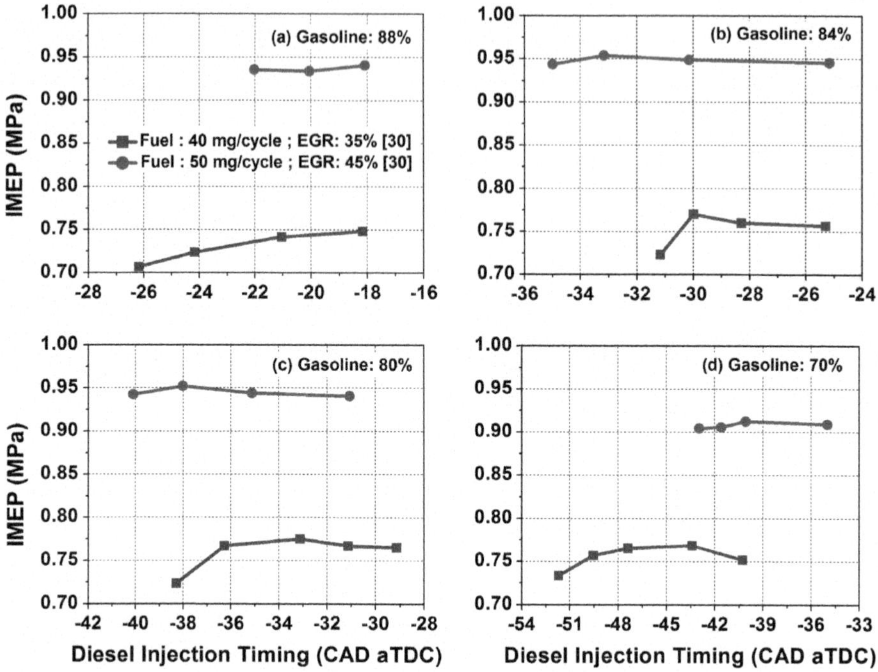


**Fig. 8** Effect of premixing ratio and injection timing on IMEP in RCCI engine (adapted from [30])

50 mg/cycle). Figure 9 indicates that MPRR decreases with advanced injection timings of the high reactivity fuel due to lower in-cylinder temperature [30]. Studies reported that in-cylinder mean temperature has direct impact on MPRR [30, 31]. It can also be observed from Fig. 9 that the MPRR decreases with an increase in the gasoline proportion. It also attributed to lower in-cylinder mean temperature with an increase in gasoline proportion [30].

Figure 10 shows the effect of low reactivity fuel proportion on the in-cylinder mean temperature in RCCI engine. It can be depicted from Fig. 10 that on increasing the proportion of gasoline, in-cylinder mean temperature decreases. The peak of in-cylinder mean temperature is different for different proportion of the fuels (Fig. 10). Mainly it is due to the varying reactivity of the fuel distribution in the combustion chamber for different proportion of low reactivity fuels (gasoline and methanol). Li et al. used cetane number distribution to characterize the reactivity of the charge [31]. They reported that ignition timing is delayed with an increase in methanol fraction because of the lower cetane number of the methanol. Moreover, with an increased fraction methanol, the specific heat of premixed charge increases during the compression stroke, which leads to reduction in the rate of cylinder temperature increase before ignition [31].

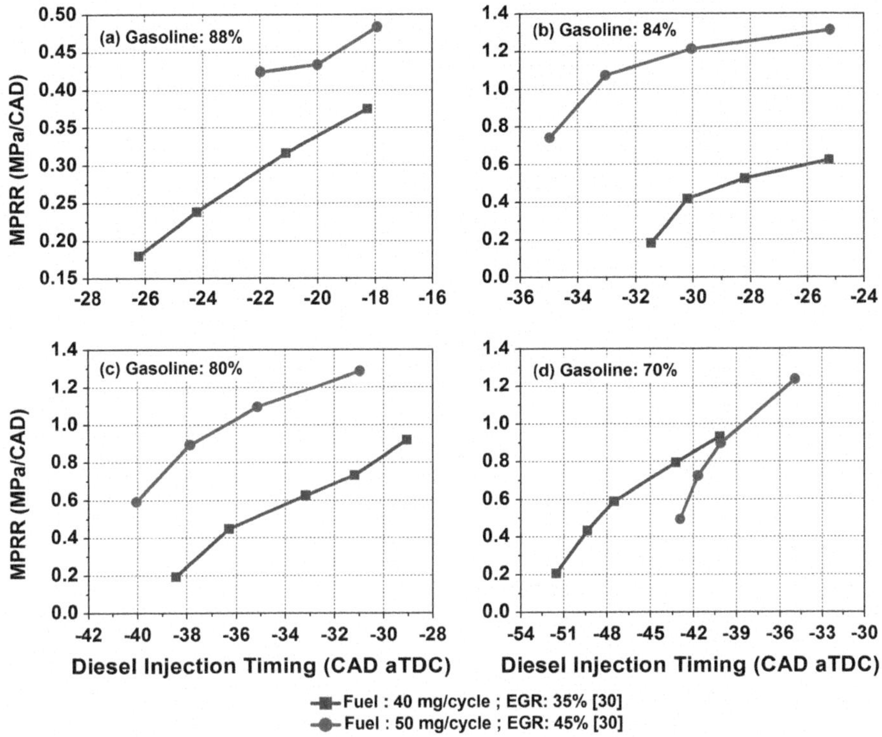


**Fig. 9** Effect of premixing ratio and injection timing on MPRR in RCCI engine (adapted from [30])

## 2.2 *Effect of Multiple Injections on Combustion Characteristics*

Injection timing of high reactivity fuel with single or multiple injections per cycle has a direct impact on the performance and combustion characteristics of the RCCI engine [32–35]. Figure 11 shows the effect of varying first injection timing (SOI-1) while keeping second injection timing (SOI-2) constant and vice versa on the performance of the RCCI engine. Figure 11 indicates that the MPRR decreases with an advanced SOI-1 timings. With an advanced injection timings more homogeneous premixed charge will be prepared and local equivalence ratio decreases because of more availability of premixing time. Combustion of lower local equivalence ratio leads to reduction in the in-cylinder combustion temperature, which results in lower MPRR [36]. Similarly, when SOI-2 retarded, the MPRR decreases because with retarded SOI-2, the ignition timing will be delayed which leads to lower in-cylinder combustion temperature. Lower in-cylinder temperature will also results into reduction in the $NO_x$ emissions [36].

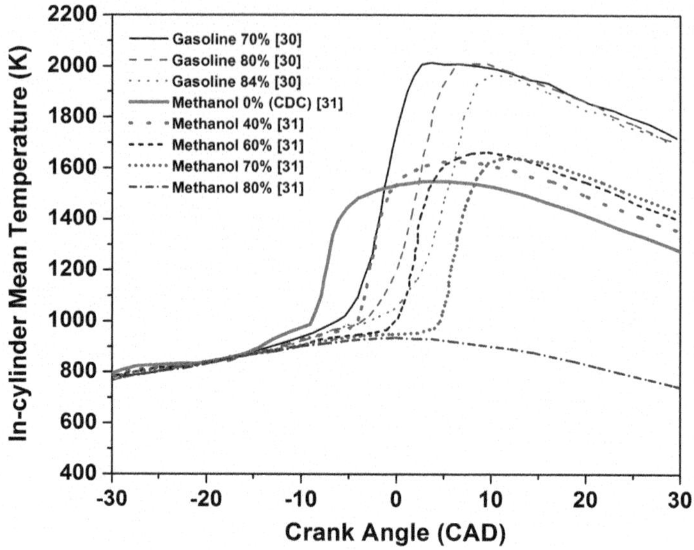


**Fig. 10** Effect of low reactivity fuel proportion on the in-cylinder mean temperature in RCCI engine

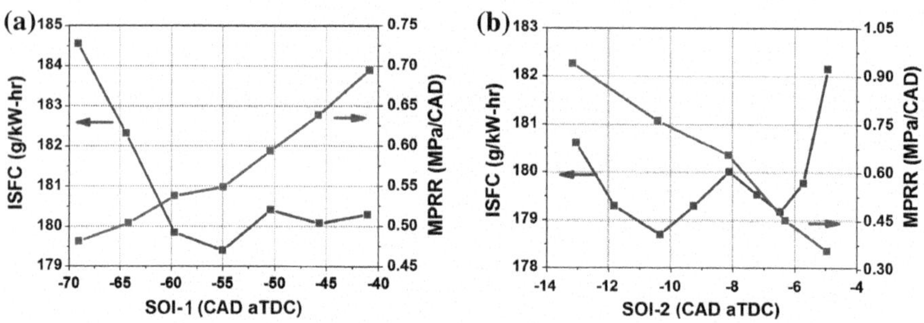


**Fig. 11** Effect of SOI-1 and SOI-2 timing on performance for late SOI2 strategy in RCCI engine (adapted from [36])

With an early single SOI timing with higher ratio of premixed charge, high efficiency along with ultra-low $NO_x$ and soot emission can be achieved in RCCI engine [36]. Early single injection and early SOI-2 show similar trend for high temperature HRR with a single stage combustion while late SOI-2 shows two stage HRR, which results in lower peak of heat release rate and increases the combustion duration. Zhang et al. determined the optimal combination of parameters including EGR rate, diesel injection timing with single and double injections, ratio of single and double injections, A/F ratio, premixing ratio and compression ratio for the

RCCI engine [33]. They reported that higher combustion efficiency could be achieved using double injection strategy as compared to single injection strategy but double injection strategy has higher MPRR. Splitter et al. [34] computationally investigated the injection effect at lower load condition in the RCCI engine. Their results revealed that with early injection timing best thermal efficiency could be achieved while late injection of high reactivity fuel increases the pressure rise rate. Study also reported that multiple (two injections per cycle) injection strategy can decrease CO and HC emissions by 40% and increase the thermal efficiency by 1% (if liner impingement could be avoided).

Figure 12 indicates the effect of variations of CA50 on the performance of the RCCI engine for early single, early SOI-2 and late SOI-2. Figure 12 depicts that lower indicated specific fuel consumption (ISFC) is achieved with early SOI-2 and lowest ISFC when CA50 is at 6 CAD aTDC. It means that higher indicated thermal efficiency obtained at CA50 of 6 CAD aTDC. However, lowest MPRR and combustion efficiency is observed in case of late SOI-2 operating conditions. Study concluded that early SOI-2 is suitable for the medium operating load condition while late SOI-2 operating conditions can be used to increase the higher engine load operations.

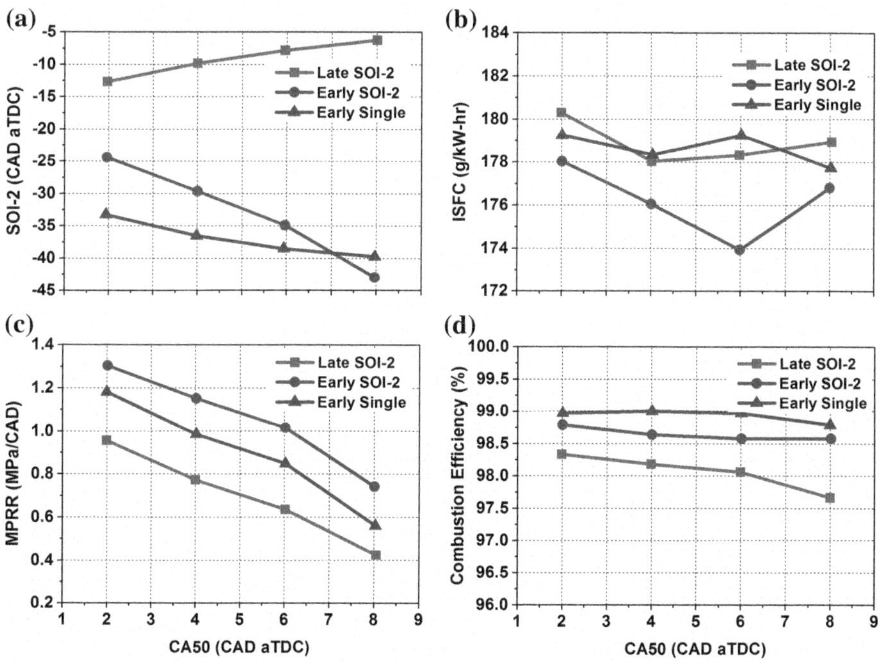


**Fig. 12** Comparison of performance characteristics of the RCCI engine for three different injection strategies (adapted from [36])

## 3 Emission Characteristics of RCCI Engine

In this section, the impact of the fuel premixing ratio and the SOI timings on emissions characteristics of the RCCI engine are discussed. The CDC engines are most favorable for heavy-duty vehicles applications because of their higher thermal efficiency as compared to spark-ignition engines. The CDC engines typically emit higher soot and $NO_x$ emissions and there exists a trade-off between the $NO_x$ and soot emissions [37]. The main motivation for RCCI combustion strategy is due to simultaneous reduction of $NO_x$ and soot emissions to a very low level while maintaining similar or higher thermal efficiency than CDC strategy. Studies confirmed that the RCCI engine has the capability to reduce the $NO_x$ and soot emissions simultaneously [38–40]. Figure 13 shows the variations of $NO_x$ and soot emissions with IMEP in a RCCI engine. Figure 13 depicts that the $NO_x$ and soot emissions are lower than the 2010 EPA HD limits in RCCI engine without any post-treatment device.

### 3.1 *Effect of Fuel Premixing Ratio*

The RCCI engine is a very promising approach to achieve lower $NO_x$ and soot emissions within the Euro VI limits. However, RCCI engine generates higher CO and HC emissions [41, 42]. Figure 14 reveals the effect of varying fuel mass fraction on the emission characteristics of RCCI engine. The fuel mass fraction plays a major role in emission formation in the RCCI engine, especially in the case of CO and HC emissions (Fig. 14). It can be observed that HC emission decrease with an increase in the methanol fraction up to 60% (Fig. 14a). due to the high temperature region of the combustion chamber [31]. A higher temperature can also

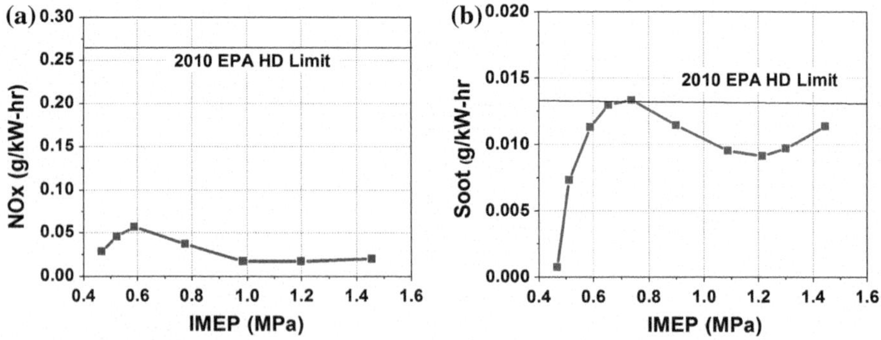


**Fig. 13** Variation of $NO_x$ and soot emission with IMEP in a RCCI engine (adapted from [43])

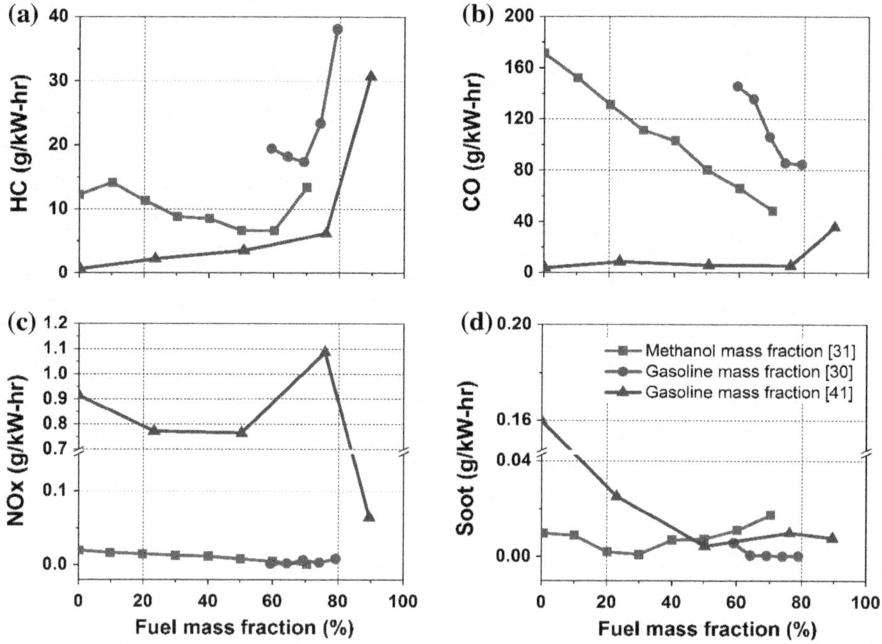


**Fig. 14** Effect of varying fuel mass fractions on the **a** HC, **b** CO, **c** $NO_x$ and **d** Soot emissions characteristics of a RCCI engine

be noticed from the Fig. 10. High temperature region attributes to preparation of more homogeneous charge with an increase in the fraction of methanol during premixing, which leads to reduction in the local fuel rich zones and more charge will be burned in a larger area, which results into higher temperature [31]. However, Fig. 14a also indicates that HC emission increases with further increase in the methanol fraction (above 60%). It is because of the combined effect of longer ignition delay and shorter reaction time, which results into lower combustion temperature. Figures 14a-b also reveals that with 100% fraction of diesel (zero low reactivity fuels), CO and HC emission are very low. However, with a decrease in the fraction of diesel or the increase in the fraction of gasoline, the global reactivity of charge decreases which retards the oxidation rate. Consequently, HC and CO emission increases in RCCI engine. When gasoline fuel fraction increased from 75 to 90%, HC and CO emission increases rapidly due to the deterioration in the combustion process (Fig. 14a–b). Thus, because of the poor combustion process the adiabatic flame temperature reduces which leads to reduction in the $NO_x$ and soot emission (Fig. 14 c–d) [41].

It can also be observed from Fig. 14b that the CO emission decreases monotonically with an increase in the fraction of methanol. Decrease in CO emission can be possibly because of higher temperature in the combustion chamber (Fig. 10). It may also be due to a higher percentage of oxygen molecule in methanol, which enhances the oxidation of CO [31]. The effect of varying fuel mass fraction on the $NO_x$ and soot emissions are presented in the Figs. 14c,d respectively. $NO_x$ and soot emissions are very low in RCCI combustion strategy (Fig. 14c,d). It can be noticed from Fig. 14c that $NO_x$ emission decreases with an increase in the fraction of the methanol. The formation of $NO_x$ emission depends upon the in-cylinder temperature and reaction time. With an increase in the fraction of methanol, the in-cylinder temperature increases (Fig. 10), which enhances the formation of $NO_x$. However, with a higher fraction of methanol, the delay period increase, which retards the ignition timing. With retard ignition timing, the residence time of charge in the high temperature decreases which decreases the chances of the formation of $NO_x$ [31]. Formation of soot mainly depends upon the local equivalence ratio of the charge. With an increase in the fraction of methanol, the local rich pockets of the charge decreases which results in lower soot formation.

## *3.2 Effect of Start of Injection*

The effect of variation of SOI timings on the emission characteristics of the RCCI engine at different premixing ratio of fuel is presented in the Fig. 15. SOI timings effect the formation of emissions in RCCI engine because the variation in SOI timings leads to change the air-fuel charge reactivity gradient distribution in the cylinder. Figure 15a indicates that the HC emission decreases with advanced SOI timings due to more homogeneous charge preparations with advanced SOI timings. In case of diesel and 50% methanol- diesel, it is observed that CO emission initially increases from SOI 7 to 32 CAD bTDC while decreases from 32 to 37 CAD bTDC (Fig. 15b). In case of 25 and 35% diesel, the CO emission decreases with an advance SOI timings due better oxidation and higher in-cylinder temperature.

Formation of $NO_x$ is mainly dependent on the local adiabatic flame temperature while soot formation depends upon the local equivalence ratio of charge. Figure 15c–d shows that the $NO_x$ and soot (soot emission in the case of 25 and 35% diesel) emissions slightly increases with an advance SOI timing. $NO_x$ emission increases because more premixed charge will burn, which leads to increase the adiabatic flame temperature. In addition, shorter ignition delay of delayed injection timings implies the richer local equivalence ratios. High adiabatic flame temperatures and rich local equivalence ratios lead to increase $NO_x$ and soot formation respectively [41]. Moreover, since there is a trade-off relation between the $NO_x$ and soot, thus with an advance SOI timing, longer ignition delay avoided the formation of rich fuel cores, which results into lower soot emissions (in the case of neat diesel and 50% methanol/50% diesel).

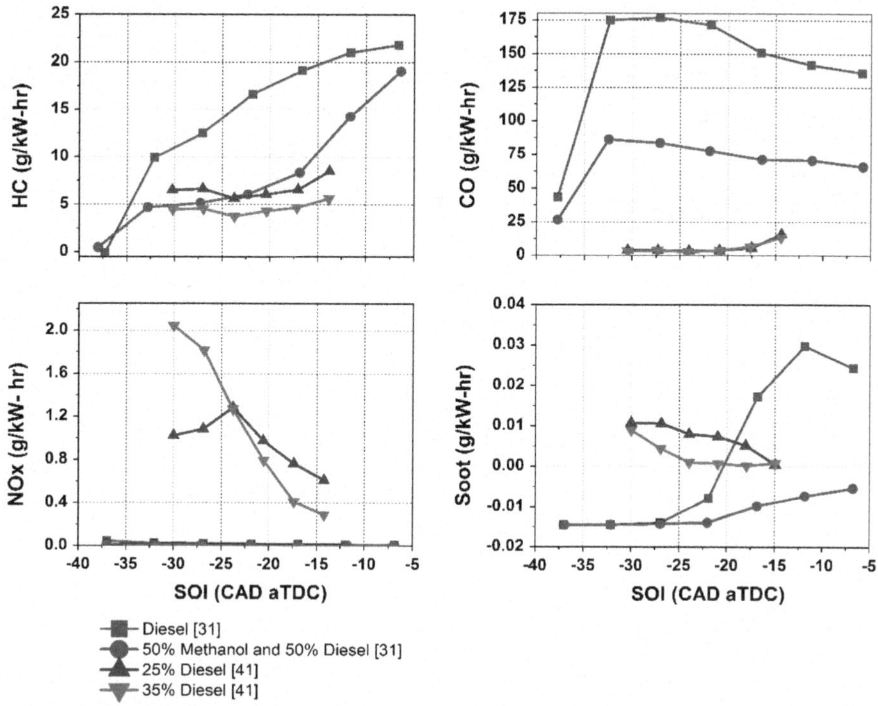


**Fig. 15** Effect of variation of SOI timings on the **a** HC, **b** CO, **c** $NO_x$ and **d** Soot emissions characteristics of a RCCI engine

# 4 Conclusions

This study demonstrates the concept of RCCI engine, including its advantages over the other combustion strategies. The RCCI combustion strategy has potential to meet the current and future emission mandates while maintaining higher thermal efficiency. RCCI strategy has superior ignition control in comparison to other LTC strategies. However, RCCI engine has lower engine operating range as compared to CDC engines, which can be increased to some extent by using bio-fuels and optimized injection events and premixing ratio of fuels. Moreover, further investigation is required to extend the operating range of the RCCI engine for operating over a wide range driving cycle. RCCI strategy has a higher thermal efficiency while maintaining the $NO_x$ and soot emission under Euro VI limits without any post treatment devices. In addition, RCCI strategy has higher HC and CO emissions. RCCI strategy has a lower combustion efficiency as compared to CDC strategy. Injection events and the premixing ratio of low and high reactivity fuel has a significant effect on the performance and emission characteristics of RCCI engine. Advanced SOI timings, increases the air-fuel charge reactivity in the cylinder while retard injection of high reactivity fuel leads to increase the air-fuel charge stratification in the cylinder. Multiple injection strategy leads to a reduction in CO and HC emissions. The premixing ratio of low and high reactivity fuel has a significant effect on the emissions of RCCI engine. With increasing in the fraction of gasoline/methanol, $NO_x$ and soot emissions are decreasing. Moreover, CO and HC emissions decrease with an increase in the fraction of methanol, but still need to be reduced in order to meet the current and future norms.

## References

1. Mathivanan K, Mallikarjuna JM, Ramesh A (2016) Influence of multiple fuel injection strategies on performance and combustion characteristics of a diesel fuelled HCCI engine–An experimental investigation. Exp Thermal Fluid Sci 77:337–346
2. Maurya RK, Agarwal AK (2014) Experimental investigations of performance, combustion and emission characteristics of ethanol and methanol fueled HCCI engine. Fuel Process Technol 126:30–48
3. Maurya RK, Agarwal AK (2015) Combustion and emission characterization of n-butanol fueled HCCI engine. J Energy Res Technol 137(1):011101
4. Wang Y, Liu H, Huang Z, Liu Z (2016) Study on combustion and emission of a dimethyl ether-diesel dual-fuel premixed charge compression ignition combustion engine with LPG (liquefied petroleum gas) as ignition inhibitor. Energy 96:278–285
5. Park H, Kim J, Bae C (2015). Effects of hydrogen ratio and EGR on combustion and emissions in a hydrogen/diesel dual-fuel PCCI engine (No. 2015–01-1815). SAE Technical Paper
6. Maurya RK, Agarwal AK (2015) Experimental investigations of particulate size and number distribution in an ethanol and methanol fueled HCCI engine. J Energy Res Technol 137 (1):012201
7. Li H, Neill WS, Chippior WL (2012) An Experimental Investigation of HCCI Combustion Stability Using n-Heptane. J Energy Res Technol 134(2):022204
8. Liu B, Cheng X, Qiu L, Dong, S, Wang X (2015). Numerical simulation and experimental research on PPC combustion and emissions characteristics of diesel engine (No. 2015–01-0401). SAE Technical Paper
9. Hanson R, Splitter D, Reitz R. D (2009). Operating a heavy-duty direct-injection compression-ignition engine with gasoline for low emissions (No. 2009–01-1442). SAE Technical Paper
10. Monteagudo AP, Guijarro CP (2015). Combustion homogeneity with partially premixed combustion a light duty optical engine (Master of Technology dissertation, Lund University) ISRN LUTMDN/TMHP-15/5350-SE
11. Manente V, Johansson B, Cannella W (2011) Gasoline partially pre-mixed combustion, the future of internal combustion engines? Int J Engine Res 12(3):194–208
12. Kalghatgi GT, Risberg P, Angstrom HE (2007). Partially pre-mixed au-to-ignition of gasoline to attain low smoke and low NOx at high load in a compression ignition engine and comparison with a diesel fuel (No. 2007–01-0006). SAE Technical Paper
13. Paykani A, Kakaee AH, Rahnama P, Reitz RD (2016) Progress and recent trends in reactivity-controlled compression ignition engines. Int J Engine Res 17(5):481–524
14. Han M, Assanis DN, Bohac SV (2009) Sources of hydrocarbon emissions from low-temperature premixed compression ignition combustion from a common rail direct injection diesel engine. Combust Sci Technol 181(3):496–517

15. Aoyama T, Hattori Y, Mizuta JI, Sato Y (1996). An experimental study on premixed-charge compression ignition gasoline engine (No. 960081). SAE Technical paper
16. Zehni A, Saray RK, Poorghasemi K (2017) Numerical comparison of PCCI combustion and emission of diesel and biodiesel fuels at low load conditions using 3D-CFD models coupled with chemical kinetics. Appl Therm Eng 110:1483–1499
17. Curran S, Prikhodko V, Gao Z, Parks J, Wagner R (2013) High efficiency clean combustion in multi-cylinder light-Duty engines. DOE hydrogen pro-gram and vehicle technologies annual merit review, May 14, 2013
18. Splitter, DA, Wissink, ML, Hendricks, TL, Ghandhi, JB., Reitz, RD (2012, September). Comparison of RCCI, HCCI, and CDC operation from low to full load. In THIESEL 2012 conference on thermo-and fluid dynamic processes in direct injection engines
19. Reitz RD, Duraisamy G (2015) Review of high efficiency and clean re-activity controlled compression ignition (RCCI) combustion in internal combustion engines. Prog Energy Combust Sci 46:12–71
20. Splitter DA, Reitz RD (2014) Fuel reactivity effects on the efficiency and operational window of dual-fuel compression ignition engines. Fuel 118:163–175
21. Inagaki, K, Fuyuto, T, Nishikawa, K, Nakakita, K, Sakata, I (2006). Dual-fuel PCI combustion controlled by in-cylinder stratification of ignitability (No. 2006–01-0028). SAE Technical Paper
22. Dempsey, AB (2013). Dual-fuel reactivity controlled compression ignition (RCCI) with alternative fuels (Doctoral dissertation, The University of Wisconsin-Madison) (2013)
23. Curran, S, Gao, Z, Szybist, J, Wagner, R. "Fuel effects on RCCI combustion: Performance and drive cycle considerations", presentation, CRC Workshop on advanced fuels and engine efficiency, February 2014
24. Molina S, García A, Pastor JM, Belarte E, Balloul I (2015) Operating range extension of RCCI combustion concept from low to full load in a heavy-duty engine. Appl Energy 143:211–227
25. Lim, JH. (2015). High power output operation of RCCI combustion (Doctoral dissertation, The University Of Wisconsin-Madison)
26. Curran, S, Gao, Z, Szybist, J, Wagner, R. "Fuel effects on RCCI combustion: Performance and drive cycle considerations", presentation, CRC Workshop on Advanced Fuels and Engine Efficiency, February 2014
27. Splitter, D, Wissink, M, Del Vescovo, D, Reitz, RD (2013). RCCI engine operation towards 60% thermal efficiency (No. 2013–01-0279). SAE Technical Paper
28. Dempsey, AB, Reitz, RD (2011). Computational optimization of reactivity controlled compression ignition in a heavy-duty engine with ultra low compression ratio. SAE International Journal of Engines, 4(2011–24-0015), 2222-2239
29. Dempsey AB, Walker NR, Gingrich E, Reitz RD (2014) Comparison of low temperature combustion strategies for advanced compression ignition engines with a focus on controllability. Combust Sci Technol 186(2):210–241
30. Wang Y, Yao M, Li T, Zhang W, Zheng Z (2016) A parametric study for enabling reactivity controlled compression ignition (RCCI) operation in diesel engines at various engine loads. Appl Energy 175:389–402
31. Li Y, Jia M, Liu Y, Xie M (2013) Numerical study on the combustion and emission characteristics of a methanol/diesel reactivity controlled compression ignition (RCCI) engine. Appl Energy 106:184–197
32. Wissink, ML, Lim, JH, Splitter, DA, Hanson, RM, Reitz, RD (2012, September). Investigation of injection strategies to improve high efficiency RCCI combustion with diesel and gasoline direct injection. In ASME 2012 Internal Combustion Engine Division Fall Technical Conference (pp. 327–338). American Society of Mechanical Engineers
33. Zhang, Y, De Ojeda, W, Wickman, D(2012). Computational study of combustion optimization in a heavy-duty diesel engine using in-cylinder blending of gasoline and diesel fuels (No. 2012–01-1977). SAE Technical Paper

34. Splitter, D, Hanson, R, Kokjohn, S, Wissink, M, Reitz, RD (2011).Injection effects in low load RCCI dual-fuel combustion (No. 2011–24-0047). SAE Technical Paper
35. Lim JH, Reitz RD (2014) High load (21 bar IMEP) dual fuel RCCI combustion using dual direct injection. J Eng Gas Turbines Power 136(10):101514
36. Ma S, Zheng Z, Liu H, Zhang Q, Yao M (2013) Experimental investigation of the effects of diesel injection strategy on gasoline/diesel dual-fuel combustion. Appl Energy 109:202–212
37. Heywood, JB (2011). Internal combustion engine fundamentals. TMH Publications: ISBN: 1–25-900207-1
38. Kokjohn, S, Hanson, R, Splitter, D, Kaddatz, J, Reitz, RD (2011). Fuel reactivity controlled compression ignition (RCCI) combustion in light-and heavy-duty engines. SAE International Journal of Engines, 4(2011–01-0357), 360-374
39. Kalsi SS, Subramanian KA (2016) Experimental investigations of effects of EGR on performance and emissions characteristics of CNG fueled reactivity controlled compression ignition (RCCI) engine. Energy Convers Manag 130:91–105
40. Pohlkamp, K, Reitz, RD (2012). Reactivity controlled compression ignition (RCCI) in a single-cylinder air-cooled HSDI diesel engine (No. 2012–32-0074). SAE Technical Paper
41. Benajes J, Molina S, García A, Belarte E, Vanvolsem M (2014) An investigation on RCCI combustion in a heavy duty diesel engine using in-cylinder blending of diesel and gasoline fuels. Appl Therm Eng 63(1):66–76
42. Curran, S, Hanson, R, Wagner, R, Reitz, RD (2013). Efficiency and emissions mapping of RCCI in a light-duty diesel engine (No. 2013–01-0289). SAE Technical Paper
43. Kokjohn SL, Hanson RM, Splitter DA, Reitz RD (2011) Fuel reactivity controlled compression ignition (RCCI): a pathway to controlled high-efficiency clean combustion. Int J Engine Res 12(3):209–226
44. Hanson, RM (2014). Experimental investigation of transient RCCI combustion in a light duty diesel engine (Doctoral dissertation, The University of Wisconsin-Madison)

# Exergy Analysis of Hydrogen-Fueled Spark Ignition Engine Based on Numerical Investigations

**Saket Verma, S.C. Kaushik and L.M. Das**

**Abstract** Hydrogen fuelled IC engines ($H_2$ICEs) have been considered as one of the most promising systems for pollution free transportations and their performance and combustion merits have been extensively discussed in the literature. However, studies related to these discussions have largely been linked to first-law analysis. On other hand, second-law of thermodynamics coupled with first-law, also known as exergy analysis, can give better insight into the engine performances. Bearing it in mind, this work presents second-law quantification of hydrogen engine processes and sub-processes, which helps to understand its true potential to deliver the output and simultaneously estimates various losses. This study quantifies different process inefficiencies in terms of irreversibilities thereby identifying the gaps to be addressed for further improvements. A computational fluid dynamics model has been prepared to simulate hydrogen-fueled spark-ignition engine ($H_2$SIE) operations and second-law equations have been coupled to ascertain different exergy terms. Present study shows that combustion process is the biggest source of irreversibility in IC engines. It has also been found that the level of irreversibility for a hydrogen-operated engine is substantially lesser as compared to that with gasoline engine under identical ranges of operating conditions. Combustion irreversibilities for $H_2$SIE and gasoline engine were found to be 15% and 23.6% of the total input fuel exergy respectively. Moreover, significant increase in second-law efficiency for $H_2$SIE as 44.4% compared to 36.8% that for gasoline engines has been found. Another important conclusion from this work includes exergy distribution for

S. Verma (✉) · S.C. Kaushik · L.M. Das
Centre for Energy Studies, Indian Institute of Technology Delhi, New Delhi, India
e-mail: saketverma@hotmail.com

S.C. Kaushik
e-mail: kaushik@ces.iitd.ac.in

L.M. Das
e-mail: lmdas@ces.iitd.ac.in



A.K. Agarwal et al. (eds.), *Combustion for Power Generation and Transportation*, DOI 10.1007/978-981-10-3785-6_14

$H_2$SIE, which is considerably diverse from gasoline engine operation. It indicates that optimization and improvements of different $H_2$SIE processes require specific attentions; nevertheless, show much better ability to deliver.

**Keywords** Exergy · Hydrogen · SI engine · Irreversibility · Combustion

## Nomenclature

### Abbreviations

| | |
|---|---|
| BDC | Bottom dead center |
| CI | Compression Ignition |
| CFD | Computational fluid dynamics |
| $CO_2$ | Carbon dioxide |
| EVO | Exhaust valve open |
| $H_2$SIE | Hydrogen fueled Spark ignition engine |
| ICE | Internal combustion engine |
| IVC | Intake valve close |
| LHV | Lower heating value |
| MBT | Maximum brake torque |
| $NO_x$ | Oxides of nitrogen |
| SI | Spark Ignition |
| TDC | Top dead center |

### Symbols

| | |
|---|---|
| $c_p$ | Specific heat at constant pressure, J/Kg K |
| e | Specific exergy, J/Kg |
| g | Gravitation constant, $m^2/s$ or specific gibbs free enthalpy, J/Kg |
| H | Enthalpy, J |
| h | Specific enthalpy, J/Kg or height from datum, m |
| $h_w$ | Heat transfer coefficient, $W/m^2K$ |
| I | Irreversibility, J |
| $\dot{m}$ | Mass flow rate, Kg/s |
| N | Speed, rpm |
| P | Absolute pressure, Pa |
| Q | Heat, J |
| R | Specific gas constant, J/Kg K |
| $\overline{R}$ | Universal gas constant J/mole K |
| S | Entropy, J/K |
| s | Specific entropy, J/Kg K |
| T | Absolute temperature, K |
| t | Time, s |

| | |
|---|---|
| U | Internal energy, J |
| u | Specific internal energy, J/Kg |
| V | Volume, $m^3$ |
| v | Velocity, m/s |
| W | Work, J |
| X | Exergy or Availability, J |
| x | Mole fraction |

**Subscripts**

| | |
|---|---|
| 0 | Dead state |
| 1 | Initial state |
| 2 | Final state |
| I | First-law terms |
| II | Second-law terms |
| ch | Chemical |
| cv | Control volume |
| cyl | Cylinder |
| fuel | Incoming fuel |
| ICE | Internal combustion engine |
| in | Incoming |
| i, j | Arbitrary number/species |
| K | Kinetic |
| loss | Losses through cylinder walls |
| out | Outgoing |
| P | Potential |
| ph | Physical |
| Q | Heat |
| surr | Surrounding |
| tm | Thermomechanical |
| W | Work |

**Superscripts**

| | |
|---|---|
| 0 | Reference state |

**Greek Symbols**

| | |
|---|---|
| $\theta$ | Crank angle |
| $\phi$ | Equivalence ratio |
| $\eta$ | Efficiency |

# 1 Introduction

In the pursuit of finding a non-depleting and clean burning fuel for internal combustion engines (ICEs), hydrogen has always been an attractive option. In fact, hydrogen perfectly fits in the idea of a zero emission vehicles in order to tackle the twin challenges of energy crisis and environmental degradation. Hydrogen is a form of secondary energy, which can be produced from available primary energy sources; and its basic resource (water) is plentiful and naturally recyclable [9]. Provided, hydrogen is produced from renewable primary energy sources, it can be proved to be absolutely clean, and also efficient energy source for IC engine applications[8]. This is because combustion of hydrogen with oxygen produces only water as engine exhaust product, albeit, with air, it also produces oxides of nitrogen ($NO_x$). Nevertheless, technological developments have shown that $NO_x$ emissions can be effectively controlled [5]. Furthermore, utilization of hydrogen in ICEs has many inherent advantages owing to some unique properties of hydrogen. A comparison of various properties of hydrogen with methane and gasoline is given in Tables 1 and 2. Higher calorific value of hydrogen with exceptionally high burning velocity as compared to gasoline, allows for much quicker and complete combustion leading to better energy conversion efficiency. This effect is assisted by high diffusivity of hydrogen, which helps in good mixing with air producing homogenous air-fuel mixture. Additionally, very high diffusivity of gaseous hydrogen and small emissivity of hydrogen flame make it safer in cases of leakage and fire respectively. With high flammability limit of hydrogen in air, much wider range of ICE operations with respect to equivalence ratios is possible. This allows for very lean operation of the engine, which brings many combustion merits also with emission benefits [4, 14, 22]. Yet, very high auto ignition temperature of hydrogen indicates that its utilization is much easier in spark ignition (SI) engines compared to compression ignition (CI) engines [23].

In the present study, an effort has been made to enhance the understanding related to the merits of $H_2$SIEs over conventional gasoline SI engines. In a view to achieve this objective, operations of $H_2$SIE have been studied based on second-law

**Table 1** Thermodynamic properties of hydrogen, methane and gasoline (generally accepted values from the literature) [4]

| Properties | Hydrogen | Methane | Gasoline |
|---|---|---|---|
| Molecular weight | 2.016 | 16.043 | 107.0 |
| Density of gas at NTP (g $m^{-3}$) | 83.764 | 651.19 | 4400 |
| Heat of combustion (low) (kJ $g^{-1}$) | 119.93 | 50.02 | 44.5 |
| Heat of combustion (high) (kJ $g^{-1}$) | 141.86 | 55.53 | 48 |
| Specific heat ($c_p$) of NTP gas (J $g^{-1}K^{-1}$) | 14.86 | 2.22 | 1.62 |
| Viscosity of NTP gas (g $cm^{-1}s^{-1}$) | 0.0000875 | 0.000110 | 0.00005 |
| Specific heat ratio ($\gamma$) of NTP gas | 1.383 | 1.308 | 1.05 |
| Gas constant (R) ($cm^2$ atm $g^{-1}$ $K^{-1}$) | 40.7030 | 5.11477 | 0.77 |
| Diffusion coefficient in NTP air ($cm^2$ $s^{-1}$) | 0.61 | 0.16 | 0.005 |

**Table 2** Combustion properties of hydrogen, methane and gasoline (generally accepted values from the literature) [4]

| Properties | Hydrogen | Methane | Gasoline |
|---|---|---|---|
| Limits of flammability in air (volume %) | 4.0–75.0 | 5.3–15.0 | 1.0–7.6 |
| Stoichiometric composition in air (volume %) | 29.53 | 9.48 | 1.76 |
| Minimum energy for ignition in air (mJ) | 0.02 | 0.29 | 0.24 |
| Auto-ignition temperature (K) | 858 | 813 | 501–744 |
| Flame temperature in air (K) | 2318 | 2148 | 2470 |
| Burning velocity in NTP air (cm $s^{-1}$) | 265–325 | 37–45 | 37–43 |
| Quenching gap in NTP air (cm) | 0.064 | 0.203 | 0.2 |
| Percentage of thermal energy radiated from flame to surrounding (%) | 17–25 | 23–32 | 30–42 |
| Diffusivity in air ($cm^2$ $s^{-1}$) | 0.63 | 0.2 | 0.08 |
| Normalized flame emissivity (2000 K, 1 atm) | 1.00 | 1.7 | 1.7 |
| Limits of flammability (equivalence ratio) | 0.1–7.1 | 0.53–1.7 | 0.7–3.8 |

of thermodynamics. It is well accepted that although energy analysis based on first-law of thermodynamics apprises of overall engine performance, it lacks in identifying specific areas of losses. On the other hand, exergy analysis based on second-law of thermodynamics allows for quantification of individual process inefficiencies and therefore presents the opportunity for improvements [3, 7]. Application of second-law of thermodynamics enables to identify individual locations of irreversibilities and hence apprises us about the scope of improvements. However, few such works are available concerning studies of $H_2$SIE with the help of exergy analysis. Early studies on this were done by Rakopoulos and Kyritsis [17], who studied the effect of hydrogen enrichment on natural and landfill gases combustion in ICEs. Based on computational study using zero-dimensional model, they found that there was monotonous decrease in combustion irreversibility as hydrogen enrichment was increased in both the fuels. This was related to the mechanism of entropy generation during combustion of hydrogen, which was found significantly reduced. In another study, Rakopoulos et al. [18] studied the effect of hydrogen addition on biogas fueled SI engine based on quasi-dimensional multi-zone combustion model. It was found that with 15% (by volume) of hydrogen addition in biogas, second-law efficiency was increased from 40.85 to 42.41%, which was accompanied by the reduction of combustion irreversibilities from 18.25 to 17.18% of the total availability. In another study similar results were observed with hydrogen-natural gas mixture [19]. Ozcan [15] found increased second-law efficiency with simultaneous decrease in irreversibility, and this effect was found to be pronounced at lean limit operating conditions. Besides, exergy losses to the cylinder walls were also increased with higher level of hydrogen supplementations. Operation of the neat hydrogen operated SI engine was examined by Fathi et al. [6] with main focus on the study of the effect of initial charge temperature. Based on their computational study, it was found that by increasing the initial charge temperature, combustion irreversibility, work availability and entropy generation were decreased, however, both heat transfer and exhaust availability were increased. Effects of equivalence ratio and injection timing were studied by Barzegar et al. [2] and it was found that retarded injection timings for direct injection $H_2$SIE showed better results from second-law perspectives. In a similar study conducted by Nemati et al. [13], it was found that variation in equivalence ratio has strong effects on combustion irreversibility and exhaust gas availability of direct injection $H_2$SIE.

Present work is motivated by the encouraging results obtained by previous researchers and it is an effort to enhance the understanding in areas of exergy analysis of $H_2$SIE. This analysis gives better insight of IC engine operations with hydrogen as a fuel, which shows ways for its efficient utilization in future prime movers.

## 2 Methodology

### 2.1 *Energy Analysis Based on First-Law of Thermodynamics*

First-law of thermodynamics is based on the principle of conservation of energy, which states that for any thermodynamic system, energy can neither be created nor be destroyed. Energy analysis deals with the energy balance for the system under consideration, essentially equating all incoming energies to all the outgoing energies. Therefore, energy analysis is a quantitative analysis, which shows the distribution of incoming fuel energy to the various IC engine processes. In IC engines operation, incoming fuel energy is distributed to output shaft work, energy transfer by heat losses through cylinder walls, energy transfer with exhaust gases and unaccounted energy losses (mechanical friction, leakage loss etc.). Following equations are used to calculate these terms:

Rate of incoming fuel energy $\dot{Q}_{fuel}$ is given by:

$$\dot{Q}_{fuel} = (\dot{m}_{fuel} \times LHV_{fuel}) \tag{1}$$

where $\dot{m}_{fuel}$ is the mass flow rate and $LHV_{fuel}$ is the lower heating value of the incoming fuel.

Rate of work transfer $\dot{W}_{cv}$ is given by:

$$\dot{W}_{cv} = \int_{IVC}^{EVO} (P - P_0) \frac{d\dot{V}_{CV}}{d\theta} d\theta \tag{2}$$

where $P$ and $P_0$ are instantaneous cylinder pressure and environmental pressure respectively. $d\dot{V}_{CV}/d\theta$ shows rate of change of cylinder volume with crank angles (CAs).

Rate of heat transfer losses $\dot{Q}_{loss}$ is given by Heywood [7]:

$$\dot{Q}_{loss} = h_w A \left[T_w - T_g\right] \tag{3}$$

where, $h_w$ and $A$ represent average heat transfer coefficient and cylinder area of heat transfer respectively. $T_w$ and $T_g$ represent cylinder wall temperature and instantaneous gas mixture temperature respectively.

First-law efficiency $\eta_{I,\,ICE}$ of the present IC engine system is given by Heywood [7]:

$$\eta_{I,\,ICE} = \frac{\dot{W}_{cv}}{\dot{Q}_{fuel}} \tag{4}$$

## 2.2 Description of Computational Model

In the present work, $H_2$SIE processes have been simulated using the computational fluid dynamics methodology. This simulation work has three important steps: (a) preparation of geometry (b) generation of computational mesh (c) computational calculations. Geometry of the system was prepared with the experimental engine specifications as shows in Table 3. Subsequently, meshing of the system geometry was performed. Present work requires modeling of transient in-cylinder combustion with piston motion; consequently, a moving mesh technique has been adopted. Furthermore, selective meshing was achieved by blocking and O-grid techniques [1]. This mesh mainly consists of tetrahedral elements except at the deforming cylinder walls, which consists of hexahedral elements assuring better reliability against deformation and reconstruction of mesh elements. Sensitivity analysis was performed to study the grid dependency of the computational results and finally a mesh consisting approximately of 620000 numbers of total elements was selected for computation. In the present work, computations were performed for the closed

**Table 3** Experimental engine specifications

| Engine parameter | Dimension (mm) | Valve parameter | Value |
|---|---|---|---|
| Bore | 86 | Intake valve open | −355° |
| Stroke | 68 | Intake valve close | −155° |
| Connecting rod length | 120 | Exhaust valve open | 150° |
| Inlet valve diameter | 37 | Exhaust valve close | 355° |
| Exhaust valve diameter | 30 | Valve overlap | 0° |
| Maximum valve lift | 7.65 | Compression ratio | 9.5 |

part of the cycle between intake valve closure (IVC at 25° after BDC) to the exhaust valve opening (EVO at 30° before BDC). Working conditions were chosen as equivalence ratio = 0.48, engine speed = 2000 rpm, spark timing = 12° BTDC (MBT timing). These numerical calculations were performed with the help of ANSYS Fluent CFD codes, which solve basic governing equations of conservation of mass, momentum and energy, and found reasonably accurate to represent in-cylinder combustion phenomena. In this modelling, RNG k-epsilon equations were used for viscous turbulent flow and partially premixed approach helps distinguish between burned and unburned zones. Additionally, other equations were solved e.g. species formation and chemical reactions, mean mixture fraction and flame front transport to represent the combustion model [1]. It has been reported that laminar burning speed of hydrogen is significantly different from that of gasoline; therefore, in present work correlation presented by Verhelst [21]was used to model increased laminar speed of hydrogen.

## 2.3 Exergy Analysis Based on Second-Law of Thermodynamics

There are subtle differences between energy and exergy analysis, and a comprehensive comparison is presented in Table 4. Exergy analysis is a qualitative measure to understand the performance of the system, and therefore governed by the second-law of thermodynamics. The most generalized definition of exergy was given by Szargut et al. [20] as:

"The maximum amount of work obtained when an energy carrier is brought from its initial state to a state of thermodynamic equilibrium (an inert state) with the common substances of the natural environment by means of reversible processes, involving interaction only with the above-mentioned substances of nature."

In other words, exergy is the maximum amounts of work that can be obtained from a given form of energy using environmental parameters as reference state. Therefore, high-grade energies (work, electricity, kinetic energy and potential energy) are equivalent to their exergy potentials. Exergy is a composite state function and, must be calculated by means of only reversible processes. Thus, exergy is a function of both thermodynamic states of the system and thermodynamic states of the reference elements in natural environment. It must be noted that when initial state is fixed, exergy shows the work potential of an energy source or system with reference to its dead state. However, any amount of reversible work can be extracted for an energy source or a system for its different final states, and when this final state corresponds to dead state, maximum reversible work is termed as exergy of the system.

**Table 4** Comparison between energy and exergy analysis

| Energy analysis | Exergy analysis |
|---|---|
| It is governed by the first-law of thermodynamics | It is governed by the second-law of thermodynamics |
| Energy is always conserved in any analysis | Exergy is also conserved for a reversible process; however, it can be destroyed during an irreversible process due to irreversibilities |
| It is based on the quantity of energy | It is based on the quality of energy |
| Energy is a function of states of the system under consideration | Exergy is a function of the states of the system and the surrounding as well |
| Energy of the system increases with the temperature rise | Temperature shows the quality of energy; higher the temperature higher the exergy |
| Energy analysis does not give any information about the performances of individual processes of an IC engine | Exergy analysis can be used to find out the irreversibilities associated with individual processes and hence indicates the scope for improvements |
| There are many processes (e.g. throttling etc.) for which first law analysis shows no energy loss | However, there is always some loss in terms of quality of energy, which can be ascertained by second law analysis |
| The datum for the energy analysis is arbitrary chosen | The datum for exergy is scientifically chosen such that at the datum (dead state) system cannot produce any useful work |

### 2.3.1 Environment, Dead State and Equilibrium

Natural environment can be considered as a very large medium with infinite source of chemical elements constituting its composition, where all the elements are in thermodynamic equilibrium (thermal equilibrium, chemical equilibrium and mechanical equilibrium). Therefore, no potential difference exists between the elements of this natural environment, and consequently no work can be extracted by interaction between these elements. However, any system outside the environment differing in any parameters such as pressure, temperature and/or concentration from the environmental state conditions will have some work potential with respect to the environment. Thus, it becomes vital to first conceptualize a realistic environment for exergy evaluation. Environment comprises of atmosphere, water and earth, which largely depends on the geographical region and even varies seasonally. Therefore, it is very difficult to choose a universal environment with unique standard species at their standard states. However, for equitable exergy based comparison of different engineering devices and concepts, standard environment composition and states have been adopted throughout the world [10, 12, 20].

A system is said to be in equilibrium with environmental state when the conditions for thermal equilibrium and mechanical equilibrium are satisfied, which requires that temperature and pressure of the system are equal to the temperature and pressure of the environment. However, exchange of matter between system and environment is not permitted, which means system is not in chemical equilibrium with the environment. That is why this equilibrium is also called *restricted equilibrium.* A system is said to be in equilibrium with its dead state, when it is in thermodynamic equilibrium with the environment, and hence conditions for thermal, mechanical and chemical equilibrium are satisfied. This implies that the system equals its temperature and pressure with the environment; moreover, there is no chemical potential difference between the substances of the system and that of the environment. This system is in *dead state* because it cannot change its state by any kind of energy interactions with its environment.

### 2.3.2 Exergy Transfer

In thermodynamic processes, exergy can be transferred by means of work transfers, heat transfers and mass transfers. As we have discussed, work transfer is a high-grade energy transfer, and therefore manifests equivalent exergy transfer in both magnitude and direction. Exergy associated with work transfer $X_W$ is given by [3]:

$$X_W = W_{CV} - W_{Surr} \ (For\, boundary\, work) \tag{5}$$

where $W_{CV}$ and $W_{Surr}$ show work done in the control volume and work done against the surrounding respectively.

However, heat being the low-grade energy, rest on the second-law of thermodynamics to represent its equivalence of exergy. Based on the definition, exergy associated with heat transfer is the maximum amount of work that can be obtained when it comes in the equilibrium with the surrounding. Exergy potential $X_Q$ of this heat content $Q$ at temperature $T$, with respect to the environment at temperature $T^0$ will be [3]:

$$X_Q = \left(1 - \frac{T^0}{T}\right) Q \tag{6}$$

It can be seen that exergy of heat transfer $X_Q$ depends on Carnot efficiency, for which environment is considered as heat sink.

Exergy transfer associated with mass transfer is defined by Kotas [10] as:

"Exergy of a steady stream of matter is equal to the maximum amount of work obtainable when the stream is brought from its initial state to the dead state by processes during which the stream may interact only with the environment."

Classification of exergy transfer associated with mass transfer is given in the Fig. 1. It is shown that this steady stream of matters can possess energies both in the form of low-grade energies (chemical energy and physical energy) and high-grade energies (potential energy and kinetic energy). Therefore, potential and kinetic energies (also called mechanical exergies) also show equivalent potential and kinetic exergies associated with the stream of matters and as given by the Eqs. 7 and 8. [3]

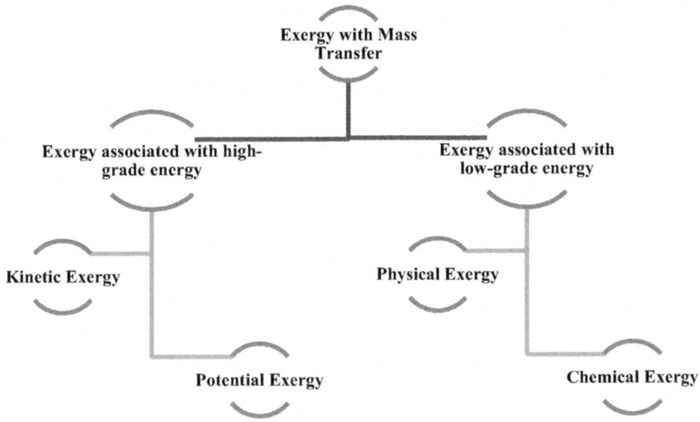


**Fig. 1** Classification of exergy transfers associated with mass transfer

$$X_K = \frac{1}{2} m v^2 \tag{7}$$

$$X_P = mgh \tag{8}$$

where $m$ is the mass of the fluid stream, $g$ is the gravitational acceleration, $h$ is the altitude of the flow and $v$ is the velocity of flow stream.

While, physical exergy is defined as the maximum amount of obtainable work when mass of stream is allowed to come in thermal and mechanical equilibrium with the environmental state defined by $P^0$ and $T^0$, by physical processes involving only thermal interaction with the environment. Therefore, for a mass of stream at state 1, physical exergy $X_{ph}$ is given by [3]:

$$X_{ph} = m\left[(h_1 - h_0) - T^0(s_1 - s_0)\right] \tag{9}$$

where $m$, $h$, and $s$ are flow mass, specific enthalpy and specific entropy of the stream of matter respectively.

It has been found convenient to show both physical exergy and mechanical exergy (potential exergy and kinetic exergy) in a single term called thermomechanical exergy $X_{tm}$. Consequently, for a steady flow stream, a specific exergy function $\psi$ can be defined as [3]:

$$e_{tm} = \psi = h - T^0 s + \frac{v^2}{2} + gz \tag{10}$$

It can be noted that in specific applications to IC engine processes, mechanical exergy term is commonly neglected being relatively small compared to physical exergy of the streams.

Chemical exergy is defined as:

"Chemical exergy is equal to the minimum amount of work necessary to synthesise, and to deliver in the environmental state, the substance under consideration from environmental substance by means of processes involving heat transfer and exchange of substances only with the environment." [10]

To find the chemical exergy of fuels, a reversible process has to be selected in the form of reversible chemical reaction to change the fuel into one or more reference substances. Following chemical reaction can be chosen involving combustion of hydrocarbon fuel with oxidant:

$$C_xH_yO_z + \left(x + \frac{y}{4} - \frac{z}{2}\right)O_2 \rightarrow xCO_2 + \frac{y}{2}H_2O \tag{11}$$

where, $C_xH_yO_z$ is a hydrocarbon fuel; then specific chemical exergy of this hydrocarbon fuel can be calculated as [10]:

$$e^0_{ch,\,fuel} = -\left[\left(xg^0_{CO_2} + \frac{y}{2}g^0_{H_2O}\right) - \left(g^0_{C_xH_yO_z} + \left(x + \frac{y}{4} - \frac{z}{2}\right)g^0_{O_2}\right)\right] + \left[xe^0_{CO_2} + \frac{y}{2}e^0_{H_2O} - \left(x + \frac{y}{4} - \frac{z}{2}\right)e^0_{O_2}\right] \tag{12}$$

where, $e^o_{ch,\,fuel}$ is molar chemical exergy of the fuel and $g^o$ is standard free enthalpy. Noticeably, in case of calculating chemical exergy of hydrogen as fuel, x = z = 0.

Finally, exergy transfer associated with mass transfer can be expressed as [3]:

$$e_{mass} = e_{tm} + e_{ch} \tag{13}$$

### 2.3.3 Exergy Analysis of IC Engine System

Considering IC engine as an open thermodynamic system undergoing heat, work and mass transfer with its surrounding, exergy balance equation on time basis can be written as follows [11]:

$$\underbrace{\Sigma\left(1 - \frac{T^0}{T}\right)\dot{Q}}_{\text{Exergy transfer by heat}} - \underbrace{\left(\dot{W}_{CV} - P^0\frac{dV_{CV}}{dt}\right)}_{\text{Exergy transfer by work}} + \underbrace{\sum_{in}\dot{m}_{in}\,.\,e_{in} - \sum_{out}\dot{m}_{out}\,.\,e_{out}}_{\text{Exergy transfer by mass}} - \dot{X}_{destroyed} = \frac{dX_{cv}}{dt} \tag{14}$$

It is more convenient to represent this equation on crank angle basis, which then takes the form of Eq. 15 [16]:

$$\frac{dX_{fuel}}{d\theta} + \frac{dX_Q}{d\theta} - \frac{dX_W}{d\theta} + \frac{\dot{m}_{in}e_{in} - \dot{m}_{out}e_{out}}{6N} - \frac{dI}{d\theta} = \frac{dX_{cyl}}{d\theta} \tag{15}$$

where $dX_{fuel}/d\theta$ shows the rate of change of chemical exergy of incoming fuel with crank angles. Incoming exergy of fuel is given by following equation [7, 16]:

$$\frac{dX_{fuel}}{d\theta} = \frac{dm_{fuel}}{d\theta} e^{0}_{fuel} \tag{16}$$

Here, burning rate of fuel mass $dm_{fuel}/d\theta$ is obtained using computational combustion model and specific fuel exergy is evaluated from Eq. 12.

As it has been already discussed in the previous section that exergy associated with heat transfer $dX_Q/d\theta$ and work transfer $dX_W/d\theta$ were calculated with the Eq. 5 and Eq. 6 respectively.

$\dot{m}_{in}e_{in} - \dot{m}_{out}e_{out}/6N$ represents the net exergy exchange of IC engine system associated with mass transfers. $\dot{m}_{in}e_{in}$ is the incoming exergy with air-fuel mixture, $\dot{m}_{out}e_{out}$ shows outgoing exergy with the exhaust gases and $N$ represents rotational speed of the crank in rpm. Here, specific exergy terms only include physical exergy of the flow stream. As present work concerns only with the closed part of the cycle (IVC to EVO), exergy terms associated with mass transfer are eliminated. Rate of change of exergy of the system (cylinder contents) with crank angles is given by [16]:

$$\frac{dX_{cyl}}{d\theta} = \frac{dU}{d\theta} + p_0\frac{dV}{d\theta} - T_0\frac{dS}{d\theta} - \frac{dG_0}{d\theta} \tag{17}$$

where $dU/d\theta$, $dV/d\theta$, $dS/d\theta$ and $dG_0/d\theta$ represent rate of change of internal energy, cylinder volume, entropy and free enthalpy of the gas mixture with crank angle respectively. These terms are calculated based on the results obtained from the first-law analysis of the computational engine model discussed above.

The last term $dI/d\theta$ represents rate of change of irreversibility generation with crank angle and is obtained by balancing the Eq. 15. Cumulative exergy terms are obtained by integrating these rate terms over the crank angles as: $\int_{\theta_i}^{\theta_f} (dX/d\theta)\, d\theta$.

Second-law efficiency for the present IC engine system is defined as the ratio of output work exergy over the incoming fuel exergy as given by Eq. 18. [3, 7]:

$$\eta_{II,\,ICE} = \frac{\dot{X}_W}{\dot{X}_{fuel}} \tag{18}$$

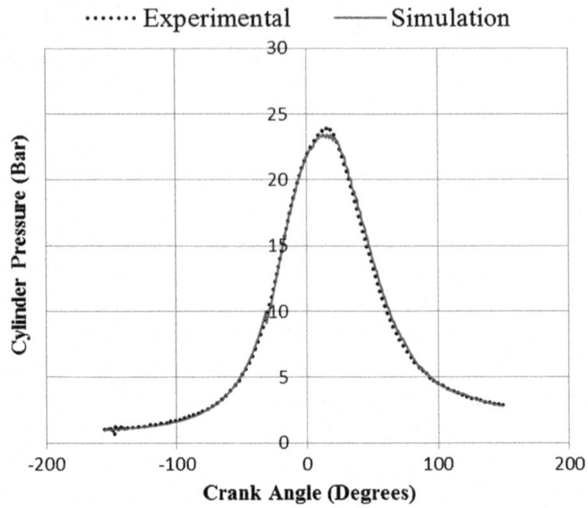


**Fig. 2** Comparison of experimental and simulation data obtained for cylinder pressure V/s crank angle. (speed = 2000 rpm, equivalence ratio = 0.48, spark timing = 12° BTDC)

## 3 Results and Discussion

Firstly, in-cylinder pressure data obtained from simulation of the present computational model were validated against the experimental results from a single cylinder, modified hydrogen-fueled SI engine, for which specifications are given in the Table 3. Figure 2 shows comparison between experimental and simulation data of in-cylinder pressure variation against crank angles for equivalence ratio of 0.48, engine speed of 2000 rpm and spark timing of 12° BTDC. Model was found reasonably accurate to predict compression, initial pressure rise, pressure peak and expansion processes for present $H_2$SIE operations. The chosen crank angle resolution of 0.25° was also found satisfactory showing insignificant variations for generated data.

Figure 3 shows variations of rate of change of exergy and Fig. 4 shows cumulative exergy values against crank angles. Based on the discussion on exergy analysis, variations of following exergy terms have been shown: (a) input-fuel exergy (b) work exergy (c) heat loss exergy (d) cylinder contents exergy (e) irreversibility. From Figs. 3 and 4, it can be seen that input fuel exergy values (both rate and cumulative terms) remain zero up to the point of start of ignition. This is because before the start of combustion, no fuel mass is burned and hence no fuel exergy is liberated inside the engine cylinder. However, as the ignition starts, there is drastic increase in input fuel exergy rate term and consequently in cumulative term due to very fast combustion of hydrogen as fuel in the combustion chamber. Peak value of rate term was observed at 3° CA BTDC and it was also found that up to 90% of fuel is burned at CA of 21° ATDC showing cumulative exergy value of 676.2 J. After this CA position, fuel exergy rate term becomes very small and

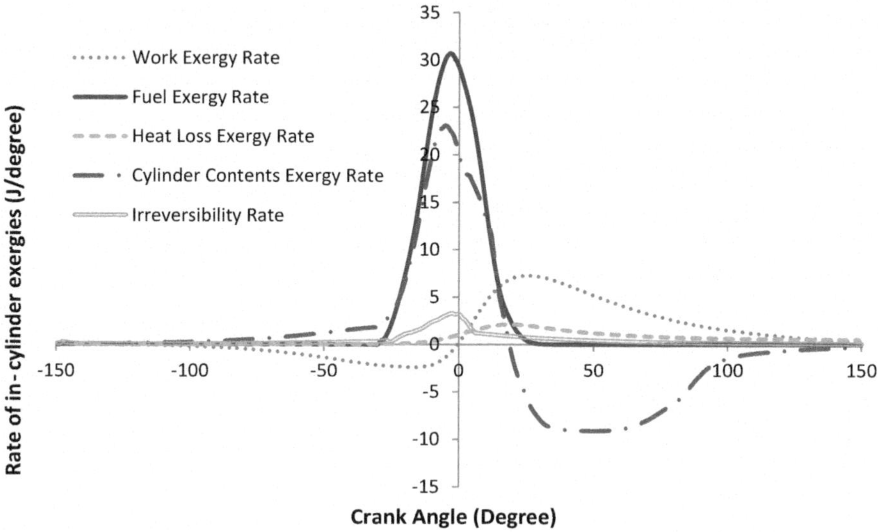


**Fig. 3** Development of exergy rate terms for closed part of the engine cycle

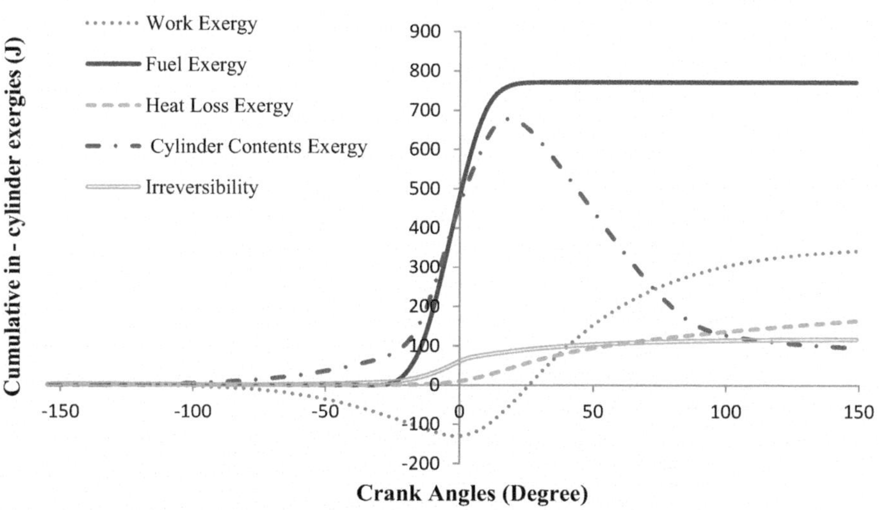


**Fig. 4** Development of cumulative exergy terms for closed part of the engine cycle

finally terminates. Work exergy rate term shows negative values during compression process as work is done on the system. However, after TDC, as the change in cylinder volume becomes positive, work exergy rate also becomes positive and starts increasing. Consequently, cumulative work exergy term also increases and positive work output is obtained. After expansion in the cylinder, both pressure and temperature of the cylinder contents decrease and therefore rate of work exergy also decreases and become zero.

Variation of heat loss exergy was found insignificant (though positive) during initial compression process. This is because during compression process, charge temperature is smaller than the average wall temperature, consequently,[1] heat transfer takes places from walls to the system. However, as the combustion proceeds, temperature of the system increases and heat is lost to the environment from the system causing heat loss rate term to become negative. This term gradually subsided by the end of the expansion process. Cylinder content exergy closely follows work exergy (but in opposite manner) before the combustion starts, and thereafter drastically increases. This is the consequence of combustion process, which significantly increases in-cylinder pressure and temperature increasing exergy of system contents. However, as the temperature starts decreasing in the expansion process, both cylinder content exergy rate and cumulative term start decreasing, which represent conversion of this exergy into useful work.

Irreversibility is the most important term and the focus of this study. It was found that rate of irreversibility is negligible up to the point of start of combustion, however, it drastically increases in the early phase of combustion and in expansion stroke again start decreasing. During initial compression process, there is negligible heat loss from the system and therefore, irreversibility generation is also negligible. However, as rate of combustion increases, irreversibility generation also increases. This conforms that combustion process is the biggest source of irreversibility generation in IC engines. Peak irreversibility rate of 3.34 J/degree was found at 3° CA BTDC and cumulative irreversibility value was 115.7 J.

Work was extended to find out various exergy terms for gasoline-fueled SI engine and similar trends as that of $H_2$SIE were obtained. Nevertheless, relative distributions of these exergy terms were found very different. Comparison of work transfer exergy, heat transfer exergy and combustion irreversibility as percentages of input fuel exergy is shown in Fig. 5. Work transfer exergy for $H_2$SIE was found to be 44.4% compared to 35% for gasoline operation. This significantly higher work transfer exergy attributed to much higher second-law efficiency of $H_2$SIE showing overall better performance of the engine. One of the main reasons for this high second-law efficiency of $H_2$SIE can be attributed to significantly lower combustion irreversibility of the engine. Combustion irreversibilities for $H_2$SIE was found to be 15% compared to 29% for gasoline operation. This significant decrease in combustion irreversibility with hydrogen operation of the engine is due to the unique nature of hydrogen combustion, where smaller molecules are combined to form a relatively heavier molecule in contrast of hydrocarbon combustion, where heavier molecules are broken down to the smaller ones [17]. Consequently, entropy generated by hydrogen combustion is much lower compared to that of hydrocarbon combustion. Additionally, heat transfer exergy with $H_2$SIE was found slightly

[1]It should be noted that present evaluation considers heat loss from the system as a negative term.

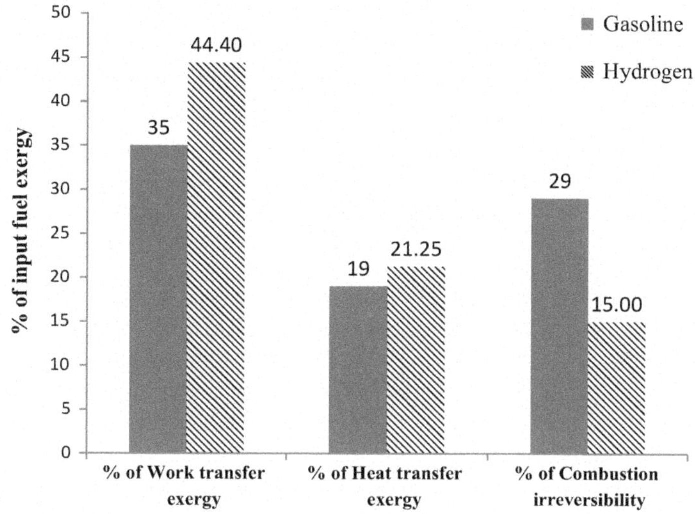


**Fig. 5** Comparison of different exergy terms between gasoline and hydrogen fueled IC engines

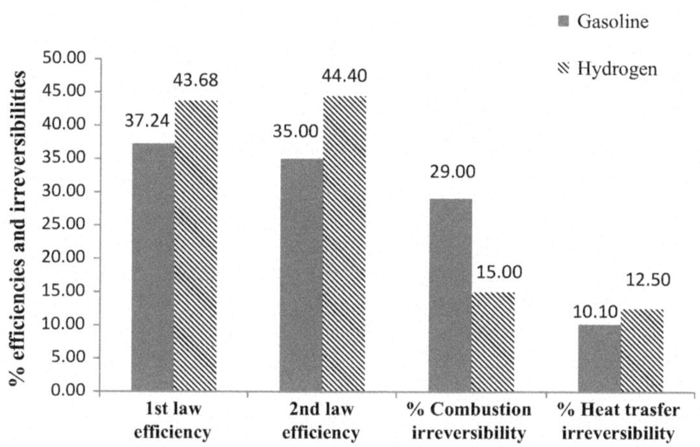


**Fig. 6** Efficiencies and irreversibilities for gasoline and hydrogen fueled SI engine

higher than that with gasoline operation. This is due to relatively higher average temperature during combustion with hydrogen engines. In addition to that high flame velocity and smaller quenching distance of hydrogen flames travel closer to cylinder walls causing higher rate of wall heat losses [14, 22].

First and second-law efficiencies of $H_2$SIE is shown in Fig. 6, which were found to be 43.68% and 44.4% of the input fuel exergy respectively. It can be noted that first-law efficiency of $H_2$SIE is always lower than second-law efficiency owing to the differences in lower heating value (LHV) and chemical exergy of hydrogen as fuel. It has been found that chemical exergy of hydrogen is lower than its LHV, however, opposite is true for hydrocarbon fuels. Figure 6 also shows heat transfer (from cylinder walls to atmosphere) irreversibility from $H_2$SIE, which was found to be 12.5% of the input fuel exergy.

## 4 Conclusions

In the present work exergy analysis of $H_2$SIE is carried out based on CFD methodology and its various exergy terms have been compared to that with conventional gasoline operation of the engine. This comparative exergy analysis between gasoline and $H_2$SIE has indicated that a hydrogen-fueled engine is more efficient than a gasoline fuelled engine from second-law perspective. One of the main reasons to explain the greater second-law efficiency of a hydrogen-fueled engine is efficient combustion of hydrogen, which enables it to convert higher amount of chemical potential into work causing lower specific fuel consumption. Following results were obtained from present analysis:

1. It was found that work exergy with hydrogen engine is around 9.4% higher compared to that with gasoline operation.
2. Most noticeable results show significantly reduced combustion irreversibility with $H_2$SIE which was found to be 15% of input fuel exergy. It is due to simple molecular structure of hydrogen.
3. Due to reduction in combustion irreversibility, overall exergetic performance of $H_2$SIE is better than gasoline engine.
4. However, heat transfer exergy and exergy loss with exhaust is much higher than gasoline engine. It requires careful attention for overall utilization.
5. First and second-law efficiencies were found to be 43.68% and 44.4% respectively for $H_2$SIE. It should be noted that heat transfer irreversibility is also more pronounced at around 12.5%.

**Acknowledgements** We would like to acknowledge the support of Council for Scientific and Industrial Research (CSIR) from where the author (Saket Verma) receives research fellowship.

## Appendix

See Table 5.

**Table 5** Various exergy terms used in the analysis

| Exergy terms | Equations | Remarks |
|---|---|---|
| Fuel exergy rate | $\frac{dX_{fuel}}{d\theta} = \frac{dm_{fuel}}{d\theta} e^0_{fuel} = m_{fuel} . e^0_{fuel} \cdot \frac{dx_b}{d\theta}$ | $e^0_{fuel}$ is fuel exergy |
| Wall heat transfer exergy rate | $\frac{dX_Q}{d\theta} = \sum \left(1 - \frac{T_0}{T}\right) \dot{Q}$ | Represents loss of exergy due to heat from the engine to the cylinder walls |
| Work exergy rate | $\frac{dX_W}{d\theta} = (P_{Cyl} - P_0) \frac{dV_{CV}}{d\theta}$ | Represents rate of exergy transfer with respect to crank angle by boundary work transfer |
| Mass exchange exergy rate | $\frac{\dot{m}_{in} e_{in} - \dot{m}_{out} e_{out}}{6N}$ | Represents net exergy exchange by incoming and outgoing masses in the engine cylinder |
| Irreversibility rate | $\dot{X}_{destroyed} = \frac{dI}{d\theta}$ | Irreversibility represents the loss of available work due to increase in the entropy of universe during a process |
| Control volume exergy rate | $\frac{dX_{cyl}}{d\theta}$ | $\int_0^{720} \frac{dX_{cyl}}{d\theta} d\theta = 0$ |

# References


1. Ansys Fluent (2011). 14.0, Theory Guide
2. Barzegar R, Khalilarya S, Fathi V, Nemati A (2013) The exergy analysis in a hydrogen fueled DISI engine under the effects of equivalence ratio and fuel injection timing. Tech J Eng Appl Sci 3(21):2803–2811
3. Cengel YA, Boles MA (2015) Thermodynamics: an engineering approach, 18th edn. McGraw-Hill Education, New York
4. Das LM (1990) Hydrogen engines: a view of the past and a look into the future. Int J Hydrogen Energy 15(6):425–443
5. Das LM (2002) Hydrogen engine: research and development (R&D) programmes in Indian Institute of Technology (IIT), Delhi. Int J Hydrogen Energy 27:953–965
6. Fathi V, Nemati A, Khalilarya Sh, Jafarmadar S (2011) The effect of the initial charge temperature under various injection timings on the second law terms in a direct injection SI hydrogen engine. Int J Hydrogen Energy 36:9252–9259
7. Heywood JB (1988) Internal combustion engine fundamentals. McGraw-Hill series in mechanical engineering, McGraw-Hill inc
8. Hudson MSL, Dubey PK, Pukazhselvan D, Pandey SK, Singh RK, Raghubanshi H, Shahi RR, Srivastava ON (2009) Hydrogen energy in changing environmental scenario: Indian context. Int J Hydrogen Energy 34:7358–7367
9. Kelly NA (2014) 6—hydrogen production by water electrolysis. In: Basile A, Iulianelli A (eds) Advances in hydrogen production, storage and distribution. Woodhead Publishing, pp 159–185
10. Kotas TJ (1995) The exergy method of thermal plant analysis. Krieger Publishing Company Malabar, Florida
11. Moran MJ (1982) Availability analysis: a guide to efficient energy use. Prentice Hall, New Jersey

12. Morris DR, Szargut J (1986) Standard chemical exergy of some elements and compounds on the planet earth. Energy 11(8):733–755
13. Nemati A, Fathi V, Barzegar R, Khalilarya S (2013) Numerical investigation of the effect of injection timing under various equivalence ratios on energy and exergy terms in a direct injection SI hydrogen fueled engine. Int J Hydrogen Energy 38:1189–1199
14. Niu R, Yu X, Du Y, Xie H, Wu H, Sun Y (2016) Effect of hydrogen proportion on lean burn performance of a dual fuel SI engine using hydrogen direct-injection. Fuel 15:792–799
15. Ozcan H (2010) Hydrogen enrichment effects on the second law analysis of a lean burn natural gas engine. Int J Hydrogen Energy 35:1443–1452
16. Rakopoulos CD, Giakoumis EG (1997) Development of cumulative and availability rate balance in a multi-cylinder turbocharged indirect injection diesel engine. Energy Convers Manage 38(4):341–369
17. Rakopoulos CD, Kyritsis DC (2006) Hydrogen enrichment effects on the second law analysis of natural and landfill gas combustion in engine cylinders. Int J Hydrogen Energy 31:1384–1393
18. Rakopoulos CD, Michos CN, Giakoumis EG (2008a) Studying the effects of hydrogen addition on the second-law balance of a biogas-fuelled spark ignition engine by use of a quasi-dimensional multi-zone combustion model. Proc. IMechE J. Automob Eng 222 (D):2249–2268. doi:10.1243/09544070JAUTO947
19. Rakopoulos CD, Scott MA, Kyritsis DC, Giakoumis EG (2008b) Availability analysis of hydrogen/natural gas blends combustion in internal combustion engines. Energy 33:248–255
20. Szargut J, Morris DR, Steward FR (1988) Exergy analysis of thermal, chemical, and metallurgical processes. Hemisphere Publishing Corporation, New York, London
21. Verhelst S (2005) A study of the combustion in hydrogen-fuelled internal combustion engine. PhD Thesis, Ghent University, Belgium
22. Verhelst S, Sierens R (2001) Hydrogen engine-specific properties. Int J Hydrogen Energy 26 (90):987–990
23. Verhelst S, Wallner T (2009) Hydrogen-fueled internal combustion engines. Prog Energy Combust Sci 35:490–527

# Thermodynamic Modelling of Combustion Process in a Spark Ignition Engine and its Numerical Prediction

**Vijaya Shree and V. Ganesan**

**Abstract** This chapter deals with the details of thermodynamic modelling of combustion process in a spark ignition engine. Because of the advent of fast digital computers, various modelling techniques have evolved over the period. They can be classified as thermodynamic, phenomenological and multi-dimensional models. Depending upon the complexities involved appropriate techniques are chosen for predicting the various engine processes. Of the three modelling techniques mentioned, thermodynamic modelling is the simplest one and can provide first cut solutions in the engine development. It may be noted that whatever modelling procedure used combustion is not yet completely amenable to mathematical analysis. In this chapter attempts have made to bring the combustion process under the ambit of thermodynamics. Ten thermodynamic combustion models available in the literature have been assembled and the details are explained. For the simulation of various engine processes from suction to exhaust appropriate thermodynamic equations have been employed and solved. Towards this a software have been developed and is named as 'GANESH'. The acronym 'GANESH' stands for Graphical and Numerical Software Hub. Using the software pressure-volume and pressure-crank angle variations are predicted and compared with experiments. Various engine performance parameters are predicted and analysed using different combustion models. Where ever possible predicted results are compared with experimental results. It is found that Wiebes law with m = 2 along with Woschni's equation for heat transfer is found to give good results. It is concluded that the developed software can be used with confidence for getting fast and reliable results

V. Shree (✉) · V. Ganesan
Department of Mechanical Engineering, Indian Institute of Technology Madras, Chennai 600036, India
e-mail: vijaya.shree.iitm@gmail.com

V. Ganesan
e-mail: vganesan.iitm@gmail.com



A.K. Agarwal et al. (eds.), *Combustion for Power Generation and Transportation*, DOI 10.1007/978-981-10-3785-6_15

# 1 Introduction

Having undergone a sustained development for well over a century, the modern internal combustion engine has become a very successful and fairly sophisticated machine. Ever since Otto made his first engine in 1876, a worldwide effort, encouraged by the multitudinous uses in stationary and also automotive applications to which the reciprocating engines can be put, has been made to improve the engine performance in all aspects. This development has proceeded along with the advances made in material science and fuel technology.

The internal combustion engine system presents many difficulties to an accurate analysis. These difficulties are 'a natural corollary' of the complexities of its operating cycle. For example, during the gas exchange phase of the engine operation the nature of unsteady flow through the intake/exhaust passage with moving boundaries poses a difficult task. Further, heat transfer and wave-propagation, and addition of volatile liquid fuel in spark ignition engines, complicates the situation beyond a straightforward analytical description. The difficulties relating to the combustion and expansion phases of the engine operation are equally formidable. These arise because of the extreme complexity of the physical situation brought about by many simultaneous events in a medium, which is in a complex state of thermal and mechanical agitation. Among them are (i) air-fuel mixing, (ii) piston motion, (iii) heat transfer to and from the combustion chamber, and (iv) chemical reaction with heat release including flame motion in a spark ignition engine. To understand the effect of above mentioned simultaneous processes, there is an immediate need for a workable mathematical model. The efficacy of this model to yield complete and accurate results depends to a large measure on the quantitative understanding of the process it models.

The most significant difference in the operating cycle of spark ignition (SI) engines and compression ignition (CI) engines occurs during the combustion phase. In a CI engine, combustion of fuel is brought, about by injecting a metered quantity of fuel in the form of a fine spray, into the air heated by compression—the temperature of the air being sufficient to cause ignition of fuel droplets. In an SI engine, the combustion is initiated by an electric spark discharge in the compressed mixture of air and fuel vapour. The flame then propagates outwards until the whole of the charge is consumed. Basically, therefore, the CI engine combustion process is characterized by a diffusion flame around each of the numerous fuel droplets, whereas the SI engine combustion process is characterized be deflagration initiated at the spark plug position and sweeping through the combustion chamber.

Because of the large number of very fine fuel particles in a relatively small combustion volume in an engine, it can be considered as a zone of homogeneous reaction. Rate of heat release depends on the spray characteristics, the air motion, the chemical properties of fuel, and the pressure and temperature of air. This approach has been found to yield fairly good results in the CI engine cycle calculations. Such a simplification, however, is not possible for the SI engine cycle calculations where, (at any instant during combustion) the charge is divided by the flame front into a burnt charge portion and an unburnt charge portion—almost all the heat release occurring at the flame front. The present work deals with the thermodynamic modelling of combustion process in a spark ignition engine and its numerical predictions.

The system of equations should use correlate time derivatives of such variables as the mass, volume and temperature of the burned and the unburned charge, and the cylinder pressure. The possibility of representing the effects of instantaneous heat transfer and heat release rates, the thermodynamic properties of fuel vapour and air, and the geometry of the combustion chamber by separate functional relationships in the system of equations used in modelling, offers many advantages. Primarily, such a formulation of the problem would provide great flexibility to the calculation procedure allowing several different models, of the phenomena mentioned above to be individually evaluated for their relative usefulness and accuracy.

## 2 Review of Literature

The advantages of the ability to simulate the operating cycle of the IC engine are always recognized after computers came into vogue. A continuous effort is being made by researchers to use the available information to obtain an accurate picture of the engine processes. In this section we will review the literature available pertaining to the topic chosen.

### *2.1 Thermodynamic Modelling of Combustion in SI Engines*

First law of thermodynamics provides the key equation in the incremental procedure used in thermodynamic models. Cylinder charge is divided into a number of zones in these models. Each zone has uniform pressure, temperature and composition. These models have been developed for forward and reverse analysis of engine cycles. The forward model is used for improvement of engine design and to predict engine behaviour at different conditions. The reverse operation is used to extract information additional to those directly measured experimental engine data.

Abu-Nada et al. [1] have developed a theoretical thermodynamic model of air-standard Otto cycle having temperature dependent specific heats analysis. They found that there were significant variations between the results obtained by using temperature dependent specific heats with those obtained at constant specific heats especially at high engine speeds. Hence, they inferred that use of temperature dependent specific heat is much better while studying the temperature variation in the actual cycles.

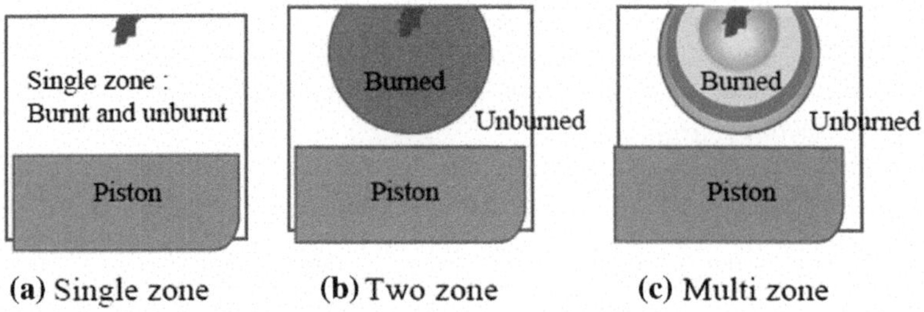


**Fig. 1** Pictorial representations of single-, two- and multi-zone models [70]

Jelić and Ninić [33] have done a review on the work done on analysis of thermodynamics of the combustion process in an engine cylinder in SI and CI engines and also on second law of thermodynamics analysis of IC engine. Relation between all thermodynamic data with entropy has been achieved by applying the second law of thermodynamics in analysis of the combustion process. The model provided by them can be considered as a useful tool for better insight and optimization of SI and CI engines to achieve lower fuel consumption and lower emissions.

These models can be further subdivided into two categories according to the number of zones used, viz. single- or two-zone (zero-dimensional or phenomenological), and multi-zone (quasi-dimensional) models. These models are described and reviewed in the following sub-sections. The details are shown pictorially in Fig. 1.

### 2.1.1 Single or One Zone Models

In single- or one-zone models, the working fluid in the engine is assumed to be a thermodynamic system, which undergoes energy and/or mass exchange with the surroundings. The energy released during the combustion process is obtained by applying the first law of thermodynamics to the system. Single zone models, mostly, assume uniform pressure, temperature and charge concentration at different crank angles. Heat addition and duration of combustion are given as input as per some mathematical law. Some of the models widely referred to in the literature are proposed by Lyn [47], Borman [10], Raine [61], Sezer and Bilgin [68], Scappin et al. [66].

Cylinder pressure, temperature, and composition are assumed uniform throughout the combustion chamber and the flame propagation process is neglected in single-zone models. Heat losses to the cylinder walls and heat addition as functions of time are included in the model. Heat release rate of engine cycle can be extracted using pressure-crank angle data in the reverse form of these models [16, 24].

Ramos [63] predicted that the one-zone thermodynamic models mainly depend on the assumptions of instantaneous mixing of burned and unburned gases. Further, such a model assumes uniform temperature and composition throughout the combustion chamber, and heat transfer loss and heat release were taken care of by sub-models. Following were assumed in the study:

1. heat release related to the enthalpy of combustion and independent of the temperature
2. cylinder mixture temperature, pressure and composition were uniform throughout the cylinder
3. cylinder gases behaved as an ideal gas mixture.

Heat transfer losses were calculated by Woschni correlation [74], and the internal energy of the cylinder gases was computed by empirical correlation dependent on the mixture temperature and composition. Further, the model assumes instantaneous mixing between the burned and unburned gases. The first law of thermodynamics is represented as:

$$m\Delta e = -p\Delta V - \dot{Q}_w\Delta\theta + \dot{Q}_c\Delta\theta \tag{1}$$

where $m$ is the mass contained in the cylinder, $e$ is the sensible internal energy per unit mass, $V$ is the combustion chamber volume, $\theta$ is the crank angle, $\dot{Q}_w$ is the heat loss through the solid walls, and $\dot{Q}_c$ is the heat released by the combustion process. The one-zone thermodynamic model represented by Eq. (1) can be used either as a diagnostic or as a predictive tool. Amount of heat release per unit time has to be specified when the model is used a predictive tool. Using the heat transfer losses calculation of the previous pressure and temperature values, the model gives the pressure and temperature as functions of time or crank angle. As a diagnostics tool, experimental pressure-crank angle curve has to be given as input, and the model gives the heat release curves and mixture temperature as functions of time.

***Heat release rate evaluation in single or one-zone models***—Merker et al. [55] have described in detail the single zone model to find heat release rates from measured pressure. Applying first law of thermodynamics to the system (single zone):

$$\frac{\delta Q_{ch}}{d\theta} - \frac{\delta Q_{ht}}{d\theta} - \frac{\delta W}{d\theta} + \frac{dM_f}{d\theta}h_f = \frac{dU}{d\theta} \tag{2}$$

where

| | |
|---|---|
| $\frac{\delta Q_{ch}}{d\theta}$ | Apparent rate of chemical energy (or heat release) |
| $\frac{\delta Q_{ht}}{d\theta}$ | Rate of heat transfer out of the system |
| $\frac{\delta W}{d\theta}$ | Rate of work transfer out of the system |
| $\frac{d(Mfh_f)}{d\theta}$ | Rate of enthalpy inflow with the fuel |
| $\frac{dU}{d\theta}$ | Rate of change of internal energy of the system. |

All the energy rates are expressed with respect to the crank angle, $\theta$.

After mathematical manipulation and simplification of the above Eq. (2) they have arrived at

$$\frac{dQ_n}{d\theta} = \frac{dQ_{ch}}{d\theta} - \frac{dQ_{ht}}{d\theta} = \left(\frac{\gamma}{\gamma - 1}\right) p \frac{dV}{d\theta} + \left(\frac{1}{\gamma - 1}\right) V \frac{dp}{d\theta} \tag{3}$$

where $\frac{dQ_n}{d\theta}$ is net apparent rate of heat release [55]. Using Eq. (3) they have modelled the heat release in SI engine. They have obtained the heat release from the measured $p$-$\theta$ variation.

### 2.1.2 Two-Zone Models

In two-zone models the working fluid is imagined to consist of two zones, an unburned zone and a burned zone. These zones are actually two distinct thermodynamic systems with energy and mass interactions between themselves and their common surroundings, the cylinder walls. The mass-burning rate or the cylinder pressure as a function of crank angle has been numerically computed by solving the simplified equations by applying the first law to the two zones. These models have been traditionally used in two different directions as shown in Fig. 2.

1. In one way, both these models have been used to predict the in-cylinder pressure as a function of crank angle from an assumed energy release or mass burned profile (as a function of crank angle).
2. Another use of these models lies in determining the energy release/mass burning rate as a function of crank angle from experimentally obtained in-cylinder pressure data.

Two-zone thermodynamic combustion model assumes that an infinitesimally thin flame propagates throughout the engine cylinder and calculates the mass and

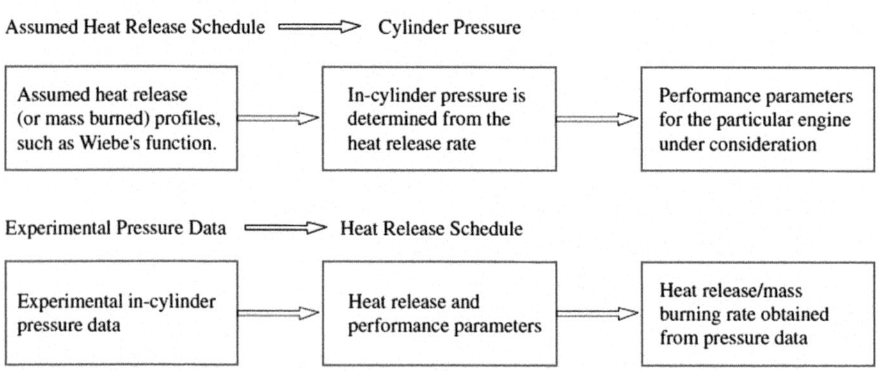


**Fig. 2** Directions followed by different variations of the single and two-zone models for different purposes

energy of the burned and unburned gases as a function of time [63]. The flame divides the mixture into unburned and burned gas regions having homogeneous temperature and composition. The burned gases at the existing instantaneous temperature are assumed to be in chemical equilibrium.

The gas mixture in both the burned and unburned zones is assumed to behave as an ideal gas and pressure as uniform throughout the combustion chamber. Heat transfer losses to the solid walls are considered [74]. However, no heat transfer is considered between the burned and unburned zones. The governing mass and energy equations of the burned and unburned gases are given by:

$$m = m_u + m_b \tag{4}$$

$$V = V_u + V_b \tag{5}$$

$$\frac{d}{dt}(m_u e_u) = -p\frac{dV_u}{dt} + \dot{Q}_u + \dot{m}_u h_u \tag{6}$$

$$\frac{d}{dt}(m_b e_b) = -p\frac{dV_b}{dt} + \dot{Q}_b + \dot{m}_b h_b \tag{7}$$

$$pV_u = m_u R_u T_u \tag{8}$$

$$pV_b = m_b R_b T_b \tag{9}$$

where subscripts $u$ and $b$ denote the unburned and burned gases respectively. The heat transfer losses $\dot{Q}_u$ and $\dot{Q}_b$ are calculated by Woschni correlation applied to each zone. Instantaneous composition of the burned gases is assumed to be in chemical equilibrium and calculated by the method proposed by Way [71].

The internal energy and enthalpy are evaluated by means of empirical correlations as functions of the temperature. Flame propagation can be determined by assuming that the flame front to be spherical. This type of model has been introduced by Patterson and Van Wylen [59] and modified and improved by Krieger and Borman [41], Lancaster et al. [42] and Mattavi et al. [50]. The two-zone model can be used to determine the turbulent flame speed as:

$$S_1 = \frac{\dot{m}_b}{\rho_u A_F} \tag{10}$$

where $A_F$ is the flame front surface area. Using the expressions proposed by Metghalachi and Keck [56] the laminar flame speed $S_1$ can be calculated and used to calculate the flame speed ratio (*FSR*), $S_t/S_1$.

Ramachandran [62] developed a thermodynamic combustion model for the simulation of a spark ignition engine running on hydrocarbon fuel. The model was based on the classical two-zone approach. Heat transfer from the cylinder, blowby energy loss and heat release rate have been considered. The general fuel has been specified by C–H–O–N values. He has employed curve-fit coefficients to simulate air and fuel data along with frozen composition and practical chemical equilibrium routines to plot various thermodynamic parameters with respect to crank angle.

El-Awad et al. [19] modelled a computer-based thermodynamic model of SI engines. The model could analyse the engine's performance with alternative fuels. Chow and Wyszynski [15] have given an overview of engine systems modelling by first and second law analysis.

Krieger and Borman [41], Fathi et al. [20], and Guezennec and Haurama [26] have studied the two-zone reverse thermodynamic models. They have developed models to derive the temperature and the volume of burned and unburned gases, heat release and mass fraction burned based on the pressure time history measured during combustion.

Jensen and Schramm [34] investigated the effects of crevices and cyclic variations on unburned hydrocarbon emissions ignoring blowby phenomenon. They developed a three-zone heat release model by assuming the crevice volume as a third zone.

Similarly, a number of researchers have studied and modelled forward simulation for two-zone reversed models [13, 31, 40, 43, 44, 48, 53, 54].

***Heat release rate evaluation in two-zone models***—A simple thermodynamic model, which incorporates two zones, viz. an unburned zone and a burned zone is desirable to perform parametric simulation studies on engine combustion. Therefore, it is important to understand and choose a suitable two-zone model to predict heat release rates in SI and CI engine combustion. Relatively well-defined and continuous flame propagation might occur in spark-ignited engines whereas there may be several distinct burned and unburned regions scattered throughout the cylinder in diesel engine combustion as shown in Fig. 3. Though there is difference in the nature of combustion in spark-ignited and diesel engine, two-zone model would be applicable for both the engines since it is a zero dimensional model,

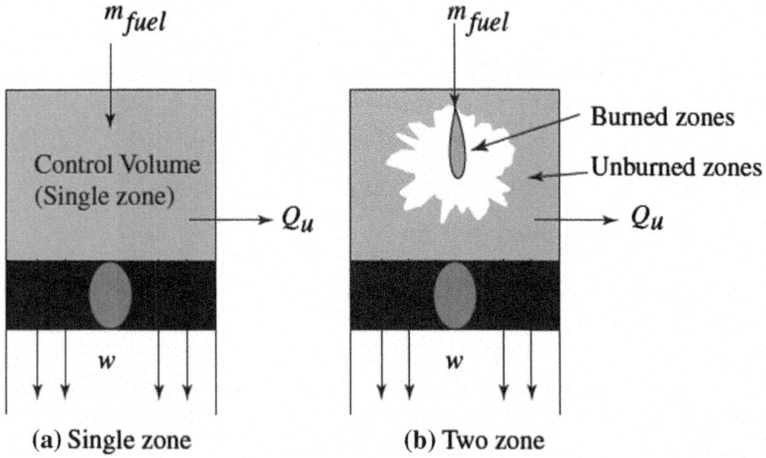


**Fig. 3** Representation of combustion in one- and two-zone models [70]

which considers only unburned and burned zones without any consideration of the spatial location of such zones.

Reason for considering two-zone model is to utilise the ideal gas equations in obtaining an "apparent" mass burned fraction curve or an "apparent" rate of heat release curve and also the conservation of mass and energy by first law of thermodynamics.

The heat release or mass burned fraction are apparent since they are obtained indirectly from measuring in-cylinder pressure. Its accuracy is limited by both the assumptions of the pressure data and the model. The corresponding equations for unburnt and burnt zones can be shown as

$$-Q_{ht,u} - PV_u + m_u h_u = \overline{m_u u_u} \tag{11}$$

$$Q_{ch} - Q_{ht,u} - PV_b - m_u h_u = \overline{m_b u_b} \tag{12}$$

The governing equations for the two-zone model are Eqs. (11) and (12). These equations are solved simultaneously each time (or crank angle) step to determine the unknown quantities $T_u$, $T_b$, $V_u$, $V_b$, $m_u$, $m_b$, and heat release rate from the known cylinder pressure $p$ and total volume $V$.

### 2.1.3 Multi-zone Models

In multi-zone models, the charge in the chamber is divided into more than one zone usually two zones, burned and unburned. The two zones are separated from each other by a discontinuity surface, viz. flame front. Flame propagation is usually governed by a turbulent burning velocity expression in such models. Uniform temperature and pressure are assumed in the burned and unburned regions and heat transfer between the two zones are not taken into consideration. The unburned gases are assumed to be of frozen composition and the burned gases in thermo-chemical equilibrium.

Blizard and Keck [7] introduced an eddy entrainment model assuming a spatially homogeneous volume of eddies. They stated that the entrained eddies were burned by diffusive transport of radicals between neighbouring eddies at a laminar burning velocity over a characteristic time. Such models are termed as "entrainment models". They have the advantage of linking with experimental flame visualisation techniques, viz. natural light and Schlieren photography. The models usually incorporate two differential equations, one for mass entrained and the other for mass burned. Using the turbulent entrainment and eddy burn-up model, several thermodynamic models have been constructed and developed [31] (Liu et al. 1999; Zhao et al. 1999; Reid and Douglas 1994; Wahiduzzaman et al. 1993; Chen et al. 1992). Multi-zone models consider energy and mass balances over several zones thus obtain results that are closer to reality.

AI-Himyary and Karim [2] studied a simple diagnostic combustion model based on pressure-time data for spark-Ignition engines. The model considers the charge to be made up of two zones, viz. burnt products and unburnt reactants. Each of which underwent a series of continuously varying yet distinctly different polytropic processes. The two zones exchanged mass across the flame front due to combustion. This approach did not utilise any correlations. The empirical formulae produced the results for rate of burning of the reactants, the rate of change of volume of each of the two zones and mean temperature histories of each of the two zones throughout the combustion period.

Kaprielian et al. [37] worked for fast and accurate automated design and predictive capabilities of quasi-dimensional combustion models of spark engines. The models have been based on mass and energy conservation principles and were supplemented by sub models based on experimental correlations. Initially they generated a three-zone model by introducing a reacting zone near the walls. The gases burnt at a lower temperature in the third zone than in the main reacting zone due to heat losses to the walls. Then, a multi-zone model has been built by dynamically adding new reacting zones at given crank-angle intervals. The multiple zones took into account the temperature and concentrations gradients in the flame. Energy release rates and pressures time histories predicted by the three-zone and by the multi-zone models have been compared with experimental data to validate the models.

Žák et al. [77] have dealt with the utilization of multi-zone model results in the thermodynamic cycle modelling of a four-stroke spark ignition engine. The required outputs based on available geometrical and experimental data have been obtained using inverse form of the model.

Buttsworth [12] has developed a multizone model for the thermodynamic simulation of internal combustion engines. Both burned and unburned zones could be tracked throughout the entire simulation. Heat transfer areas for burned and unburned zones could be specified as a function of engine geometry. Leakage of gas past piston rings could be simulated using a flow rate model driven by pressure differences. Further, the injection of fuel during the compression stroke and charge stratification across the zones could also be simulated. An initial assessment of the pressures, temperatures, and NO concentrations derived from the model has been performed by comparing the results obtained from the model.

## 2.2 Combustion in SI Engines

The cylinder mixture is in a complex turbulent flow field. It is ignited by a spark at late in the compression stroke in spark ignition engines. A burned flame kernel is created because of electrical discharge of the spark. The flame kernel propagates due to heat, combustion of unburned charge and turbulent diffusion between unburned and burned gases. Pressure, temperature and the flow field of the unburned mixture vary during the flame propagation because of burning gas expansion, piston motion and turbulence decay. The cylinder charge experiences complex chemical reactions, viz. heat transfer (with chamber wall) and work transfer (with piston) during the combustion process. Chamber geometry, gas motion, auto-ignition, mixture stratification and mass flow to/from crevices further complicate the situation.

A basic understanding of physics and chemistry of the phenomena is a must. It can be obtained by conducting simplified experimental studies. However, much fundamental research are being conducted in Bunsen burners under laminar conditions and constant volume bombs to understand the effects of pressure, temperature, diluents on laminar burning velocity, fuel type and equivalence ratio. Engine combustion chamber as an open thermodynamic system is schematically represented in Fig. 4.

Many investigations are being done to characterise the turbulent flow field in motored engines and well-stirred bombs in the absence of a combustion event [4, 28, 32].

Several investigators have carried out quite a few studies on heat exchange between the cylinder charge and chamber walls [3, 57, 74]. Investigations have been carried out on ring-piston-cylinder pack crevice flow, which is important for engine efficiency and unburned hydrocarbon emissions [27, 38, 52, 67].

Large number of research investigations have been conducted to understand combustion chamber geometry such as standard piston and multi-chambered piston [14, 75] and auto-ignition phenomena [17, 36], combustion chamber shape [21].

Several semi-empirical and empirical correlations and theoretical models have been presented to study the flame speed or turbulent burning velocity in

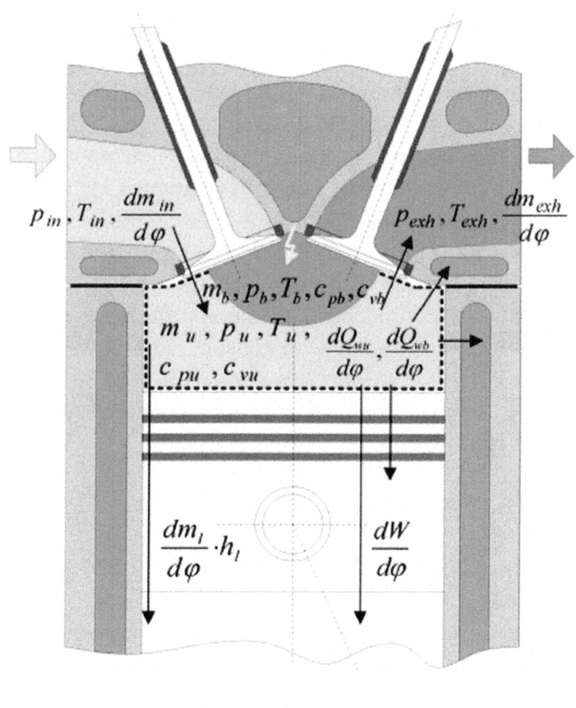


**Fig. 4** Engine combustion chamber as an open thermodynamic system [60]

homogenous charge in terms of laminar burning velocity, turbulent intensity, length scale, Lewis number, effective turbulent intensity and flame radius etc. [46]. Though various simplifications and assumptions are usually required to model engine combustion processes, they can be modelled using physics and chemistry equations of the phenomena.

## 2.3 Mass Fraction Burned During Combustion

The energy conversion during a combustion cycle at a specific crank angle degree (CAD) can be described by the Mass Fraction Burned (MFB). In IC engines, the MFB depends on engine geometry, engine speed, ignition angle, air-fuel ratio, residual mass etc.

Mass fraction burned (MFB) describes the process of chemical energy release as a function of crank angle in each individual engine cycle. It is a normalised quantity with a scale of 0–1. MBF is commonly determined based on burn rate analysis—a procedure developed by Rassweiler and Withrow [64, 65]. There are three approximate approaches described by Marvin [49], McCuiston et al. [51] and Rassweiler and Withrow [64, 65].

Marvin assumed that a particular amount of heat energy liberated by combustion would produce a proportional increase in pressure under isochoric conditions. However, combustion occurs at varying volumes in the real engine. Therefore, he allowed for this to reference conditions by a polytropic expansion. Young and Lienesch [76] implemented Marvin's approach to the approximate method developed by them within the EDPAC system. They have implemented a linear switch over between compression and expansion lines.

McCuiston et al. [51] deduced their approach from the adiabatic isochoric form of the first law of thermodynamics. Their approach has a similarity with Marvin [49]. It can be seen that Marvin's method is identical if the specific heat ratio is kept constant throughout combustion. Hence, the differences between the results of these methods must be due to the selected variation of specific heat ratio [73].

Rassweiler and Withrow's technique [64, 65] is widely used than the other approximate approaches. All the approaches based their method on two empirical rules from experiments with constant volume bombs. However, the final calculation of mass fraction burnt were similar in form to Marvin's, except that the pressure changes due to combustion being referenced to isochoric conditions and are separated from the effects of piston motion.

## 2.4 Formulation of the Method for Combustion Analysis

The main disadvantage of the thermodynamic analysis of engine combustion is the requirement of processing time. These techniques needed cycle by cycle calculations of the burn period, which are highly impractical. Therefore, fast methods for the analysis are desired.

Rassweiler and Withrow [64, 65] approach is based on observations from experiments with combustion bombs of constant volume. The fraction of the charge mass that has burnt at any instant in the combustion bomb is directly proportional to the fractional pressure rise. Fractional pressure rise is defined as the pressure at the particular instant divided by the final pressure attained from all the mixture burning. In addition, the final pressure attained from the combustion of a unit mass of mixture is inversely proportional to the volume of the bomb. Rassweiler and Withrow [64, 65] found a procedure to make use of the experimental observations where the combustion within the spark ignition engine could be considered analogous to that within a combustion bomb. Thus, the burnt fraction of the charge could be deduced.

The difference between the combustion in bombs and engines is that the pressure attained in the engine is due to the compression or expansion of the mixture by the motion of piston plus pressure change due to combustion. Hence, Rassweiler and Withrow [64, 65] hypothesized that the measured pressure in the engine cylinder was produced by the independent effects of combustion and the movement of the piston and therefore they are additive.

Therefore, the pressure increments due to combustion alone can be found by subtracting the pressure induced by the piston motion from the actual cylinder pressure. If all the increments of combustion pressure at different crank location is calculated, i.e. at different volumes and are referenced to as a common volume. By this analogous conditions between combustion within the engine and combustion within a single bomb can be considered to be similar. Hence, if the changes in combustion pressure are determined, the fractional pressure rise at any crank location is known and thereby the mass fraction burnt.

Rassweiler and Withrow [64, 65] suggested that the effects of combustion could be separated from the effects of piston motion by assuming that the measured pressure in the cylinder in a short interval of crank angle is obtained by the piston movement alone during the interval, followed by the instantaneous combustion of fuel at the end of the interval. To put it mathematically

$$\Delta p = \Delta p_p + \Delta p_c \tag{13}$$

According to Wiseman [73] the points worth to be noted about the approximate method suggested by Rassweiler and Withrow [64, 65] are:

(i) The isolation of the effects of the piston motion from the cylinder pressure are each interval and the initial pressure for the polytropic process is the actual cylinder pressure that was measured at the end of the previous interval
(ii) The increments of combustion pressure must ideally be zero before spark time and after the end of combustion. There is inaccuracy in the calculation of the polytropic process if there is deviation from these ideal pinpoints.

(iii) The mass fraction burnt is the ratio of burnt mass at a particular crank angle to the mass of all the mixture, which burns during the cycle.
(iv) The reference volume acts as a scaling factor on the combustion pressure. However, Rassweiler and Withrow [64, 65] suggest using the cylinder volume at spark time as the reference. If so, then the final combustion pressure obtained will be a function of spark timing and burn efficiency. Therefore, measure of the amount of fuel burnt will be lost.
(v) Calculation of heat energy liberation by thermodynamic analysis as a route to determine the fuel burning rate.
(vi) The mass fraction burnt calculated by approximate and thermodynamic approaches will differ if the lower heating value of the fuel changes significantly with crank angle.

Another functional form often used to represent the mass fraction burned $(x_b)$ versus crank angle is the Wiebe's function [25, 72]

$$x_b = 1 - \exp\left[-\alpha\left(\frac{\theta - \theta_0}{\Delta\theta}\right)^{m+1}\right] \tag{14}$$

where $\theta$ is the crank angle, $\theta_0$ is the angle where the start of combustion occurs, $\Delta\theta$ is the total combustion duration and $\alpha$ and m are adjustable parameters. Actual mass fraction burned curves have been fit with $\alpha = 5$ and $m = 2$. Hence, the mass fraction that remains unburned is given by $(1 - x_b)$.

Borg and Alkidas [35] have employed Wiebe function for simulating the fuel burning in IC engines. They have examined the various forms of the Wiebe's functions to present a methodology that optimizes the Wiebe parameters, which could simultaneously best fit the cumulative heat release and the rate of heat release histories. They have compared the accuracy of the results of the Wiebe function by solving for two parameters and for four parameters. The four parameters methodology was done using optimization routine and applied for the simulation of combustion in SI engines under normal and knocking conditions. Various researchers [9, 22] have employed Wiebe's function to determine the MFB.

Bonatesta [8] have investigated a S-shaped MFB profile from combustion initiation, assumed to coincide with spark timing to termination, i.e. 100% MFB using the independent parameters of the Wiebe function. The total burning angle has been taken as the 0–90% burn interval. Lindström et al. [45] preferred to 0–99.9% burn interval used in their work. However, 90% MFB could be determined experimentally with greater certainty.

Table 1 gives the comparison between approximate and thermodynamic methods of combustion analysis in a comprehensive manner.

**Table 1** Comparison between thermodynamic and approximate methods of combustion analysis

| Contributory factor | Approximate analysis | Thermodynamic analysis |
|---|---|---|
| Data input | Only pressure-volume data | Extensive: Pressure-volume data, thermodynamic tables, heat transfer correlation |
| Analysis time | Very short<br>No iterations | Long. Runge-Kutta method iteration required at each step |
| Heat transfer | Implicit in selection of polytropic index. Not problematic as mass fraction burnt is not sensitive to heat transfer | Heat transfer correlation used. Uncertainty as heat flux varies partially within cylinder, coefficient requires adjusting to each test rig<br>Effects absolute heat release, but not mass fraction burnt |
| Output | Mass fraction burnt only | Mass fraction burnt. Absolute heat release. Temperature profiles<br>Emissions calculations |
| Applicability | Good for comparing the combustion at each operating condition. Ideal for cycle by cycle analysis | Analysis of averaged pressure histories<br>Not very useful for cycle by cycle analysis |

## 2.5 Conclusions Drawn from the Review of the Literature

The following general remarks may be appropriate based on the review of literature presented in this chapter.

The need for an accurate procedure for analysing the processes of an IC engine from the standpoint of design optimisation is well recognised. It is a known fact that efforts towards a detailed description of the physical processes of the IC engine will yield a deeper and better insight into, and control of the IC engine operation.

Hitherto, conventional approach to cycle analysis using burned mixture charts have been carried out. Against this, the common modern approach to cycle analysis consists of applying the appropriate thermodynamic equations based on a specific model description of the IC engine processes. The equations are then solved step by step using a small time (or crank angle) increment taking into account the gas exchange, heat transfer, and heat release, as described by the model. The procedure followed for moving from one step to another, generally, consists of a series of sub-steps. Each sub-step refers either to a simple thermodynamic path or to an exchange process. Often an iterative procedure is used to arrive at the final step properties. This procedure is, however, unnecessary when a straight forward solution can be obtained by direct integration of the system of equations representing the model description.

Several attempts have been made to model the various processes of IC engines and to obtain performance results from the solution of these models. Digital computers are particularly helpful for the complex and lengthy calculations involved. These 'models' differ from the conventional cycle analysis approach (and from each other) in three broad aspects:

1. Treatment of intake and exhaust process,
2. Treatment of heat transfer analysis, and
3. Application of combustion models.

The present knowledge of these phenomena is, however, inadequate for a complete analytical description. This fact, coupled with the necessity to limit the complexity of the calculation procedure vis-a-vis the specific aspect of the operating cycle being investigated has led to several newer model descriptions of each of the three phenomena mentioned above. Models for the gas exchange processes are to date not adequate to predict an accurate value for the volumetric efficiency.

The relative suitability of a bunch of available heat transfer correlations is open to discussion. Theoretical advantages may be claimed for some of these. However, for all practical purposes, it appears that comparable results can be obtained by using the earlier correlation proposed by Eichelberg [18].

It is well known that combustion inside an SI engine is a complex phenomenon. It involves flame propagation through a highly turbulent gas mixture with simultaneous increase in temperature and pressure. It may be safely mentioned that no experimental or theoretical investigations regarding combustion in such a situation has been made till date. Consequently, models for treating combustion in SI engines have in general been based on empirical reasoning and often are not realistic in detail. Almost in all simulations, either an a priori rate of heat release is assumed or the flame front is assumed to propagate according to an arbitrarily assumed function. Such assumptions reduce the utility of the analysis for actual engine performance calculations. Realistic assumptions for the shape of the flame front and its rate of propagation as determined by the physical and chemical state of the charge are the need of the hour. For the cycle analysis to be useful and accurate for engine design the combustion process should be so modelled that the heat release is obtained from instantaneous conditions obtained inside the engine cylinder.

## 3 Piston Engines

A brief introduction of various nomenclatures associated with a spark ignition engine is presented in this section.

A schematic sketch of a piston cylinder arrangement of an engine is shown in Fig. 5, in which TDC and BDC refer to the top dead center and bottom dead center respectively. Further, $V_c = V_2$ is the clearance volume, $V_{disp}$ is the displacement volume, and the sum of the two, $V_t = V_{disp} + V_c = V_1$is the total volume. The diameter of the cylinder is $b$, known as the bore; the length of the cylinder that is traversed by the piston is the stroke length, or stroke, $s$. Thus, the displacement volume is the product of the stroke length and cross sectional area of the cylinder. In a

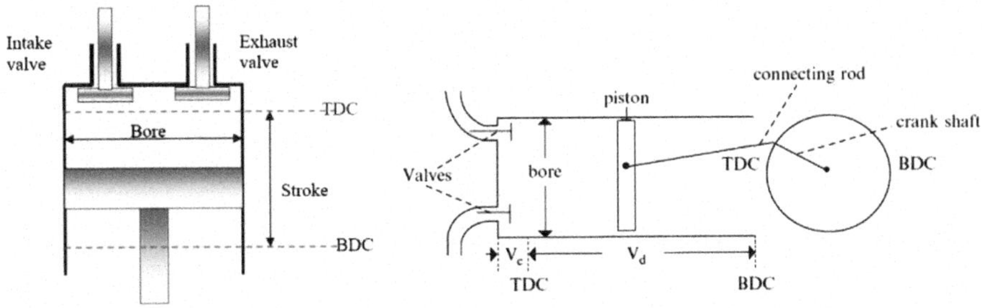


**Fig. 5** A schematic sketch of a piston engine [70]

four-stroke SI engine, the valve opens and allows the fuel-air mixture inside the cylinder and withdraws the piston toward BDC creating a vacuum inside the cylinder.

At the appropriate moment, the valve is closed and the piston pushing forward towards TDC compresses the trapped fuel-air mixture along with the products of combustion of the previous cycle. The fuel is ignited and the resultant withdrawal of the piston gives the effective work (power stroke), and, finally the exhaust valve opens and the product of combustion are pushed out. Thus, in a four-stroke engine, there are four different strokes in two complete rotation of the crankshaft for each power stroke.

Following relations among the three volumes are obtained with simple algebraic manipulations, (Fig. 6)

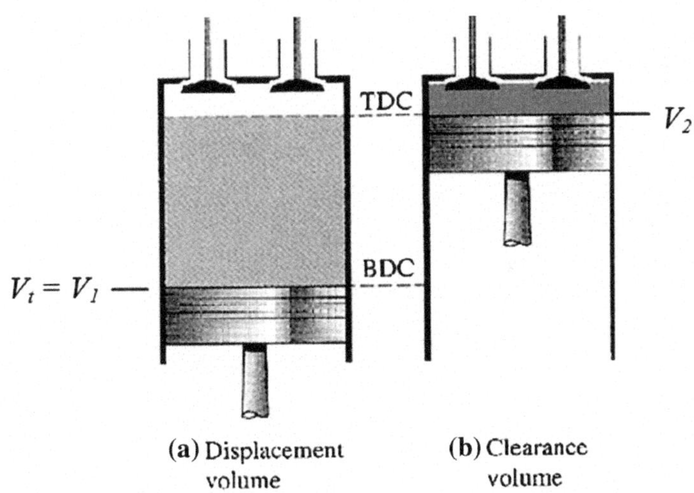


**Fig. 6** A schematic sketch of displacement and clearance volume [70]

$$V_c = \frac{V_t}{CR} = V_2;$$

$$V_{disp} = V_t\left(\frac{CR-1}{CR}\right)$$

$$V_t = V_1$$

Ideal cycle involves isentropic processes (compression and expansion) without any heat exchange with the wall and the isochore (constant volume) process involves instantaneous change in the pressure at a constant volume due to heat addition.

## 4 SI Engine Simulation with Progressive Combustion (PCS)

Instantaneous combustion at constant-volume is assumed in usual Fuel-air cycle simulation (FCS) analysis. However, in actual SI engines the combustion is spread over certain degrees of crank angle, which will have a significant effect on the performance of the engine.

All naturally aspirated engines exhibit a power peak with respect to speed. Hence, adiabatic combustion analysis carried out must be modified suitably to achieve closer approximation to study the behaviour of actual engines. Modifications can be attempted in the following three processes.

1. the combustion process,
2. the gas exchange process, and
3. the heat transfer process.

Combustion in spark ignition engine is not instantaneous. The spark is passed well before TDC. Combustion depends on many factors, such as

(a) the composition of the reactant mixture,
(b) the engine speed,
(c) the spark plug location,
(d) combustion chamber size and geometry.

Two events in the cycle and the $p$-$V$ trace during combustion have to be defined to calculate the net work done by the gases between the start of ignition and the end of combustion. The volume, $V$, of the cylinder with respect to the crank angle, $\theta$ has to be found out first for the analysis. The relation between $V$ and $\theta$ is governed by the engine geometry and kinematics.

## 4.1 Progressive Combustion

The details of the progressive combustion process are shown in Fig. 7. Ignition starts before the piston reaches top dead centre and combustion is not completed until the piston has moved beyond top dead centre. It is quite clear that the work done on the gases during the compression process will increase due to progressive burning while the work done by the gases during the expansion process will decrease. Hence, there will be a decrease in the power output. Equations to calculate the pressure-volume variation during the combustion process and to compute the power output will be described in this section.

The essential idea underlying the analysis of progressive combustion is expressed in the equation [24].

$$\Delta p = \Delta p_p + \Delta p_c \tag{15}$$

where

$\Delta p$ pressure change during a small interval of time,
$\Delta p_p$ pressure change during that time interval as a result of piston movement,
$\Delta p_c$ pressure change during that time interval as a result of combustion.

From Eq. (15), it is clear that $\Delta p_p$ and $\Delta p_c$ to the total pressure change, $\Delta p$ is merely additive [64, 65]. Compression, combustion and expansion are assumed to proceed without heat transfer to or from the gases in the working space. Hence,

$$pV^k = C \tag{16}$$

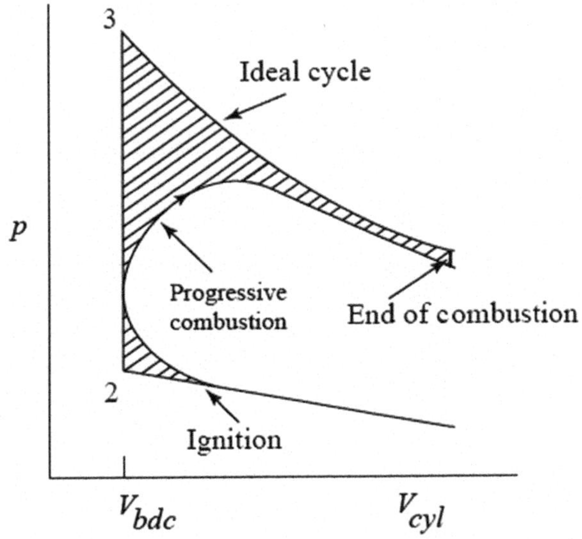


**Fig. 7** Details of progressive combustion [23]

from which

$$\Delta p_p = -pk\frac{\Delta V}{V} \tag{17}$$

where $k = C_p/C_v$.

In order to solve $\Delta p_c$, events that accompany combustion in a closed vessel have to be examined. Consider a combustible mixture is contained in a closed vessel, which is rigid and insulated under known conditions of pressure and temperature. The mixture is ignited at some point. Immediately a flame front is established and subsequently moves outward in all directions from the point of ignition. When the combustion is complete, the whole mixture is in a burned state at a pressure and temperature can be calculated using an adiabatic combustion process.

$$n = \frac{(a-b)\left(\frac{p}{p_i}\right)^{\frac{k_r-1}{k_r}} + b\left(\frac{p}{p_i}\right) - a}{(a-b)\left(\frac{p}{p_i}\right)^{\frac{k_r-1}{k_r}} + b\left(\frac{p_e}{p_i}\right) - a} \tag{18}$$

For a typical combustion engine, reactant mixture $a \approx 0.7$ and $b \approx 0.9$ are reasonable values. This means that *a–b* can be taken to be small for all practical purposes and can be considered to be very nearly equal to zero in the limiting case. Hence, by eliminating terms containing *a–b* in Eq. (18), *n* can be written as

$$n = \frac{bp - a\,p_i}{b\,p_e - a\,p_i} \tag{19}$$

If *a–b* can be approximated to zero, then it is nearly legitimate to assume $a \approx b$; therefore, Eq. (19) can be further simplified to

$$n = \frac{p - p_i}{p_e - p_i} \tag{20}$$

From Eq. (20), $\Delta p_c$ can be substituted into Eq. (15) to give

$$\Delta p_c = (p_e - p_i)\Delta n \tag{21}$$

Note that, in the analysis of combustion at constant-volume, $p_i$ and $p_e$ are the initial and final pressures. In the fuel-air cycle analysis, where combustion is assumed to be instantaneous and at constant volume, $p_i = p_2$ and $p_e = p_3$. Consequently, when combustion proceeds at some volume *V*, or when a fraction *n* of the reactants burns at volume *V*, the relation between $p_e$ and $\Delta n$ is given as

$$\Delta p_c = (p_3 - p_2)\frac{V_{tdc}}{V}\Delta n \tag{22}$$

$p_2$ and $p_3$ are calculated following the adiabatic combustion process. Hence, pressure change during combustion is

$$\Delta p = \Delta p_p + \Delta p_c = -pk\frac{\Delta V}{V} + (p_3 - p_2)\frac{V_{tdc}}{V}\Delta n \tag{23}$$

Equation (23) relates the change in cylinder pressure, $\Delta p$, to the volume change, $\Delta V$, and the mass fraction of gas that burned during the time interval required for $\Delta V$. Since the difference between the ratios of the specific heats of reactants and products, $k_r$ and $k_p$ respectively, is not insignificant, the integration of Eq. (23) can be carried out with the relation

$$k = k_r + (k_p - k_r)n \tag{24}$$

Note that Eq. (24) merely provides a simple and convenient transition from reactants to products as the combustion process proceeds.

In order to obtain $p$ and $V$ during the progressive combustion process, Eq. (23) should be integrated. It will then be possible to compute the net work done by the working gases during the combustion period. Let

$\theta_i$ is crank angle at the start of ignition,
$\Delta\theta_c$ is combustion duration in terms of crank angle,
$\theta_c$ is crank angle when combustion ends ($= \theta_i + \Delta\theta_c$).

An easy way to carry out the integration is to divide the combustion period into $N$ steps, each occupying a uniform interval of time. Then

$$\theta = \frac{\theta_c}{N} \tag{25}$$

The volume change $\Delta V$ can be written as

$$\Delta V = V\ (\theta + \theta) - V(\theta) \tag{26}$$

where $V(\theta)$ is given by

$$V(\theta) = V_{disp} \times \left[\frac{CR}{CR-1} - \frac{(1-\cos\theta)}{2} + \frac{L}{S} - \frac{1}{2}\sqrt{\left(\frac{2L}{S}\right)^2 - \sin^2\theta)}\right] \tag{27}$$

where $\theta = 0$ when the piston is at BDC.

The main problem is to know the variation of mass fraction, $n$ in the burned state with $\theta$. It is to be noted that thermodynamics deals only with equilibrium states. The dependence of $n$ on $\theta$ involves a series of equilibrium states that change with time. As thermodynamics cannot help, there are two possibilities to take care of the situation.

1. propose hypothetical relations between $n$ and $\theta$ which can be reasonable and possible, or
2. examine the events that take place in real engines as best as can. Then propose an empirical relation from the experimental data.

The second one is more appealing but meets with great difficulties when one has to pursue it actively in the laboratory. Combustion is a rapid chemical reaction of enormous complexity. It takes place in a relatively small space in engines that is not easily accessible to measuring apparatus and in terms of milliseconds.

Lot of investigations has been carried out in examining the happenings when the spark passes in as reciprocating engine by high-speed photography. These photographs are taken through quartz cylinder heads. They clearly reveal the happenings. A rapid development of flame front, which moves away from the spark plug in radial directions, could be seen. As the flame front progresses reactants are converted to products. The rate of conversion depends on two main factors.

1. The momentary surface area of the flame front, and
2. Speed with which the flame front moves relative to the adjacent unburned gas.

The first factor can be closely associated with the geometry of the combustion chamber and the location of the spark plug. Therefore, various models to take care of the situation are presented by means of different models proposed by different researchers.

(i) ***Uniform rate law*** [23]

Even though a uniform rate assumption is only academic, it recommends itself if only for its simplicity. With ingenuity, it might be possible to develop a combustion chamber that can produce a uniform rate of reaction. Whether it bears any relationship to real events in a real engine is largely a matter of geometry. However, a close analysis can reveal that it is not realistic during the early stages of combustion. During the early stage of flame development, a flame front starts as a tiny sphere between the spark electrodes and, therefore, $\frac{\Delta n}{\theta}$ should be very small for $\theta \approx \theta_i$. Equations (28) and (29) does not fulfill that requirement.

$$n = \frac{\theta - \theta_i}{\theta_c} \tag{28}$$

$$\Delta n = \frac{\theta}{\theta_c} = \frac{1}{N} \tag{29}$$

where $N$ is the number of steps, each occupying a uniform interval of time through which the combustion is completed (Fig. 8).

**Fig. 8** Variation of *n* with θ for uniform rate law

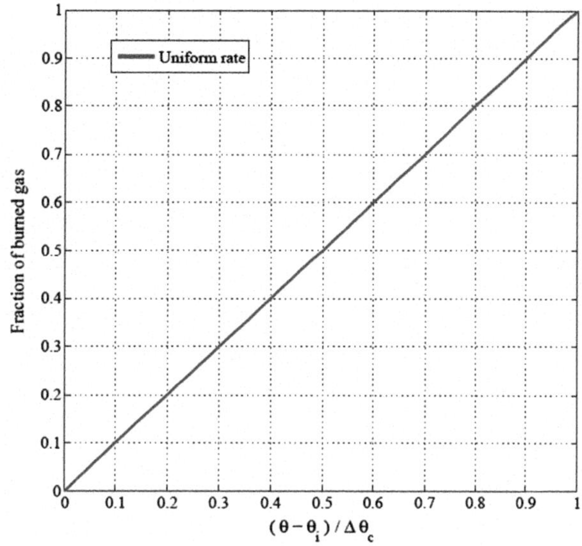


(ii) ***Square law*** [23]

A second possibility is a square law given by (Fig. 9)

$$n = \left(\frac{\theta - \theta_i}{\theta_c}\right)^2 \tag{30}$$

$$\Delta n = 2\left(\frac{\theta - \theta_i}{\theta_c}\right)\frac{\theta}{\theta_c} = \frac{2}{N}\left(\frac{\theta - \theta_i}{\theta_c}\right) \tag{31}$$

**Fig. 9** Variation of *n* with θ for square law

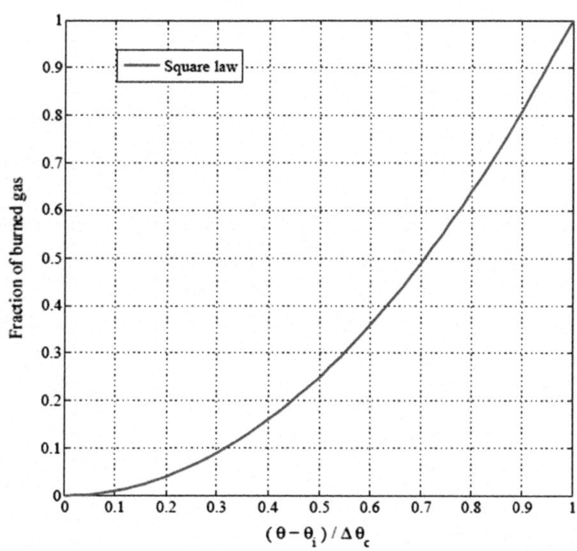

Fig. 10 Variation of $n$ with θ for cubic law

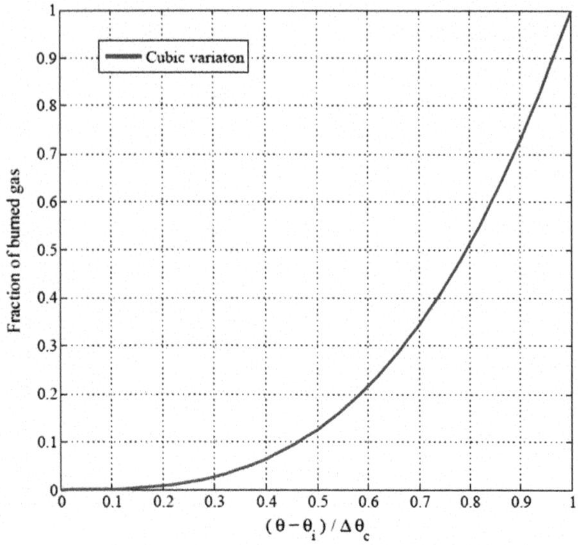


Equations (30) and (31) satisfies the condition of slow initial growth of the flame front. However, it predicts continuous growth of the flame front. Actually, there can be no continuous increase of flame front area. It should be noted that in piston engines, advancing flame front would reach the cylinder wall and the piston face over a period of time. This will cause an abrupt decrease in the rate of flame front area change with time (Fig. 10).

(iii) ***Cubic law*** [23]

$$n = \left(\frac{\theta - \theta_i}{\theta_c}\right)^3 \tag{32}$$

$$\Delta n = 3\left(\frac{\theta - \theta_i}{\theta_c}\right)^3 \left(\frac{\theta}{\theta_c}\right) \tag{33}$$

However, instead of square law if the cubic variation is applied it will reveal good slow initial growth but it depicts continuous growth of the flame. The difference between the square and cubic variation is that only first 10% burning happens slowly in square law and then pickups whereas in cubic variation 20% burning takes place slowly and the growth of the flame is comparatively slower (Fig. 11).

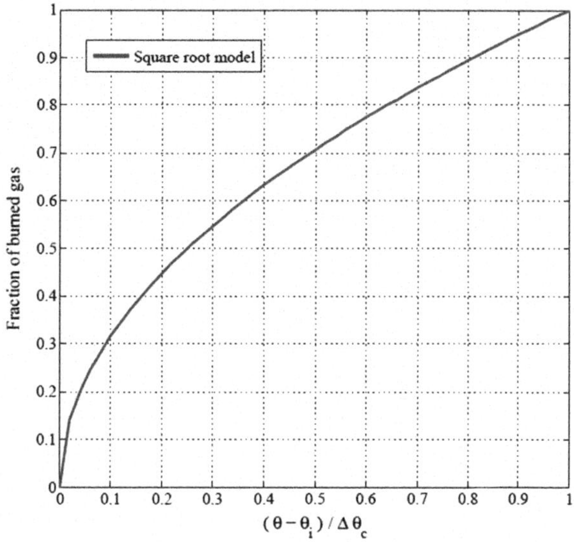


**Fig. 11** Variation of *n* with θ for square root law

(iv) ***Square root law*** [23]

$$n = \left(\frac{\theta - \theta_i}{\theta_c}\right)^{0.5} \tag{34}$$

$$\Delta n = 0.5 \times \left(\frac{\theta - \theta_i}{\theta_c}\right)^{0.5} \left(\frac{\theta}{\theta_c}\right) \tag{35}$$

Square root model does not predict the burning correctly since as shown in the figure, first 10% burning happens very fast and then pickups. As it is not bringing out the progressive combustion correctly, it may be discarded.

(v) ***Empirical cosine law*** [23]

Therefore, a fifth possible relation (model 5) can be written as an empirical law in the form. The relationship has been derived from engine studies and satisfies the condition that $\frac{\Delta n}{\theta}$ is small both at the beginning and the end of combustion (Fig. 12).

$$n = \frac{1}{2}\left[1 - \cos\left(\frac{\theta - \theta_i}{\theta_c}\pi\right)\right] \tag{36}$$

$$\Delta n = \pi 2N \sin\left(\frac{\theta - \theta_i}{\theta_c}\pi\right) \tag{37}$$

**Fig. 12** Variation of $n$ with θ for empirical cosine law

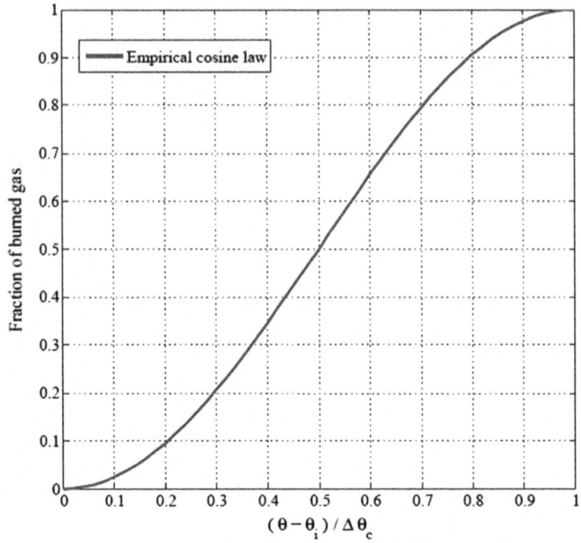


(vi) ***Wiebe's law (m = 2)*** [25]

A functional form often used to represent the mass fraction burned versus crank angle is the Wiebe's function:

$$\mathrm{n} = 1 - \exp\left[-\alpha\left(\frac{\theta-\theta_i}{\Delta\theta}\right)^{m+1}\right] \tag{38}$$

$$\Delta\mathrm{n} = \frac{15}{\Delta\theta}\left(\frac{\theta-\theta_i}{\Delta\theta}\right)^2 \exp\left[-5\left(\frac{\theta-\theta_i}{\Delta\theta}\right)^3\right] \tag{39}$$

where θ is the crank angle, $\theta_i$ is the angle where the start of combustion occurs, Δθ is the total combustion duration and α and m are adjustable parameters. Actual mass fraction burned curves have been fitted with α = 5 and $m$ = 2, where α and $m$ are tuning parameters. Coefficient α represents the combustion efficiency. Obviously, the mass fraction that remains unburned is given by (1 − $n$). Most of the researchers have used Wiebe's law and therefore, in this investigation also this law has been employed (Fig. 13).

(vii) ***Wiebe's Law (m = 3)*** [62]

Burning rate mostly depends on the position of the spark plug and combustion chamber shape [62]. The Wiebe's function represents the mass fraction burned, $n$ versus crank angle and defined as $n = 1 - \exp\left[-\alpha\left(\frac{\theta-\theta_i}{\Delta\theta}\right)^{m+1}\right]$. The value of α and $m$ are taken as 5 and 2 respectively [30]. The model is capable of representing many combustion chamber shapes with different positions of spark plugs by

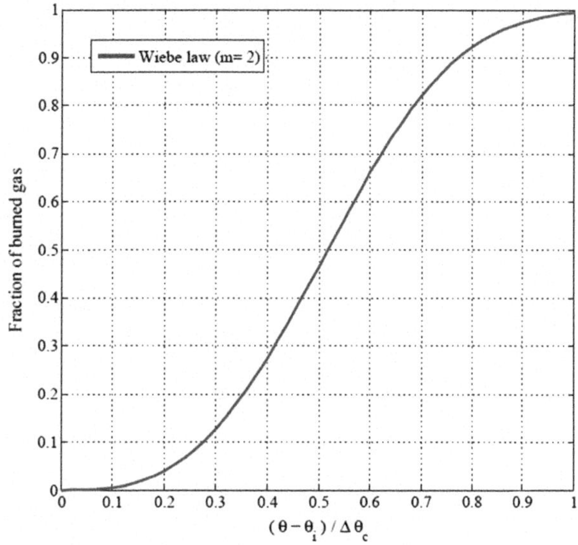


**Fig. 13** Variation of $n$ with θ for Wiebe's law for $m = 2$

adjusting α and $m$. The constant α is responsible for maximum value of $n$ which is unity [39]. Hence, α may represent the positions and/or number of spark plugs. It is observed that by varying $m$ the starting point and end point of $n$ are independent of $m$. However, for a particular α, its rate closely depends on $m$. Therefore, as $m$ increases, $n$ decreases and $m$ represent combustion chamber shape-factor (Fig. 14).

$$n = 1 - \exp\left[-5\left(\frac{\theta - \theta_i}{\Delta\theta}\right)^{3+1}\right] \quad (40)$$

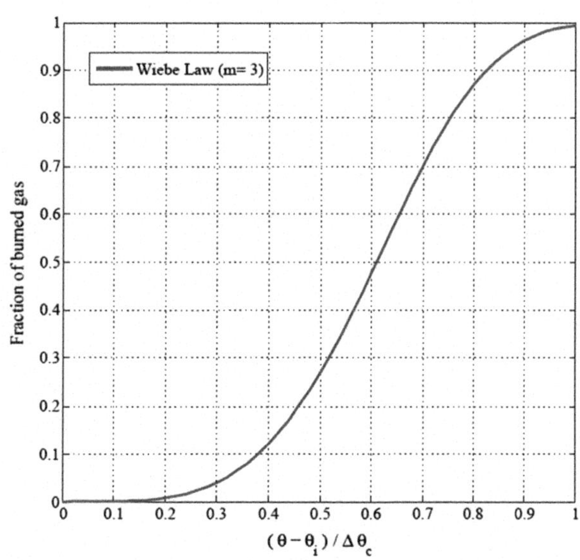


**Fig. 14** Variation of $n$ with θ for Wiebe's law for $m = 3$

$$\Delta n = -\exp\left[-5\left(\frac{\theta-\theta_i}{\Delta\theta}\right)^{3+1}\right] - 5\times 4\times\left(\frac{\theta-\theta_i}{\Delta\theta}\right)^{3}\times\frac{1}{\theta_i} \tag{41}$$

(viii) ***Exponential law*** [70]

$$n = xr + (1-xr)\times\left(1-\exp\left[-\alpha\left(\frac{\theta-\theta_i}{\Delta\theta}\right)^{m+1}\right]\right) \tag{42}$$

$$\Delta n = (1-xr)\frac{15}{\Delta\theta}\left(\frac{\theta-\theta_i}{\Delta\theta}\right)^{2}\times\exp\left[-5\left(\frac{\theta-\theta_i}{\Delta\theta}\right)^{3}\right] \tag{43}$$

where $xr$ is the mass fraction burnt during preflame reaction (Fig. 15).

Even though the variation is similar to Wiebe's law, the flame does not start from zero flame front but assume a small preflame reaction.

(ix) ***Experimentally fitted law for rich mixtures*** [29]

Harish Kumar and Antony [31] have modified Taylor's equation for predicting the overall combustion duration and flame development angle and ignition delay. They have formed a new set of logistic equation with conditional variability in the mass fraction burnt. The equation takes into account the influence of engine geometry, swirl, spark plug location etc. With available experimental data for rich mixture (Fig. 16),

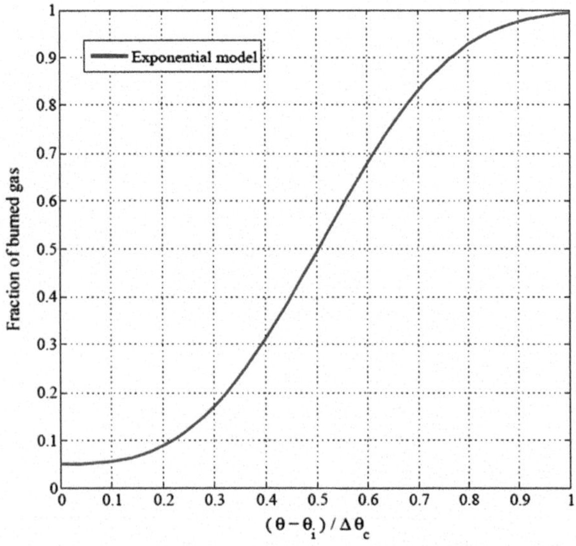


**Fig. 15** Variation of $n$ with θ for experimental law

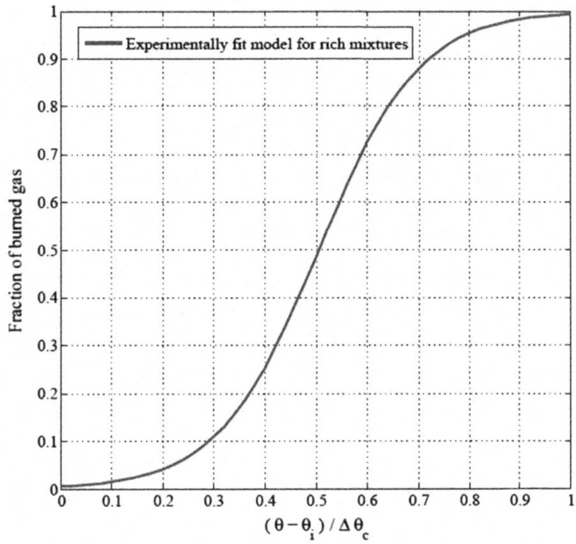


**Fig. 16** Variation of $n$ with θ for experimentally fitted law for rich mixtures

$$n = 1/\left(1 + 176.2 \times \exp\left[-10.23\left(\frac{\theta - \theta_i}{\Delta\theta}\right)\right]\right) \tag{44}$$

$$\Delta n = n^2 \times 176.2 \times 10.23 \times \exp\left[-10.23\left(\frac{\theta - \theta_i}{\Delta\theta}\right)\right] \tag{45}$$

(x) ***Polynomial law*** [5]

The Wiebe's function is commonly used to model the heat released by the combustion in a SI engine. When α = 1 and $m$ = 2, then

$$n = 1 - e^{-5y\theta_i^3} \tag{46}$$

$$\Delta n = 15\theta_i^2(1 - n) \tag{47}$$

The starting point of this function is $(\theta_i = 0;\ n = 0)$ and for $\theta_i = 1$, $n$ is very close to unit (0.99). A more simple function is $\theta_i = n$ and $\Delta n = 1$. The constant derivative i.e. constant combustion velocity is not plausible in a spark ignition engine; hence, a better approximation is obtained by the following polynomial function [5] (Fig. 17)

$$n = 3 \times \theta_i^2 \times \left(1 - \frac{2}{3} \times \theta_i\right) \tag{48}$$

$$\Delta n = \theta_i \times \frac{(1 - \theta_i) \times \pi}{180} \tag{49}$$

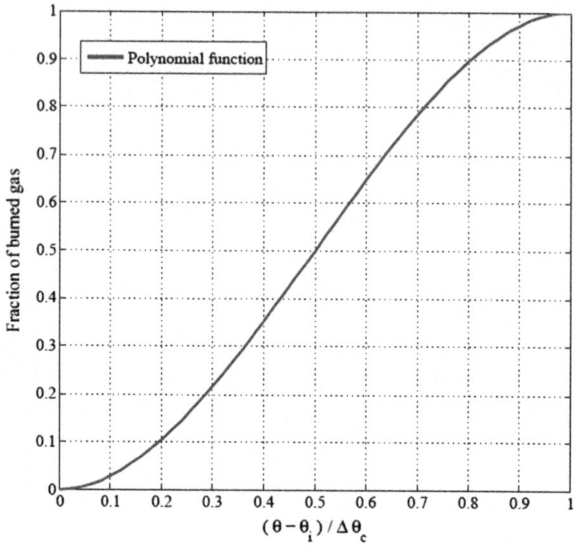


**Fig. 17** Variation of $n$ with θ for polynomial law

Figure 18 shows comparison of these relationships. The relationship between n and θ is only a part of a conjecture about the combustion process. Each relationship contains a term Δθc, the duration of combustion in terms of crank angle in degrees.

Cycle calculations for an engine operating with progressive combustion can be carried out as follows [23]:

Start the cycle at BDC. Prior to compression assume values for temperature $T_1$ and residual exhaust gas mole number $N_x$

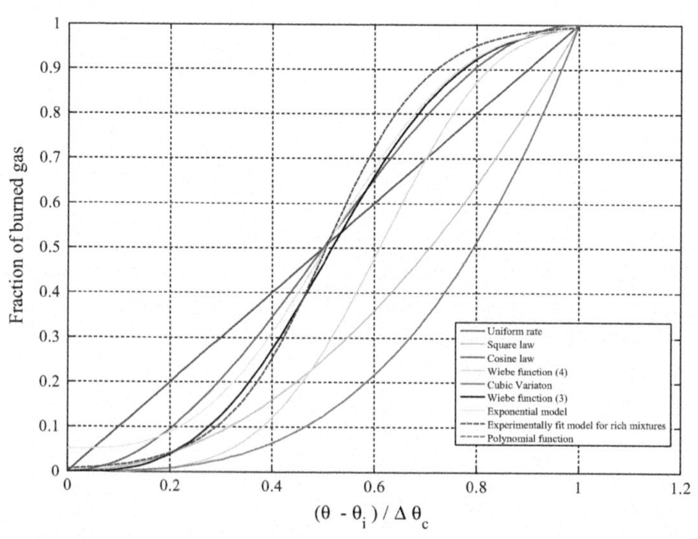


**Fig. 18** Variaton of $n$ with θ for 9 models

Specific heat ratio of the reactant mixture, $k_r$, can be obtained with known $N_x$.

The analysis can be carried till the end of constant-volume adiabatic combustion by obtaining the values for $p_2$ and $p_3$ in

$$\Delta p_c = (p_3 - p_2)\left(\frac{V_{tdc}}{V}\right)\Delta n \tag{50}$$

The ratio of specific heats for the product gases, $k_p$, can be obtained using $T_3$.

Choose a spark advance angle, $S_{adv}$, in degrees before TDC. With subscript $i$ denoting the values at ignition, find

$$\theta_i = 180 - S_{adv} \tag{51}$$

$$V_i = V(\theta_i) \tag{52}$$

$$P_i = p_1\left(\frac{V_{bdc}}{V_i}\right)^{k_r} \tag{53}$$

$$T_i = T_1\left(\frac{V_{bdc}}{V_i}\right)^{k_r - 1} \tag{54}$$

The work required for compression stroke is given by

$$W_{comp} = (N_m + N_x)C_{vr}(T_i - T_1) = -\sum\left(p + \frac{\Delta p}{2}\right)\Delta V \tag{55}$$

where $C_{vr}$ denotes the constant-volume specific heat of the reactant mixture ($N_x + N_m$). The summation is carried out from ignition to TDC using equal increments in crank angle with which the corresponding increments in $V$, $p$ and $n$ is calculated, with

$$\Delta p = \Delta p_p + \Delta p_c = -pk\left(\frac{\Delta V}{V}\right) + (p_3 - p_2)\left(\frac{V_{tdc}}{V}\right)\Delta n \tag{56}$$

$$k = k_r + \left(k_p - k_r\right)n \tag{57}$$

Continuing beyond TDC, with subscript $e$ identifying the end of combustion, $p_e$ and $V_e$ are known, since the end of combustion is fixed by the assumed value of $\Delta\theta_c$. Then

$$T_e = \frac{p_e V_e}{R\ N_p} \tag{58}$$

$$T_4 = T_e \left( \frac{V_e}{V_{bdc}} \right)^{k_p - 1} \tag{59}$$

where $N_p$ is the moles of product gas and $T_4$ is the temperature at BDC, on completion of the expansion stroke. The work of expansion is given by

$$W_{exp} = \sum \left( p + \frac{\Delta p}{2} \right) \Delta V + N_p \{ u_p(T_e) - u_p(T_4) \} \tag{60}$$

The summation extends from TDC to the end of combustion. Equation (56) is integrated numerically in stepwise fashion. The accuracy of the calculation depends on the number of steps, *N*, selected. Several values for *N* should be tried to acquire the size that will be adequate but not excessive. The rest of the simulation is completed by the method for adiabatic combustion for full throttle operation. The cycle must be repeated as done for the previous analysis and stopped when successive cycles repeat one another.

## 5 SI Engine Simulation with Gas Exchange Process

The gas exchange process that takes place in an actual engine is analysed in this section. Further, heat transfer and friction will be taken into account to simulate an actual cycle.

Friction calculations are also taken into account by empirical equations. In the analysis of exhaust and intake processes in actual engine where the piston moves at finite speeds and can become large. When large their motion exerts a controlling influence on engine performance. Since the events are considered as the piston moves, the variation of pressure with respect to time can be expressed as [23].

For exhaust process

$$\frac{dp}{dt} = k_p p \left( \frac{1}{M} \frac{dM}{dt} - \frac{1}{V} \frac{dV}{dt} \right)_{exhaust} \tag{61}$$

For intake process

$$\frac{dp}{dt} = k_r \left( \frac{R\,T_m}{V} \frac{dM}{dt} - \frac{p}{V} \frac{dV}{dt} \right)_{intake} \tag{62}$$

Both Eqs. (61) and (62) require mass flows rates $\frac{dM}{dt}$, which are found from the equations of fluid mechanics. When a gas at *p, T* with a heat capacity ratio *k* flows through an opening with effective area *A* into a region at pressure $p_O$, the flow should be considered in two regimes, (i) subsonic and (ii) supersonic. The effective area, $A$ is usually the product of the actual area opening multiplied by a suitable discharge coefficient.

The critical pressure ratio that defines the two regimes is called $PR_{crit}$ and it depends only on the value of $k$ of the gas.

$$PR_{crit} = \left(\frac{k+1}{2}\right)^{\frac{k}{k-1}} \tag{63}$$

When $\frac{p}{p_0}$ less than PRcrit, the flow is subsonic and the mass flow rate is given by

$$\frac{dM}{dt} = Ap_0\sqrt{\frac{2k}{RT(k-1)}\left(\frac{p}{p_0}\right)^{\frac{k-1}{k}}\left[\left(\frac{p}{p_0}\right)^{\frac{k-1}{k}} - 1\right]} \tag{64}$$

The flow will be supersonic when $\frac{p}{p_0}$ exceeds $PR_{crit}$ and $\frac{dM}{dt}$ is given by

$$\frac{dM}{dt} = Ap\sqrt{\frac{k}{RT}\left(\frac{2}{k+1}\right)^{\frac{k+1}{k-1}}} \tag{65}$$

The $\frac{dM}{dt}$ of supersonic equation does not contain $p_O$, which is characteristic of supersonic flow. In supersonic flow what matters is where the gas is coming from, and not where it is going.

On the exhaust stroke, $p$ is the cylinder pressure and $p_O$ is the exhaust manifold pressure, whereas on the intake stroke, $p$ is the intake manifold pressure, and $p_O$ is the cylinder pressure. By suitably integrating the Eqs. (61) and (62), the variation of pressure with respect to time or crank angle can be estimated.

Volumetric efficiency is the ratio of the volume occupied by the fresh charge drawn into the engine on the intake stroke, measured at intake manifold conditions, to the displacement volume.

The mass of gas in the cylinder at TDC, $M_{tdc}$ is known. By summing the dM amounts of fresh charge that enter, the mass in the cylinder at BDC, $M_{bdc}$ is known:

$$M_{bdc} = M_{tdc} + \sum dM \tag{66}$$

Then the volumetric efficiency is given by

$$\eta_v = \frac{(M_{bdc} - M_{tdc})R\ T_m}{p_m V_{disp}} \tag{67}$$

In order to carry out cycle calculations that incorporate details of the gas exchange process, it is necessary to specify the variation of the effective valve area $A$, that appears in Eqs. (64) and (65) as a function of crank angle. The valve area

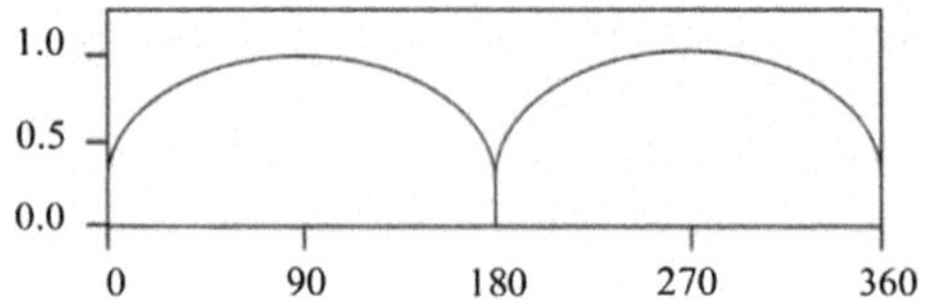


**Fig. 19** Variation of valve area ratio $A/A_{ex}$ or $A/A_{in}$ with crank angle

calculation depends upon the engine geometry and should be specified. However, a simple scheme that can be used is given below:

*Exhaust valve*:

$$A = A_{ex} (\sin\theta)^{\frac{1}{3}} \tag{68}$$

$$\text{opens at BDC}, \theta = 0^\circ \qquad \text{closes at TDC}, \theta = 180^\circ$$

*Intake valve*:

$$A = A_{in} (|\sin\ \theta|)^{\frac{1}{3}} \tag{69}$$

$$\text{opens at TDC}, \theta = 180^\circ \qquad \text{closes at BDC}, \theta = 360^\circ$$

where $A_{ex}$ and $A_{in}$ are the wide-open valve areas for $A$ and is assumed identical for both valves. Generally, the exhaust valve area is larger than the intake valve area since the specific volume of the exhaust gas is much higher due to its high temperature. This can be very easily incorporated in the simulation.

Figure 19 shows $A/A_{ex}$ or $A/A_{in}$ against crank angle as per Eqs. (68) and (69). It may be noted that provisions must be made in the computer program to incorporate the relevant area variation of the valves of the engine concerned.

The final item required to complete the cycle calculation is $W_{loop}$, the work associated with the exhaust and intake strokes:

$$W_{loop} = -\sum\left(p + \frac{\Delta p}{2}\Delta V\right) \tag{70}$$

where $\Delta p$ denotes the change in pressure accompanying the volume change $\Delta V$. The summation extends from the point in the cycle at which the exhaust valve opens to the point at which the intake valve closes.

Heat transfer process can be modelled by using the available equations in the literature. The most often used equations are given below.

## 5.1 Heat Transfer Correlations

A number of heat transfer correlations are available in the literature. A few of them are reproduced in this section. Which correlation should be used depends upon the geometry and the type of engine.

Considerable efforts are being expended to find a single equation to suit all engines, taking into account the various factors. This is, indeed, a very difficult job and till such an equation is arrived at, one has to largely depend on empirical formulae only for the determination of $h_c$.

The process of heat transfer in simulation is quiet important. Mainly heat transfer reduces the power output. It reduces the expansion work and increases the compression work. During the exhaust stroke, heat transfer from the working fluid will lead to an increase in the quantity of residual gases in the cylinder thus affecting scavenging efficiency. The heat transfer to the working fluid during the intake stroke will decrease the quantity of fresh charge inhaled, thus affecting volumetric efficiency.

### 5.1.1 Woschni's Equation [74]

$$h_c = 0.13B^{0.12}P^{0.8}T^{-0.5}Z^{0.8}\,\mathrm{kJ\,s^{-1}\,m^{-2}\,K^{-1}} \tag{71}$$

$h_c$ is heat transfer coefficient; $B$ is cylinder bore in m; $p$ is cylinder pressure in atm, $T$ is working fluid temperature in K, $Z$ is working fluid velocity in m/s.

### 5.1.2 Nusselt's Equation [58]

$$h_c = 0.99(1 + 1.24C_m)\left(p^2T\right)^{\frac{2}{3}}\mathrm{kcal\,h^{-1}\,m^{-2}\,{}^{\circ}C^{-1}}) \tag{72}$$

where $C_m$ is the average piston speed in m/s.

### 5.1.3 Brilling's Equation [11]

$$h_c = 0.99 + (3.5 + 0.185C_m)\left(p^2T\right)^{\frac{2}{3}}\,\mathrm{kcal\,h^{-1}m^{-2}K^{-1}} \tag{73}$$

### 5.1.4 Eichelberg's Equation [18]

$$h_c = 2.1(C_m)^{\frac{1}{3}}(PT)^{\frac{1}{2}}\,\mathrm{kcal\,h^{-1}m^{-2}K} \tag{74}$$

### 5.1.5 Annand's Equation [3]

$$h_c = a\left(\frac{\lambda}{B}\right)Re^{0.7} + \frac{C}{T - T_w}\left[\left(\frac{T}{100}\right)^4 - \left(\frac{T_w}{100}\right)^4\right]\mathrm{kcal\backslash h^{-1}\backslash m^{-2}K^{-1}} \tag{75}$$

where $a$ is 0.35–0.8 increasing with increased speed, $C$ is 0 for compression stroke of SI engine, $C$ is 0.37 for other strokes and 2.81 for CI engines, λ is thermal conductivity of gas in kcal $h^{-1}$ m °$C^{-1}$, Re is Reynold's number $= \frac{\rho D C_m}{\mu}$, $C_m$ is mean piston speed.

### 5.1.6 Van Tyen's Equation [69]

$$h_c = 5.46 \times 10^{-4} P^{\frac{2}{3}} T_g^{\frac{1}{3}} (3.19 + 0.885 C_m) \quad (76)$$

where $T_g$ is in-cylinder gas mean temperature (K); $P$ is in-cylinder pressure (Pa); $C_m$ is mean piston speed (m/s).

## 5.2 *Friction Calculations*

Part of the power developed in the cylinder of an IC engine is absorbed to overcome friction in the various parts, inlet and throttling losses and pumping losses. Empirical relations proposed by Bishop [6] have been used to calculate the power absorbed in friction and the various expressions are as given below:

### 5.2.1 Mean Effective Pressure (MEP) Lost to Overcome Friction Due to Gas Pressure Behind Rings

$$F_{mep\,1} = 0.42 \times (p_a - p_{imf}) \times \frac{S}{B^2} \times \left(0.0888CR + 0.182CR^{1.33 - 0.394\frac{C_m}{100}}\right) \quad (77)$$

where $p_a$ is atmospheric pressure (kg/cm$^2$), $p_{imf}$ is manifold vacuum (kg/cm$^2$), $S$ is stroke (cm), $B$ is bore (cm), $CR$ is compression ratio, $C_m$ is mean piston speed (m/min).

### 5.2.2 Mean Effective Pressure Absorbed in Friction Due to Wall Tension of Rings

$$F_{mep\,2} = 10 \times \frac{0.377S\ n_{pr}}{b^2} \quad (78)$$

where $n_{pr}$ is number of rings.

### 5.2.3 MEP Absorbed in Friction Due to Piston and Rings

$$F_{mep\,3} = 128.6 \times \frac{P_{sl}}{BS} \times \frac{C_m}{1000} \quad (79)$$

where $P_{sl}$ is piston skirt length (cm).

### 5.2.4 Blow-by Loss

$$F_{mep4} = \sqrt{(p_a - p_{imf})} \times \left[0.121\, r^{0.4} - (0.0345 + 0.001055r) \times \left(\frac{N}{1000}\right)^{1.185}\right] \quad (80)$$

where $N$ is speed in rpm.

### 5.2.5 MEP Lost in Overcoming Inlet and Throttling Losses

$$F_{mep5} = \frac{p_e}{2.75} + p_{imf} \quad (81)$$

where $p_e$ is exhaust backpressure (kg/cm$^2$).

### 5.2.6 MEP Absorbed to Overcome Friction Due to Valve Gear

$$F_{mep6} = 226 \times \left(30 - \frac{4N}{1000}\right) \times \frac{GH^{1.75}}{B^2 S} \quad (82)$$

where $G$ is number of intake valves/cylinder; $H$ is intake valve diameter (cm).

### 5.2.7 MEP Lost in Pumping

$$F_{mep7} = 0.0275 \times \left(\frac{N}{1000}\right)^{1.5} \quad (83)$$

### 5.2.8 MEP Absorbed in Bearing Friction

$$F_{mep8} = 0.85 \times \frac{B}{S} \times \frac{N}{1000} \quad (84)$$

### 5.2.9 MEP Absorbed in Overcoming Combustion Chamber and Wall Pumping Losses

$$F_{mep9} = \sqrt{\left(\frac{p_{imep}}{11.45}\right)} \times 0.0915 \times \left(\frac{N}{1000}\right)^{1.7} \quad (85)$$

### 5.2.10 Total MEP Lost in Friction

$$F_{\text{Total mep}} = F_{\text{mep1}} + \cdots + F_{\text{mep9}} \quad (86)$$

$$\text{Net brake MEP} = \text{Indicated MEP} - F_{\text{Total mep}} \quad (87)$$

$$\text{Mechanical efficiency, } \eta_{\text{mech}} = \frac{\text{bmep}}{\text{imep}} \quad (88)$$

However, when gas exchange process is included, some of the frictional losses can be neglected in friction calculations. The above nine losses are only indicative and should be applied with caution. The expressions are given in order to make the reader familiar with friction calculations. The appropriate friction model or mechanical efficiency if available for the engine under consideration then that model or value should be used.

## 6 Typical Results

Large number of parametric studies have been carried out using the software developed. One set of typical results obtained through the present simulation are presented and discussed in the following sections. Wherever possible, the predicted results are compared with experiments. In all the figures, symbols refer to experimental data whereas the full lines represent the predicted values.

The various performance parameters, viz., indicated mean effective pressure (*imep*), brake mean effective pressure (*bmep*), indicated power (*ip*), brake power (*bp*), indicated thermal efficiency ($\eta_{ith}$), brake thermal efficiency ($\eta_{bth}$) and mechanical efficiency ($\eta_m$) are presented and discussed for various loads. The test matrix for data collection is given in Table 2.

The developed software for simulation is named as "GANESH". The acronym "GANESH" stands for Graphical And Numerical Software Hub. It should be mentioned here that different combustion models explained in the text, have been incorporated in the software "GANESH". However, for parametric studies the

**Table 2** Test matrix for data collection

| Compression ratio, (CR) | Speed, (*N*) [rpm] | Load, (*MAP*) [kPa] | Spark advance, (CAD) [degrees] | Measured values |
|---|---|---|---|---|
| 8.3 | 1000 | 48 | 159 | *imep* |
| 8.3 | 2000 | 58 | 159 | *bmep* |
| 8.3 | 3000 | 68 | 147 | *ip* |
| 8.3 | 4000 | 78 | 145 | *bp* |
| 8.3 | 4800 | 88 | 143 | $\eta_{ith}$ |
| | | 98 | | $\eta_{bth}$ |
| | | | | *T* |

results presented in this chapter are for only one combustion model viz. Wiebe's law with $m = 2$. However, the details regarding burning rate variation for various combustion models have already been presented and discussed in Sect. 4.1.

# 7 Effect of Compression Ratio on *p*-*V* and *p*-*θ* Variation

In an actual engine, the compression ratio (*CR*) plays a vital role in the performance characteristics. In order to demonstrate the capability of the software "GANESH" predicted variation of *p*-*V* and *p*-*θ* for different compression ratios are shown in Figs. 20 and 21 respectively for the engine under consideration.

The engine chosen for experiments is a popular two wheeler engine in India the specifications of which are given in Appendix. Experiments were carried out on this engine using an eddy current dynamometer. The experimental engine has a compression ratio of 8.3. The engine is equipped with all the necessary sensors to obtain the required data. The pressure-volume and pressure-crank angle were measured using pressure pickup. from the measured *p*-*θ* variation the required performance parameters can be calculated.

As can be seen (Figs. 20 and 21) that with increase in compression ratio the peak pressure increases. For example when the compression ratio is increased by 63.5% (6.3–10.3) the peak pressure almost doubles This will increase the mean effective pressure of the engine which will improve the power output and efficiency. The improvement in power output and efficiency due to increase in compression ratio (CR) can be attributed to two factors:

(i) with increase in CR the expansion ratio also increases increasing power.
(ii) with increase in CR the exhaust gas contamination reduces increasing η.

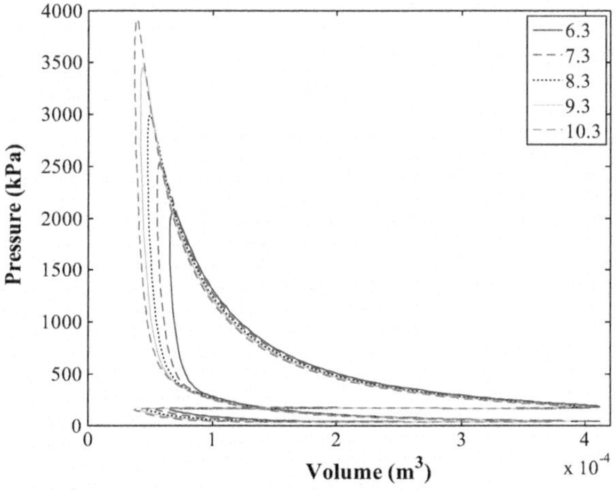


**Fig. 20** Effect of compression ratio on *p*-*V* variation of the engine

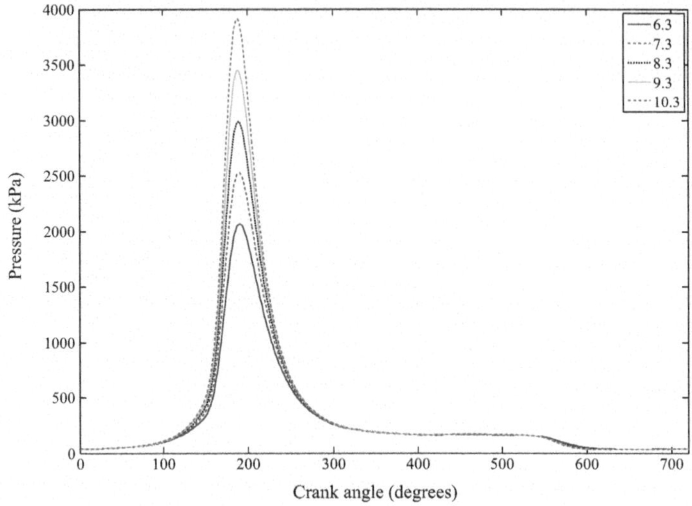


**Fig. 21** Effect of compression ratio on $p$-$\theta$ variation of the engine

In order to validate the predicted results, experiments were carried out on a two wheeler engine and the comparison between the two is shown in Figs. 22 and 23.

The results presented are for a *CR* of 8.3 at 4500 rpm, which are the rated values of the engine under consideration and the results presented are for part throttle operation since two wheelers operate mostly under part throttle conditions on road. For prediction, various combustion models discussed in Sect. 4.1 can be used. However, the results presented here are for Wiebe's law with $m = 2$, since it is found to predict the $p$-$\theta$ variation quite well compared to others. As can be seen from Figs. 22 and 23, the predicted results agree quite well with experiments. The maximum deviation is less than 2% during the expansion stroke. Therefore, it may

**Fig. 22** Pressure-volume variations

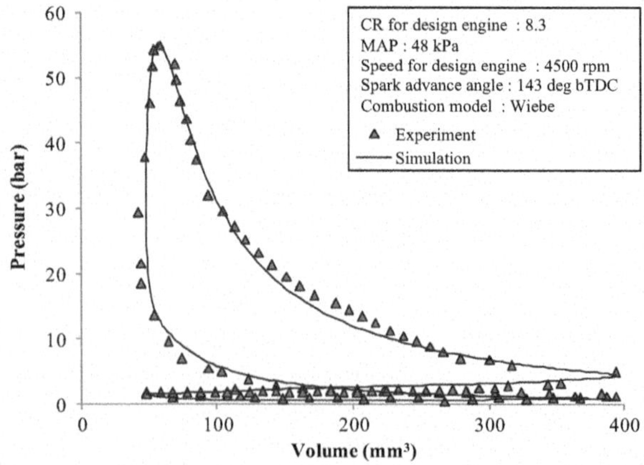

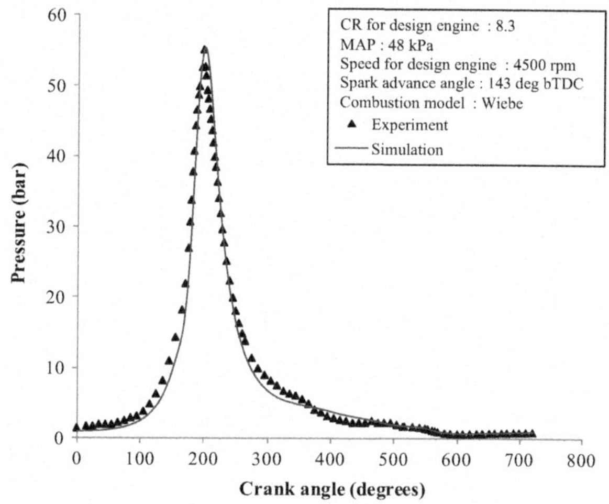


**Fig. 23** Pressure-crank angle variations

be argued with confidence that the thermodynamics modelling procedure as well as Wiebe's combustion model used in this study is able to predict *p-V and p-θ* variation quite well.

# 8 Effect of Speed on the Performance

Another important parameter for studying the performance is the speed of the engine. The load on the engine can be represented by the inlet manifold pressure, which is a function of throttle opening. The inlet manifold pressure increases as we open the throttle, thereby admitting more amount of air. This being a port fuel injected (PFI) engine the fuel quantity is adjusted by injection duration. In the following sections the effect of load on the various performance characteristics are presented and discussed.

## *8.1 Effect of Speed on* imep *and* bmep *at Different Loads*

Figures 24 and 25 show the variations of *imep* and *bmep* with respect to speed for different manifold absolute pressure MAP which represents the load. As can be seen, with respect to speed both the *imep* and *bmep* decreases. This is due to the decrease in the volumetric efficiency with increase in speed. However, increase in load (MAP) at a particular speed both *imep* and *bmep* increases. This is due to induction of more air that enhances combustion which in turn increases the peak pressure and mep.

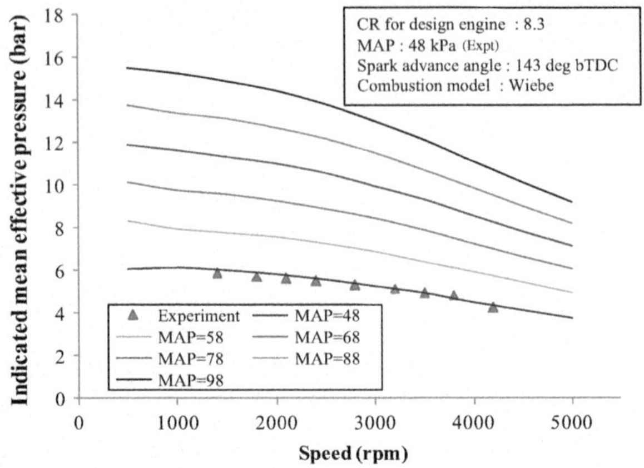


**Fig. 24** Variation of indicated mean effective pressure with load for different speeds

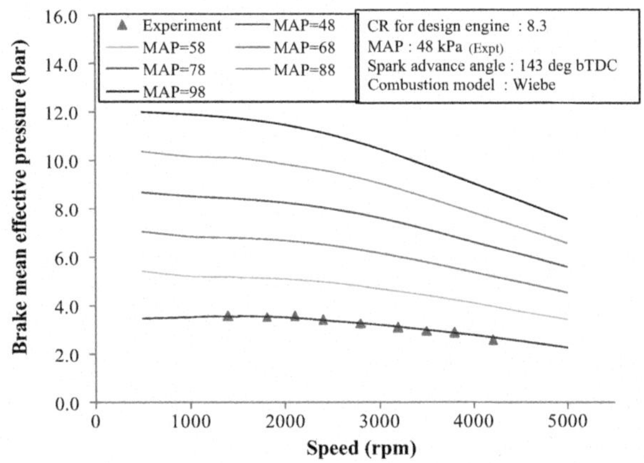


**Fig. 25** Variation of brake mean effective pressure with load for different speeds

At *MAP* = 48 kPa at which the experiment have been conducted the predicted values have been compared with experiments. As can be seen agreement is found to be quite good.

## 8.2 *Effect of Speed on ip and bp for Different Loads*

Figure 26 shows the effect of speed on the indicated power for various load (48–98 kPa). In order to achieve higher speeds at constant load more fuel should be injected. As power is directly proportional to speed the indicated power increases with speed. This is clearly seen at all loads in Fig. 26.

Further, with increase in load the indicated power increases. As seen, the variation of indicated power with speed at constant load (*MAP*) is non-linear. Initially the rate of increase of *ip* is higher up to 3000 rpm and rate gradually decreases

**Fig. 26** Variation of indicated power with load for different speeds

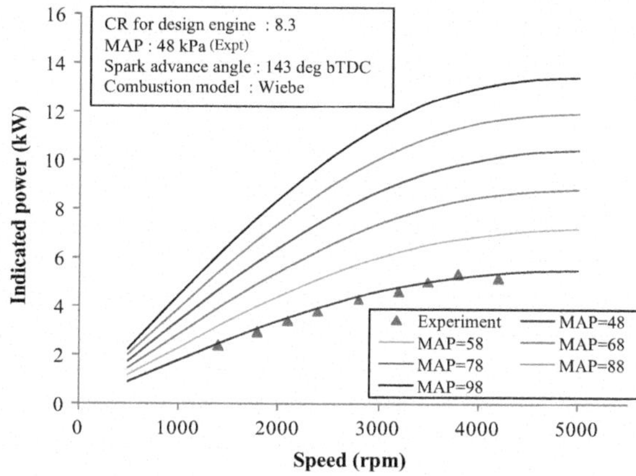


thereafter. It is known that at each load the engine will have a power peak. At full load the power peak occurs at the rated speed. As can be seen this trend is very well predicted. After the rated speed the indicated power will fall due significant reduction in volumetric efficiency.

At lower speed (say 750 rpm), the increase in indicated power is comparatively smaller with increase in load. From no load to full load the increase in power is around 122% whereas at full load and rated speed the increase in power is around 333%. This is achieved by injecting more fuel to sustain the load. In case of SI engines used for two wheelers at full load and rated rpm the air-fuel ratio is set to have good drivability. This trend is well predicted by the Weibe's model. As expected Fig. 27 shows similar variation for brake power. The reasons explained for indicated power holds good for brake power also. As can be seen the agreement between predicted and experimental results are quite good at MAP = 48 kPa at which experiments have been conducted.

**Fig. 27** Variation of brake power with load for different speeds

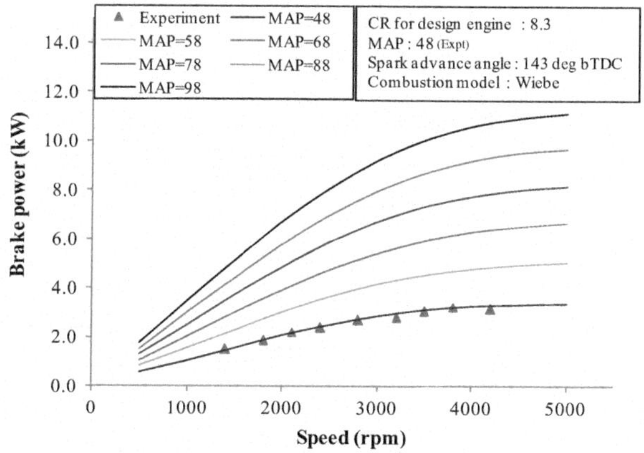

## 8.3 Effect of Speed on Indicated and Brake Thermal Efficiencies at Different Loads

Figure 28 shows the variation of indicated thermal efficiency with respect to speed for different loads. As can be seen, with increase in load at a constant speed the indicated power increases due to the induction of more air which can augment combustion. This increases the indicated thermal efficiency. At full load and 4500 rpm, the engine develops a maximum indicated thermal efficiency of about 44.4%. As can be seen at all speeds the predicted and experimental values agree closely for the part throttle operation of $MAP$ = 48 kPa. It may be seen that brake thermal efficiency does not vary much with speed indicating good combustion at all speeds (Fig. 29). The trend is similar for all loads (MAP: 48 kPa to Map: 98 kPa).

## 8.4 Effect of Speed on Mechanical Efficiency at Different Loads

Mechanical efficiency is the ratio of *brake thermal efficiency* to *indicated thermal efficiency*. From no load to full load and at all speeds the mechanical efficiency increases with load. The engine has a mechanical efficiency of around 80% at full load whereas at no load it is only 60%. As shown in Fig. 30, there is a good agreement between predicted and experimental results at MAP = 48 kPa at which experiments have been conducted.

Similarly, with "GANESH" various parametric studies can be carried out and the results can be obtained almost dynamically. It is hoped that the developed software will be a helping tool for the researchers and engineers.

In this study, for validation only Wiebe's model has been used. However, to understand the capability of various models, nine models (except square root

**Fig. 28** Variation of indicated thermal efficiency with load for different speeds

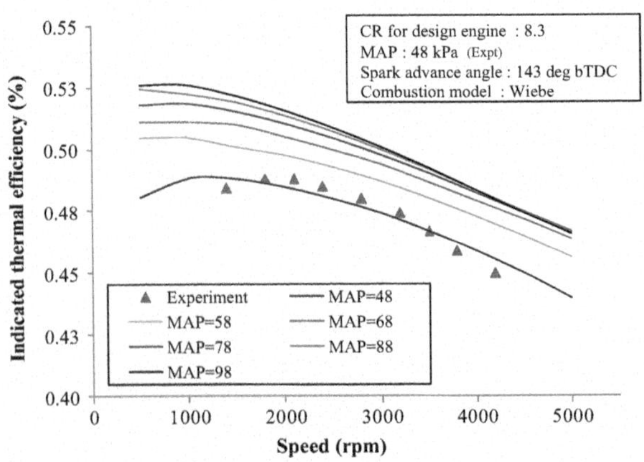

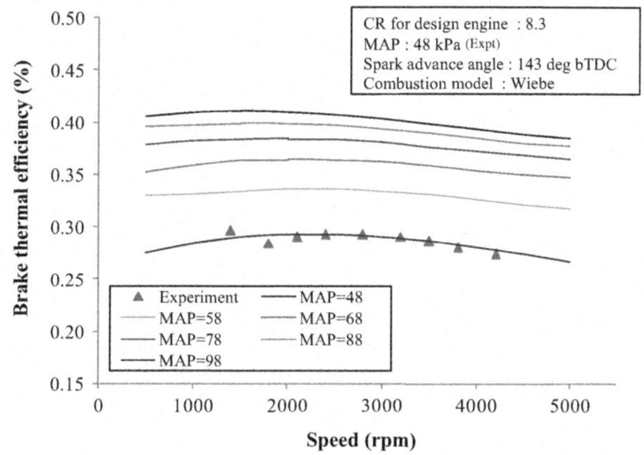


**Fig. 29** Variation of brake thermal efficiency with load for different speeds

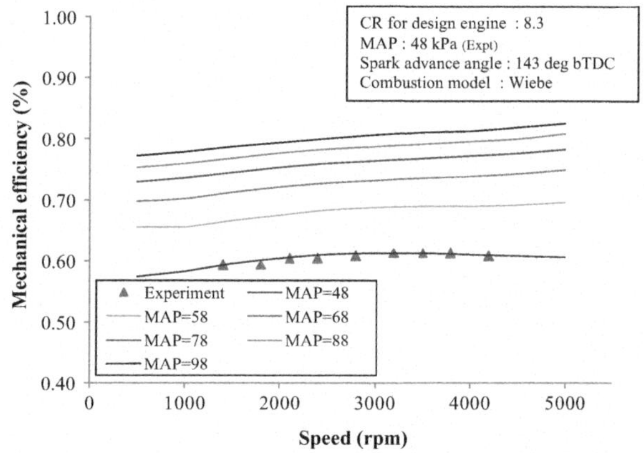


**Fig. 30** Variation of mechanical efficiency with load for different speeds

model) are compared in Fig. 31 during the combustion process, viz. from start of ignition to end of combustion. If one closely observes the curves one can come to the conclusion that Wiebe's model is comparatively predicting variation of pressure satisfactorily during the combustion process. As we have already shown, it compares favourably with experiments. However, in the software "GANESH", all 10 models are incorporated and any model can be invoked with ease. However, it may be mentioned here that the weak link is the friction model where there is scope for improvement since Bishop's equations may not be applicable for all the engines. Another important area for improvement is the calculation of valve area with respect to crank angle. What is given is only a simple equation and therefore such an equation may not be valid for all engines. With the various heat transfer model given Woshni's and Anand's model correlates well for the present engine. The selection of model depend upon the engine and therefore one should take care in selecting appropriate model for the engine selected.

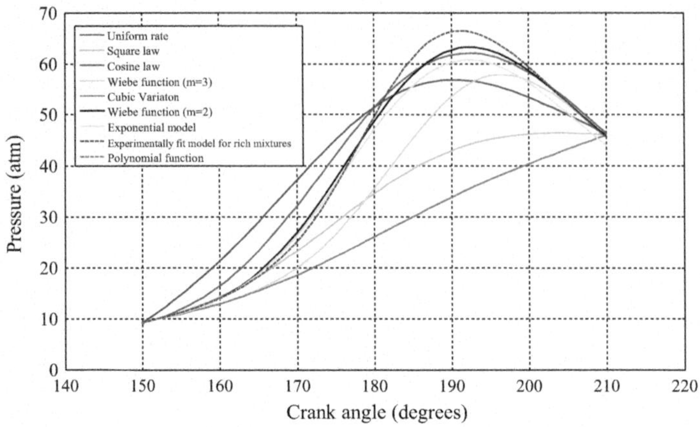


**Fig. 31** Pressure-crank angle diagram for different combustion models

# 9 Summary and Conclusions

1. Thermodynamic models for compression, combustion, expansion, exhaust and intake have been presented.
2. The heat transfer and friction effects have been taken into account using various models available in the literature.
3. After assembling the models a software code "GANESH" has been developed to simulate all the processes of a spark-ignition engine.
4. In order to see the validity of the models, experiments have been carried out to compare the predictions. The agreement is found to be quite satisfactory.
5. Burn rate of various combustion models have been presented and discussed and it was found that Wiebe's combustion model is found to work well
6. Among the various heat transfer models, Woschni's heat transfer correlation is found to be good.
7. The friction model could be improved as Bishop's empirical equations may not be applicable for all the engines.

Thus, it is concluded that the objectives of the work, viz. modelling, simulation and validation have been achieved. However, no attempt has been made to predict the emission characteristics of the engine and this could be the future scope of the work.

# Appendix—Specifications of the Engine

| | |
|---|---|
| Make | Royal Enfield Ltd. |
| Type | Single cylinder, air-cooled, four-stroke, PFI engine |
| Bore | 70 mm |
| Stroke | 90 mm |

| | |
|---|---|
| Displacement | 346 $cc^3$ |
| Connecting Rod Length | 160 mm |
| Compression Ratio | 8.3:1 |
| Rated Power | 14.6 kW |
| Rated Speed | 5200 rpm |

*Valve Timing*

| | |
|---|---|
| Inlet valve opening | 40° bTDC |
| Valve closing | 35° aTDC |
| Exhaust valve opening | 45° bTDC |
| Exhaust valve closing | 30° aTDC |

## References


1. Abu-Nada E, Al-Hinti I, Al-Sarkhi A, Akash B (2006) Thermodynamic modelling of spark-ignition engine: effect of temperature dependent specific heats. Int Commun Heat Mass Transf 33:1264–1272
2. Al-Himyary T, Karim G (1988) A diagnostic two-zone combustion model for spark-ignition engines based on pressure-time data. SAE Technical Paper 880199. doi:10.4271/880199
3. Annand WJD (1963) Heat transfer in the cylinder of reciprocating internal combustion engines. Inst Mech Eng 177(36):973–990
4. Atashkari K (1997) Experimental study of flow and turbulence in a V-flame burner and a SI engine. PhD thesis, University of Leeds, UK
5. Beccari A, Beccari S, Pipitone E (2010) An analytical approach for the evaluation of the optimal combustion phase in spark ignition engines. J Eng Gas Turbines Power 132/032802-1
6. Bishop IN (1965) Effect of design variable on friction and economy. SAE Trans 73
7. Blizard NC, Keck JC (1974) Experimental and theoretical investigation of turbulent burning model for internal combustion engines. SAE paper 740191
8. Bonatesta F (2013) Premixed combustion in spark ignition engines and the influence of operating variables. In: Dr Ng HK (ed) Advances in internal combustion engines and fuel technologies. doi:10.5772/55495
9. Bonatesta F, Waters B, Shayler PJ (2010) Burn angles and form factors for Wiebe function fits to mass fraction burned curves of a spark ignition engine with variable valve timing. Int J of Engine Research 11(2):177–186
10. Borman GL (1964) Mathematical simulation of internal combustion engine processes and performance including comparisons with experiments. PhD thesis, University of Wisconsin, USA
11. Brilling NR (1931) Investigating of working process and heat transfer in diesel engine, Moscow, pp 210–256
12. Buttsworth D (2009) Multizone internal combustion engine modelling: initial assessment of a simulation tool developed in Matlab. In: ICEE 2009 3rd international conference on energy and environment, Malacca, Malaysia
13. Cairns A (2001) Turbulent flame development in a spark ignition engine. PhD thesis, University of Leeds, UK
14. Channappagoudra MN, Thaned S, Ramesh K, ManavendraG (2013) Optimization of combustion chamber geometry for a direct injection diesel engine: a review. Int J Eng Res Technol 2(11)

15. Chow A, Wyszynski ML (1999) Thermodynamic modelling of complete engine systems—a review. Inst Mech Eng Part D: J Automobile Eng 213:403
16. Chun KM, Heywood JB (1987) Estimating heat-release and mass-of mixture burned from spark-ignition engine pressure data. Combust Sci Tech 54:133–143
17. Eckert P, Kong S, Reitz R (2003) Modelling auto-ignition and engine knock under spark ignition conditions. SAE paper 2003-01-0011
18. Eichelberg G (1939) Some new investigations on old combustion engine problems. Engineering 148(463–446):547–560
19. El-Awad MM, Kheiralla AF, Chai CM (2011) A computer-based air-fuel model for analyzing the performance of spark-ignition internal combustion engines. Univ Khartoum Eng J (UOFKEJ) 1(2):40–46
20. Fathi M, Saray RKh, Checkel MD (2010) Detailed approach for apparent heat release analysis in HCCI engines. Fuel 89(9):2323–2330
21. Floch A, Dupont A, Baby X (1998) In cylinder flow investigation in a gasoline direct injection four valve engine: bowl shape piston effects on swirl and tumble motions. Fédération Internationale des Sociétés, d'Ingénieurs des Techniques de l'Automobile, World automotive congress, Paper F98T049, Paris, France, 1998
22. Galindo JH, Climent B, Plá VD, Jiménez (2010) Correlations for Wiebe function parameters for combustion simulation in two-stroke small engines. Appl Thermal Eng. doi:10.1016/j.applthermaleng.2010.12.020
23. Ganesan V (1996) Computer simulation of spark ignition engine process. Universities Press (I) Ltd., Hyderabad, India
24. Gatowski JA, Balles EN, Chun KM, Nelson FE, Ekchian JA, Heywood JB (1984) Heat release analysis from pressure data. SAE paper 841359, SAE Trans 93
25. Ghojel JI (2010) Review of the development and applications of the Wiebe function: a tribute to the contribution of Ivan Wiebe to engine research. Int J Engine Res 11:297
26. Guezennec YG, Haurama W (1999) Two heat release analysis of combustion data and calibration of heat transfer correlation in an I.C. engine. SAE paper 1999-01-0218
27. Hadjiev K (2008) Influences of SI engine parameters on crevice fuel losses. Ann Fac Eng Hunedoara J Eng Tome VI
28. Hall MJ, Bracco FV (1987) A study of velocities and turbulence intensities measured in firing and motored engines. SAE paper 870453
29. Harish Kumar R, Antony A (2008) Progressive combustion in SI-engines-improved empirical models for simulating and optimizing engine performance. SAE technical paper 2008-01-1630. doi:10.4271/2008-01-1630
30. Heywood JB (1975) Pollutant formation and control in spark-ignition engines. Prog Energy Combust Sci 1:135–164
31. Hynes J (1986) Turbulence effects on combustion in spark ignition engines. PhD thesis, University of Leeds, UK
32. Jakubik T (2002) Engine LDA technique. Progress Report, The University of Leeds, UK
33. Jelić M, Ninić N (2008) Analysis of internal combustion engine thermodynamic using the second law of thermodynamics. Strojarstvo 50(2):85–94
34. Jensen TK, Schramm J (2000) A three-zone heat release model for combustion analysis in a natural gas SI engine-effects of crevices and cyclic variations on UHC emissions. SAE paper 2000-01-2802
35. Jonathan MB, Alkidas AC (2009) On the application of Wiebe functions to simulate normal and knocking spark-ignition combustion. Int J Veh Des 49(1/2/3):52–69
36. Kalghatgi GT, Bradley D (2012) Pre-ignition and 'super-knock' in turbocharged spark-ignition engines. Int J Engine Res 13(4):399–414. doi:10.1177/1468087411431890
37. Kaprielian L, Demoulin M, Cinnella P, Daru V (2013) Multi-zone quasi-dimensional combustion models for spark-ignition engines. SAE paper 2013-24-0025. doi:10.4271/2013-24-0025
38. Kim S, Min KIB (2003) Piston crevice hydrocarbon oxidation during expansion process in an SI engine. KSME Int J 17(6):888–895

39. Kodah ZH, Soliman HS, Abu Qudais M, Jahmany ZA (2000) Combustion in a spark-ignition engine. Appl Energy 66:237–250
40. König G (1993) Autoignition and knock aerodynamics in engine combustion. PhD thesis, University of Leeds, UK
41. Krieger RB, Borman GL (1966) The computation of apparent heat release for internal combustion engines. ASME paper 66-WA/DGP-4
42. Lancaster DR, Krieger RB, Sorenson SC, Hull WL (1976) Effects of turbulence on spark-ignition engine combustion. SAE Trans 85:689–710
43. Langridge S (1995) Imaging and thermodynamic analysis of autoignition and knock in S.I. engines. PhD thesis, University of Leeds, UK
44. Lee AP (1995) The effects of bulk motions and turbulence oil combustion in Sl engines. PhD thesis, University of Leeds, UK
45. Lindström F, Angstrom H, Kalghati G, Moller C (2005) An empirical S.I. combustion model using laminar burning velocity correlations. SAE paper 2005-01-2106
46. Lipatnikov AN, Chomiak J (2002) Turbulent flame speed and thickness: phenomenology, evaluation and application in multi-dimensional simulations. Prog Energy Combust Sci 28: 1–74
47. Lyn WT (1963) Study of burning rate and nature of combustion in diesel engines. In: Ninth symposium (international) on combustion. Academic, New York, pp 1069–1082
48. Mahmud ZA (1999) End gas autoignition and knock in spark ignition engines. PhD thesis, University of Leeds, UK
49. Marvin CF (1927) Combustion time on the engine cylinder and its effect on engine performance. NACA Technical report 276
50. Mattavi JN, Groff EG, Lienesch JH, Matekunas FA, Noyes RN (1980) Engine improvements through combustion modelling. In: Mattavi JN, Amann CA (eds) Combustion modelling in reciprocating engines Plenum Press, New York, pp 537–587
51. McCuiston F, Lavoie GA, Kauffman CW (1977) Validation of a turbulent flame propagation model for SI engine. SAE 770045, vol 86, pp 200–221
52. Meng FM, Wang JX, Xiao K (2010) A study of the influences of particles in the gas flow passage of a piston ring pack on the tribological performances of the piston ring. Inst Mech Eng Part C J Mech Eng Sci 203–210 01/2010, 224(1):201–215
53. Merdjani S (1996) Leeds University spark ignition engine computer model, version 6.1, User Manual
54. Merdjani S, Sheppard CGW (1993) Gasoline engine cycle simulation using the Leeds turbulent burning velocity correlations. SAE paper 932640
55. Merker GP, Schwarz C, Stiesch G, Otto F (2006) Simulating combustion—simulation of combustion and pollutant formation for engine-development. Springer, USA
56. Metghalachi M, Keck JC (1982) Burning velocities of mixtures of air with methanol, iso-octane and indolene at high pressure and temperature. Combust Flame 48:191–210
57. Morel T, Keribar R (1985) A model for predicting spatially and time resolved convective heat transfer in bowl-in-piston combustion chambers. SAE paper 850204
58. Nusselt W (1928) Graphische bestimming des winkelverh ̈altnisses bei der w ̈armestrahlung. VDI Zeitschrift 72:673
59. Patterson DJ, Van Wylen GJ (1963) A digital computer simulation for spark-ignited engine cycles. SAE paper 633F
60. Popovic S, Tomic M (2014) Possibilities to identify engine combustion model parameters by analysis of the instantaneous crankshaft angular speed. Thermal Sci 18(1):97–112
61. Raine RR (1990) A performance model of the dual fuel (diesel/natural gas) engine. SAE paper 900387
62. Ramachandran S (2009) Rapid thermodynamic simulation model of an internal combustion engine on alternate fuels. In: International multi conference of engineers and computer scientists, vol II, IMECS 2009, Hong Kong, p 2146
63. Ramos JI (1986) Comparisons between thermodynamic and one-dimensional combustion models of spark-ignition engines. Appl Math Model 10(6):409–422

64. Rassweiler GM, Withrow L (1938) Motion pictures of engine flames correlated with pressure cards. SAE Trans 83, pp 184–204. Reissued as SAE paper 800131,1980
65. Rassweiler GM, Withrow L (1938) Motion pictures of engine flames correlated with pressure cards. SAE paper 800131
66. Scappin F, Stefansson SH, Haglind F, Andreasen A, Larsen U (2012) Validation of a zero-dimensional model for prediction of NOx and engine performance for electronically controlled marine two-stroke diesel engines. Appl Therm Eng 37:344–352
67. Senzer EB (2007) Piston ring pack design effects on production spark ignition engine oil consumption: a simulation analysis. MS thesis, Massachusetts Institute of Technology, USA
68. Sezer I, Bilgin A (2008) Mathematical analysis of spark ignition engine operation via the combination of the first and second laws of thermodynamics. Proc R Soc A 464:3107–3128
69. Van Tyens W (1962) Methode zur Berechung der Zylinderbuchsentemperaturen in Dieselmotoren. Dissertation thesis, U. Delft, Netherlands
70. Vijayashree (2014) Thermodynamic simulation of spark ignition engine processes and optimization using artificial neural network. PhD thesis, Anna University, Chennai, India
71. Way RIB (1976) Methods for determination of composition and thermodynamic properties of combustion products for internal combustion engine calculations. Proc Inst Mech Eng 190 (60/76):687–697
72. Wiebe JJ (1960) Das Wiebe-Brenngesitz, ein Fortschnitt in der Thermodynamik der Kreigspozessc von Verbrennungsmotoren translated from Russian by Joute A, Dresen TD. Kraftakrzeugtechnik 9, pp 1–11
73. Wiseman MW (1990) Spark ignition engine combustion process analysis. Ph.D. thesis, University of Nottingham, UK
74. Woschni G (1967) A universally applicable equation for the instantaneous heat transfer coefficient in the internal combustion engine. SAE paper 670931. doi:10.4271/670931
75. Yadollahi B, Boroomand M (2012) A numerical investigation of combustion chamber geometry effects on natural gas direct injection properties in a SI engine with centrally mounted multi-hole injector. Paper ICES2012-81153, pp 797–808. doi:10.1115/ICES2012-81153
76. Young MB, Lienesch JH (1978) An engine diagnostic package (EDPAC)—software for analysing cylinder pressure-time data. SAE paper 780967
77. Žák Z, Hvězda J, Emrich M, Macek J, Červenka L (2012) Utilization of multi-zone model results in SI engine modelling. In: MECCA 02, pp 23–20

# Gasoline Direct Injection—Challenges

**Srinibas Tripathy, Sridhar Sahoo and Dhananjay Kumar Srivastava**

**Abstract** The demand for reduced fuel consumption and low-emission spark ignition engine is rising, in response to concerning energy conservation and preservation of the global environment. Air-fuel mixture formation plays a significant role in controlling exhaust emissions and engine efficiency. The primary purpose of fuel injection system is to provide the engine with correct air-fuel ratio according to engine operating condition. Fuel injection system of spark ignition engine has developed in last thirty years monotonically from the carburetor to throttle-body injection, and then to port fuel injection (PFI). Although, the current port fuel injection technology is highly evolved, but it is difficult to achieve further stringent emission norms and demand for lower fuel economy. Gasoline direct injection (GDI) system is a promising next generation of technology for the automotive engine to improve fuel economy, performance and emissions control. GDI engines are characterized by injecting fuel directly into the combustion chamber at high pressure during the compression stroke. Air-fuel mixture formation takes place inside the combustion chamber as in a diesel engine. GDI engine offers a significant potential for improving the fuel economy and reduction of exhaust emissions from spark ignited engine, while maintaining or improving the high specific power output of current PFI engines. GDI engines has several potential advantages over PFI system, however, it does have a number of inherent problems that need to be address. These constraints are controlling the stratified charge combustion over wide range of engine operation, complexity in injection strategy for seamless load changes, injector deposits, increased particulate emissions etc. In this paper, several difficulties associated with GDI development and its probable solution will be discussed.

**Keywords** GDI · Air-fuel mixing · Homogeneous combustion · Stratified combustion · Injector deposit · Emissions

---

S. Tripathy (✉) · S. Sahoo · D.K. Srivastava
Indian Institute of Technology, Kharagpur, India



A.K. Agarwal et al. (eds.), *Combustion for Power Generation and Transportation*, DOI 10.1007/978-981-10-3785-6_16

## 1 Introduction

Gasoline and diesel engines are most widely used engines in the world's transportation sector. Concerning energy conservation and stringent emission norms; the prime goal of the automobile industry is to achieve an optimized engine technology with improved fuel economy, enhanced power and reduced emissions over the wide range of engine operating condition. SI engine have high specific power output and less thermal efficiency; however, due to high compression ratio and un-throttled operation, CI engine have higher thermal efficiency, but NOx and PM emission is still a major concern. Due to lack of variability of charge preparation for a wide range of engine operation in carburetor system, port fuel injection (PFI) system has been developed, where fuel is injected into the intake manifold at much less pressure compared to injection in CI engine. The air motion is restricted by a venturi in carburetor system which reduces the volumetric efficiency. Hence PFI system is introduced for higher volumetric efficiency to overcome the above problem. Thus increasing the volumetric efficiency increases the breathing ability of the engine which increases the combustion efficiency. Along with that, it has more uniform fuel distribution, precise control of air-fuel ratio and lower engine exhaust emissions [7]. Due to the fuel accumulation behind the intake valve in PFI engine causes inaccurate fuel metering. Moreover, during cranking and cold start, due to low intake manifold temperature, some fuel in the form of liquid droplet injected into the cylinder and remains there resulting soot formation [8, 38]. Hence, concerning the future emission norms with improved fuel economy, researchers are focused on gasoline direct injection engine (GDI) to merge the specific power output of PFI engine to that of diesel engine efficiency. This can be achieved by directly injecting gasoline into the engine cylinder by generating a combustible air-fuel mixture around the spark gap. As power output is varied by fuel injection, hence pumping loss is minimized. The prime motive of the engine is to operate in stratified mode at part load operation by maintaining rich mixture near the spark plug and ultra-lean mixture in remaining combustion chamber. This is done by injecting the fuel during the end of compression stroke. As turbulence increases at full load, it is challenging to obtain stratification; hence homogeneous mixture is prompted throughout the engine cylinder by injecting the fuel during the intake stroke [37]. The switching of homogeneous to stratified mode and vice versa is controlled by early and late injection depending upon the engine speed. Direct injection of gasoline provides in-cylinder charge cooling benefit which increases the volumetric efficiency, allowing a high compression ratio of engine and thus improving combustion efficiency, reduced pumping loss by the un-throttled operation, more precise A/F ratio control, fuel cut off during deceleration, reduced cold-start UBHC, and $CO_2$ emissions are major advantages of gasoline direct injection engine [17, 37]. Considering these advantages, GDI engine has various limitations which suppress its productivity. Hence, these limiting areas must be taken into consideration for further improvements. Fuel is injected directly into the engine cylinder results in piston and cylinder wall impingement which produces

UBHC and PM emissions. As time availability between injection and ignition is very short hence the level of atomization of fuel must be high to obtain vaporization. HC emission due to incomplete combustion at ultra-lean mixture zone and soot formation at over rich region near spark plug limits to achieve the stringent emission norms. In addition to the above, controlling the stratified combustion, injector deposits, increased electrical power for injectors and drives, high noise level, high load UBHC and NOx emissions are major limitations of gasoline direct injection engine [39]. To meet these challenges, leading automobile companies makes the mechanical modifications to the engine. The major modification is the wide spacing of spark plug and the fuel injector that helps in reducing injector deposits and spark plug fouling [19, 26, 27]. The shape of the piston is modified so that charge can be transported from the injector to the spark plug with proper mixture formation. The upright straight intake port avoids wall impingement by reverse tumble mechanism introduced by Mitsubishi [29]. Swirl fuel injector is used for optimized spray geometry [14, 29]. Based on the above changes, leading automobile companies like Toyota (Lexus GS 300), Ford (1.8 L Duratec SCi), BMW (N73 V12) and Ferrari (FF V12) have launched their GDI engine in the European and American market. But still, its production and operation are limited in many countries due to lack of engine development. Hence the future work in the development of GDI engines must be focused on following aspects: reproducibility of mixture distribution, the stability of stratification, tuning of in-cylinder gas motion, suitable fuel injection systems and optimized injection and ignition parameters for the entire engine operating range [39]. For this purpose a lot of both experimental and theoretical investigations are necessary.

Gasoline Direct Injection engine is more viable in recent years in combating stringent emission norms, and better fuel economy and more power output. GDI engine has a capability of working in all load conditions obtaining more fuel economy and better performance in comparison to diesel and PFI engines. Automobile companies have taken the initiative making it successful and Mitsubishi Motor Corps being the first to introduce GDI engine powered vehicle in the market [39]. Many experiments were done by researchers, where improvement and optimization in the injection system, intake system, air-fuel mixing, combustion performance and emissions of GDI engine were considered.

## 2 Air-Fuel Mixture Formation

Keeping air-fuel mixture within flammability limits is an important parameter for the optimum combustion process. This, in turn, depends on flow velocity parameters and loading conditions, to obtain a stable flame kernel throughout the combustion chamber. The critical parameters that affect air-fuel mixing are stability, flow components of airflow, turbulence induced throughout compression and velocity distribution near spark plug during ignition time [26].

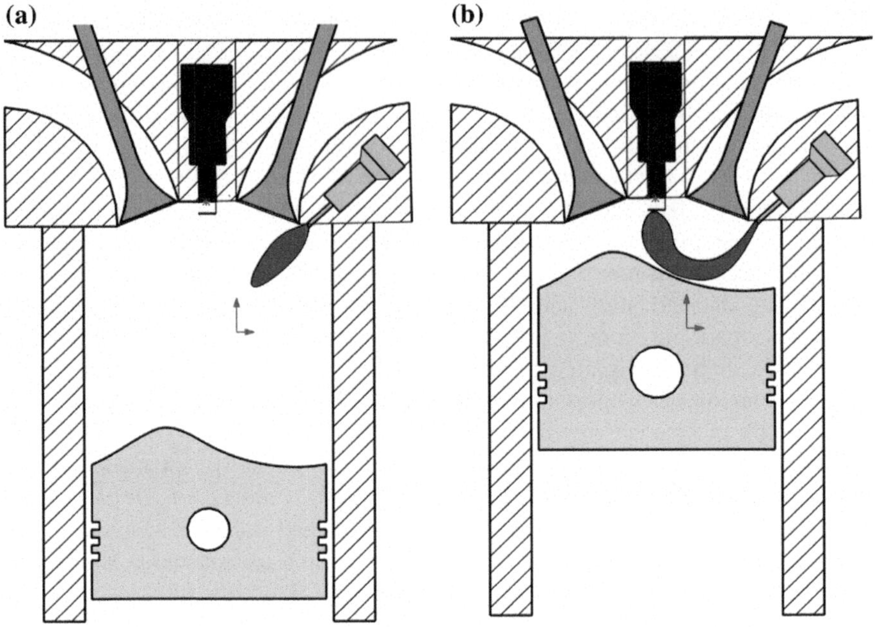


**Fig. 1** Modes of GDI combustion. **a** Homogeneous combustion mode. **b** Stratified combustion mode

Two modes of combustion are present covering all conditions w.r.t. air-fuel ratio needed inside combustion chamber which is shown in Fig. 1.

(a) Homogeneous Mode: This mode applicable in high-load high-speed engine operating conditions. At full load, the uniform air fuel mixture is required for high power over the cylinder volume. Hence fuel is introduced during the intake stroke of the engine operation achieving a homogeneous mixture at the time of spark formation.
(b) Ultra lean burn or Stratified Mode: This mode is applicable in low load conditions, reducing engine speed with no need of acceleration. Due to low load condition, a small amount of fuel is injected towards the end of compression stroke, achieving a rich combustible mixture near spark plug and very lean mixture at other portions of the combustion chamber. The combustion takes place in the narrow cavity above piston surface, enabling ultra-lean mixture combustion at low load.

For homogeneous charge in PFI or GDI engine, high turbulence and low velocity near spark plug are needed for sufficient air-fuel mixing and spark formation for a stable combustion process. Whereas for a stratified charge, high stability of flow and velocity field near spark plug are required, turbulence has to be kept low in this case [27]. Mitsubishi employed reverse tumble facility to incorporate a deflected motion to fuel taking it towards spark plug making it

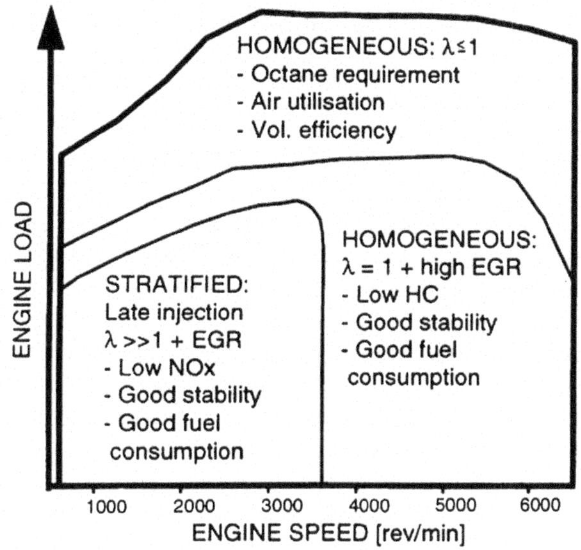


**Fig. 2** GDI engine operating condition [27] (Copyright by SAE International)

rich/stoichiometric zone. Swirl effect was implemented to lower the penetration length making it suitable for easy evaporation and reducing fuel impingement on piston crown. A high amount of swirl tends to centrifuge the higher mass droplets towards boundary increasing the chance of UBHC formation. In stratification mode, low flame speed on account of lean mixture is compensated by tumble effect resulting in faster combustion thereby reducing any type of abnormal combustion to take place [26]. Figure 2 shows the target strategy of operation with the different load and speed conditions using GDI engine [27].

Early injection at 90–120°ATDC in the intake stroke, increases the chance of wetting the piston surface increases emission level. This has to be sorted out by optimizing ignition timing, engine speed, and spray flow field. The regions having a high swirling effect inside the combustion chamber is to be encountered in combustion stability and smoke formation. Early injection has a very high effect than late injection in the intake stroke, due to having high-density and more drag in air formed inside the combustion chamber. Homogeneous charge combustion gets an advantage of charge cooling in direct fuel injection, which in turn improves the volumetric efficiency of the cylinder by increasing the air density; in cylinder charge cooling also increases torque output and knock limited compression ratio compromising the high-end gas temperature. An optimization between early and late injection has to be done for achieving high volumetric efficiency and no fuel drops on surface wall.

Piston head shape plays a great role in mixture formation within the combustion chamber. The GDI piston head is made up of bowl shape over the conventional flat piston head for homogeneous and stratified mixture preparation. It directs the incoming fresh charge coming through the straight intake port can be directly transported towards the spark plug for better combustion achievement [40].

## 3 Engine Management System

Engine management system (EMS) is electronic controlled systems which precisely monitor and control the fuel injection unit to improve the efficiency and drivability of the engine. The components of EMS are input sensors, electronic control unit, and output actuators [5]. The input sensors are air intake temperature sensor, mass air flow sensor, engine temperature sensor, intake manifold pressure sensor, throttle position sensor, camshaft position sensor, rail fuel pressure sensor, accelerator pedal position sensor, knock sensor, lambda sensor upstream of primary catalytic converter, exhaust gas temperature sensor, lambda sensor downstream of main and primary catalytic converter, engine speed sensor [7]. The output actuators are fuel pressure control valve, EGR valve, ignition coils, fuel injectors, electric fuel pump, fuel-vapor retention system valve, throttle valve positioned and fan control [1]. All the input sensors detect the engine operating conditions in the management system and the information is sent to the engine ECU, which in turn controls the actuators electronically to operate the engine at optimal conditions. Once the engine is started, EMS collects data from different sensors and makes a judgment about the engine operating conditions. High precision air-fuel ratio feedback, injection timing, ignition timing, knocks and emission control are major functions of EMS [16].

The crankshaft position sensor measures the engine speed that plays a key role for fuel injection, spark event and valve timing [1]. Direct injection spark-ignition has various operating modes depending on the engine operating condition for a stable engine operation. The GDI engine operates in two modes: stratified and homogeneous combustion. In stratified combustion, ultra-lean combustion is achieved by maintaining rich mixture near the spark plug and ultra-lean mixture in the remaining combustion chamber thereby improving fuel economy. However, with increasing the engine speed, it is difficult to maintain the stratification in the combustion chamber due to increase in turbulence. So that engine combustion mode has to switch from stratified to homogeneous combustion when the operating range changes from low load to the full load condition. In between homogeneous charge mode and stratified charge mode the engine runs in homogeneous lean charge mode where the lean mixture is homogeneously distributed throughout the combustion chamber. The engine management system must focus to control the switching modes to avoid large torque fluctuations. The engine should generally operate in stratified combustion mode. When the conditions for stratified mode are fulfilled, the controller detects and performs the switching into the homogeneous mode [15].

The engine control unit chooses the engine operating mode condition depending on power, torque, and safety requirements. In torque-based engine management system, target torque is implemented by the injected fuel quantity in stratified charge mode where intake air mass and ignition advance angle doesn't have a significant effect. However, torque requirements are controlled by ignition point and air mass in case of homogeneous charge mode and homogeneous lean charge mode. The injection quantity is predefined by the air mass hence not used for optimal torque condition [1]. Hence the function of EMS in gasoline direct injection engine is to control the amount of fuel, injection timing and ignition timing for all load conditions at very high pressures.

Maintaining the stratified combustion and switching of combustion charge modes within very short duration are the challenging task of EMS in GDI engine. The focus of GDI EMS system is to improve the stratified charge mode over a wide range of engine operating condition [15]. Automobile researcher claimed that a stable stratified charge could be achieved up to the medium speed range and engine load operated at a lambda of 1.6 to 3 [1]. Park et al. [28] performed an experiment in a spray guided GDI engine with piezo type gasoline injector to investigate the stratified lean combustion operation. Results show a well-established stable stratified combustion is achieved up to 2000 rpm by eliminating the wall wetting phenomenon. Optimized ignition timing with EGR can maintain an ultra-lean combustion of lambda 3 by maintaining sufficient ignitable rich mixture near spark plug. Sementa et al. [32] investigated the combustion stability for both gasoline and ethanol fuel for different injection strategy at urban driving operating condition (1000 and 1500 rpm) in a DI engine. He found that ethanol stratified combustion shows higher combustion stability with reduced emissions as compared to gasoline direct injection engine. Jung et al. [22] conducted an experiment by taking gasoline and n-butane as fuel for DI engine and found a more stable stratified combustion for n-butane at low injection pressure. This is due to the higher evaporation rates of n-Butane spray due to its higher vapor pressure. Zeng et al. [36] experimentally investigated the effect of in-cylinder swirl on combustion stability of a spray-guided direct injection spark ignition engine. It was found that swirl creates a strong repeatable vortex near the piston bowl Centre thereby enhancing the flame in multiple directions. Hence stable stratified combustion can be achieved at an elevated temperature of 2000 rpm. Tamura et al. found the stable stratified combustion can be achieved by using a high-pressure fuel pump and high responsive fuel injector.

Considering the above challenges Delphi, Bosch, Magneti Marelli and Denso are the leading industries implemented high processing ECU with precise algorithms in the engine management system of gasoline direct injection engine and the vehicles are successfully implemented achieving the future emission norms.

## 4 Emissions Characterization

In current market two major variant available for gasoline operated engines: gasoline direct injection (GDI) and port fuel injection (PFI). GDI engine offers better fuel economy and lower $CO_2$ emission, because of its independent control over injection timing and the mass of fuel injection. Due to its limited time for air-fuel mixing, GDI engine has been reported producing the higher particulate matter (PM) emission with its counterpart PFI engine. GDI engine needs concern on Lean $NO_x$ catalyst for lean burn mode deactivating sulfur content and PM

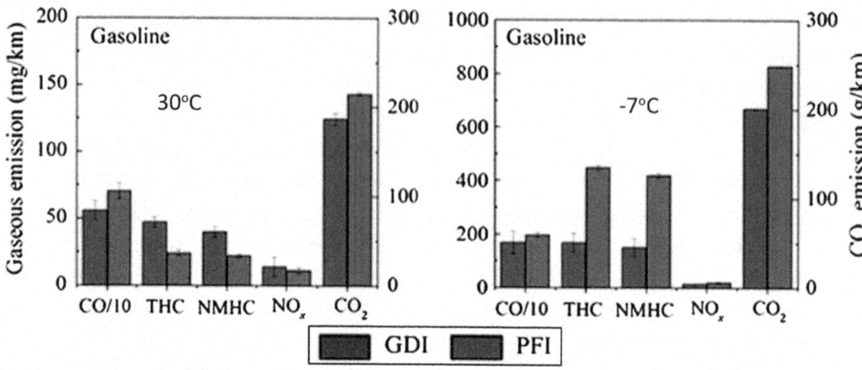


**Fig. 3** Comparison of gaseous emissions from the GDI and PFI vehicles with gasoline in the cold-start tests [42] (Copyright by ELSEVIER)

emission. A brief difference between exhaust tailpipe emissions from GDI and PFI engine is shown in Fig. 3 for different ambient temperature condition.

Although it was expected that GDI having rich combustion zones might lead to high CO emission, CO emission generated by PFI engine is much higher than GDI at both high as well as low ambient temperature. This may be due to the inefficient performance of TWC as an after-treatment at the low-temperature condition. CO emission during cold start increases 2.5 to threefold due to more injection of fuel and light off condition of TWC. During high ambient temperature condition, GDI engine has high THC and NMHC emission. This trend reverses for the cold operating condition with a factor of 3.6–3.7. PFI engine shows much higher emission in cold start condition due to inefficient fuel evaporation inside the cylinder. $NO_X$ emission is too small and not very different than PFI engine, due to the presence of after-treatment. $CO_2$ emission is less than PFI engine at all condition due to low fuel consumption advantage of GDI over PFI engine [42]. Fan et al. [12] have reported that PFI has lower CO and HC emission at the low-speed and low-load condition and GDI has lower emission at high load and high speed condition.

Aromatic compounds that present in the fuel have a significant role in PM emissions. With increased in the number of carbon present in aromatics enhances the chance of PM emission for the same level of aromatic test fuels. This is due to carbons present in aromatic compound sticks on the cylinder wall and does not take part in combustion process [18]. Fuel impingement, wall wetting, and inhomogeneity in mixture formation are the key reasons for high PM and PN emission in GDI engines. Jang et al. [20] found particulate emission by GDI emission higher than PFI engine.

Reducing PM emission is the major challenge that is needed to be taken care of for new upcoming stringent norms. Affording a suitable technology and retrofitting in GDI engine is a challenging task. Engine design modification or adjustment in combustion process for optimum combustion process can be a better solution from pretreatment point of view. There are other viable options for PM reduction by changing in fuel properties, or some after-treatment arrangement has to be installed in exhaust tailpipe. High carbon number aromatic compound present in the fuel has a higher boiling point, larger density, and lower vapor pressure. This property makes it easy to stick to the combustion wall and avoid taking part in combustion thereby results in PM emission. Sulfur free fuel with low aromatics having higher carbon can be a solution for the reduction of emission. This is because the high sulfur content in the exhaust gas reduces the activity of $NO_X$ reduction catalyst at low-speed condition [18]. Oxygenated blend fuel is an alternate solution towards reducing unburnt hydrocarbon and helps in improving combustion. Therefore, CO production due to the presence of rich air-fuel mixture is compensated by oxygen present in the ethanol or butanol blend fuel. Oxygenated fuel blend has a high latent heat of vaporization creating charge cooling effect and increase in volumetric efficiency. High charge cooling effect parameters of oxygenated fuel blend suppress the in-cylinder temperature during combustion, thereby reducing $NO_X$ emissions at great level [24]. An addition of oxygenated fuel with gasoline reduces aromatic hydrocarbon present in the fuel, hence this unavailability of carbon in the blended fuel reduces the chance of soot formation. Engine modification is also a feasible solution for reducing HC emission by reducing crevice volume present and reformulating lubricants. Variable valve timing with variable spark timing can be used for controlling cold start emissions [30].

## 5 Injector Deposit

Gasoline direct injection (GDI) system is a promising next generation of technology for the automotive engine to improve fuel economy, performance and emissions control. There is two basic variant of combustion mode in GDI technology currently of interest: "Stratified combustion mode" and "Homogeneous combustion mode". In the case of stratified combustion mode, fuel is injected during the late compression stroke in such a way that a stratified-charge cloud is transported to the spark plug at the time of ignition. In the case of Homogeneous Combustion, Fuel is injected during intake stroke inside the combustion chamber to provide a homogeneous mixture. Thus, the spray patterns are of great importance to the engine performance and combustion in the GDI engine. Injectors are directly mounted in the cylinder head. Compared with PFI system, the GDI injector works at a much harsh environment because it is mounted in the high-pressure, high-temperature combustion chamber rather than being placed in the intake manifold. Due to harsh environment condition inside the engine cylinder, a deposit may form on the injector tip or even inside the injector hole that might cause the significant change in spray patterns and the fuel flow rate. This would consequentially affect the engine combustion and emissions. The spray pattern distortion due to deposit will drastically influence the stratified combustion mode because stratified combustion is highly sensitive to spray geometry.

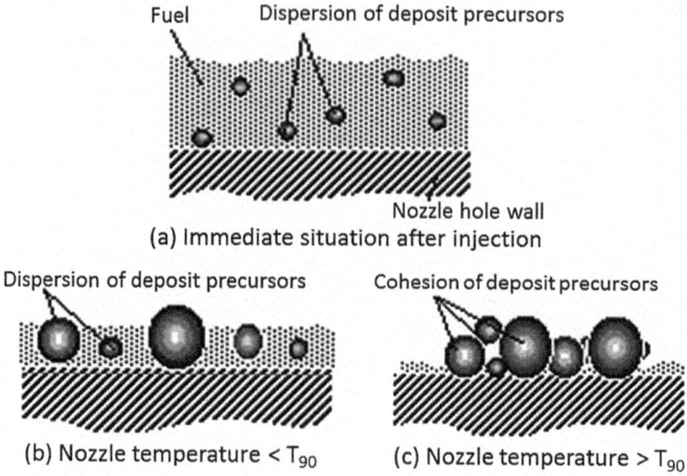


**Fig. 4** Mechanism of deposit formation [25] (Copyright by SAE International)

Kinoshita et al. [25] has proposed the mechanism of deposit formation in the injector hole. Deposit formation is related to nozzle temperature and 90% distillation temperature of the fuel ($T_{90}$). They proposed that a small amount of fuel remains trapped in the nozzle hole after the fuel injection. Initially, deposit precursors are homogeneously dispersed in the liquid fuel as shown in Fig. 4a. When the nozzle temperature is lower than $T_{90,}$ deposit precursors are remains in the state of dispersions in the fuel (Fig. 4b). In this condition, the deposit precursors are easily washed away in the following injection event. However, when the nozzle temperature is higher than $T_{90}$ then most of the fuel in the nozzle evaporates resulting in the deposit precursors are strongly adhering to the nozzle hole wall (Fig. 4c). In this state, the deposit precursors are not easily washed away thus progresses the deposit formation in the nozzle hole. Fuel composition and engine operating condition also influenced the deposit formation. Several researchers [3, 4, 9, 13, 35] have studied the impact of fuel composition on injector deposit in the GDI engine.

Injector deposit can be characterized by the spray visualization, fuel droplet size distribution, fuel flow rate measurement, change in the injector pulse width, endoscopy measurement, chemical analysis by GCMS, energy dispersive analysis (EDS), and deposit morphology by Scanning Electron Microscopy (SEM). The spray visualization, droplet size distribution, fuel flow rate can be measured with clean and fouled injector. However, to visualize the deposit across the nozzle hole, it is required to access the internal part of the injector. The morphology and chemical structure of deposit inside the nozzle hole need to understand in great detail to determine the deposit formation mechanism. Dearn et al. [11] has proposed the mechanical cracking technique of the injector to access the deposit across the GDI injector fuelled with gasoline and blends of 85% ethanol. Mechanical cracking method has an advantage that deposit inside the injector was not disturbed. It was possible to access deposit on the internal and external hole of the injector, deposit on the ball and seat. They found that deposits form in the region of nozzle tip and its size and quantity reduces as distance increases away from the combustion chamber.

The effect of injector deposit on the engine performance and exhaust emissions have been studied by several researchers [6, 10, 21, 31]. David et al. [10] found that HC and CO emissions increased with injector deposit. Sandquist et al. [31] reported that 80% increase in HC emission after 60 h of the dirty-up test due to injector fouling. Joedicke et al. [21] also found an increase in HC, CO emissions and fuel consumption after 55 h of the dirty-up test. Particulate emissions are another concern related to GDI engine. Berndorfer et al. [6] found that intense flame radiation in the vicinity of injector tip towards the end of combustion. They have correlated with this injector diffusion flame with particulate emissions. Diffusion flame acts as a source of particulate emissions.

The injector deposit is a one of the major challenge in the GDI engine. Deposit formation can be minimized by adding additives in the fuel and design consideration of injector and engine. Aradi et al. [2] studied the effect of fuel additive (Mannichs and Polyetheramines) with regard to injector deposit control. The Mannichs additive show the superior performance compares to Polyetheramines additives.

## 6 Conclusions

High PM emission and low fuel economy is the major challenge in present era ahead of having lowered fuel consumption and high power output in GDI engine. Stratified charge, lean mode operation endorses the formation of soot particles inside cylinder chamber having rich fuel zones. GDI alone has much complexity in handling with gasoline fuel since it shows very high emission and low fuel economy. GDI engine lack in electronic management system as well as electronic fuel injector improvements. Emission cannot be compromised; hence research is needed in reducing after treatment process. Implementing dual fuel could be one of the alternative idea eliminating fossil fuels that degrades environment. Engine downsizing is also another prospective towards increasing efficiency. Split injection strategy will help in achieving high volumetric efficiency, better combustion and low emission. Engine fouling is the major problem need to be explored.

## References

1. Anon (2002) Bosch motronic MED7 gasoline direct injection. Volkswagen Self Study Program 253, Volkswagen AG
2. Aradi AA, Colucci WJ, Scull HM, Openshaw MJ (2000) A study of fuel additives for direct injection gasoline (DIG) injector deposit control. SAE 2000-01-2020

3. Aradi AA, Imoehl B, Avery NL, Wells PP, Grosser RW (1999) The effect of fuel composition and engine operating parameters on injector deposits in a high-pressure direct Injection gasoline (DIG) research engine. SAE 1999-01-3690
4. Ashida T, Takei Y, Hosi H (2001) Effects of fuel properties on SIDI fuel injector deposit. SAE 2001-01-3694
5. Ashok B, Ashok SD, Kumar CR (2016) A review on control system architecture of a SI engine management system. Ann Rev Control 41:94–118
6. Berndorfer A, Breuer S, Piock W, Bacho PV (2013) Diffusion combustion phenomena in GDI engines caused by Injection process. SAE 2013-01-0261
7. Celik MB, Ozdalyan B (2010) Gasoline direct injection, fuel injection. InTech. http://cdn.intechopen.com/pdfs-wm/11336.pdf
8. Cheng WK, Hamrin D, Heywood JB, Hochgreb S, Min K, Norris M (1993) An overview of hydrocarbon emissions mechanisms in spark-ignition engines. SAE 9332708
9. China P, Rivere JP (2003) Development of a direct injection spark ignition engine test for injector fouling. SAE 2003-01-2006
10. David CA, Malcolm JM (2000) The effect on vehicle performance of injector deposits in a direct injection gasoline engine. SAE 2000-01-2021
11. Dearn K, Xu J, Ding H, Xu H, Weall A, Kirkby P, Cooper B, Edington I, Venus JK (2014) An investigation into the characteristics of DISI Injector deposits using advanced analytical methods. SAE 2014-01-2722
12. Fan Q, Li L (2013) Study on first-cycle combustion and emissions during cold start in a TSDI gasoline engine. Fuel 103:473–479
13. Fernandes H, Braga LC, Martins AR, Braga SL, Braga CVM (2012) Fuel sulfate content influence in the formation of inorganics components deposits in the engine injectors with technologies of gasoline direct injection. SAE 2012-36-0314
14. Fraidl G. K.; Piock W. F.; Wirth M (1996) Gasoline direct injection: actual trends and future strategies for injection and combustion systems. SAE 960465
15. Gafvert M, Arzen KE, Pedersen LM, Bernhardsson B (2004) Control of GDI engines using torque feedback exemplified by simulations. Control Eng Pract 12:165–180
16. Grahn M, Johansson K, Mckelvey T (2013) A diesel engine management system strategy for transient Engine operation. In: 7th IFAC symposium on advances in automotive control, vol 46, pp 1–6
17. Harada J, Tomita T, Mizuno H, Mashiki Z, Ito Y (1997) Development of direct injection gasoline engine. SAE 970540
18. Iizuka M (2007) Effect of fuel properties on emissions from direct injection gasoline vehicle. In: 5th Asian petroleum technology symposium, Jakarta, Indonesia
19. Iwamoto Y, Noma K, Nakayama O, Yamauchi T, Ando H (1997) Development of gasoline direct injection engine. SAE 970541
20. Jang J, Lee J, Kim J, Park S (2015) Comparisons of the nanoparticle emission characteristics between GDI and PFI vehicles. J Nanoparticle Res 17:486
21. Joedicke A, Krueger VJ, Bohr P, Cracknell R, Doyle D (2012) Understanding the effect of DISI injector deposits on vehicle performance. SAE 2012-01-0391
22. Jung J, Park S, Choongsik B (2016) Combustion characteristics of gasoline and n-butane under lean stratified mixture conditions in a spray-guided direct injection spark ignition engine. Fuel 187:146–158
23. Kano M, Saito K, Basaki M, Matsushita S, Takeshi G (1998) Analysis of mixture formation of direct injection gasoline engine. SAE 980157
24. Karavalakis G, Short D, Vu D, Villela M, Asa-Awuku A, Durbin TD (2014) Evaluating the regulated emissions, air toxics, ultrafine particles, and black carbon from SI-PFI and SI-DI vehicles operating on different ethanol and iso-butanol blends. Fuel 128:410–421
25. Kinoshita M, Akinori S, Matsushita S, Shibata H, Niwa Y (1999) A method for suppressing formation of deposits on fuel injector for direct injection gasoline Engine. SAE 1999-01-3656
26. Kume T, Iwamoto Y, Lida K, Murakami M, Akishino K, Ando H (1996) Combustion control technologies for direct injection SI engine. SAE 960600

27. Lake TH, Stokes J, Whitaker P, Crump J (1998) Comparison of direct injection gasoline combustion systems. SAE 980154
28. Park C, Kim S, Kim H, Moriyoshi Y (2012) Stratified lean combustion characteristics of a spray-guided combustion system in a gasoline direct injection engine. Energy 41:401–407
29. Queiroz C, Tomanik E (1997) Gasoline direct injection engines—A bibliographical review. SAE 973113
30. Sanchez FP et al (2012) Estimated cost of emission reduction technologies for light duty vehicles. A report by International council of clean transportation (ICCT)
31. Sandquist H, Denbratt I, Owrang F, Olsson J (2001) Influence of fuel parameters on deposit formation and emissions in a direct injection stratified charge SI engine. SAE 2001-01-2028
32. Sementa P, Vaglieco BM, Catapano F (2012) Thermodynamic and optical characterizations of a high performance GDI engine operating in homogeneous and stratified charge mixture conditions fueled with gasoline and bio-ethanol. Fuel 96:204–219
33. Spicher U, Kolmel A, Kubach H, Topfer G (2000) Combustion in spark ignition engines with direct injection. SAE 2000-01-0649
34. Tamura H (1998) Engine management system for direct injection gasoline engine. Adv Microsyst Autom Appl 98:199–212
35. Taniguchi S, Yoshida K, Tsukasaki Y (2007) Feasibility study of ethanol applications to a direct injection gasoline engine journal. SAE 2007-01-2037
36. Zeng W, Sjoberg M, David LR, Zongjie H (2016) The role of spray-enhanced swirl flow for combustion stabilization in a stratified-charge DISI engine. Combust Flame 168:166–185
37. Zhao FQ, Lai MC, Harrington DL (1997) A review of mixture preparation and combustion control strategies for spark-ignition direct-injection gasoline engines. SAE 970627
38. Zhao F, Lai MC, Harrington DL (1995) The spray characteristics of automotive port fuel injection—a critical review. SAE 950506
39. Zhao F, Lai MC, Harrington DL (1999) Automotive spark-ignited direct-injection gasoline engines. Prog Energy Combust Sci 25:437–562
40. Zhao H (2014) Overview of gasoline direct injection engines. Advanced direct injection combustion engine technologies and development. Volume 1: Gasoline and Gas Engines, Chapter 1, pp 01–18
41. Zhao H, Ladommatos N (1998) Optical diagnostics for in-cylinder mixture formation measurements in IC engines. Prog Energy Combust Sci 24:297–336
42. Zhu R, Hu J, Bao X, He L, Lai Y, Zu L, Li Y, Su S (2016) Tailpipe emissions from gasoline direct injection (GDI) and port fuel injection (PFI) vehicles at both low and high ambient temperatures. Environ Pollut 216:223–234

# Syngas as SI Engine Fuel: Combustion Perspective

**Atmadeep Bhattacharya and Amitava Datta**

**Abstract** Depletion of conventional energy resources and the adverse impact of fossil fuel combustion on environment threaten the sustainable development potential in the energy sector. Biomass can be an alternative, renewable fuel to overcome these negativities. Secondary fuels, derived from biomass, can be used to run the transportation engines. Syngas, which is obtained from biomass gasification, is a potential fuel for the spark ignition engines. However, certain combustion characteristics of the fuel are needed to be studied for the efficient and smooth operation of the engines with the alternative fuel. Laminar burning velocity and ignition delay are two such important characteristics, which influence flame propagation and abnormal combustion, like detonation and surface ignition, inside the engine cylinder. Determination of these combustion characteristics shows considerable deviation between conventional gasoline fuel and syngas (50% $H_2$ and 50% CO). However, blending a small amount of conventional hydrocarbons bring the characteristic parameters with syngas fuel close to those of gasoline. Both physical properties and chemical characteristics are attributed as the reasons behind such variations in the combustion behaviour. Use of syngas is found to decrease the emission levels compared to gasoline in many respect.

## 1 Introduction

The concept of sustainability lies in the ability to satisfy the necessity of the present without compromising the resources for the future. As energy is a key component in the development, sustainability in energy usage finds a lot of interests in the scientific and technical community. In this context, it is worth mentioning that the

A. Bhattacharya
Department of Mechanical Engineering, National Institute of Technology, Jamshedpur 831014, India

A. Datta (✉)
Department of Power Engineering, Jadavpur University, Salt Lake Campus, Kolkata 700098, India



A.K. Agarwal et al. (eds.), *Combustion for Power Generation and Transportation*, DOI 10.1007/978-981-10-3785-6_17

worldwide productions of oil, coal and gas are expected to decrease exponentially after reaching their respective peaks in the years 2015, 2025 and 2035 respectively [1]. Hence, the contemporary energy sector has been exploring more sustainable resources for the use in future. The government of India has adopted and signed the United Nation's development agenda for 2016–2030 with the motto—'Transforming our world: the 2030 agenda for sustainable development'. The agenda comprises of 17 sustainable development goals (SDG) and 169 related targets [2]. Among these, there are some serious concerns regarding the targets which are related to the energy sector. It has been observed that with time the Indian energy sector has relied more heavily on the fossil fuels. The power sector in India is primarily dependent upon coal as fuel. It may be seen from Fig. 1 that the coal consumption in the country has increased in the span of 2000–2013 [3]. This increase can be attributed to the increase in the electricity demand and installation of new coal based power plants. On the other hand, the transportation sector in India is mostly dependent on oil and natural gas as fuels. The combined percentage consumption of these fuels has remained almost unchanged over the time shown in Fig. 1. Considering the growth in the primary energy consumption during this period, this indicates a corresponding growth in the usage of these fuels. This is consistent with the growth in the number of vehicles on Indian roads, as is evident from the increase in car sales in the past years (Fig. 2) [4]. Therefore, it may easily be said that accomplishing the sustainability targets within the desired time is difficult to achieve if the fossil fuels continue to dominate the transportation fuel demand.

Although biomass is considered as a carbon neutral renewable energy source, the high share of the biomass fuel in the overall fuel demand in India, as shown in Fig. 1, is misleading from the clean energy point of view. The cause of such high demand of biomass fuel is due to the fact that people in rural India still use biomass as a traditional cooking fuel. This type of direct burning of several biomass feedstock (like agricultural wastes, dried cow dung etc.) in traditional ovens creates

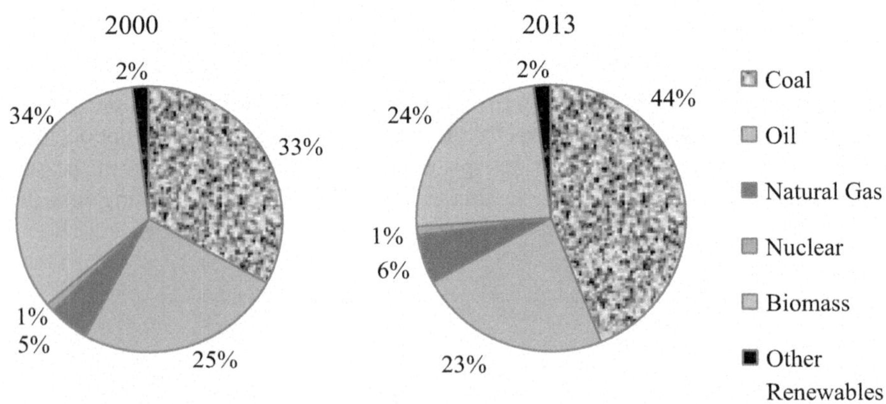


**Fig. 1** Fuel demand in India in the year 2000 and 2013 [3]

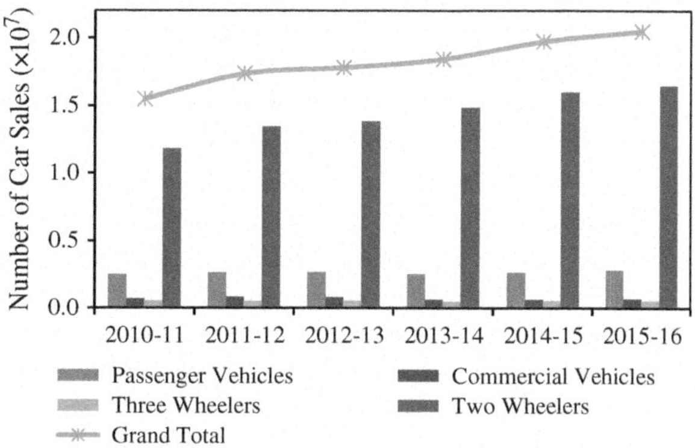


**Fig. 2** Number of car sales in the past six financial years [4]

severe health hazards. Recently, Government of India has initiated programs to extend clean cooking fuel, like LPG, to every household in the country, by which the dependence on biomass for cooking fuel will see a waning. However, biomass is the 4th most abundant energy source after coal, oil and natural gas throughout the world, contributing almost 14% of the total energy demand globally [5]. Therefore, in the future setting with increased dependence on alternative and renewable resources, biomass should play an important part.

Some global energy scenario studies have indicated that biomass may contribute up to 30% of the energy supply by 2100 in India [6]. Though such claims may draw a debate, in an agriculture-based economy the role of biomass towards sustainability in the energy sector is undeniable. Biomass derived fuels of different forms can replace the petro-fuels in the transportation sector as well. For example, ethanol from corn and sugarcane, and biodiesel from soy, rapeseed, and oil palm have been in the international market as biofuels for quite some time now [7]. However, these biomass feedstock are edible and their abundant use gives rise to the food versus fuel crisis. The present challenge is to produce good quality transportation fuel from non edible biomass resources like crop fuels, agricultural wastes, saw dust, municipal solid wastes etc. Direct combustion of solid biomass in engines is not possible and therefore it requires their conversion into liquid or gaseous forms. Such conversion of biomass into other liquid or gaseous biofuels can be performed through either thermochemical or biochemical routes. The thermochemical conversion process includes three sub-categories, viz. pyrolysis, gasification, and liquefaction. The biochemical route is considered to be less energy intensive but requires a large reactor volume due to its low yield.

In thermochemical processes (i.e. liquefaction, pyrolysis and gasification), high temperature is used to break down the lignocellulosic structure of biomass. In the liquefaction and the pyrolysis processes, the biomass is heated in absence of oxidizer. On the other hand, a sub-stoichiometric oxidation environment is used for gasification [8, 9]. Many studies have proved the superiority of gasification over other biomass conversion processes [10, 11]. Therefore, biomass derived syngas (primarily a mixture of $H_2$ and CO) produced via gasification may also be considered as a promising option for replacing the use of conventional fossil fuels in transportation. There are four different types of gasifiers available in the literature, viz. fixed bed, fluidize bed, entrained flow and transport reactor, that are suitable for the gasification of biomass [12]. The fixed bed gasifiers can further be categorized into updraft and downdraft configurations. All these types have certain benefits and drawbacks with respect to fuel type, application and operation. A useful description of these aspects is available in the review made by Samiran et al. [13]. During gasification process of biomass, unwanted by-products such as tar, impurities and ash are produced. Each type of gasifier has different parametric conditions and consequently different tar compositions and syngas production rates. The tar is a mixture of heavier hydrocarbons. It needs to be removed or further processed due to the fact that it tends to condense at a downstream location, thereby causing serious problems including fouling of engines and deactivation of catalysts. The tar yield is the highest for the updraft gasifiers [13]. The tar is primarily produced in the pyrolysis zone of the gasifier. The tar yield is higher if the gasification temperature is lower and the vice versa [14]. It has been seen by Öhrman et al. [15] that the oxygen blown entrained flow gasifier produces almost tar free syngas among all the gasification processes due to the occurrence of high temperature inside the reactor. The required tar quantity has to be limited to less than 50 $mg.Nm^{-3}$ for spark ignition engine applications [16].

Apart from the tar, there are other hurdles in the use of biomass derived syngas in the vehicular engines as well. It is well accepted that due to the high auto-ignition temperature of the components gases, syngas alone is not a suitable fuel in compression ignition engines. Therefore, such engines need to be operated in spark ignition mode. The composition and heating value of syngas change abruptly with the change in biomass feedstock type in the gasifier [17, 18]. Since the volumetric energy content of syngas is lower than the conventional hydrocarbon fuels, the volumetric efficiency as well as the engine output decrease when syngas is used as fuel in SI engine [19]. Moreover, Li et al. [20] have reported higher NO concentration in the exhaust when the syngas is used in the SI engine. They attributed the cause of such increase in the NO formation to the high adiabatic flame temperature of the fuel compared to conventional gasoline. On the other hand, there are certain advantages of using syngas as an SI engine fuel as well. The fuel has very high resistance to detonation in the engine. Therefore, the loss in volumetric efficiency and power output of the engine can be compensated by operating the engine at higher compression ratio. Sridhar et al. [21] have demonstrated smooth SI engine operation with syngas as fuel at a compression ratio as high as 17 without any abnormality. However, the use of syngas in S.I. engine needs to sort out certain related combustion issues.

# 2 Combustion in SI Engine

## *2.1 Stages of Combustion*

There are three stages of combustion in SI Engine:

(1) Preparation stage or ignition lag
(2) Flame propagation stage
(3) After burning stage

Following the ignition initiated by the spark (at point A in Fig. 3), the flame front propagates very slowly at the beginning due to additional heat loss to the relatively cold spark plug. Therefore, cylinder pressure fails to build up sufficiently during the initial stages up to point B in the pressure diagram. This delay in building up of pressure is called the ignition lag. The ignition lag is generally expressed in terms of crank angle (θ) and the pressure rise is generally detected at about 6° of crank rotation after spark. The fraction of fuel burned during the first stage is only very little and can be within 5% of the total fuel burned in one working cycle.

Once the ignition lag is over, the flame front propagates through the unburned reactant mixture in the cylinder at a rapid pace burning the fuel. It results in heat release and corresponding increase in temperature and pressure (as shown in Fig. 3 from B to C). The rate of pressure rise is rapid and depends upon the rate of flame propagation. Under the ideal condition of Otto cycle, the flame propagation is considered to be infinitely fast causing heat addition at constant volume. With the finite rate of flame propagation in practice, the rate of pressure rise will be less and the correspondingly the power output reduces. The peak pressure occurs after the top dead centre (TDC), but close to it, so that high pressure can work on the piston for a considerable part of the expansion stroke. The premixed charge within the

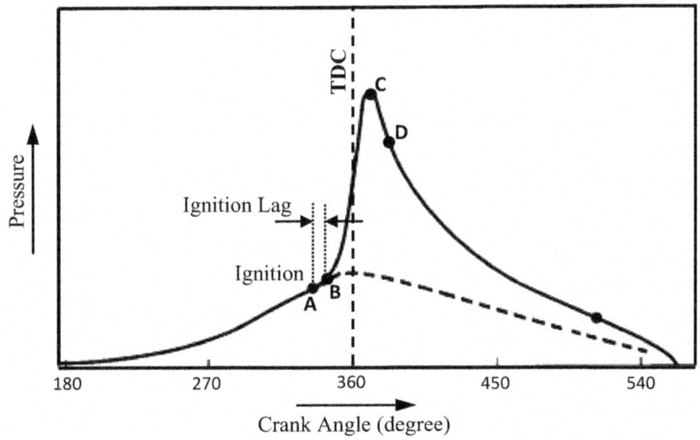


**Fig. 3** Variation of cylinder pressure with crank angle. The *dotted line* indicates the cylinder pressure in absence of combustion [22]

engine cylinder maintains a turbulent flow structure in which the flame propagates as a wrinkled laminar flame. The turbulent flame propagation velocity depends on the intensity of turbulence in the fluid flow and the laminar burning velocity of the fuel-air mixture.

Majority of the fuel fired into the cylinder burns during the second stage. A small fraction of about 5% of the fresh charge is burnt in the third stage, known as after burning stage. The third stage takes place during the expansion stroke. Therefore, pressure is observed to be dropping during after burning (C–D), though there may be a rise in temperature due to additional heat release.

## 2.2 *Effect of Spark Timing*

The spark timing has an important effect on the performance of spark ignition engine. The spark is given near the end of the compression stroke in a four stroke engine at an optimum time that produces the maximum work output from the engine. If the spark is advanced too much, the pressure builds up too early inside the cylinder. Therefore, the work transfer from the piston to the gas in the compression stroke increases. On the other hand, if the ignition timing is retarded beyond the optimum point, the flame propagation phase proceeds well into the expansion stroke. Therefore, there is a considerable decrease in the peak cylinder pressure and also the peak pressure shifts away from the TDC location. The optimization of the spark timing is done by balancing these two mutually opposite effects of advancing and retarding the spark timing. Thus flame propagation in the cylinder and the resulting rate of pressure rise has important effects in optimizing the spark location.

## 2.3 *Abnormal Combustion Phenomena in SI Engine*

The two major abnormal combustion phenomena that are of prime importance are detonation and surface ignition. During the flame propagation stage, as the flame proceeds across the chamber, the pressure and temperature of unburned charge increase due to compression by the burnt portion of the charge. These compressed unburnt gases are further heated by the heat loss from the advancing flame. Detonation occurs when the unburnt end charge inside the combustion chamber auto ignites after spending a specific delay period beyond auto-ignition temperature. When detonation takes place, there is an extremely rapid release of the chemical energy in the end-gas, causing very high local pressures and the propagation of pressure waves. These pressure waves ultimately collide with the normal pressure wave inside the cylinder, resulting pressure oscillation, which may create engine damage.

Another abnormal combustion phenomenon is the surface ignition. It is the ignition of the air-fuel mixture by a hot spot on the combustion chamber wall. The hot spots can be generated by an overheated valve or spark plug, or glowing combustion chamber deposits; i.e., by any means other than the normal spark discharge. It can occur before the normal ignition by the spark (pre-ignition) or after it (post-ignition). Following surface ignition, a turbulent flame develops at each surface-ignition location and starts to propagate across the chamber in the same manner to what occurs with normal spark ignition. The problem of surface ignition often leads to detonation as well. Both these abnormal situations can be avoided if there is sufficient ignition delay. When the ignition of the end charge is delayed by the time it is consumed by the normal flame, the problem of knocking as well as surface ignition can be prevented.

## 3 Methodology

From the above discussion on SI engine combustion, it is clear that the laminar burning velocity, which is directly related to the turbulent burning velocity, and ignition delay are the two parameters of prime importance regarding SI engine combustion. If the current engine designs are to be retained for a new fuel like syngas, it is necessary to keep these parameters nearly unaltered from the existing value with conventional fuels. Therefore, the viability of syngas-air mixture needs to be evaluated from the perspective of the laminar burning velocity and the ignition delay for conventional gasoline based SI engine applications. The syngas has been assumed to be as equimolar mixture of $H_2$ and CO. The laminar burning velocities have been simulated using the PREMIX code [23] of the CHEMKIN PRO package. The assumptions for the code include steady, one-dimensional, adiabatic, freely propagating flame in a doubly infinite domain. The simulations have been done by taking thermal diffusion (Soret effect) and mixture averaged transport properties into consideration. For the evaluation of the ignition delay the closed homogenous batch reactor model of the CHEMKIN PRO package has been adopted for the numerical simulations. The physical model utilizes the transient, spatially homogeneous form of the conservation equations for mass, species and energy. The fuel–oxidizer-diluent mixture is assumed to be in a closed adiabatic system under constant pressure conditions. Since the system of ordinary differential equations shows large disparity in time scales for the evolution of different species in the reactor system, an implicit time integration scheme has been adopted for solving the set of equations [24]. The ignition is assumed to occur when the hydroxyl radical (OH) concentration reaches its maximum during the reaction process [25]. A skeletal mechanism, following Yoo et al. [26], constituting 143 species and 643 reaction steps has been adopted for the simulations of laminar burning velocity and ignition delay for isooctane, syngas as well as isooctane-syngas blends. This mechanism was earlier reduced from a detailed isooctane-air kinetic mechanism of Mehl et al. [27] and validated against experimental auto-ignition delay and extinction residence time data over a wide parameter range of pressure (1–50 atm), equivalence ratio (0.3–2.0), and initial temperature (750–1800 K).

## 4 Results and Discussion

The commercial gasoline is a complex mixture of many hydrocarbons generally in the range of $C_4$–$C_{10}$, including linear and branched paraffins, naphthenes, olefins, and aromatics. The concentrations of these individual constituents vary depending upon the geographical origin of the fuel. Therefore, it is impossible to model any combustion phenomena using commercial gasoline unless a suitable surrogate is chosen. It has been shown by many researchers that isooctane can successfully be taken as a surrogate for the real world gasoline [28, 29]. In Fig. 4, the experimental data on laminar burning velocity of gasoline-air mixture is compared with the simulated values of isooctane-air mixture at three different pressures and two unburnt gas temperatures. The experimental data at 1 bar is of Bhattacharya et al. [30] and at 10 and 20 bar are taken from Jerzembeck et al. [31]. Bhattacharya et al. [30] adopted the heat flux burner method for the experimental determination of the laminar burning velocity at atmospheric pressure. On the other hand, Jerzembeck et al. [31] used constant volume spherical bomb setup for the determination of the laminar burning velocity at higher pressure. It may be seen from the figure that the agreement is very good in the lean side of stoichiometry. However, on the rich side, there is a slight amount of mismatch. The maximum deviation is about 9% at $\varphi = 1.2$ for $T_u = 423$ K and $P = 1$ bar. At atmospheric pressure, the peak laminar burning velocity occurs at $\varphi = 1.1$ for the experimental data. However, the simulation predicts the maximum laminar burning velocity at $\varphi = 1.2$. This mismatch is not seen for the higher pressures where both the experimental as well as the simulation results predict the peak laminar burning velocity to occur at $\varphi = 1.1$. The deviation in the predicted and experimental values of laminar burning velocities indicate that the adopted chemical mechanism for isooctane somewhat fails to capture the conditions for gasoline on the rich side of stoichiometry beyond a particular equivalence ratio.

In order to validate the numerical simulations further, the predicted laminar burning velocity data are compared with the experimental data for isooctane-air mixture at three different pressures (Fig. 5). It may be seen from the figure that the

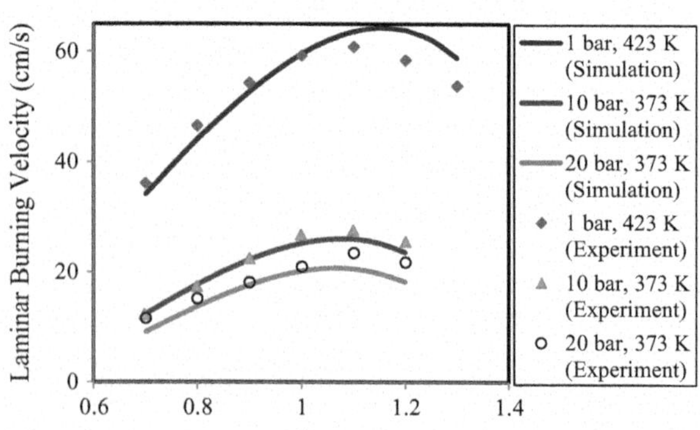


**Fig. 4** Variation of laminar burning velocity of commercial gasoline-air mixture with equivalence ratio at different unburnt gas temperatures and pressures. The experimental data at 1 bar is taken from Bhattacharya et al. [30] and at 10 and 20 bar from Jerzembeck et al. [31]

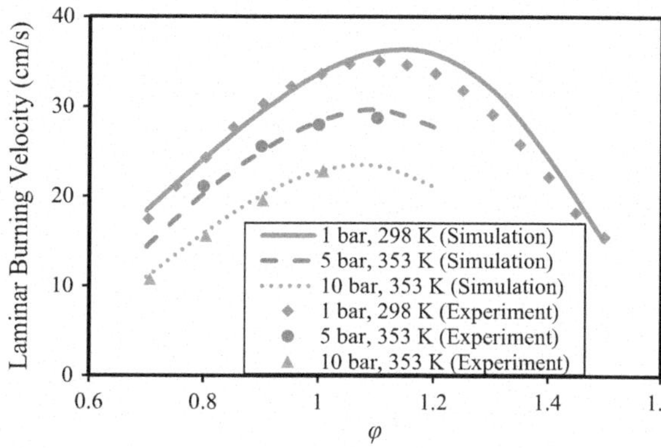


**Fig. 5** Variation of laminar burning velocity of isooctane-air mixture with equivalence ratio at different unburnt gas temperatures and pressures. The experimental data at 1 bar is taken from Dirrenberger et al. [32] and at 5 and 10 bar from Kelley et al. [33]

experimental laminar burning velocity data is well predicted by the present set of simulation results. The simulation at the rich side of stoichiometry beyond $\varphi = 1.2$ has not been done for higher than atmospheric pressures due to the unavailability of experimental data. As the pressure increases, the laminar burning velocity is found to reduce primarily due to the increase in density, and therefore decrease in thermal diffusivity. For all the simulation values in Figs. 4 and 5, the maximum deviation is observed to be within 10% of the experimental data.

In Fig. 6, the ignition delay for the stoichiometric isooctane–air mixture from the simulation has been compared with the ignition delay time data obtained from experiments of Sakai et al. [34], Oehlschlaeger et al. [35], Minetti et al. [36] and correlation from Goldsborough [37] at 10 atm pressure. The correlation of Goldsborough [37] has already been validated with a large number of experimental data points. It is evident from the figure that the ignition delay time is quite accurately predicted by the simulation results in the higher temperature region where the ignition delay time is found to decrease with the increase in temperature. However, as the temperature of the reactant mixture is reduced further, it is found that the ignition delay time shows a decreasing trend with the reduction in

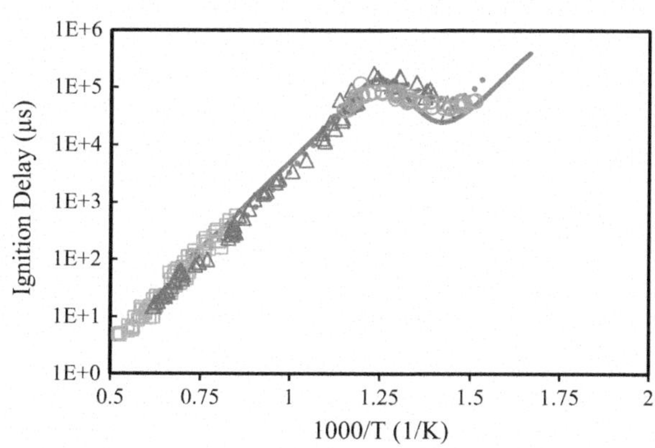


**Fig. 6** Variation of ignition delay with temperature for stoichiometric isooctane–air mixture at 10 bar pressure. The simulation data (—) are compared with the experimental data taken from Sakai et al. [34] (Δ), Oehlschlaeger et al. [35] (□), Minetti et al. [36] (○) and correlation from Goldsborough [37] (…)

temperature within a certain temperature range. This region is called the negative temperature coefficient (NTC) region [38].

The NTC behaviour in the variation of ignition delay of isooctane-air mixture can be explained through the low temperature oxidation pathway of hydrocarbon species as observed in a cool flame [39]. The reaction mechanism involves the following reaction steps for a generic alkane (RH) molecule:

$$RH + O_2 \rightarrow \mathbf{R} + HO_2 \tag{R1}$$

$$\mathbf{R} + O_2 + M \rightarrow \mathbf{RO_2} + M \tag{R2}$$

$$\mathbf{R} + O_2 \rightarrow \text{Olefin} + \mathbf{HO_2} \tag{R3}$$

$$\mathbf{RO_2} + RH \rightarrow RO_2H + \mathbf{R} \tag{R4}$$

$$RO_2H + M \rightarrow \mathbf{RO} + \mathbf{OH} + M \tag{R5}$$

$$\mathbf{HO_2} + RH \rightarrow H_2O_2 + \mathbf{R} \tag{R6}$$

In the mechanism shown above the bold faced species represent radicals. The initiation reaction in the mechanism proceeds with reaction (R1) forming **R** radical. The radical can react with $O_2$ following two alternative reaction paths–(R2) and (R3). At low temperature, reaction path (R2) is favoured over (R3) based on their respective kinetic parameters. The exothermic chain propagation path (R2) followed by reaction (R4) generates the radical **R** again. It further produces $RO_2H$, which breaks up following the degenerate branching reaction (R5) to produce more radicals. The increased radical population leads to ignition under a low temperature and establishes a cool flame. When the temperature is increased, reaction path (R3) prevails over (R2). The $\mathbf{HO_2}$ radicals formed in the process further reacts through path (R6) to produce peroxide species $H_2O_2$. $H_2O_2$ is a metastable species, unless the temperature goes high (beyond 750 K), it does not contribute to chain branching. As a result, ignition gets inhibited and ignition delay increases again giving the NTC behaviour.

In Fig. 6, the NTC behaviour of isooctane-air flame is seen to be predicted in the proper temperature range with marginal under-prediction. It is seen From Figs. 4, 5 and 6 that the chemical mechanism adopted for the model is able to predict the engine relevant key combustion properties of gasoline and isooctane-air mixture with acceptable accuracy. However, to compare the same properties of syngas-air mixture with gasoline, this chemical mechanism has to be validated against experimental data for syngas as well.

For this purpose, simulations have been run with syngas-oxidizer mixtures under different pressure and temperature conditions to validate the predictions of combustion characteristics. The same chemical mechanism, as is used for isooctane before, is also used for syngas to predict the parameters. The variation of laminar burning velocity of syngas has been plotted against $\varphi$ in Fig. 7a, b and c for $P$ = 1, 5 and 10 bar, respectively. The unburnt gas temperature is 298 K for all the three

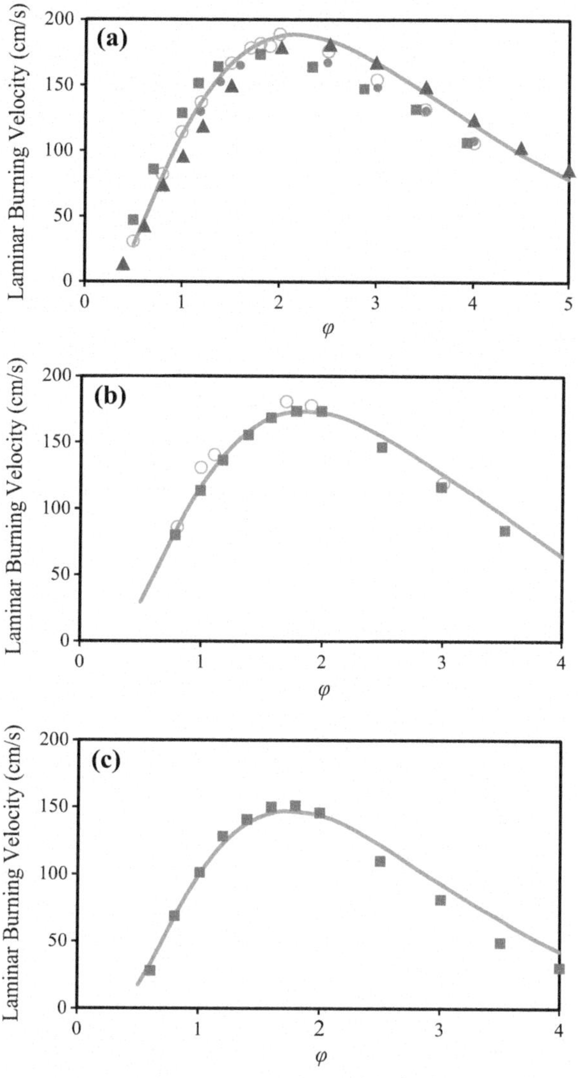


**Fig. 7** Variation of laminar burning velocity of **a** syngas-air mixture with equivalence ratio at $T_u = 298$ K and $P = 1$ bar **b** syngas-oxidizer ($O_2$: He::1:7) mixture with equivalence ratio at $T_u = 298$ K and $P = 5$ bar **c** syngas-oxidizer ($O_2$: He::1:7) mixture with equivalence ratio at $T_u = 298$ K and $P = 10$ bar. The simulation results (—) are compared with the experimental data taken from Krejci et al. [40] (○), Sun et al. [41] (■), Bouvet et al. [42] (▲) and Frassoldati et al. [43] (•)

cases. The predicted results are compared with the experimental data of Krejci et al. [40], Sun et al. [41], Bouvet et al. [42] and Frassoldati et al. [43]. The above results show that for all the pressures studied, the peak laminar burning velocity for syngas occurs at a richer fuel-air mixture ($\varphi = 1.75 - 2.5$) compared to that for isooctane. The peak value of laminar burning velocity is also found to be much higher than that for gasoline and isooctane. The increase in pressure results in the drop in the burning velocity, which is clearly evident from the three different pressure cases. Moreover, the variation curve of the laminar burning velocity with the equivalence ratio is rather asymmetrical on lean and rich sides of stoichiometry. The rise of the laminar burning velocity to the peak value is sharper than its decline as "$\varphi$" is increased. The classical models of laminar burning velocity proposed by Mallard and Le Chatelier and Zeldovich, Frank-Kamenetskii and Semenov, show the thermal diffusivity of the reactant mixture and the rate of reaction as the key parameters influencing the burning velocity of a particular air-fuel mixture [44]. The reaction rate has a dependence on the adiabatic flame temperature for a particular fuel-air mixture. Since the thermal diffusivities of $H_2$ and CO are higher than air, the thermal diffusivity for the syngas-air mixture increases monotonically with the increase in the equivalence ratio. On the other hand, the adiabatic flame temperature increases with the increase in fuel quantity in the fuel-air mixture up to the stoichiometric point. Hence, in the lean side of stoichiometry, both the adiabatic flame temperature and thermal diffusivity increase with the increase in equivalence ratio. Therefore, the laminar burning velocity of the syngas-air mixture increases sharply in this region. Beyond the stoichiometric region ($\varphi = 1$), as the equivalence ratio increases further, the adiabatic flame temperature starts decreasing due to the drop in volumetric heat generation rate as a result of incomplete combustion. But due to the dominating influence of thermal diffusivity, the laminar burning velocity still continues to rise. After a certain amount of fuel enrichment, the thermal diffusivity loses its dominance over flame temperature and reaction rate. Therefore, the laminar burning velocity starts to decline with equivalence ratio after showing a peak in the region of $\varphi = 1.75 - 2.5$.

It is also observed from Fig. 7 that the comparison of predicted and experimental results show excellent agreement over a much wider range of equivalence ratio and for all the three pressure cases. Both the magnitude and equivalence ratio for the peak laminar burning velocity are excellently predicted at all the pressure cases. It happens because the kinetic parameters for the hydrogen and carbon monoxide oxidation reactions are much well defined, unlike those parameters for higher hydrocarbon fuels.

Figure 8 shows the comparison of the predictions of ignition delay time of syngas-air mixture with model predictions of Jain et al. [38] at $P = 15$ bar and $\varphi = 0.4$. It may be seen from the figure that the numerical predictions with the

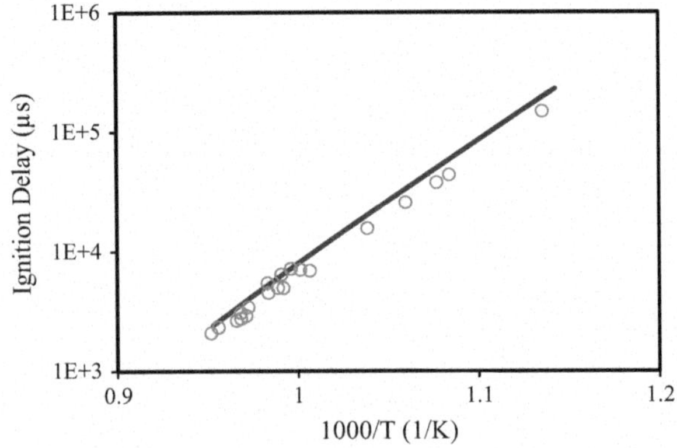


**Fig. 8** Variation of ignition delay of syngas-air mixture with temperature at 15 bar pressure and $\varphi = 0.4$. The line represents the present simulation data and the symbols represent the simulation data using the chemical mechanism of Jain et al. [38]

adopted mechanism are very close to the ignition delay times published in the literature. With the increase in temperature, the ignition delay for the syngas-air mixture decreases monotonically and, unlike isooctane, no NTC behaviour is seen in the ignition delay variation. This is because, in absence of the alkane hydrocarbon molecules, the low temperature oxidation reactions leading to ignition and establishment of cool flame do not occur.

Figure 9a shows the variation in laminar burning velocity with varying equivalence ratio for different syngas mole fractions in the isooctane-syngas-air mixture. As observed from Figs. 5 and 7, the laminar burning velocity of syngas is much higher than that of isooctane and the peak burning velocity of the former occurs at a richer fuel-air mixture compared to that for isooctane-air. However, it may be observed from the Fig. 9a that when a small amount of isooctane is blended with syngas, the laminar burning velocity reduces drastically compared with the pure syngas-air laminar burning velocity and becomes closer to the laminar burning velocity of pure isooctane-air mixture. It may be seen from the figure that the blending of 10% isooctane causes the laminar burning velocity of syngas to drop almost 60% at stoichiometric conditions. Finally, for a mixture of 30% isooctane and 70% syngas, the laminar burning velocity is within 10% of the pure isooctane-air mixture.

**Fig. 9** **a** Variation of laminar burning velocity with equivalence ratio ($\varphi$) for syngas-isooctane-air mixture at $T_u$ = 298 K and $P$ = 1 bar. **b** Comparison of ignition delay of the 30% isooctane-70% syngas, 100% isooctane and 100% syngas-air mixture for different temperatures at $P$ = 10 atm and $\varphi$ = 1

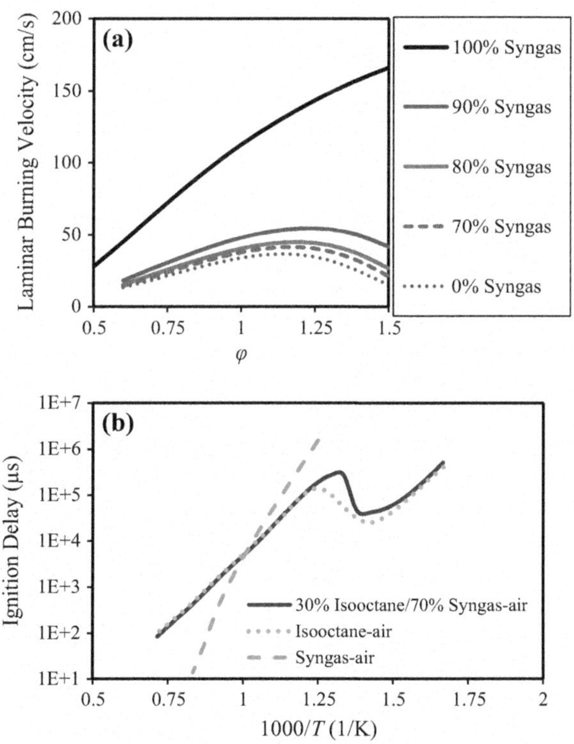

This change in laminar burning velocity with addition of isooctane in syngas can be explained from the perspective of the change in thermal diffusivity and adiabatic flame temperature of the reactant mixture. As discussed earlier, the thermal diffusivity of isooctane-air mixture is much lower than the corresponding value of syngas-air mixture for all fuel-air equivalence ratios. Therefore, the addition of isooctane reduces the thermal diffusivity of the reactant mixture and the dominance of the adiabatic flame temperature increases over the laminar burning velocity with the increase in isooctane mole fraction in the syngas-air mixture. From the combustion chemistry perspective, with the blending of the higher hydrocarbon like isooctane to syngas, the H radical concentration in the flame drops and comes closer to the H radical concentration of isooctane fuel. Moreover, the maximum point of peak H radical mole fraction shifts towards the stoichiometric value with the blending of isooctane in syngas as well. It has been reported that the effect of hydrogen radical (H) concentration on the laminar burning velocity is much greater than the influence of other radicals for syngas combustion [25]. Therefore, the shape of the laminar burning velocity versus equivalence ratio curve for the 30% isooctane-70% syngas mixture is similar to that for 100% isooctane fuel.

The ignition delay of the 30% isooctane-70% syngas blend is shown in Fig. 9b along with the ignition delay of 100% isooctane and 100% syngas for different temperatures at $P = 10$ atm and $\varphi = 1$. The curves show that the ignition delay of the syngas is much higher than isooctane when the temperature is lower than 1000 K. Due to this fact, the knocking tendency of the syngas is much lower than gasoline under the same operating conditions in an SI engine. The variation in the ignition delay time of isooctane shows the NTC behaviour, which may also facilitate autoignition and detonation. However, for the syngas, the NTC pattern is not seen. However, with the blend of 30% isooctane to 70% syngas, the NTC behaviour may again be observed. It may be seen from the figure that the ignition delay times of isooctane and the 30% isooctane-70% syngas blend are almost equal until the NTC region is reached.

In order to consider syngas as a fuel of the SI engines, the consequences in terms of pollutant emissions should also be taken into account. In the recent past, Shah et al. [16] have compared the emissions from syngas-air mixture with gasoline-air mixture when applied to an SI engine under similar load conditions. They reported a decrease in the emission of CO in case of syngas compared to gasoline. This observation was further corroborated by Homdoung et al. [45]. The primary reason behind this reduction is the more homogeneous mixing of the gaseous fuel with air avoiding rich burning in the cylinder. The $NO_x$ emission primarily depends upon the adiabatic flame temperature for a particular fuel-air mixture [30]. The biomass derived syngas contains large quantity of $N_2$ with it [19]. As a result, the adiabatic flame temperature of biomass derived syngas is comparatively low despite the fact that the adiabatic flame temperature of $H_2$ present in the syngas is quite high. Therefore, the $NO_x$ emissions have also been reported to be lower for the case of syngas compared to gasoline by the authors [16]. Even in case of syngas containing little or no nitrogen (e.g. due to oxygen based gasification) the $NO_x$ emission can be low. This is because due to the high laminar burning velocity of syngas, the spark timing needs to be retarded considerably. As a result, the residence time of the gas in the high temperature combustion chamber will be reduced, thus reducing NO formation [46]. The unburnt hydrocarbon emission has also been reported to be low due to the lack of presence of hydrocarbon species in the syngas. The $CO_2$ emission is higher from the engine operating on biomass derived syngas compared to gasoline [16, 47]. The primary reason behind such increase is the oxidation of the CO molecule along with the presence of $CO_2$ in the fuel itself. However, as biomass is a carbon-neutral fuel the increased $CO_2$ in the exhaust does not result in carbon loading to the atmosphere. Thus the gasoline-syngas blends will only favour the engine performance in terms of emission.

## 5 Conclusion

Laminar burning velocity and ignition delay are two important combustion parameters for smooth and efficient operation of SI engine. The laminar burning velocity and ignition delay characteristics of commercial gasoline can be simulated with isooctane. Major modification in the engine design is required in order to incorporate syngas as an SI engine fuel due to its large difference in combustion properties with gasoline. However, if syngas is blended with conventional gasoline, the laminar burning velocity and the ignition delay characteristics are almost similar to that of pure gasoline. It has been found from the present study that the gasoline, being the heavier hydrocarbon, dominates the combustion properties of the syngas-gasoline-air mixture. This dominating effect of gasoline can be attributed to the reduction in thermal diffusivity of the reactant mixture compared to that for pure syngas after mixing. From the combustion chemistry point of view, the reduction in laminar burning velocity can be attributed to the reduction in the H radical mole fraction in the flame. An optimum blend of 30% gasoline and 70% syngas has shown almost the same laminar burning velocity and ignition delay patterns compared to pure gasoline in air.

## References

1. Höök M, Tang X (2013) Depletion of fossil fuels and anthropogenic climate change—a review. Energy Policy 52:797–809
2. NITI AAYOG. http://niti.gov.in/content/overview-sustainable-development-goals. Accessed 04 Nov 2016
3. India Energy Outlook (2015) World energy outlook special report
4. Society of Indian Automobile Manufactures. http://www.siamindia.com/statistics.aspx?mpgid=8&pgidtrail=14. Accessed 04 Nov 2016
5. Pudasainee D, Paur HR, Fleck S et al (2014) Trace metals emission in syngas from biomass gasification. Fuel Process Technol 120:54–60

6. Hamelinck CN, Faaij APC (2006) Outlook for advanced biofuels. Energy Policy 34 (17):3268–3283
7. Agarwal AK (2007) Biofuels (alcohols and biodiesel) applications as fuels for internal combustion engines. Prog Energy Combust Sci 33:233–271
8. Bhattacharya A, Bhattacharya A, Datta A (2012) Modeling of hydrogen production process from biomass using oxygen blown gasification. Int J Hydrog Energy 37(24):18782–18790
9. Bhattacharya A, Das A, Datta A (2014) Exergy based performance analysis of hydrogen production from rice straw using oxygen blown gasification. Energy 69:525–533
10. EUCAR-CONCAWE-JRC. http://iet.jrc.ec.europa.eu/about-jec/sites/iet.jrc.ec.europa.eu.about-jec/files/documents/wtw3_wtw_report_eurformat.pdf. Accessed 04 Nov 2016
11. Kırtay E (2011) Recent advances in production of hydrogen from biomass. Energy Converse Manag 52(4):1778–1789
12. Mondal P, Dang GS, Garg MO (2011) Syngas production through gasification and cleanup for downstream applications—recent developments. Fuel Process Technol 92(8):1395–1410
13. Samiran NA, Jaafar MN, Ng JH et al (2016) Progress in biomass gasification technique–with focus on Malaysian palm biomass for syngas production. Renew Sustain Energy Rev 62:1047–1062
14. Bhattacharya SC, Hla SS, Pham HL (2001) A study on a multi-stage hybrid gasifier-engine system. Biomass Bioenergy 21(6):445–460
15. Öhrman OG, Weiland F, Pettersson E et al (2013) Pressurized oxygen blown entrained flow gasification of a biorefinery lignin residue. Fuel Process Technol 115:130–138
16. Shah A, Srinivasan R, To SD et al (2010) Performance and emissions of a spark-ignited engine driven generator on biomass based syngas. Bioresour Technol 101(12):4656–4661
17. Roy PC, Datta A, Chakraborty N (2010) Assessment of cow dung as a supplementary fuel in a downdraft biomass gasifier. Renew Energy 35(2):379–386
18. Roy PC, Datta A, Chakraborty N (2013) An assessment of different biomass feedstocks in a downdraft gasifier for engine application. Fuel 106:864–868
19. MartínezJD Mahkamov K, Andrade RV et al (2012) Syngas production in downdraft biomass gasifiers and its application using internal combustion engines. Renew Energy 38(1):1–9
20. Li H, Karim GA (2005) Exhaust emissions from an SI engine operating on gaseous fuel mixtures containing hydrogen. Int J Hydrog Energy 30(13):1491–1499
21. Sridhar G, Paul PJ, Mukunda HS (2001) Biomass derived producer gas as a reciprocating engine fuel—an experimental analysis. Biomass Bioenergy 21(1):61–72
22. Pulkrabek WW (1997) Engineering fundamentals of the internal combustion engine. Prentice Hall, New Jersey
23. Kee RJ, Grcar JF, Smooke MD et al (1985) PREMIX: a Fortran program for modeling steady laminar one-dimensional premixed flames. Sandia National Laboratories Report SAND85-8249
24. Li S, Petzold L (2000) Software and algorithms for sensitivity analysis of large-scale differential algebraic systems. J Comput Appl Math 125(1):131–145
25. Bhattacharya A, Datta A, Wensing M (2016) Laminar burning velocity and ignition delay time for premixed isooctane–air flames with syngas addition. Combust Theor Model 1–20
26. Yoo CS, Luo Z, Lu T et al (2013) A DNS study of ignition characteristics of a lean iso-octane/air mixture under HCCI and SACI conditions. Proc Combust Inst 34(2):2985–2993
27. Mehl M, Pitz WJ, Sj berg M et al (2009) Detailed kinetic modeling of low-temperature heat release for PRF fuels in an HCCI engine. SAE Technical Paper No. 2009-01-1806
28. Baloo M, Dariani BM, Akhlaghi M et al (2015) Effect of iso-octane/methane blend on laminar burning velocity and flame instability. Fuel 144:264–273
29. Liu K, Fu J, Deng B et al (2014) The influences of pressure and temperature on laminar flame propagations of n-butanol, iso-octane and their blends. Energy 73:703–715
30. Bhattacharya A, Banerjee DK, Mamaikin D et al (2015) Effects of exhaust gas dilution on the laminar burning velocity of real-world gasoline fuel flame in air. Energy Fuels 29:6768–6779

31. Jerzembeck S, Peters N, Pepiot-Desjardins P et al (2009) Laminar burning velocities at high pressure for primary reference fuels and gasoline: experimental and numerical investigation. Combust Flame 156(2):292–301
32. Dirrenberger P, Glaude PA, Bounaceur R et al (2014) Laminar burning velocity of gasolines with addition of ethanol. Fuel 115:162–169
33. Kelley AP, Liu W, Xin YX et al (2011) Laminar flame speeds, non-premixed stagnation ignition, and reduced mechanisms in the oxidation of iso-octane. Proc Combust Inst 33(1): 501–508
34. Sakai Y, Ozawa H, Ogura T et al (2007) Effects of toluene addition to primary reference fuel at high temperature. SAE Technical Paper No. 2007-01-4104
35. Oehlschlaeger MA, Davidson DF, Herbon JT et al (2004) Shock tube measurements of branched alkane ignition times and OH concentration time histories. Int J Chem Kinet 36(2): 67–78
36. Minetti R, Carlier M, Ribaucour M et al (1996) Comparison of oxidation and autoignition of the two primary reference fuels by rapid compression. In: Paper presented at symposium (international) on combustion, December 1996, vol 31
37. Goldsborough SS (2009) A chemical kinetically based ignition delay correlation for iso-octane covering a wide range of conditions including the NTC region. Combust Flame 156(6):1248–1262
38. Jain S, Li D, Aggarwal SK (2013) Effect of hydrogen and syngas addition on the ignition of iso-octane/air mixtures. Int J Hydrog Energy 38(10):4163–4176
39. Law CK (2006) Combustion physics. Cambridge University Press, Cambridge
40. Krejci MC, Mathieu O, Vissotski AJ et al (2013) Laminar flame speed and ignition delay time data for the kinetic modeling of hydrogen and syngas fuel blends. J Eng Gas Turbines Power 135:21503–21509
41. Sun H, Yang SI, Jomaas G et al (2007) High-pressure laminar flame speeds and kinetic modeling of carbon monoxide/hydrogen combustion. Proc Combust Inst 31(1):439–446
42. Bouvet N, Chauveau C, Gökalp I et al (2011) Experimental studies of the fundamental flame speeds of syngas ($H_2$/CO)/air mixtures. Proc Combust Inst 33(1):913–920
43. Frassoldati A, Faravelli T, Ranzi E (2007) The ignition, combustion and flame structure of carbon monoxide/hydrogen mixtures. Note 1: detailed kinetic modeling of syngas combustion also in presence of nitrogen compounds. Int J Hydrog Energy 32(15):3471–3485
44. Glassman I, Yetter RA, Glumac NG (2014) Combustion. Academic press, London
45. Homdoung N, Tippayawong N, Dussadee N (2015) Performance and emissions of a modified small engine operated on producer gas. Energy Convers Manag 94:286–292
46. Dasappa S, Paul PJ, Mukunda HS et al (2004) Biomass gasification technology—route to meet energy needs. Curr Sci 87(7):908–916
47. Mustafi NN, Miraglia YC, Raine RR et al (2006) Spark-ignition engine performance with 'Powergas' fuel (mixture of CO/$H_2$): a comparison with gasoline and natural gas. Fuel 85(12): 1605–1612

# Part V
# Environmental Aspects

# Carbon Sequestration and Optimization of Enhanced Oil and Gas Recovery

**Ramesh K. Agarwal, Razi Safi, James Biagi and Subhodeep Banerjee**

**Abstract** There has been a strong emphasis on the development of safe and economical carbon capture, utilization, and storage (CCUS) technologies in recent years because of rising concerns about carbon dioxide ($CO_2$) emissions from fossil fueled power plants. Two technologies that show promise for CCUS application are enhanced oil recovery (EOR) and enhanced gas recovery (EGR), where $CO_2$ is used as a working fluid to extract oil and natural gas respectively from depleted reservoirs. Permanent carbon sequestration is achieved as a byproduct due to subsurface fluid losses throughout the life of the system. In this chapter, numerical simulations of subsurface flow in EOR are conducted using the multiphase flow solver COZ-Sim. For EGR simulations, the TOUGH2 (Transport of Unsaturated Groundwater and Heat) code is employed. An optimization code based on a multi-objective genetic algorithm is combined with COZSim and TOUGH2 and modified for EOR and EGR application respectively. Using GA-COZSim and GA-TOUGH2, the $CO_2$ injection rates are optimized for both constant mass and constant pressure injection scenarios to manage the production of methane or oil for EGR and EOR respectively to ensure high output for the entire life of the system thus allowing for more efficient use of $CO_2$. The results of this study highlight the scope of EGR and EOR technologies along with $CO_2$ sequestration for consideration of deployment on a commercial scale. This chapter presents a review of the authors' previous work reported in the literature on EOR (Safi et al., Chem Eng Sci 144:30–38, 2016 [1]) and EGR (Biag et al., Energy 94:78–86, 2016 [2]) simulations and optimization.

**Keywords** CCUS · Enhanced oil recovery · Enhanced gas recovery · Genetic algorithm · Optimization

---

R.K. Agarwal (✉) · R. Safi · J. Biagi · S. Banerjee
Department of Mechanical Engineering and Materials Science,
Washington University in St. Louis, 1 Brookings Drive, St. Louis, MO 63130, USA
e-mail: rka@wustl.edu

 
A.K. Agarwal et al. (eds.), *Combustion for Power Generation and Transportation*, DOI 10.1007/978-981-10-3785-6_18

# 1 Introduction

Since the beginning of the 20th century, the fossil fuel consumption worldwide has grown at an alarming rate leading to a marked increase in Greenhouse Gas (GHG) emissions. $CO_2$ emissions were close to 25,000 teragrams in early 1900; current levels are greater than 30,000 teragrams [3]. As a result, atmospheric $CO_2$ levels have increased dramatically from nearly 315 ppm in 1960 to current levels of approximately 400 ppm [4]. Carbon Capture and Geological Sequestration (CCGS) offers a potential solution to reduce the direct $CO_2$ emissions from stationary sources into the atmosphere. In a typical CCGS project, $CO_2$ is captured from power plants or other industrial sources, for example the cement factories and steel mills, and is permanently sequestered in nearby geological formations such as the saline aquifers, oil and gas fields, and unmineable coal seams. These geological formations have a highly permeable reservoir capped with a relatively impermeable layer of rock which prevents the leakage of $CO_2$ from the reservoir. If fully implemented, geological sequestration of $CO_2$ has the capacity to reduce $CO_2$ emissions from power plants by as much as 80–90% [4]. Once the $CO_2$ is injected into the geological formation, it undergoes four distinct trapping mechanisms to ensure permanent storage within the reservoir. These trapping mechanisms are categorized as structural trapping, residual trapping, solubility trapping, and mineral trapping [5, 6]. A similar approach that is more lucrative financially is Carbon Capture, Utilization and Storage (CCUS) because the captured $CO_2$ can be used as the working fluid for enhanced oil recovery, enhanced gas recovery, or enhanced geothermal systems to benefit further energy production [7].

$CO_2$ enhanced oil recovery (EOR) can be used in the tertiary stage of oil production to extract the 50–80% of the original oil in place (OOIP) that remains in the oil field at the end of the primary and secondary (water flood) production stages [8]. $CO_2$ injected into deep oil reservoirs becomes supercritical and miscible with oil, leading to a reduction in the oil viscosity and the surface tension between the oil and the surrounding rock, and a swelling effect in the oil. This results in increased mobility of the oil allowing for easier extraction from the recovery well. In addition, using $CO_2$ for the oil recovery process may help maintain pore pressure in the reservoir and reduce the risk of induced seismicity that is otherwise a concern in large-scale geological storage of $CO_2$ in saline formations [9, 10]. EOR is used regularly across the US and around the world and the results are well-documented and appear promising. Today, there are 114 $CO_2$ floods underway in the United States, of which 64 are in the Permian Basin reservoir in West Texas and southeast New Mexico [11, 12]. In this chapter, numerical simulations of subsurface flow in an EOR system are presented using the multiphase flow solver package COZSim/COZView developed by Nitec, LLC [13]. The $CO_2$ injection is optimized for both constant rate and pressure-limited injection scenarios using a genetic algorithm based optimizer integrated with COZSim. A detailed mathematical description of the numerical modeling process for EOR can be found in the authors' previous work [1].

The concept of carbon sequestration with enhanced gas recovery (EGR) was originally proposed by van der Burgt et al. [14] and Blok et al. [15] in the 1990s. The idea did not receive much attention at the time due to the high price of $CO_2$ and concerns about $CO_2$ contamination in the recovered $CH_4$ gas. With advancements in carbon capture technologies and possible government mandates to limit $CO_2$ emissions, EGR is likely to become more economically viable in the future. Like with EOR, the dimensions of the reservoirs can extend to several kilometers in area and can be of hundreds of meters in depth while the $CO_2$ interactions with the in situ $CH_4$, brine, or minerals often occur at the microscopic level. Moreover, the life spans of EGR projects range from a few years to many decades with carbon trapping mechanisms extending to thousands of years [16]. Due to the large physical dimensions and time periods of industrial-scale EGR projects, it is difficult to analyze them in small laboratory-scale or pilot projects. Numerical simulations provide the only a priori alternative for assessing the viability and planning of the industrial scale EGR projects since field testing can only be done a posteriori. The U.S. Department of Energy computational code TOUGH2 is a CFD solver capable of modeling the three-dimensional multiphase, multicomponent flows in porous or fractured media [17]. TOUGH2 was employed by the authors to develop an EGR model which was then optimized by employing a genetic algorithm to maximize the $CH_4$ recovery with a more efficient $CO_2$ injection. Details of the modeling setup for the numerical simulation and optimization of EGR were reported in a previous paper [2].

## 2 Numerical Simulation Methodology

$CO_2$ becomes supercritical above temperatures of 31 °C and pressures of 72.8 atm. While $CO_2$ behaves like any gas at standard temperature and pressure, in its supercritical phase, $CO_2$ adopts properties of both a gas and a liquid [18]. In particular, supercritical $CO_2$ typically has a density close to that of a liquid but viscosity and diffusivity similar to that of a gas. $CO_2$ injection in CCUS systems occurs at elevated temperature and pressure due to the characteristics of the individual oil or natural gas reservoir in which the $CO_2$ is injected. The reservoirs typically occur at depths between 2000 ft. and 15,000 ft., and have temperatures ranging from 100 °F to 250 °F and pressures ranging from a few hundred psi to 5,000 psi. It is therefore expected that the injected $CO_2$ will be supercritical for both EOR and EGR operation.

## 2.1 Governing Equations of Multicomponent Multiphase Flow

Because of the large physical dimensions of the reservoirs and the long time periods for CCUS in industrial scale projects, it is not feasible to analyze them empirically for each oil field of interest. In order to study the effects of $CO_2$ injection in a large reservoir over a long period of time, numerical simulations provide the only feasible alternative. The pertinent processes to be considered for accurately simulating EOR and EGR include the migration of $CO_2$ through the reservoir and the resulting migration of oil, gas, and/or water out of the reservoir depending upon the injection rate and pressure of $CO_2$ as well as the fracture pressure of the formation [1, 2].

The governing multicomponent, multiphase equations of fluid dynamics for subsurface flow can be expressed as

$$\frac{d}{dt}\int_{V_n} M^s dV_n = \int_{\Gamma_n} \mathbf{F}^s \cdot \mathbf{n} d\Gamma_n + \int_{V_n} q^s dV_n \tag{1}$$

where $V_n$ is any control volume, $\Gamma_n$ is the closed surface that bounds the control volume $V_n$, and $\mathbf{n}$ is the normal vector to the surface element $d\Gamma_n$ that points into the volume $V_n$. On the left-hand side of Eq. (1), $M^s$ is the rate of increase of mass or energy in the unit volume with the superscript $s$ denoting the component of mass or energy. On the right-hand side, $\mathbf{F}^s$ is the net mass or energy flux through the surface that bounds the control volume and $q^s$ represents the net mass or energy source within the control volume [19].

The mass accumulation term for each component in Eq. (1) can be written as

$$M^{\kappa} = \phi \sum_{\beta} S_{\beta} \rho_{\beta} X_{\beta}^{s} \tag{2}$$

where $\phi$ is the porosity of the medium, $S_\beta$ is the saturation of the phase $\beta$, $\rho_\beta$ is the density, and $X_\beta^s$ is the mass fraction of the component $s$.

The generalized form of the mass flux is also a sum over the phases, which is given by

$$\mathbf{F}^s = \sum_{\beta} \mathbf{F}_{\beta} X_{\beta}^{s} \tag{3}$$

where the individual phase mass fluxes, $\mathbf{F}_\beta$, are obtained using Darcy's Law for multiple phases.

$$\mathbf{F}_{\beta} = \rho_{\beta} \mathbf{u}_{\beta} = -k \frac{k_{r\beta} \rho_{\beta} g}{\mu_{\beta}} (\nabla P_{\beta} - \rho_{\beta} \mathbf{g}) \tag{4}$$

In Eqs. (3) and (4), $\mathbf{u}_\beta$ is the Darcy velocity in phase $\beta$, $k$ is the absolute permeability, $k_{r\beta}$ is the relative permeability, $\mu_\beta$ is the viscosity, $P_\beta$ is the fluid pressure, and $\mathbf{g}$ is the acceleration due to gravity. $P_\beta$ is obtained as the sum of the pressure $P$ of a reference phase pressure and the capillary pressure,

$$P_\beta = P + P_{cap\beta} \tag{5}$$

Substitution of Eqs. (2)–(4) into Eq. (1) gives the mass balance equation for a multiphase multicomponent flow in a porous media [19].

The heat accumulation term for each component in Eq. (1) can be expressed as

$$M^s = (1-\phi)\rho_R C_R T + \phi \sum_\beta S_\beta \rho_\beta u \tag{6}$$

where $\rho_R$ is the rock grain density, $C_R$ is the specific heat of the rock, $T$ is the temperature, and $u_\beta$ is the specific internal energy of the phase $\beta$.

The heat flux is expressed as the sum of conductive and convective heat flow, which is represented by the first and second term respectively in the following equation:

$$\boldsymbol{F}^s = -\lambda \nabla T + \sum_\beta h_\beta \boldsymbol{F}_\beta \tag{7}$$

where $\lambda$ is the thermal conductivity, $\nabla T$ is the temperature gradient, $h_\beta$ is the specific enthalpy of the phase $\beta$, and $\mathbf{F}_\beta$ is the heat flux of the specified phase $\beta$. As in the case of the mass equation, substituting Eqs. (6) and (7) into Eq. (1) yields the energy equation for a multiphase multicomponent flow in a porous medium [19].

## 2.2 Numerical Simulation and Optimization Software

The EOR simulation presented in this paper utilizes the CFD solver COZSim/COZView developed by Nitec, LLC sponsored by the DOE's National Energy Technology Laboratory [13]. The software is designed to accelerate EOR technical studies and is capable of simulating comprehensive three-dimensional, three-phase, four-component fluid flows in reservoirs. COZSim is the actual simulator and COZView provides a visual interface for pre- and post-processing. The equations described in Sect. 2.1 are used as the governing equations in COZSim. Built-in correlations estimate the molecular weights, fluid properties, and mole fractions based on the specific gravity of oil and hydrocarbon gas. Miscibility calculations are based on interfacial tension using black-oil data. Interfacial tension reduction is used to model the transition from immiscible to partially miscible to fully miscible conditions. Physical dispersion is determined from a Todd-Long staff type model using an interfacial tension function [20]. Simulations of the EGR process are performed in TOUGH2, a numerical program for simulating three-dimensional multicomponent multiphase flow in porous or fractured media. TOUGH2 is supplied with a series of fluid property modules that allow the software to model specific cases related to many geological applications, e.g., $CO_2$ or $N_2$ in natural gas reservoirs, geothermal reservoirs with saline fluids, $CO_2$ storage in saline aquifers, as well as other reservoir and multiphase fluid dynamics applications.

COZSim and TOUGH2 do not have any optimization capability in the original code. An optimization module based on a genetic algorithm (GA) was developed and integrated with these programs to perform optimization studies of $CO_2$ storage [1, 2]. The genetic algorithm is an optimization technique that mirrors natural biological evolution. The algorithm begins with a generation of individuals modeled as a set of vectors to form an input. These individuals represent a specific sequence of data or characteristics that help to differentiate them. Each individual is then evaluated according to a fitness value determined by the user for their reproduction. A portion of the weaker individuals (those with less desirable fitness values) is deleted from the generation and the stronger individuals (those with more desirable fitness values) are kept. New offspring is produced from the characteristics of the surviving parents and then their fitness values are again evaluated. This cycle repeats for many generations over time, much like biological evolution, until an optimal individual is found.

When all the solutions in a given generation of genetic algorithm have similar fitness values, the genetic algorithm is considered to have converged to an optimal value. The GA described in this chapter has been modeled after the research performed by Zhang [21]. The approach for the optimization methodology for EOR simulations is to create multiple input files based on fitness functions (pre-simulation processing), run COZSim, evaluate the new fitness functions resulting from the simulations (post-simulation processing), and generate a new generation of inputs based on this data [1]. A flowchart of the GA-integrated COZSim solver is shown in Fig. 1.

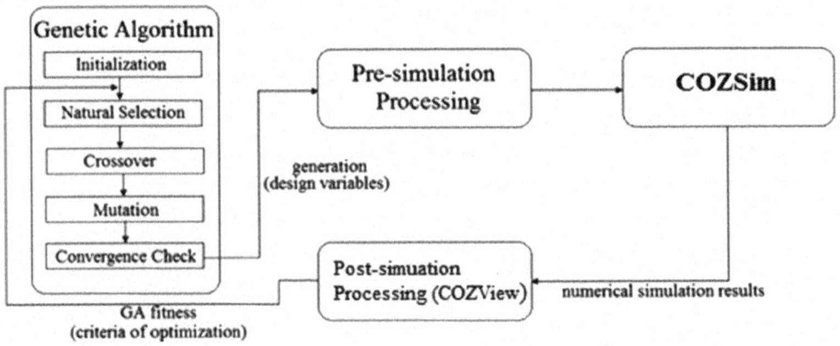


**Fig. 1** Flowchart of GA-COZSim [1]

## 3 Model Development and Optimization of a Benchmark EOR Problem

In EOR, the supercritical $CO_2$ injected into the partially depleted reservoir displaces a significant volume of oil that had not been economical to extract in the past. The fundamental problem in simulating an EOR system is to determine the total amount of liquid oil that can be recovered from a reservoir through the injection of $CO_2$ and the corresponding amount of $CO_2$ that can be injected and stored in the reservoir. The integration of the GA-based optimization methodology with the numerical solver COZSim is first demonstrated through a benchmark problem [22]. The optimization technique is later applied to a currently operational reservoir in the Permian Basin in Texas. The model for both reservoirs follows the most commonly deployed five-spot injection-production pattern. It consists of a central injection well surrounded by four production wells as shown in Fig. 2. Because of the symmetry, only one-fourth of the reservoir needs be modeled. The computational domain is divided into three layers of 25 ft. thickness each. The injection of $CO_2$ into the reservoir occurs in the lowest layer at 5,000 MSCF/day and the oil production occurs at outer corners in all three layers. This setup reduces the $CO_2$ production while attaining maximum oil production through all layers. The hydrogeological properties of the benchmark reservoir and other parameters required to generate the input file for COZSim are given in Table 1.

The simulation is terminated when gas to oil ratio (GOR), which is the ratio of the volume of gas coming out of the production wells to the volume of oil recovered exceeds 5,000, meaning that there is substantial contamination in the production stream due to outflow of $CO_2$ from the well. The production well shut-in time corresponds to this specified GOR condition. At the well shut-in time, the total oil recovery factor of the reservoir is calculated by the formula [1]:

$$\text{Recovery factor} = \frac{\text{Volume of oil recovered}}{\text{Original volume of oil place}}. \quad (8)$$

**Fig. 2** Reservoir with a five-spot injection pattern [1]

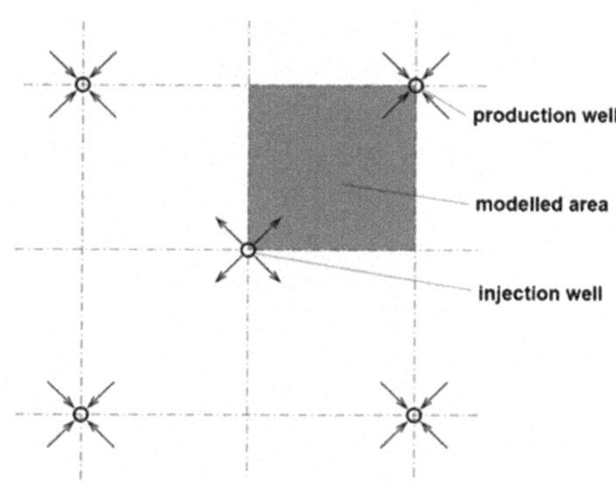

**Table 1** Physical properties and modeling parameters for the benchmark reservoir [1]

| | |
|---|---|
| Dimensions of 5-spot domain | 1320 ft. × 1320 ft. |
| Reservoir thickness | 75 ft. |
| Porosity | 0.2 |
| Reservoir temperature | 125°F |
| Initial pressure at –4500 ft. | 1500 psia |
| Initial bubble point pressure | 800 psia |
| Horizontal permeability | 50 mD |
| Vertical permeability | 5 mD |
| Initial $CO_2$ volume fraction | 0.0% |
| Initial oil volume fraction | 41.0% |
| Initial water volume fraction | 56.2% |
| Initial natural gas volume fraction | 2.8% |
| Reservoir volume | 4,636,414 bbl |
| Oil gravity | 36.3 API |
| Water salinity | 45,000 ppm |
| HC gas specific gravity | 0.7 |
| Minimum reservoir elevation | –4,513 ft. |
| Maximum reservoir elevation | –4,607 ft. |
| BHP of recovery well | 1,500 psia |
| Maximum production rate | 1,600 STB/day |
| Minimum oil rate | 5 STB/day |
| Model type | 2 phase |
| Boundary conditions | No mass flow at boundaries |

In order to increase the economic incentive to implement EOR systems in industry, it is necessary to increase the recovery factor to increase the amount of oil that can be sold for profit and to decrease the well shut-in time to reduce the cost of well operation. The genetic algorithm integrated with COZSim is employed to determine the optimal recovery factor for oil for this benchmark EOR problem for a constant volume injection rate of $CO_2$. Later, the GA code is used to optimize a variable injection scenario for $CO_2$ to maintain a constant bore-hole pressure (BHP) at the bottom of the injection well.

Before the benchmark EOR problem can be properly optimized, the effects of the injection rate on few key parameters are determined by performing a series of COZSim simulations with different $CO_2$ injection rates while keeping all other reservoir conditions the same as in Table 1. Given the low permeability of the reservoir, it is not expected that the optimal rate of $CO_2$ injection will be greater than 5,000 MSCF/day, providing the upper limit for the parametric study. Rates higher than 5,000 MSCF/day are not only economically unrealistic for a reservoir of this size, but will also cause a substantial increase in the BHP, which may possibly result in the reservoir fracture. The lower limit was established at 100 MSCF/day based on the prior literature. Figure 3 shows the effect of increasing the

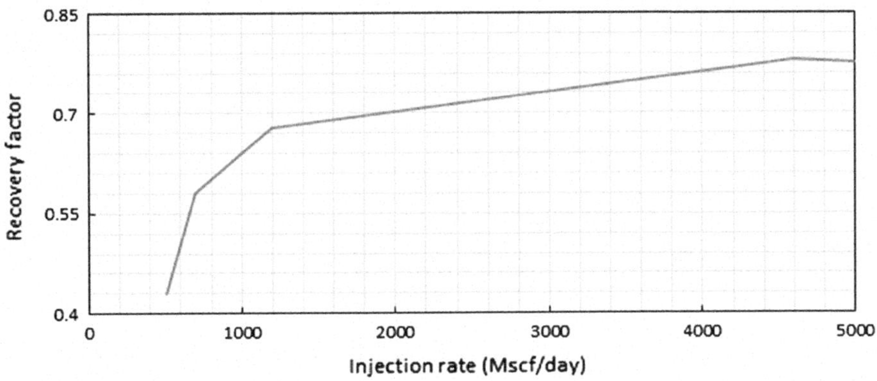


**Fig. 3** Recovery factor as a function of the $CO_2$ injection rate for the benchmark reservoir [1]

injection rate on the recovery factor. A total of five "brute force" simulations were conducted which were later used for comparison with the GA based optimization results.

From Fig. 3, it can be seen that increasing the injection rate from 1200 to 4600 MSCF/day increases the recovery factor by 6% but a further increase to 5000 MSCF/day reduces the recovery factor by 1.4%. This is because the higher injection rates have a shorter life cycle before the gas to oil ratio reaches 5,000 and the well shut-in condition is activated so that the total recovered oil is higher for the $CO_2$ injection rate of 4600 MSCF/day. It is clear from Fig. 3 that the maximum recovery peaks between 100 and 5000 MSCF/day somewhere close to 4600 MSCF/day [1].

Figure 4 shows the oil production efficiency against the total simulation time for the three $CO_2$ injection rates of 1200, 4600, and 5000 MSCF/day. The oil production efficiency is given by [1]:

$$\text{Oil production efficiency} = \frac{\text{Total volume of oil recovered}}{\text{Total volume of } CO_2 \text{ injected}}. \tag{9}$$

There are two notable characteristics demonstrated by Fig. 4. First, for all $CO_2$ injection rates, the production efficiency increases up to a point and then plateaus. The end points of the line plots in Fig. 4 correspond to the well shut-in condition. If the simulations were to continue past the established GOR limit, the production efficiency would begin to dip. Second, the efficiency for the lowest injection rate of 1,200 MSCF/day starts out lower than the higher injection rates but eventually exceeds them. This shows that the most efficient injection level in terms of oil output per unit of $CO_2$ injected does not produce the most oil over the lifecycle of the reservoir. Figure 4 also highlights the rationale behind the GOR limit of 5,000. At this ratio, there is no longer enough oil produced to warrant continued $CO_2$ injection.

The bore-hole pressures over time of the injection well for the three $CO_2$ injection rates are shown in Fig. 5. The fracture pressure of a reservoir is the

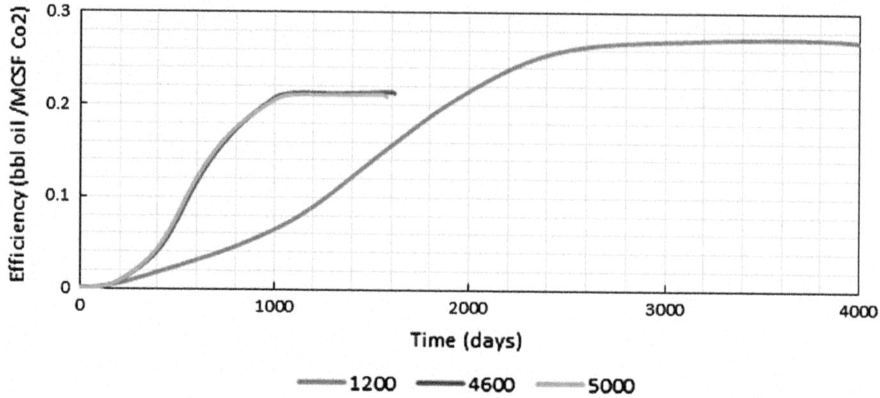


**Fig. 4** Oil production efficiency over well life cycle for three $CO_2$ injection rates [1]

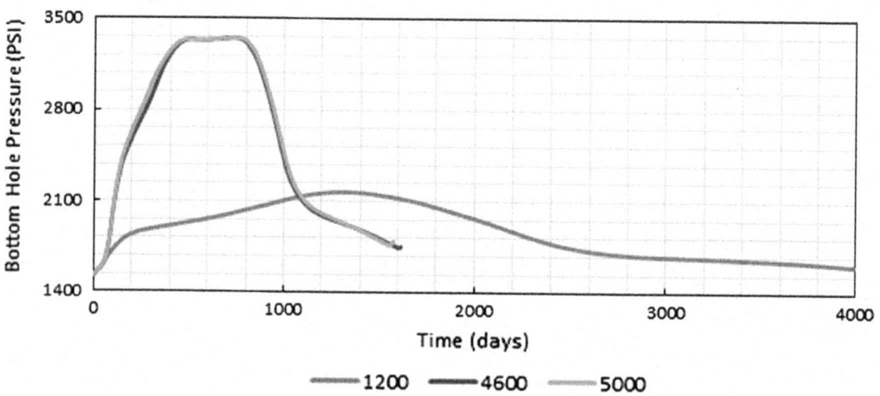


**Fig. 5** Bore-hole pressure (BHP) of injection well over well life cycle for three $CO_2$ injection rates [1]

pressure at which the rock can fracture, changing the porosity and permeability of the rock and potentially leading to $CO_2$ contamination. For the initial optimization, the BHP is not constrained. Later, constraints are added to keep the reservoir pressure below the fracture pressure.

### *3.1 Optimization of Recovery Factor for a Constant Injection Rate*

Since each injection rate results in a different $CO_2$ migration pattern, using one specific time measurement for the calculation of the recovery factor is unreliable. The concentration of $CO_2$ can be continuously monitored in the production well of

**Table 2** Parameters used in the genetic algorithm optimization [1]

| | |
|---|---|
| Individuals per generation | 4 |
| Maximum number of generations | 20 |
| Natural selection algorithm | 50% |
| Mutation rate | 8% |
| Cross-over algorithm | Semi-random combination of parents |

the reservoir in COZView. At the time of the well shut-in, corresponding to a GOR of 5,000, the approximate recovery factor of the well can be determined, and the resulting fitness function can be calculated for input into the GA. The parameters used for GA optimization for the benchmark case are given in Table 2.

The genetic algorithm achieves convergence after approximately five generations. The optimum injection rate is 4,592 MSCF/day corresponding to a recovery factor of 78.03%. This is higher than the recovery factor of 77.62% for an injection rate of 4,600 MSCF/day obtained from the "brute force" simulations for various injection rates presented in Fig. 3, indicating that the optimization was successful. For a better understanding of the $CO_2$ plume migration in the reservoir, graphical representations of $CO_2$ mass fraction at various times for the injection rate of 1,200 MSCF/day and the optimal injection rate of 4,592 MSCF/day are shown in Fig. 6 [1]. It should be noted that Fig. 6 shows 12 years of $CO_2$ injection for both injection rates for the sake of comparison; this injection time is significantly longer than necessary to achieve a GOR of 5,000 for the optimized injection rate.

## *3.2 Optimization of Recovery Factor for a Pressure-Limited System*

The GA-optimized constant injection rate can increase the recovery factor slightly. However, constant injection systems are susceptible to elevating the BHP at the injection well to a level above the formation fracture pressure. An improved optimization of the EOR system should consider a time-variant injection rate based on the injection well BHP.

The maximum pressure within the reservoir for the benchmark problem is constrained to 2,500 psia in COZSim. All other reservoir characteristics and constraints are kept the same as for the constant injection system in Sect. 3.2. The effect of varying the injection rate is investigated in the context of a pressure-limited system and the results are summarized in Table 3.

The results in Table 3 for a pressure-limited $CO_2$ injection follow the same trends as the results for the constant injection rates discussed in Sect. 3.1. As the initial injection rate increases, the well shut-in time decreases. Furthermore, there is a similar initial increase in the recovery factor as the $CO_2$ injection rate increases, followed by a slight decrease as the initial injection rate reaches the maximum value of 5000 MSCF/day. It should be noted that even though there is approximately 9%

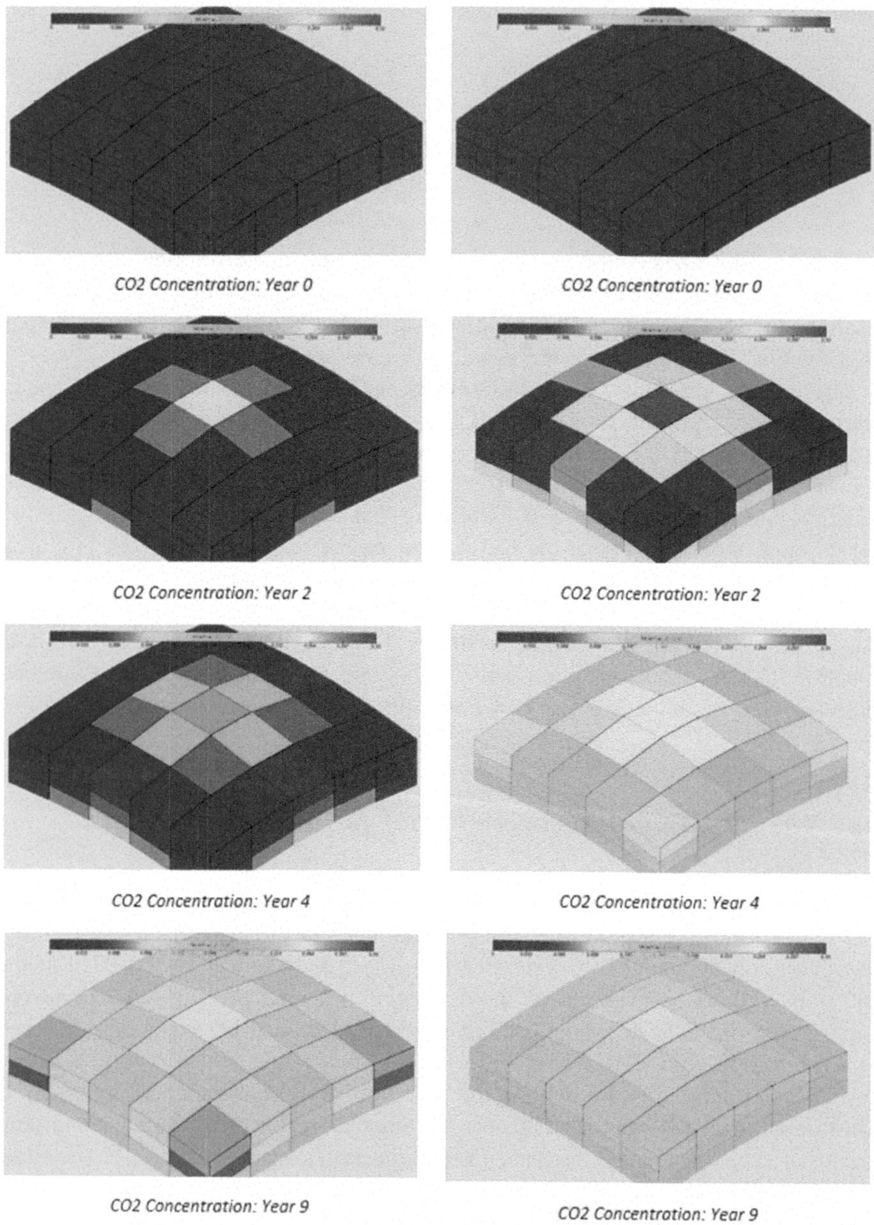


**Fig. 6** Comparison of $CO_2$ plume migration in time for the injection rates of 1200 MSCF/day (*left*) and the optimized injection rate of 4592 MSCF/day (*right*) [1]

difference between the initial $CO_2$ injection rates of 4592 and 5000 MSCF/day, the total volume of $CO_2$ injected differs by less than 1.5%, which implies that the pressure significantly constrains the subsequent injection rate in the 5000

**Table 3** Simulation results for the pressure-limited $CO_2$ injection with varying initial injection rates [1]

| Initial $CO_2$ injection rate (MSCF/day) | Oil recovered (bbl) | Recovery factor (%) | Well shut-in time (days) | Total $CO_2$ injected (MSCF) |
|---|---|---|---|---|
| 500 | 802,905 | 42.23 | 5475 | 2,750,207 |
| 1200 | 1,312,162 | 69.02 | 4015 | 4,800,075 |
| 4592 | 1,470,498 | 77.35 | 2337 | 6,278,021 |
| 5000 | 1,469,240 | 77.29 | 2325 | 6,367,122 |

**Table 4** Optimized result for the pressure-limited injection [1]

| | |
|---|---|
| Maximum $CO_2$ injection rate (MSCF/day) | 4,826 |
| Oil recovered (bbl.) | 1,471,398 |
| Recovery factor (%) | 77.40 |
| Well shut-in time (days) | 2729 |
| Maximum reservoir pressure (psia) | 2240 |
| Total $CO_2$ injected (MSCF) | 6,352,891 |
| Total $CO_2$ sequestered (MSCF) | 3,990,927 |

MSCF/day case. Therefore, it can be concluded that the optimal initial injection rate for a pressure-limited injection will fall between 4592 MSCF/day and 5000 MSCF/day [1].

In the optimization process, the injection pressure is set by limiting the BHP of the injection well to a maximum of 2500 psia to keep it consistent with the rest of the reservoir. The time-varying injection rate at any time during the simulation is determined by the simulator based on the BHP at the injection well. The simulation is run until the GOR of 5000 is achieved or after 15 years, whichever occurs first. The optimized results for the pressure-limited injection are presented in Table 4.

Figures 7 and 8 show the average reservoir pressure and the instantaneous injection rate over the well life cycle. The average pressure peaks at around 1100 days, which is compensated by a corresponding decrease in the $CO_2$ injection rate as shown in Fig. 8. It should be noted that $CO_2$ injection rate starts to decrease before the maximum average reservoir pressure reaches the peak because the injection level is constrained by the BHP at the injection well and not by the average reservoir pressure. As more oil is recovered, both the injection well BHP and the average reservoir pressure decrease and the $CO_2$ injection level starts to increase until its maximum optimized level of 4852 MSCF/day is reached. The final recovery factor is higher than the constant volume injection rate due to a higher total volume of $CO_2$ injected. Because the pressure-limited system decreases the injection to regulate pressure, the GOR is reduced and the system life cycle is extended. Despite the lower initial injection level, the pressure-limited system can produce more oil and sequester a larger amount of $CO_2$ [1].

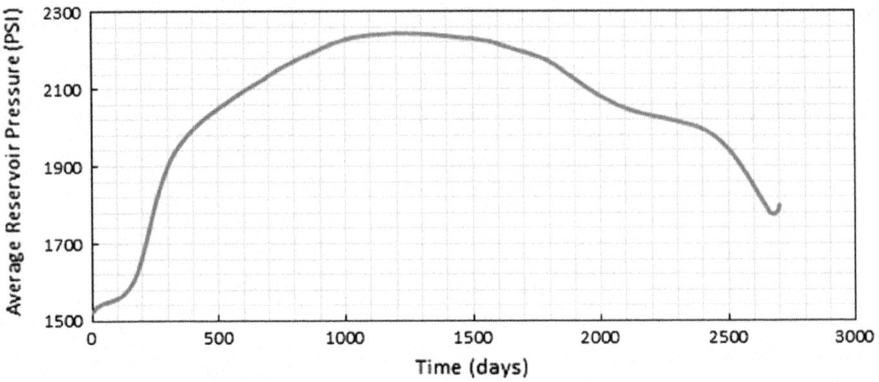


**Fig. 7** Average reservoir pressure for the optimized pressure-limited injection rate [1]

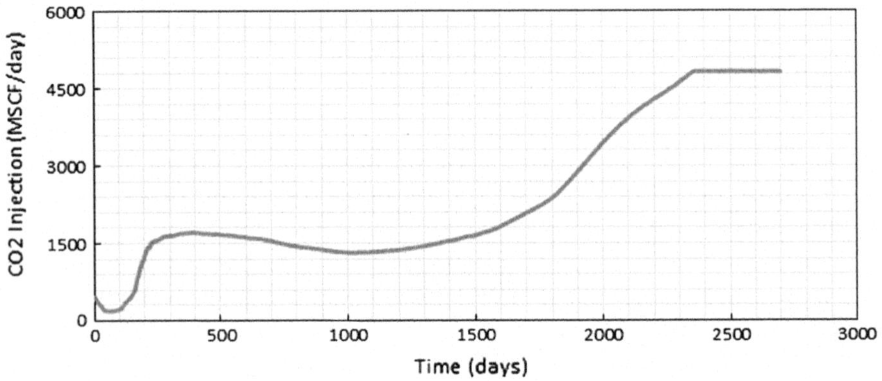


**Fig. 8** Instantaneous $CO_2$ injection rate for the optimized pressure-limited injection case [1]

# 4 Simulation and Optimization of a Permian Basin Reservoir

Using a similar method as used for the benchmark problem, a model was developed using data from a real industrial scale reservoir in the Permian Basin. The properties of the Permian basin reservoir are obtained from the work of Dai et al. [23] and are outlined in Table 5.

To determine the accuracy of the COZSim model for the Permian basin reservoir, the results are compared to the published data of Dai et al. [23]. In their study, the effect of well spacing between the injection and production well on oil recovery was examined. The data from three discrete well spacing distances are taken from their study and are compared to the results from COZSim in terms of the total oil production. In both studies, the gas is recycled so the well-shut in time is not dependent on the GOR. The results between the two studies are compared below in Table 6.

**Table 5** Physical properties and modeling parameters for the Permian Basin reservoir [1]

| | |
|---|---|
| Dimensions of 5-spot domain | 1000 ft. × 1000 ft. |
| Reservoir thickness | 100 ft. |
| Porosity | 0.12 |
| Reservoir temperature | 235°F |
| Initial pressure at −4500 ft | 1500 psia |
| Initial bubble point pressure | 800 psia |
| Horizontal permeability | 150 mD |
| Vertical permeability | 15 mD |
| Initial $CO_2$ volume fraction | 0.0% |
| Initial oil volume fraction | 46.21% |
| Initial water volume fraction | 46.73% |
| Initial natural gas volume fraction | 7.06% |
| Reservoir volume | 2,137,285 bbl |
| Oil gravity | 40.7 API |
| Water salinity | 70,000 ppm |
| HC gas specific gravity | 0.7 |
| Minimum reservoir elevation | −5600 ft. |
| Maximum reservoir elevation | −5500 ft. |
| BHP of recovery well | 1500 psia |
| Maximum production rate | 1600 STB/day |
| Minimum oil rate | 5 STB/day |
| Model type | 2 phase |
| Boundary conditions | No mass flow at boundaries |

The COZSim results in Table 6 show excellent agreement with the results of Dai et al. [23]. Slight discrepancies between the results of the two models can be attributed to the grid size limitations in COZSim, imperfect replications of the KR-GO and KR-WO saturation tables, and the heterogeneity of the reservoir rock characteristics examined by Dai et al. [23] that are not present in the COZSim model.

The COZSim model of the Permian basin reservoir examines a domain 1000 ft. by 1000 ft. to allow for a diagonal distance of approximately 1400 ft. between the wells and has a depth of 100 ft. The computational domain is discretized into four layers with a thickness of 25 ft. each and is discretized into a 5 by 5 grid of 20 ft. for a total of 125 discrete cells. An injection well and a production well are placed on opposite corners, and the wells are perforated in all four layers. The maximum

**Table 6** Comparison of total oil production for the Permian Basin reservoir from Dai et al. [23] and COZSim [1]

| Well distance (m) | 250 | 300 | 350 |
|---|---|---|---|
| Dai et al. [23] (MMbbl) | 1.4 | 1.8 | 1.7 |
| COZSim results (MMbbl) | 1.32 | 1.73 | 1.68 |

injection rate is set at 5000 MSCF/day and the simulation is allowed to run until a GOR of 5000 is reached.

## 4.1 *Optimization of the Permian Basin Reservoir*

The reservoir performance is evaluated for four discrete depletion levels: 30%, 45%, 60%, and 80% of the original oil in place (OOIP) of 1,646,850 bbl, each corresponding to a different height of the water-oil contact (WOC). It should be noted that the benchmark problem was simulated for 60% oil in place (OIP). Optimization was conducted for both constant-mass injection and a pressure-limited injection for each of the four depletion levels. The optimization process for the Permian Basin reservoir is the same as that for the benchmark EOR problem discussed in Sect. 3. The effect of recycling the $CO_2$ gas that is recovered along with the oil at the production well was also investigated. Gas recycling serves to reduce the large difference between the total volumes of $CO_2$ injected and sequestered allowing for a more effective CCUS system. It should be noted that because of differences in reservoir properties, the pressure-limited injection was set to a higher value of 3000 psia for the injection well BHP to be in line with the formation fracture pressure of this reservoir. Finally, for each of the above conditions, gas recycling levels of 0% (no recycling) and 100% were considered. A summary of each of the 16 optimized simulations and the optimal injection profiles is given in Table 7. It should be noted

**Table 7** Optimized injection profiles for the Permian Basin Reservoir [1]

| Reservoir conditions | | | Optimized injection rate |
|---|---|---|---|
| Oil in place | Injection profile | Gas recycling | $CO_2$ injection rate (MSCF/day) |
| 30% | Constant | No recycling | 4,592 |
| | | Forced 100% recycling | 2,370 |
| | Pressure-limited | No recycling | 5,000 |
| | | Forced 100% recycling | 2,486 |
| 45% | Constant | No recycling | 4,827 |
| | | Forced 100% recycling | 1,785 |
| | Pressure-limited | No recycling | 1,785 |
| | | Forced 100% recycling | 2,085 |
| 60% | Constant | No recycling | 4,592 |
| | | Forced 100% recycling | 2,947 |
| | Pressure-limited | No recycling | 4,592 |
| | | Forced 100% recycling | 2,934 |
| 80% | Constant | No recycling | 4,592 |
| | | Forced 100% recycling | 5,000 |
| | Pressure-limited | No recycling | 3,664 |
| | | Forced 100% recycling | 2,690 |

**Table 8** Optimized output results for the Permian Basin Reservoir [1]

| Reservoir conditions | | | Optimization results | | | | |
|---|---|---|---|---|---|---|---|
| Oil in place | Injection profile | Gas recycling | Cumulative oil output (bbl) | Percent OIP recovered | Percent OOIP recovered | Total $CO_2$ sequestered (MSCF/day) | Production efficiency |
| 30% | Constant | No recycling | 165,615 | 28.86 | 10.06 | 866,434 | 0.15 |
| | | 100% Recycling | 175,244 | 29.54 | 10.64 | 895,749 | 0.16 |
| | Pressure-limited | No recycling | 168,003 | 29.28 | 10.20 | 829,923 | 0.14 |
| | | 100% recycling | 168,805 | 29.42 | 10.25 | 895,489 | 0.16 |
| 45% | Constant | No recycling | 262,705 | 35.21 | 15.95 | 1,010,206 | 0.17 |
| | | 100% recycling | 273,321 | 36.63 | 16.60 | 895,489 | 0.26 |
| | Pressure-limited | No recycling | 261,116 | 34.99 | 15.86 | 1,044,824 | 0.28 |
| | | 100% recycling | 280,112 | 37.54 | 17.01 | 828,538 | 0.25 |
| 60% | Constant | No recycling | 425,594 | 43.61 | 25.84 | 1,184,214 | 0.26 |
| | | 100% recycling | 432,823 | 44.35 | 26.28 | 1,013,056 | 0.25 |
| | Pressure-limited | No recycling | 427,703 | 43.83 | 25.97 | 1,214,214 | 0.28 |
| | | 100% recycling | 432,411 | 44.31 | 26.26 | 842,321 | 0.26 |

(continued)

**Table 8** (continued)

| Reservoir conditions | | | Optimization results | | | | |
|---|---|---|---|---|---|---|---|
| Oil in place | Injection profile | Gas recycling | Cumulative oil output (bbl) | Percent OIP recovered | Percent OOIP recovered | Total $CO_2$ sequestered (MSCF/day) | Production efficiency |
| 80% | Constant | No recycling | 769,253 | 58.25 | 46.71 | 1,335,188 | 0.36 |
| | | 100% recycling | 752,786 | 57.00 | 45.71 | 1,310,207 | 0.37 |
| | Pressure-limited | No recycling | 752,786 | 57.00 | 45.71 | 1,060,022 | 0.41 |
| | | 100% recycling | 735,039 | 55.66 | 44.63 | 998,237 | 0.45 |

that the $CO_2$ injection rate for the pressure-limited injection represents the final instantaneous injection rate after it starts out at a much lower value as shown in Fig. 8 to ensure that the reservoir pressure does not exceed the provided threshold (fracture pressure). The optimal output results for the 16 cases are summarized in Table 8.

## 5 Model Development and Optimization of a Benchmark EGR Problem

The predominant reservoir fluid in an EGR system is $CH_4$ while brine or other aqueous solutions may also be present. It is expected that there will be complex interactions between $CH_4$ and $CO_2$ as well as interactions between the reservoir fluids and the surrounding rocks. As a result, it is essential to consider the dynamics of multicomponent, multiphase flow in modeling and simulation of EGR process. The pertinent processes that must be accurately modeled include the migration of $CO_2$ through a reservoir and the resulting migration of $CH_4$ out of the reservoir, as well as the injection pressure in the system. In this section, a numerical model is developed of an EGR scenario as defined by a benchmark problem proposed in a workshop at the University of Stuttgart, Germany [24]. The total amount of $CH_4$ that can be extracted from the reservoir through the injection of $CO_2$ is investigated as well as the storage of $CO_2$ in the depleted reservoir [2]. It should be noted that the displacement of $CH_4$ by $CO_2$ is assumed to be non-miscible as is common practice for such simulations [24].

The numerical model follows the common five-spot injection-production pattern as shown in Fig. 2. Because of the symmetry of the five-spot pattern, only one-fourth of the reservoir is modeled. The hydrogeological properties of this reservoir are summarized in Table 9. The computational domain is discretized into 10 layers each with a thickness of 15 ft. The injection of $CO_2$ into the reservoir occurs in the lowest layer at 0.4 kg/s, which translates into 0.1 kg/s for one-quarter of the five-spot domain; the production occurs at the opposite corner in the highest layer of the reservoir. This setup avoids substantial gas mixing of the $CO_2$ and $CH_4$ and ensures a better sweep efficiency of $CO_2$ through the reservoir. The simulation is terminated when there is 20% (by mass) $CO_2$ contamination in the production stream signifying a substantial $CO_2$ outflow from the production well. The production well shut-in time corresponds to this breakthrough condition. At the well shut-in time the total gas recovery factor of the reservoir is evaluated as [2]:

$$\text{Recovery factor} = \frac{\text{Mass of CH}_4\text{ recovered}}{\text{Initial mass of CH}_4}. \tag{10}$$

From the benchmark simulation, it was predicted that the production well shut-in time is about 1922 days with a natural gas recovery factor of about 58%. Figure 9 shows the predicted gas production. As can be seen, the steady displacement of

**Table 9** Geometric and hydrogeological properties for the generic gas reservoir [2]

| | |
|---|---|
| Dimensions of quarter five-spot domain | 201.19 m × 201.19 m |
| Reservoir thickness | 45.72 m |
| Porosity | 0.23 |
| Brine saturation | 0 |
| Reservoir temperature | 66.7 °C |
| Initial reservoir pressure | 35.5 bar |
| Coefficient of molecular diffusion | $6 \times 10^{-7}$ m$^2$/s |
| Horizontal permeability | $50 \times 10^{-15}$ m$^2$/s |
| Vertical permeability | $5 \times 10^{-15}$ m$^2$/s |
| Boundary conditions | No mass flow at boundaries |
| Initial $CO_2$ mass fraction | $X_{CO_2} = 0$ |
| Initial $CH_4$ mass fraction | $X_{CH_4} = 1$ |

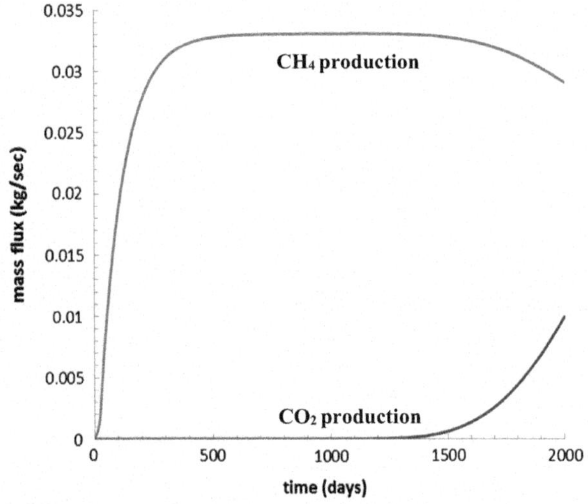


**Fig. 9** Production mass flux of $CH_4$ and $CO_2$ in the benchmark EGR simulation [2]

natural gas is established in roughly 400 days, and $CO_2$ breakthrough occurs at about 1350 days into the EGR operation. The simulation of the baseline case quantifies the extraction of $CH_4$ and the $CO_2$ storage in order to provide a reasonable estimate of the recovery factor. It is desirable to increase the recovery factor and/or delay the well shut-in time via optimization studies in order to provide a greater economic incentive for the implementation of EGR systems in industry.

The integrated GA-TOUGH2 code was used to optimize the recovery factor for a constant $CO_2$ mass-injection rate [2]. First, the key parameters that affect the recovery factor were determined by performing a series of TOUGH2 simulations at different injection rates. For these initial studies, all reservoir conditions of the benchmark problem were kept the same except the injection rate, which was varied at specified increments. Because of the low permeability and other geometric

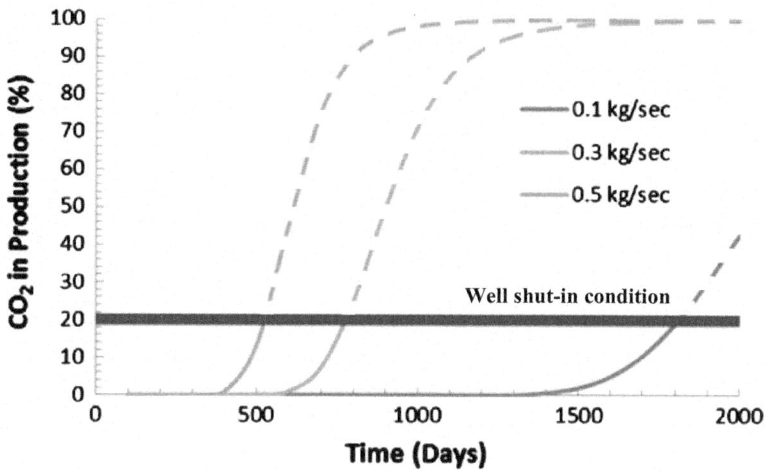


**Fig. 10** Mass percentage of $CO_2$ in production stream [2]

parameters of the reservoir, it was not expected that an optimal value of injection rate would be greater than 0.5 kg/s since a high injection rate would result in substantial pressure increase in the reservoir causing the reservoir to fracture leading to an unstable injection. A lower threshold for the injection rate was determined at 0.1 kg/s based on the parameters of the benchmark $CO_2$-EGR problem presented in Sect. 3. One additional injection rate of 0.3 kg/s was selected between the upper and lower bounds to obtain insight into the effect of change in the injection rate. These three brute force simulations were again used for comparison against the optimization results from GA-TOUGH2.

Figure 10 plots the amount of $CO_2$ contamination in the production stream against the simulation time in days. The thicker red line marks the "well shut-in" condition, which occurs at 20% $CO_2$ contamination (by mass) in the production stream. From Fig. 10, for injection rates of 0.1 kg/s, 0.3 kg/s, and 0.5 kg/s respectively, the recovery factor is 58.3%, 62.0%, and 61.3%, and the well shut-in time is 494 days, 762 days, and 1830 days respectively. That the recover factory first improved but later dropped as the injection rate was monotonically increased supports the conjecture that there is a maximum value for the recovery factor between the injection rates of 0.1 and 0.5 kg/s. Figure 10 indicates that $CO_2$-EGR operations with higher injection rates have significantly shorter life cycles than the baseline injection rate of 0.1 kg/s, which explains why the scenario with 0.5 kg/s injection rate has a lower recovery factor than that of the scenario with 0.3 kg/s injection rate [2].

Figure 11 shows a plot of the production efficiency versus the total simulation time. The production efficiency is determined by the relation [2]:

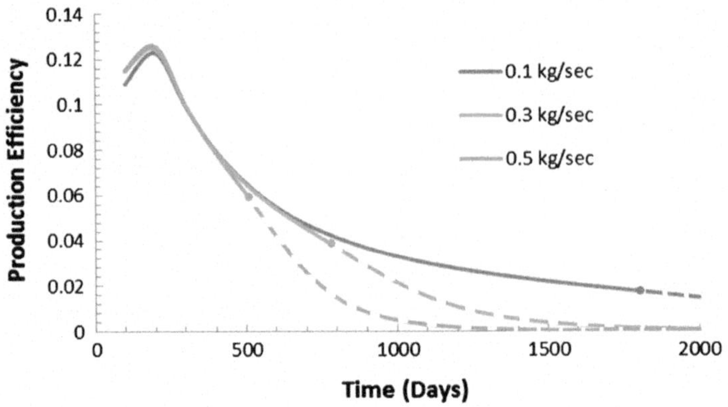


**Fig. 11** Production efficiency of $CH_4$ production [2]

$$\text{Production efficiency} = \frac{\text{Mass of } CH_4 \text{ recovered}}{\text{Initial mass of } CO_2 \text{ injected}}. \quad (11)$$

The three injection rate cases shown in Fig. 11 have very similar curves for their production efficiency until their respective well shut-in time. After well shut-in, there is little $CH_4$ produced for the amount of $CO_2$ injected and thus the production efficiency plunges. Comparing Figs. 11 to 10, the relationship between production efficiency and the well shut-in time can be clearly noted. For the 0.5 kg/s injection rate, 20% $CO_2$ contamination occurs in approximately 494 days, and as expected the production efficiency begins to decrease at this point. For the 0.3 kg/s and 0.1 kg/s injection rates, 20% $CO_2$ contamination occurs and production efficiency begins to decrease in about 762 and 1830 days respectively. Figure 11 highlights the rationale behind the decision to close the production well at 20% $CO_2$ contamination. At this stage of extraction, there is no longer enough $CH_4$ being recovered to warrant continued $CO_2$ injection [2].

In an EGR system with low injection rates, there is very limited mixing between the injected $CO_2$ and the in situ $CH_4$ [25]. A low injection rate will ensure that $CO_2$ migrates through the reservoir as a plume and will not spread in an irregular fashion due to diffusion or high pressures. Based on the results from the three injection cases shown in Fig. 11, a low injection rate also means a lower production rate of $CH_4$. On the other hand, a higher injection rate of $CO_2$ will lead to a higher production rate of $CH_4$ but the reduced well shut-in time may lead to a net reduction in the recovery factor. The higher injection rates of $CO_2$ can also cause a separation of the plume, which can lead to a premature breakthrough of $CO_2$ at the production well, further reducing the life expectancy of the extraction. These results imply that there exists a $CO_2$ injection rate between the lower and upper limits of 0.1 and 0.5 kg/s that will lead to a relatively higher recovery factor while maintaining a low $CO_2$ contamination. The GA-TOUGH2 code was employed to determine this optimum injection rate. Parameters used for this optimization study are given in

**Table 10** Parameters used in GA-TOUGH2 for the constant rate injection scenario [2]

| | |
|---|---|
| Individuals per generation | 6 |
| Maximum number of generations | 50 |
| Natural selection algorithm | 50% |
| Mutation rate | 8% |
| Cross-over algorithm | Semi-random combination of parents |

Table 10. Since each injection rate results in a different $CO_2$ migration pattern, one cannot rely on one specific time measurement for the calculation of the recovery factor. The concentration of $CO_2$ must be continuously monitored in the production cell of the reservoir. At the time of well shut-in, the approximate recovery factor of the well can be determined and the resulting fitness function can be calculated by GA-TOUGH2 [2].

Figure 12 depicts the evolution of the optimal solution in successive generations of genetic algorithm. It can be seen that GA-TOUGH2 achieves convergence after 5 generations; the corresponding injection rates are shown in the inset and can be seen to converge into the optimum value of 0.294 kg/s. The recovery factor based on the optimum injection rate is compared to the results obtained from the initial brute force TOUGH2 simulations for injection rates of 0.1, 0.3, and 0.5 kg/s in Table 11. As expected, the recovery factors for injection rates of 0.294 kg/s and 0.3 kg/s are nearly the same, confirming the validity of the GA-TOUGH2 results. Figure 13 shows plots of the $CO_2$ mass fraction for the benchmark injection rate of 0.1 kg/s and the optimal injection rate of 0.294 kg/s at various times from 100 days to 2000 days. A dramatic change in the $CO_2$ plume migration in the reservoir with an increased injection rate is evident. It is important to note that the sustained injection of $CO_2$ at 0.294 kg/s past the well shut-in time of 772 days would not be continued in actual $CO_2$-EGR operation; it is shown in Fig. 13 for the purpose of comparison only [2].

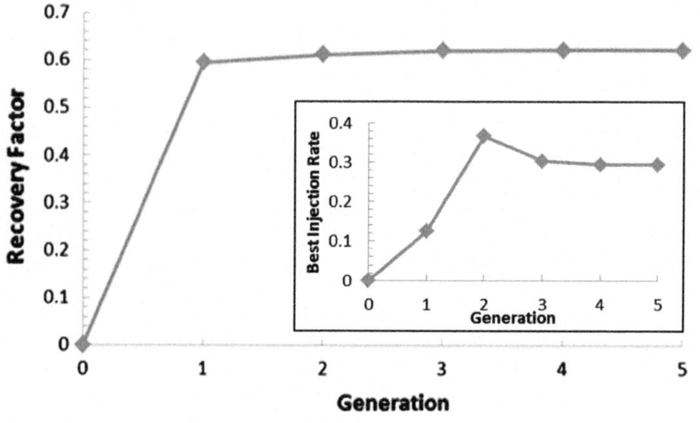


**Fig. 12** GA-TOUGH2 convergence history and the evolution of recovery factor with injection rates [2]

**Table 11** Comparison of recovery factors for different injection rates [2]

| Injection rate (kg/s) | Recovery factor (%) | Well shut-in time (days) |
|---|---|---|
| 0.100 | 58.6 | 1830 |
| 0.294 (optimum) | 62.1 | 772 |
| 0.300 | 62.0 | 762 |
| 0.500 | 61.3 | 494 |

Comparisons of the $CH_4$ production rate and the injection well pressure for the baseline $CO_2$ injection rate of 0.1 kg/s and the optimal injection rate of 0.294 kg/s are shown in Fig. 14. Based on Fig. 14, the optimal injection rate yields a significantly higher production rate of methane alongside the improved recovery factor and shorter well shut-in time as indicated in Table 11. Thus the necessity of optimization studies to improve the overall performance of CS-EGR operations is clearly demonstrated [2].

## 6 Optimization for Constant Pressure Injection Scenario for EGR

The recovery of $CH_4$ is improved significantly by optimizing the injection rate but the effect of the injection rate on the reservoir pressure is not taken into account. This is an important consideration since an injection rate that can result in reservoir pressure greater than its fracture pressure will not be acceptable due to safety concerns. The primary technique to control the bottom well pressure at the injection well is to achieve a constant pressure injection (CPI) scenario by varying the injection rate in time. The reservoir pressure for the baseline constant $CO_2$ injection rate of 0.1 kg/s in Sect. 5 reaches a maximum pressure of 38.4 kPa by the end of the injection. To increase the production of $CH_4$, an ideal injection profile would reach this maximum reservoir pressure very quickly and then taper-off to maintain this pressure throughout the life of the injection, allowing for a larger total volume of $CO_2$ injected [25].

The optimization of a CPI operation is an extension of the original GA-TOUGH2 simulation. A threshold pressure is chosen as an input into the program and the optimization adjusts the mass injection rate to best suit the pressure constraint. The fitness is defined as the difference between the threshold pressure (e.g., the fracture pressure) and the injection pressure, which is a function of the injection rate [2]:

$$\text{Fitness function} = \frac{|\text{Threshold pressure} - \text{Injection pressure}|}{\text{Mass injection rate of } CO_2} \tag{12}$$

If the entire injection life cycle is discretized and the above fitness function value is evaluated at each time increment, an ideal time-dependent injection profile can be

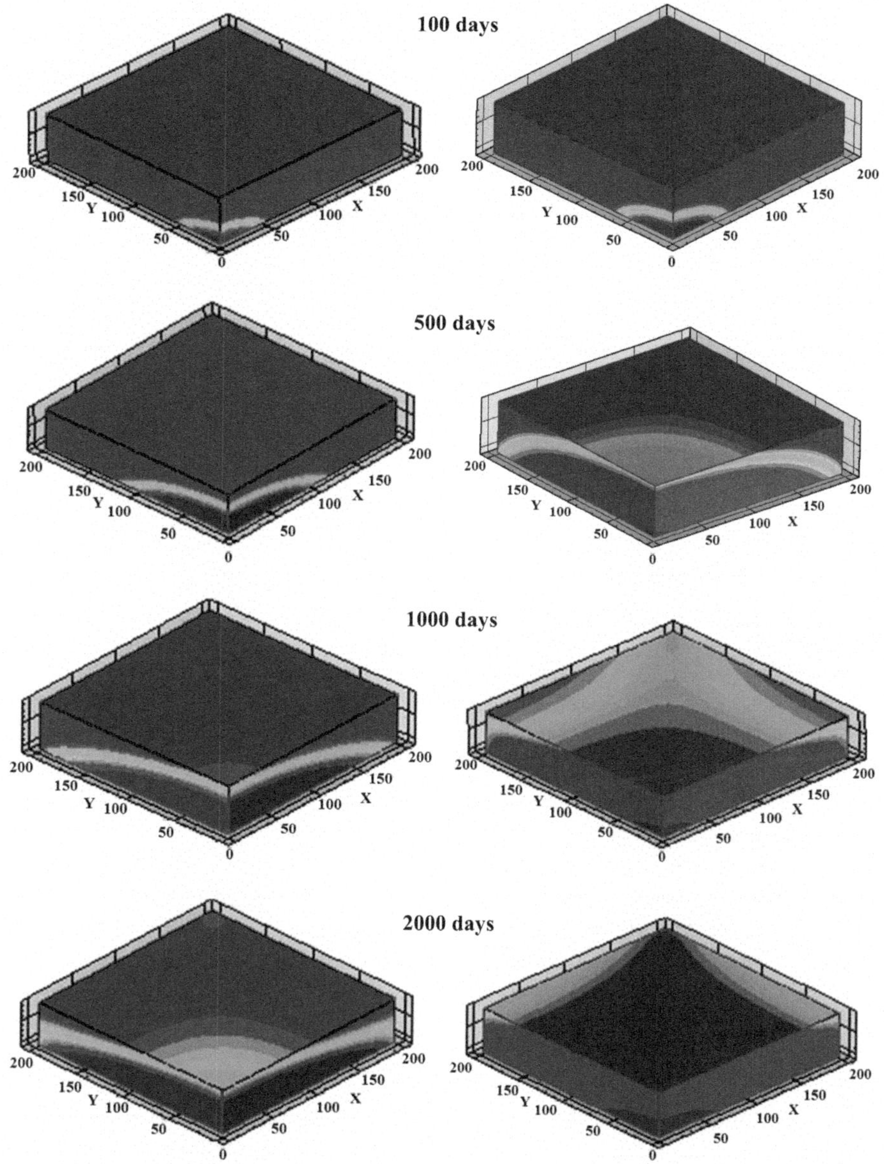


**Fig. 13** Comparison of $CO_2$ mass fraction in the reservoir for 0.1 kg/s injection rate (*left*) and optimized 0.294 kg/s injection rate (*right*) [2]

obtained as the numerator in Eq. (12) approaches zero. Although an exact zero value in the numerator of Eq. (12) may be difficult to obtain due to the unsteady nature of the injection profile, an optimal injection profile can be obtained that maintains the injection pressure within 1 bar of the threshold pressure. By taking

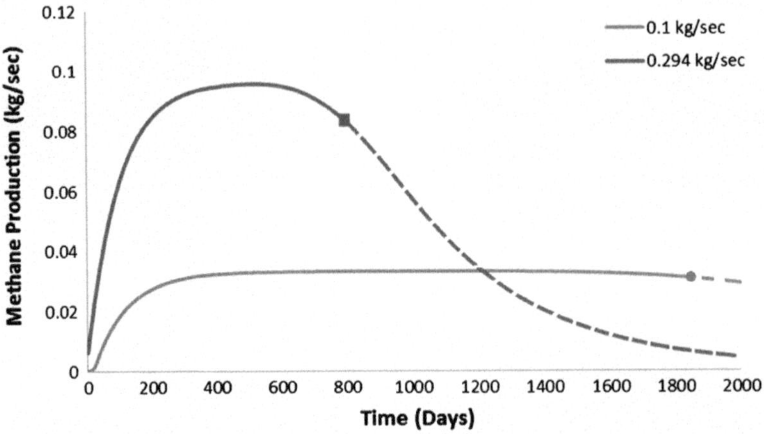


**Fig. 14** Comparison of $CH_4$ production rate for the baseline and optimum $CO_2$ injection rates [2]

the absolute value of the pressure difference, the injection pressure is allowed to overshoot or undershoot the threshold pressure when searching for the optimum. This leads to the present solution-searching optimization technique for the time-dependent injection.

The optimal solution search for EGR with constant pressure injection seeks to define a continuous time-dependent function for the injection profile. In order to accomplish this goal with relatively few design parameters, a Bézier curve is employed to define the injection profile [2]. The Bézier curve was originally utilized to design complex curved automobile parts in the early 1960s; it is now commonly used in computer graphics and simulations [26, 27]. A Bézier curve is constructed by use of a series of points defined in a Cartesian coordinate system. From these few coordinate points, a polynomial can be fitted to define a unique curve. The control points are typically denoted as $P_0$ to $P_n$ and the order of the curve is defined as (n − 1). The polynomial equation for an $n$th order curve is defined as

$$B(t) = \sum_{i}^{n} \binom{n}{i} (1-t)^{n-i} t^i P_i \tag{13}$$

For this study a cubic Bézier curve is chosen, which is a third-order polynomial dependent on four control points. The same benchmark methane gas reservoir described in Sect. 5 is used for the CPI optimization study. The reservoir conditions and all other physical parameters are the same as given in Table 9. A threshold pressure is introduced as the maximum allowable injection pressure; for simplicity, it is assumed to be the peak pressure of 38.4 bar obtained for the EGR simulation with 0.1 kg/s $CO_2$ injection rate for the benchmark case. For determining the time-dependent injection rate for CPI scenario, the search space for the injection rate is set between 0.1 and 0.5 kg/s. In order to reduce the total simulation time, the injection rate is varied only for the first 500 days of injection, which was the

**Table 12** Parameters used in GA-TOUGH2 for the constant pressure injection optimization scenario [2]

| | |
|---|---|
| Individuals per generation | 6 |
| Maximum number of generations | 50 |
| Natural selection algorithm | 50% |
| Mutation rate | 10% |
| Cross-over algorithm | Semi-random combination of parents |
| Threshold pressure | 38.4 bar |

approximate time period for the benchmark case to reach the peak injection pressure. The remaining injection from 500 to 2000 days is conducted at a gradually decreasing rate down to the baseline injection rate of 0.1 kg/s [2]. The GA-TOUGH2 parameters for the CPI optimization scenario are summarized in Table 12.

After 28 generations of GA, the optimization converged to a pressure difference of 0.13 bar between the threshold pressure and the injection pressure given in Eq. (12) for the first 500 days of injection. In comparison, the constant injection rate of 0.1 kg/s had an average pressure difference of 0.57 bar over the first 500 days of injection. Thus the optimized results reached the threshold or fracture pressure much faster. The resulting optimized time-dependent injection profile is shown in Fig. 15. The injection begins at a higher rate to establish the threshold pressure in the reservoir and then quickly decays to a nearly steady-state value of 0.1 kg/s. Figure 16 shows a comparison of the injection pressure profile under the optimized time-dependent rate and that under the non-optimized benchmark case for the first 500 days and 2000 days of operation respectively. It is evident that the optimized time-dependent injection helps the injection pressure to rises to the

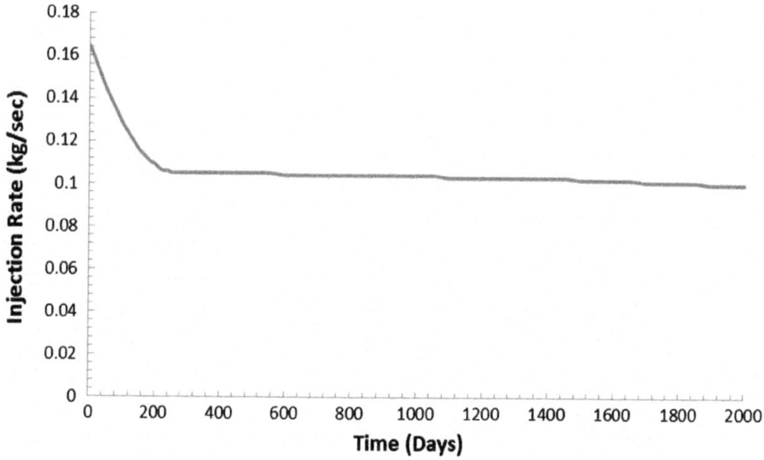


**Fig. 15** Optimal time-dependent injection rate profile for constant pressure injection [2]

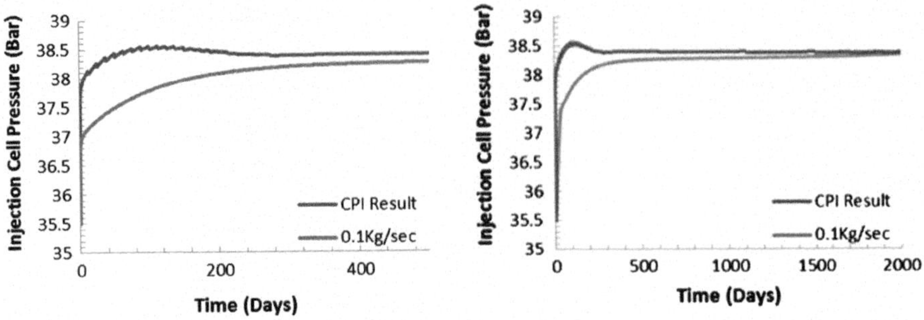


**Fig. 16** Comparison of injection cell pressures for CPI-optimized and baseline constant mass injection rates for 500 days (*left*) and 2000 days (*right*) [2]

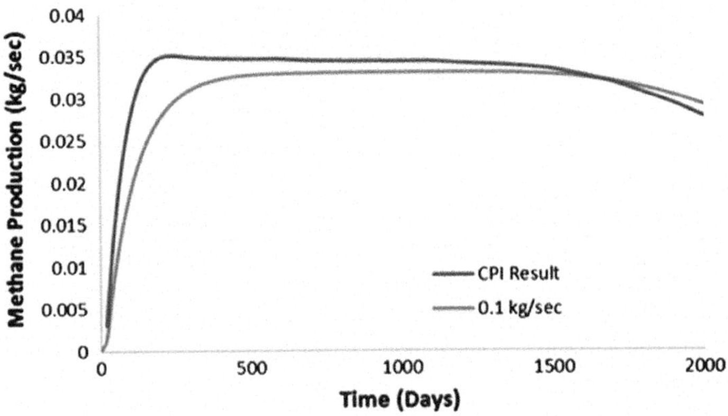


**Fig. 17** Comparison of methane production rates for CPI-optimized and baseline constant mass injection rates [2]

threshold pressure much faster compared to the scenario of non-optimized constant rate injection. The combination of a slightly greater initial injection rate and a faster pressure elevation to the threshold pressure leads to higher production rate of $CH_4$ compared to the benchmark case as shown in Fig. 17 [2].

Figure 17 can be interpreted in conjunction with Fig. 18, which shows a comparison of the $CO_2$ mass concentrations in the reservoir at various times for the baseline constant mass injection and the optimized constant pressure injection (CPI). The optimized time-dependent injection scenario ensures larger initial migration of $CO_2$ in the reservoir leading to a higher production rate of $CH_4$. Upon reaching the threshold pressure, the injection rate starts to decrease to about 0.1 kg/s. The overall $CH_4$ production thus becomes higher because of the larger initial $CO_2$ migration in the reservoir. Since the $CO_2$ plume is established more quickly in the optimized CPI case compared to the baseline case, subsequent injections have a more direct effect on extraction of $CH_4$. In summary, the

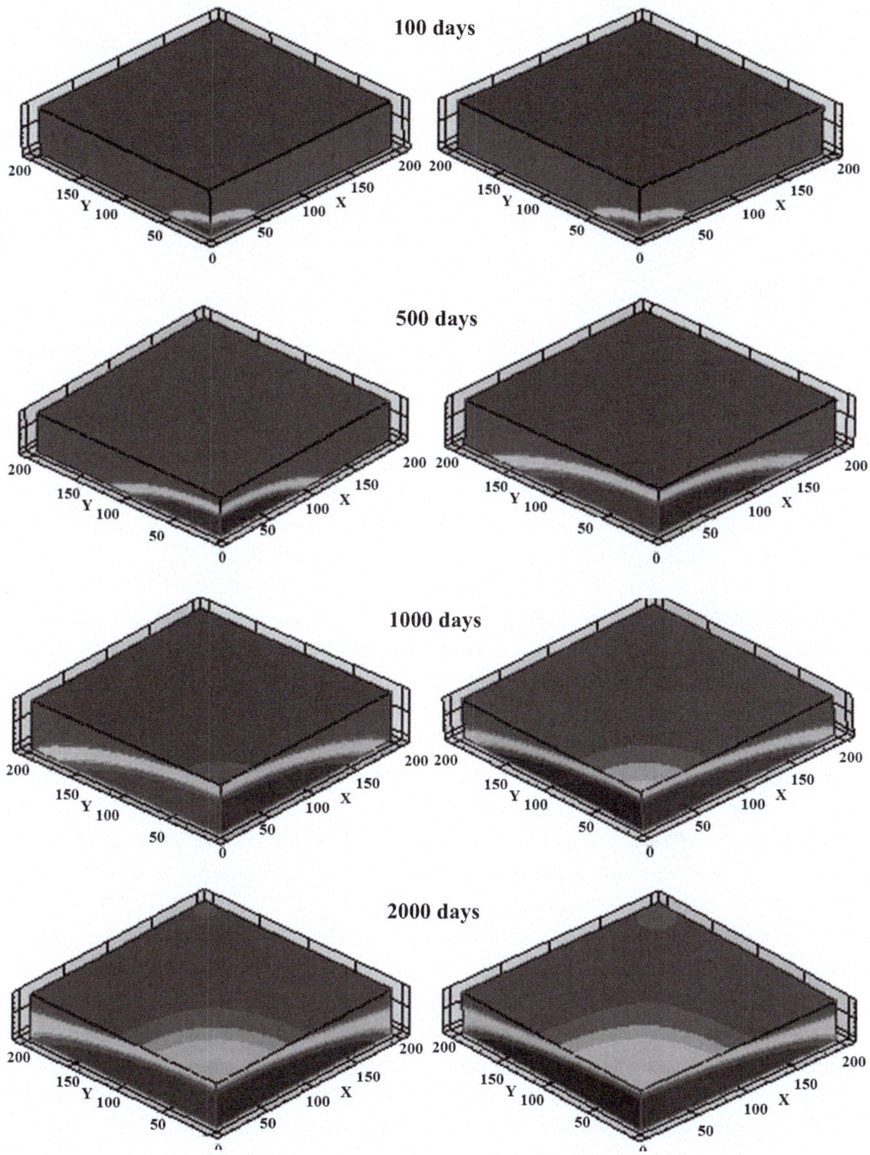


**Fig. 18** Comparison of $CO_2$ mass fraction in reservoir for 0.1 kg/s constant mass injection rate (*left*) and optimized time-dependent CPI rate (*right*) [2]

optimized injection rate for constant pressure injection (CPI) starts at a higher rate to reach the threshold pressure quickly. Once the threshold pressure is established, the injection profile decays to maintain the reservoir pressure within a very small margin of tolerance with respect to the threshold pressure. This type of pressure management leads to a higher methane production initially as well as a higher level of overall production during the life of the well [2]. The CPI approach does not affect the recovery factor significantly compared to the baseline case over the lifetime of the well, but it produces methane at a higher rate while still ensuring that the injection pressure stays below the threshold pressure at all times to maintain operational safety.

## 7 Conclusions

In this chapter, numerical simulations of $CO_2$ enhanced oil recovery (EOR) from partially depleted oil reservoirs have been reported for a benchmark reservoir and a real reservoir in the Permian Basin in Texas. For both cases, a supercritical $CO_2$ fluid was injected into the reservoir and its migration was simulated using the DOE numerical solver COZSim/COZView. The code was validated against available data for the Permian Basin reservoir and matched the previously reported results. Optimization studies were conducted on both reservoirs using a genetic algorithm based optimizer, first considering a constant injection rate to optimize the oil recovery factor, and later a pressure-limited time-varying injection rate to ensure that the reservoir pressure does not exceed the fracture pressure. The results showed that significant increase in the recovery factor and shorter well-shut in times could be obtained by optimizing the injection parameters.

The concept of $CO_2$ sequestration with enhanced gas recovery (EGR) was also investigated using numerical simulations. A high pressure $CO_2$ fluid was injected in a natural gas reservoir and its migration was simulated using the TOUGH2 numerical solver. Code validation for EGR using TOUGH2 showed good agreement compared to previous studies in the literature. The validated baseline case was then optimized considering a constant mass injection rate to optimize the recovery factor for $CO_2$-EGR and resulted in an improvement of $CH_4$ recovery factor by approximately 5% along with a shorter project life cycle. Next, an optimum injection profile was found to maintain a constant injection pressure scenario to avoid the problem of rock fracture caused by high pressures. The constant pressure injection optimization resulted in greater $CH_4$ production compared to the baseline case while ensuring safe management of the injection pressure. The results obtained from the optimization studies described in this chapter demonstrate the scope for improvements using various injection scenarios to determine the optimal $CO_2$ injection rate for optimizing the recovery factor, overall production, and shorter well shut-in time for EOR and EGR applications. Such optimization can help to make the CCUS technologies more economically viable and provide a potential use for captured $CO_2$. The efficient use of captured $CO_2$ is of vital importance for the success and proliferation of CCUS projects.

## References


1. Safi R, Agarwal RK, Banerjee S (2016) Numerical simulation and optimization of $CO_2$ utilization for enhanced oil recovery from depleted reservoirs. Chem Eng Sci 144:30–38
2. Biagi J, Agarwal R, Zhang Z (2016) Simulation and optimization of enhanced gas recovery utilizing $CO_2$. Energy 94:78–86
3. Marland G, Boden TA, Andres RJ (2000) Global, regional, and national fossil fuel $CO_2$ emissions. In Trends: a compendium of data on global change. Carbon Dioxide Information Analysis Center, Oak Ridge National Laboratory, U.S. Department of Energy, Oak Ridge, TN
4. Keeling R (2015) $CO_2$ Concentration at Mauna Loa Observatory, Hawaii, Scripps Institution of Oceanography. http://www.scrippsco2.ucsd.edu
5. Metz B, Davidson O, de Coninck H, Loos M, Meyer L (eds) (2005) IPCC special report on carbon dioxide capture and storage. Cambridge University Press, New York, NY
6. Brown D (2000) A hot dry rock geothermal energy concept utilizing supercritical $CO_2$ instead of water. In: Proceedings of 25th workshop on geothermal reservoir engineering, Stanford, CA, pp 233–238
7. Office of Fossil Energy (2014) Overview of carbon storage research, U.S. Department of Energy. http://www.energy.gov/fe/science-innovation/carbon-capture-and-storage-research/overview-carbon-storage-research
8. Institute for 21st Century Energy (2013) $CO_2$ enhanced oil recovery, U.S. Chamber of Commerce. http://www.energyxxi.org
9. Zoback MD, Gorelick SM (2012) Earthquake triggering and large-scale geologic storage of carbon dioxide. Proc Natl Acad Sci USA 109(26):10164–10168
10. National Academy of Sciences (2012) Induced seismicity potential in energy technologies. The National Academies Press, Washington, DC
11. Melzer SL (2012) Carbon dioxide enhanced oil recovery ($CO_2$ EOR): factors involved in adding carbon capture, utilization and storage (CCUS) to enhanced oil recovery. National Enhanced Oil Recovery Initiative, Midland, TX
12. Capture Carbon, Technologies Sequestration (2014) Carbon capture and sequestration project database. Massachusetts Institute of Technology, Cambridge, MA
13. Nitec, LLC (2014) COZView/COZSim User Manual. http://niteclle.com/docs/COZView_UserManual.pdf
14. van der Burgt MJ, Cantle J, Boutkan VK (1992) Carbon dioxide disposal from coal-based IGCCs in depleted gas fields. Energy Convers Manage 33(5–8):603–610
15. Blok K, Williams RH, Katofsky RE, Hendricks CA (1997) Hydrogen production from natural gas, sequestration of recovered $CO_2$ in depleted gas wells and enhanced gas recovery. Energy 22(2–3):161–168
16. Szulczewski ML, MacMinn CW, Herzog HJ, Juanes R (2012) Lifetime of carbon capture and storage as a climate-change mitigation technology. Proc Natl Acad Sci 109(14):5185–5189
17. Pruess K, Oldenburg C, Moridis G (1999) TOUGH2 user's guide, version 2.0. Report No. LBNL-43134 (revised Sep 2012), Earth Sciences Division, Lawrence Berkeley National Laboratory, University of California, Berkeley, CA
18. Borrero F, Hess FS, Hsu J, Kunze G, Leslie SA (2012) Earth science: geology, the environment, and the universe. Glencoe/McGraw-Hill, Columbus, OH
19. Peaceman DW (1977) Fundamentals of numerical reservoir simulation. Elsevier Science, New York, NY
20. Karacaer C (2013) Mixing issues in $CO_2$ flooding: comparison of compositional and extended black-oil simulators. M.S. thesis, Colorado School of Mines, Golden, CO
21. Zhang Z (2013) Numerical simulation and optimization of $CO_2$ sequestration in saline aquifers. PhD dissertation, Washington University in St. Louis, St. Louis, MO
22. Nitec, LLC (2014) COZView/COZSim Tutorial 1. http://niteclle.com/docs/Tutorial-1.pdf
23. Dai Z et al (2013) An integrated framework for optimizing $CO_2$ sequestration and enhanced oil recovery. Environ Sci Technol Lett 14(1):49–54

24. Ebigbo A et al (2007) Numerical investigations of $CO_2$ sequestration in geological formations: problem-oriented benchmarks, Problem 2, enhanced $CH_4$ recovery in combination with $CO_2$ storage in depleted gas reservoirs. In: Proceedings of the workshop at University of Stuttgart. http://www.hydrosys.uni-stuttgart.de/co2-workshop/Problem2.pdf
25. Zhang Z, Agarwal R (2013) Numerical simulation and optimization of $CO_2$ sequestration in saline aquifers. Comput Fluids 80:79–87
26. Dierckx P, Tytgat B (1989) Generating the Bézier points of a β-spline curve. Comput Aided Geom Des 6(4):279–291
27. Shao L, Zhou H (1996) Curve fitting with Bézier cubics. Graph Models Image Process 58 (3):223–232

# Development of an Indigenous Sensor for Sub-micron Aerosol Monitoring in India

**Tarun Gupta**

**Abstract** Environmental pollutants might be present in different media as well as they can easily migrate between air, water and soil. These contaminants can migrate from one media to another. Many epidemiological studies have linked particulate matter (PM) and especially the $PM_{2.5}$ (aerodynamic diameter $d_p < 2.5$ μm) mass to adverse cardiovascular and respiratory health effects. These health effects include premature mortality, asthma exacerbation, decreased lung function, and increased risk of myocardial infarction among others. These tiny inhalable particles are composed of various heterogeneously mixed chemical species including sulfate, nitrate, ammonium ions, sea salt, organic and elemental carbon, metals, and biological material. Past studies have suggested that specific toxic components like heavy metals, PAHs or biological components like endotoxins are responsible for irregular heart and lung function in animal models. Accurate measurement of contaminants in various media is the first step towards their effective control. This chapter describes the lab development and field evaluation of an inertial impaction based indigenous sensor for sub-micron aerosol monitoring developed at IIT Kanpur. Submicron particles ($PM_1$) were collected during the winter season at Kanpur, India. These samples were chemically characterized for various species like As, Ca, Cd, Cr, Cu, Fe, K, Mg, Mn, Na, Ni, Pb, Se, Ti, V, Zn, $F^-$, $Cl^-$, $NO_3^-$, $SO_4^{2-}$ and $NH_4^+$. These quantitative details were used to apportion air pollutant sources using Positive Matrix Factorization (PMF) coupled with conditional probability function (CPF) to determine the likely direction of each source from receptor site. This low cost instrument is now commercially available in the Indian market.

**Keywords** Source apportionment · PMF · Metals · $PM_1$ · CPF

---

T. Gupta (✉)
Department of Civil Engineering, Indian Institute of Technology Kanpur, 208016 Kanpur, UP, India
e-mail: tarun@iitk.ac.in

 
A.K. Agarwal et al. (eds.), *Combustion for Power Generation and Transportation*, DOI 10.1007/978-981-10-3785-6_19

# 1 Introduction

Urban air quality has been a cause of concern in recent times, primarily due to rapid industrialization, migration of people into urban centers and vehicular expansion, leading to deteriorating air quality. Due to ever increasing anthropogenic activities, the situation is getting even worse. The PM is known to cause problems on a local as well as global scale and have a pronounced effect on weather and climatology [9, 10]. Health effects due to air pollutants have been reported in various earlier studies [3, 4, 48]. A strong correlation has been found between mortality rate and concentration of fine particulate matter [48, 52, 53]. Accurate determination of PM size, concentration, morphology and chemical composition are considered logical initial steps towards their final abatement.

In the present study, PMF (US EPA PMF Ver. 3.0) was employed to get an insight into pollutant sources at a typical urban location in the Indo Gangetic Plain (IGP). The IGP is one of the most fertile areas of the world and hence highly populated. Most of the countries have adopted standards for particulate matter with aerodynamic diameter less than or equal to 10 μm ($PM_{10}$), as it constitutes the respirable fraction of the dust. However, several countries have or are looking to adopt standards for fine particulate matter ($PM_{2.5}$ or less) as it is mainly formed due to anthropogenic activities [30]. The coarser fraction ($PM_{10}$) is mainly formed from naturally occurring activities like dust storms and volcanic eruptions which are beyond our control. So, $PM_{2.5}$ is set to become a significant parameter to judge ambient air quality in urban areas. In this study, we are mainly focused on $PM_1$ at a typical urban location as $PM_1$ is a better indicator of anthropogenic sources [32, 35, 55]. Conditional Probability Function (CPF) has been employed in this work to generate concentration rose plots which point towards major pollutant sources.

Most of the source apportionment studies in India have been carried out on receptor models. However, they have not been used effectively due to paucity of India-specific source profiles [45]. In a recent study conducted by the Central Pollution Control Board, India (CPCB), source profiles for PM in six Indian cities —Bengaluru, Chennai, Delhi, Kanpur, Mumbai and Pune have been developed by Patil et al. [46]. They reported profiles for 45 $PM_{10}$ and 13 $PM_{2.5}$ sources, including leather waste burning, brick kilns, kerosene generators, power plants with low sulphur heavy stock fuel, and liquid petroleum gas (LPG) stoves which are unique to Indian conditions. A number of chemical characterization studies have been carried out in and around Kanpur [8, 12, 18, 31, 37–39, 56].

In India there are only a few studies that have used CPF. One of them being [49] carried out at Hissar, Haryana. Using PMF, they resolved 4 factors—carbonate rich aerosol, combustion rich aerosol, secondary sulphate/nitrate and an unidentified factor which was likely to be emissions from polymer industries. Carbonate rich dust flew in from Pakistan and Rajasthan towards Hissar, but CPF plots showed major contributions from North-East direction. This shows that CPF fails to locate the direction of far-off sources. The unidentified factor was assumed to be emissions from polymer industries because CPF plots showed North-East as a prominent focal point where several polymer industries are located. Bapna et al. [7] characterized 5 years of highly time resolved airborne Black Carbon (BC) concentrations in Ahmedabad. The East-North-East (ENE) direction was identified as a major source region indicating a national highway (NH-8) and two coal-fired thermal power stations as likely contributors.

## 2 Theoretical Design Considerations

According to the impaction theory, the cut point of each impaction stage can be calculated by using the dimensionless Stokes number ($S_{tk}$), which is defined as follows:

$$S_{tk} = \frac{\rho_p d_p^2 U C_c}{9\eta W} \quad (1)$$

Stokes number is the main impaction parameter which governs the collection efficiency. It is a function of particle density ($\rho_p$), particle diameter ($d_p$), jet velocity ($U$), cunningham slip correction factor ($C_c$), dynamic viscosity of air ($\eta$) and nozzle diameter ($W$ or $D_j$). $Q$ represents the air flow rate through the impactor. The theoretical $d_{50}$ (cut point diameter) can be calculated by using the equation (Hinds 1999):

$$d_{50}\sqrt{Cc} = \left[\frac{9\pi\eta D_j^3 (Stk_{50})}{4\rho_p Q}\right]^{1/2} \quad (2)$$

This equation works fairly well for flat surfaces.

***PMF Analysis and Calculation*** PMF model is a multivariate factor analysis tool used to decompose a matrix of speciated samples into two non-negative matrices-factor contributions/scores (G) and factor profiles/loadings (F). Sources are identified using source profiles obtained from PMF, emission inventories and wind data analysis. The fundamental principle behind PMF is that mass can be assumed to be conserved and hence a mass balance equation can be written to account for all $m$ chemical species in the $n$ samples as contributions from $p$ independent sources in the form of

$$x_{ij} = \sum_{k=1}^{p} f_{ik} * g_{kj} \quad (3)$$

where, $x_{ij}$ is the $i$th elemental concentration measured in the $j$th sample, $f_{ik}$ is the gravimetric concentration (ng/mg) of the $i$th element in material from the $k$th source, and $g_{kj}$ is the airborne mass concentration (mg/m$^3$) of material from the $k$th source contributing to the $j$th sample [19, 42, 44].

PMF has been very efficient in dealing with extreme values which are very common in environmental data. These values could be due to rare natural/man-made phenomenon or could be real "*outliers*" due to inaccurate measurement, reporting, etc. Such high values have a significant effect on the solution. Thus, PMF offers a "*robust*" mode. The robust factorization based on the Huber influence function [28] is a technique of iterative reweighing of the individual data values.

***PMF Model Performance*** The optimal number of factors was found out by analysing goodness of fit (Q) values, Q v's FPEAK values, the scaled residual matrices, G-space plots, the regression parameter ($r^2$) and physical meaningfulness of the factor profiles and contributions. All these factors are known as performance parameters. Besides theoretical Q ($Q_{th}$), $Q_{robust}$ and $Q_{actual}$ were also compared for judging the accuracy of the analysis.

***CPF Analysis*** CPF Ashbaugh et al. [2] analyses point source impacts from changing wind directions using the source contribution estimates from PMF coupled with the wind direction values measured on site [28]. Calm wind periods (wind speed less than 1 m/s) had been excluded from many analyses due to the isotropic behaviour of wind vane under calm winds. However, in our case satellite data obtained from GIOVANNI-NASA [15] is used, hence no data were ignored. The CPF is calculated as

$$CPF = \frac{m_{\Delta\theta}}{n_{\Delta\theta}} \tag{4}$$

where, $m_{\Delta\theta}$ is the no. of times that the source contribution exceeded certain threshold criteria when the wind blew from sector $\Delta\theta$; $n_{\Delta\theta}$ is the total no. of data points when the wind came from the same sector.

From tests with several different percentile values of the factor contributions, a threshold criterion of the upper 50th percentile was chosen. The same 8 h (sampling duration) PM contribution was assigned to each 3rd hour of a given day to match the 3 h wind data [33]. The wind direction was then split into 16 sectors of 22.5° each. Since only higher contributing source contributions are used (e.g. upper 50th percentile contributions in Buzcu-Guven [11] to upper 25th percentile contributions in Lestuari and Mauliadi [33]) in the analysis, events of smaller impacts were left out.

After getting the CPF values, a plot in polar coordinates, with the radial distance corresponding to the value of CPF and the angle corresponding to the associated wind direction was made. It visually illustrates the fraction of samples coming from a given direction that have 'high' values, and point to regions associated with these high values indicating locations of dominant sources. CPF could, however, be used for shorter distances only where the air trajectory can be assumed to follow a straight line path [2, 27].

## 3 Materials and Methods

**Sampler Design** The impactor nozzles designed using the Eqs. (1) and (2) have been tested in our laboratory. The experimental setup is shown in Fig. 1. Artificially, aerosols were generated in the laboratory using NaCl and talcum powder as wet and dry aerosols, respectively. An optical time of flight based instrument (Aerodynamic Particle Sizer, TSI Inc., USA) has been employed to test the performance of the impactor. Parametric investigation was carried out with impactor nozzles of different diameters, varying flow rates and distance between impactor nozzle and the substrate, to optimize the PM collection efficiency at the desired cut-point of 1.0 μm.

***Sampling Site*** The sampling was carried out at IITK Campus (26.51°N, 80.23°E) on the roof of Western Labs Extension building, 12 m above ground level (Fig. 2). Sampling was carried out for 10 h per day from 0700 h to 1700 h using a single-stage round nozzle, grease substrate based $PM_1$ impactor developed in our lab [16, 17]. The samples were analysed for species such as As, Ca, Cd, Cr, Cu, Fe, K, Mg, Mn, Na, Ni, Pb, Se, Ti, V, Zn, $F^-$, $Cl^-$, $NO_3^-$, $SO_4^{2-}$ and $NH_4^+$ using ICP-OES (Thermo Fisher, iCAP 6300 Duo) and IC (Metrohm compact IC 761).

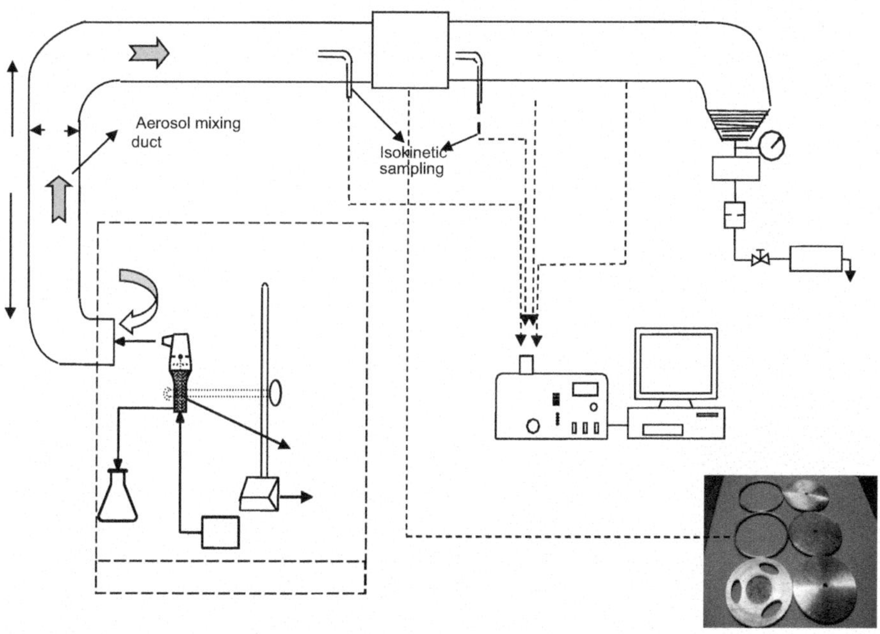


**Fig. 1** Laboratory experimental setup to evaluate $PM_1$ sampler

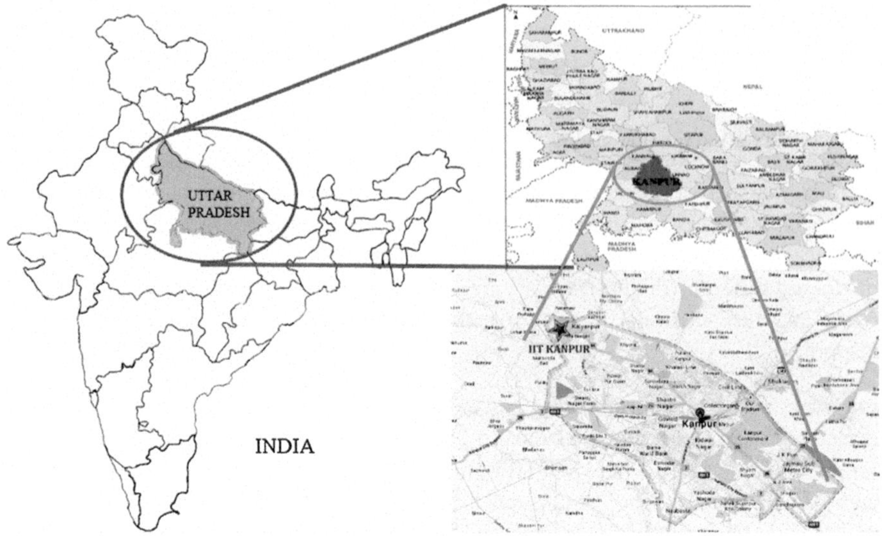


**Fig. 2** Kanpur site location map

The quality control and quality assurance (QA/QC) and the methodology have been described in our previous works [12, 18].

The IITK campus is situated at approximately 15 km N-W of the central part of Kanpur city. It is an institutional cum residential area and the dominant sources of air pollutants here usually are domestic cooking, vehicular traffic and soil and road dust [8]. Kanpur city is highly industrialized, leather industry being the most prominent here. Besides, there are several ordinance factories and brick kilns situated inside the city. Two major national highways also pass through nearby. A coal-fired thermal power plant (Panki thermal power plant) is also located 6 km to the south of our sampling site. Major pollutant sources in and around Kanpur are depicted in Fig. 3.

***Meteorological Conditions*** The mean monthly temperature and wind speed is presented in Fig. 4. Wind rose plots are shown in Fig. 5. Mean values of wind speed and temperature during winters in Kanpur were found to be 3.35 ± 0.95 m/s and 21.15 ± 2.07 °C, respectively.

Stable atmospheric conditions, calm winds and lower mixing height are generally observed during the winters when compared to summers. Mixing height varies between 200 m at midnight to 1600 m during daytime [8, 57].

Heavy fog could also be observed during the winters and both inorganic and organic aerosol get enhanced during fog time [18, 25, 51].

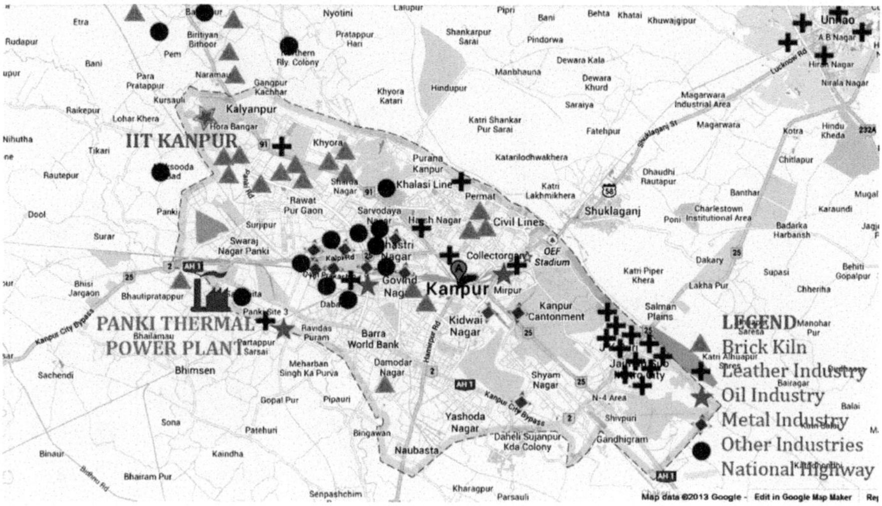


**Fig. 3** Major pollutant sources in and around Kanpur

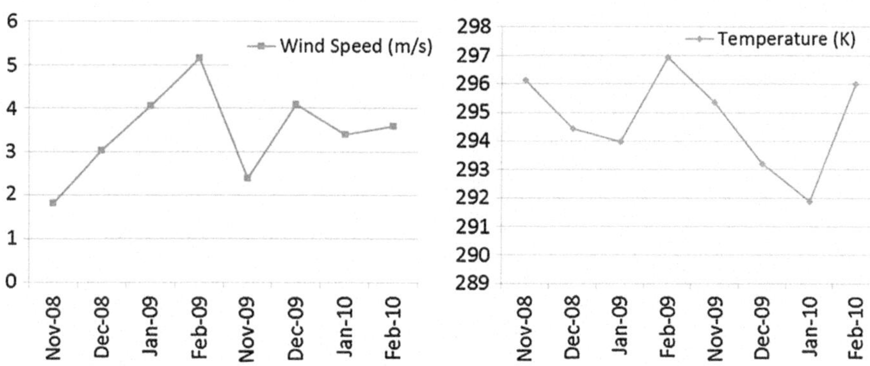


**Fig. 4** Wind speed and temperature variation at Kanpur during the sampling period

# 4 Results and Discussion

**Impactor Characteristics** Table 1 shows the salient features of the optimized $PM_1$ sampler obtained following various parametric iterations. The PM1 sampler was designed for an operational flow rate of 10 LPM and it had minimum pressure drop when used with 47 mm quartz filter paper. It used high vacuum silicone grease as an impaction substrate having significantly long particle retention capacity, thus minimizing the bounce-off of larger particles.

**Data Analysis** Species having too many missing values or below detection limit (BDL) values were rejected from the data set. Also, some extreme events could

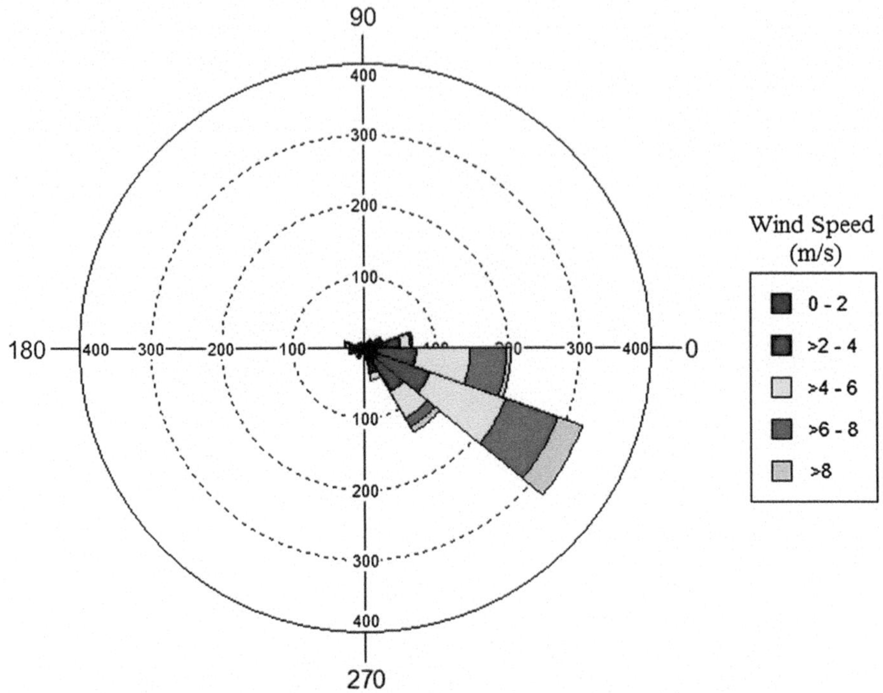


**Fig. 5** Wind rose plots during winters (2009–10)

**Table 1** Salient design and performance related features of the PM1.0 impactor

| Physical characteristics | Unit |
|---|---|
| Nozzle width (W) | 3.2 mm |
| Nozzle to plate distance (S) | 2.5 mm |
| Flow parameters | |
| Reynolds number (Re) | 4424 |
| Flow velocity (U) | 20.77 m/s |
| Experimental results | |
| d50 or cutoff diameter | 1.0 μm |
| $\sqrt{Stk}$ | 0.31 |
| σg (GSD) | 1.2 |
| ΔP (Pressure drop) of Impactor with 47 mm Teflon filter (2 μm pore size) | 18.5 cm of $H_2O$ |

either be removed or provided with a large uncertainty, so as to not influence overall result. Norris et al. [40], Polissar et al. [47] suggested the following for the concentration values and their associated error estimates:

$$x_{ij}^{k} = v_{ij}^{k} \quad \text{for determined values} \tag{5}$$

$$x_{ij}^{k} = d_{ij}^{k}/2 \quad \text{for BDL values} \tag{6}$$

$$x_{ij}^{k} = \tilde{v}_{ij}^{k} \quad \text{for missing values} \tag{7}$$

where $v_{ij}^{k}$, $u_{ij}^{k}$ and $d_{ij}^{k}$ are the measured concentration, the analytical uncertainty, and the method detection limit, respectively, for sample i, element j, and sampling site k, $\bar{d}_{ij}^{k}$ is arithmetic mean of the detection limit value for the element j and the sampling site k; and $\tilde{v}_{ij}^{k}$ is the geometric mean of the measured concentration of the element j at sampling site k. The number of BDL and missing samples are shown in Table 2.

**Uncertainty Analysis** The uncertainties for determined values are calculated using the following equation:

**Table 2** Summary table depicting the individual species concentration values and data analysis

| Species | Average concentration (μg/m$^3$) | Maximum concentration (μg/m$^3$) | Minimum concentration (μg/m$^3$) |
|---|---|---|---|
| As | 0.03 ± 0 | 0.03 | 0.03 |
| Ca | 4.11 ± 3.65 | 15.24 | 0.19 |
| Cd | 0.05 ± 0.24 | 1.72 | 0.00 |
| Cr | 0.36 ± 0.44 | 1.76 | 0.00 |
| Cu | 0.08 ± 0.15 | 1.07 | 0.01 |
| Fe | 2.6 ± 1.96 | 8.81 | 0.41 |
| K | 5.05 ± 6.5 | 39.70 | 0.06 |
| Mg | 0.76 ± 0.68 | 2.49 | 0.09 |
| Mn | 0.05 ± 0.09 | 0.62 | 0.00 |
| Na | 2.95 ± 1.84 | 7.51 | 0.09 |
| Ni | 0.57 ± 0.26 | 1.41 | 0.04 |
| Pb | 0.91 ± 2.87 | 20.05 | 0.01 |
| Se | 0.03 ± 0.04 | 0.19 | 0.01 |
| Ti | 0.06 ± 0.05 | 0.31 | 0.00 |
| V | 0.31 ± 0.42 | 2.20 | 0.00 |
| Zn | 1.01 ± 3.5 | 25.10 | 0.01 |
| $F^-$ | 0.92 ± 0.92 | 4.69 | 0.06 |
| $Cl^-$ | 2.0 ± 2.27 | 14.62 | 0.01 |
| $NO_3^-$ | 11.97 ± 10.77 | 59.75 | 0.71 |
| $SO_4^{2-}$ | 10.48 ± 8.62 | 36.55 | 1.05 |
| $NH_4^+$ | 12.37 ± 14.44 | 73.47 | 0.62 |
| $PM_1$ | 113.28 ± 71.94 | 347.59 | 18.44 |

$$S_{ij} = p_j * x_{ij} + \frac{MDL_j}{3} \tag{8}$$

where, $p_j$ is the uncertainty proportional parameter, $x_{ij}$ is the observed concentration and $MDL_j$ is the method detection limit for *j*th species [47, 50]. $p_j$ values were obtained from an earlier study [37] and systematically modified to obtain feasible solutions. Uncertainty for BDL values used was 5/6 times the species concentration, while that for missing values was 4 times the concentration. An extra modelling uncertainty of 10% had also been added to account for unknown errors. After the uncertainties have been analyzed, species are characterized as strong, weak or bad on the basis of signal to noise ratio (S/N) calculated as:

$$\left(\frac{S}{N}\right)_j = \sqrt{\frac{\sum_{i=1}^{n}(x_{ij} - s_{ij})^2}{\sum_{i=1}^{n} s_{ij}^2}} \tag{9}$$

Species were characterized as *bad* when S/N was less than 0.2; *weak* when S/N was in the range 0.2–2 and *strong* when its value was greater than 2 [43].

Marking species "weak" triples the uncertainty involved while marking "bad" excludes that species from further analysis. Four samples dated 18-11-09, 19-11-09, 28-01-10 and 29-01-10 were excluded from the analysis because they had very large deviations from the average concentrations. The $p_j$ and $MDL_j$ values are provided in Table 3.

The ability to weight specific observations with specific uncertainties allows PMF to handle data that include heterogeneous measurement uncertainty, outliers, values below measurement detection limits, and missing values. As such, PMF can often yield better results than traditional factor analysis methods [20].

***Source Profiles Obtained by PMF Analysis*** The average PM mass concentration was found to be 192.58 ± 122.29 μg/m$^3$, which varied from 347.59 to 18.44 μg/m$^3$. Mass concentrations of individual species are summarised in Table 2. Base runs were executed on each data set for 20 runs with a random initial seed. Different number of factors, varying from 4 to 10, were chosen for the analysis.

From the PMF analysis, a 7-factor solution was found to be the optimum solution. For 8 or more factors, some of the factors split into two identical factors, hence difficult to distinguish. When the factors chosen were less than 7, several factors combined into one. Hence, the 7-factor solution was most meaningful. The characterisation of species based on S/N ratio and performance parameters for a 7-factor PMF solution are shown in Table 4.

$PM_1$ along with most other species like Ca, Cr, Fe, Mn, Zn, $Cl^-$, $NO_3^-$, $SO_4^{2-}$ and $NH_4^+$ were very well modelled ($r^2 > 0.8$). The other species were also modelled satisfactorily. For a 7-factor solution, all runs converged towards a global minimum, which was quite close to the theoretical Q. The Q-robust of the present study was found equal, to be within 10% of the Q-true, indicating no outliers were impacting the Q value significantly (Table 5).

**Table 3** Table depicting instrumental method detection limits and uncertainty factor values for PMF analysis

| Species | [a]$MDL_j$ value (ng/m$^3$) | [b]$p_j$ value (%) |
|---|---|---|
| As | 2.12 | 10 |
| Ca | 0.11 | 22 |
| Cd | 1.33 | 10 |
| Cr | 4.13 | 10 |
| Cu | 10.42 | 10 |
| Fe | 6.25 | 10 |
| K | 0.11 | 22 |
| Mg | 0.1 | 10 |
| Mn | 2.08 | 10 |
| Na | 0.11 | 22 |
| Ni | 2.08 | 10 |
| Pb | 22.928 | 10 |
| Se | 2.08 | 10 |
| Ti | 2.08 | 10 |
| v | 2.08 | 10 |
| Zn | 2.08 | 10 |
| $F^-$ | 3.0 | 14 |
| $Cl^-$ | 3.0 | 20 |
| $NO_3^-$ | 8.7 | 20 |
| $SO_4^{2-}$ | 12.0 | 14 |
| $NH_4^+$ | 17.0 | 14 |

[a]*Source* [18]
[b]*Source* [37]

No F-Peak runs were required as each of the factors obtained were not correlated to each other. After a reasonable solution was found, the uncertainties in the modelled solution were estimated using a bootstrapping technique. A total of 100 bootstrap runs were performed with a minimum $r^2$-value of 0.6. Of the 100 runs, all the factors were mapped to a base factor in every run indicating a stable result. Residuals were also found to be between +3 and −3 for all species. In general, the performance of PMF on the source apportionment of PM at study site was found to be satisfactory.

The factor profiles or factor loadings obtained are shown in Fig. 6. Factor 1 had high loading on Zn (75%) and minor loadings on other species, hence this could be due to refuse burning and incineration [22, 24, 26, 29, 37].

Factor 2 had high loadings on Cr, Cu and Fe, hence identified as industrial emissions [6]; Chandra Mouli et al. [13, 31, 37, 45, 46]. In factor 3, crustal elements, Ca, Mg, Na and Ti were highly loaded along with Cd, Cu, Pb and Se and hence this factor was identified as crustal/road dust mixed with vehicular emissions. Factor 4 had high loadings on Cr, Ni and V. Hence, it was identified as leather and heavy oil combustion. Several industrial processes, including electroplating, ferrochrome production, pigment production, and leather tanning, use Cr [54, 58].

**Table 4** Species characterisation and PMF performance parameters

| Species | Category | S/N | Intrcpt | Slope | SE | $r^2$ | KS test stat | KS test P value |
|---|---|---|---|---|---|---|---|---|
| As | Bad | 0.2 | 0 | 0 | 0 | 0 | 0 | 0 |
| Ca | Strong | 7.34 | −0.11 | 1.01 | 1.04 | 0.93 | 0.17 | 0.13 |
| Cd | Weak | 0.94 | 0.01 | 0.05 | 0.01 | 0.14 | 0.12 | 0.52 |
| Cr | Strong | 7.23 | 0.02 | 0.9 | 0.14 | 0.9 | 0.18 | 0.12 |
| Cu | Weak | 0.85 | 0.02 | 0.48 | 0.04 | 0.23 | 0.19 | 0.09 |
| Fe | Strong | 7.3 | 0.27 | 0.82 | 0.83 | 0.81 | 0.16 | 0.21 |
| K | Weak | 0.89 | 0.32 | 0.67 | 1.76 | 0.55 | 0.19 | 0.08 |
| Mg | Strong | 9 | 0.14 | 0.62 | 0.29 | 0.7 | 0.2 | 0.06 |
| Mil | Weak | 1.34 | 0.01 | 0.34 | 0.02 | 0.87 | 0.19 | 0.08 |
| Na | Weak | 1.15 | 0.3 | 0.68 | 1.28 | 0.49 | 0.08 | 0.95 |
| Ni | Weak | 0.78 | 1.05 | 0.79 | 1.39 | 0.03 | 0.19 | 0.08 |
| Pb | Weak | 7.07 | 0.11 | 0.03 | 0.11 | 0.05 | 0.09 | 0.86 |
| Se | Weak | 1.13 | 0.02 | 0.26 | 0.02 | 0.37 | 0.17 | 0.15 |
| Ti | Weak | 1.06 | −0.01 | 0.69 | 0.03 | 0.65 | 0.12 | 0.55 |
| V | Weak | 1.53 | 0.06 | 0.35 | 0.13 | 0.59 | 0.2 | 0.06 |
| Zn | Strong | 4.94 | 0.12 | 0.97 | 0.66 | 0.97 | 0.18 | 0.11 |
| $F^-$ | Weak | 1.02 | 0.38 | 0.13 | 0.45 | 0.07 | 0.27 | 0.01 |
| CT | Strong | 4 | 0.42 | 0.47 | 0.7 | 0.72 | 0.1 | 0.79 |
| $NO_3^-$ | Strong | 4 | −0.88 | 1.01 | 2.8 | 0.95 | 0.11 | 0.64 |
| $SO_4^{2-}$ | Strong | 6.14 | 0.34 | 0.89 | 2.94 | 0.89 | 0.11 | 0.63 |
| $NH_4^+$ | Strong | 6.14 | 1.87 | 0.68 | 4.07 | 0.86 | 0.11 | 0.62 |
| PMI | Weak | 40414 | 7.07 | 0.91 | 24.69 | 0.89 | 0.16 | 0.2 |

**Table 5** Summary table depicting goodness-of-fit (Q) values

| Season | Factors | $Q_{th}$ | $Q_{robust}$ | $Q_{true}$ | $Q_{true}/Q_{robust}$ | $Q_{robust}/Q_{th}$ | |
|---|---|---|---|---|---|---|---|
| 2009–10 | 7 | 626 | 501.18 | 503.671 | | 1.04 | 0.804 |

Kanpur has a lot of leather tanning industries situated in the S-E direction. V and Ni mainly come from heavy oil combustion [41]. V is present in heavy fuel oils, which is released as fly ash when oils get combusted [36]. Factor 5 had the highest loading on $F^-$ and moderate loading on Cd and Se, hence this factor was identified as emissions from the brick kilns mixed with thermal power plant emissions. $F^-$ is mainly emitted from the brick kilns and ceramic industries [1, 61]. Coal combustion in modern power plants occurs at about 1500 °C and results in vaporization of most of the Cd and Se present in coal [5, 6, 12, 21, 23, 59, 60]. Factor 6 was identified as secondary sources as it had a high loading on water soluble ions ($F^-$, $NO_3^-$ and $SO_4^{2-}$) which are mainly formed from primary sources upon photo-chemical

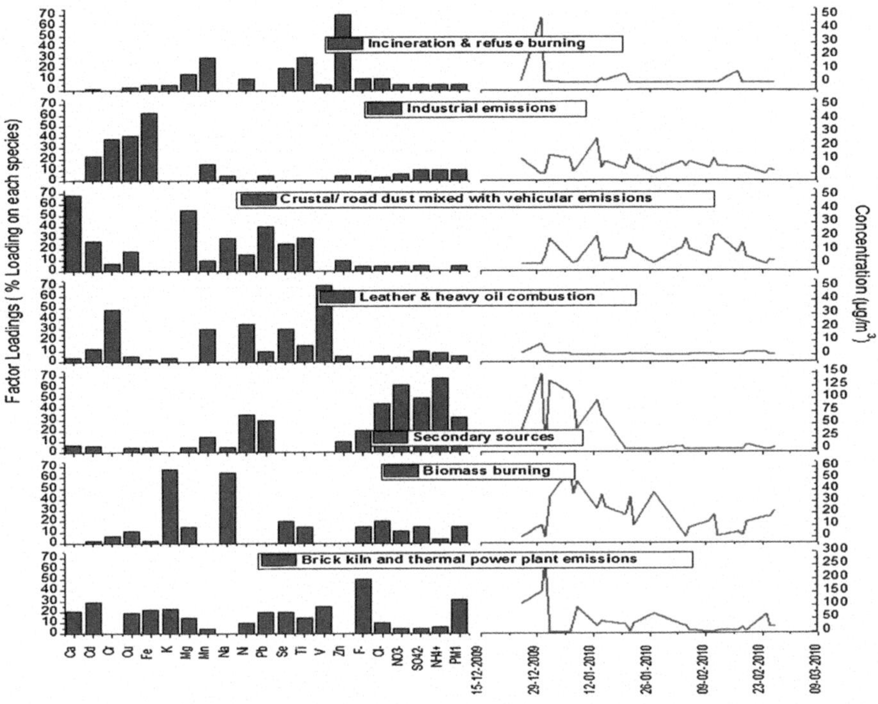


**Fig. 6** Factor profiles (*left*) and individual source contribution timeline (*right*) for PMF resolved factors during winters

reactions [12, 37]. Factor 7 had high loadings on K and Na which is attributed to biomass burning [14, 34].

***Source Contributions*** The daily variation in the contribution of a particular source towards PM is shown in Fig. 6 while the individual source contributions towards the total PM are shown in Fig. 7. These were obtained from PMF analysis. Brick kiln and thermal power plant emissions (37%), and secondary sources (33%) were the major contributors to total PM. Biomass burning (13%), industrial emissions (8%), crustal/ road dust mixed with vehicular emissions (6%), leather and oil combustion (2%), and refuse burning and incineration (1%) were minor components. Mandariya [18] had earlier used UNMIX model for the same study period and identified 5 factors. Secondary sources had the largest contribution of 57%, followed by composite motor vehicle exhausts (16%), metal smelting industries (14%), coal combustion (7%) and crustal source (7%).

Hence, UNMIX and PMF do present slightly different apportioned results for the same dataset. This could be due to the fact that the PMF is more advanced than UNMIX and can resolve more factors. In UNMIX several factors must have combined into one, which couldn't be resolved.

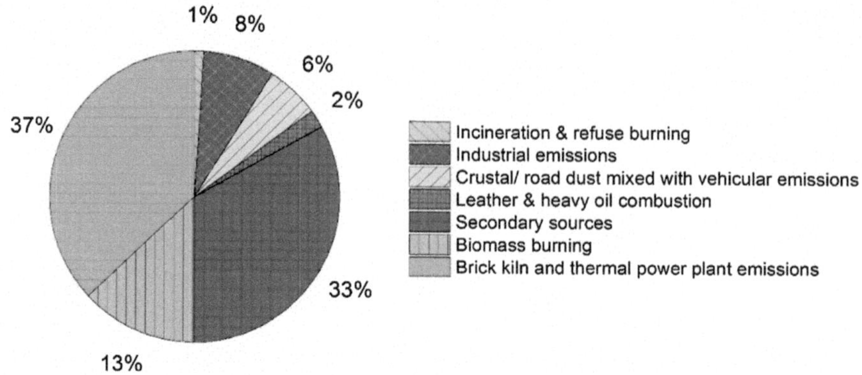


**Fig. 7** Individual source contributions for each season

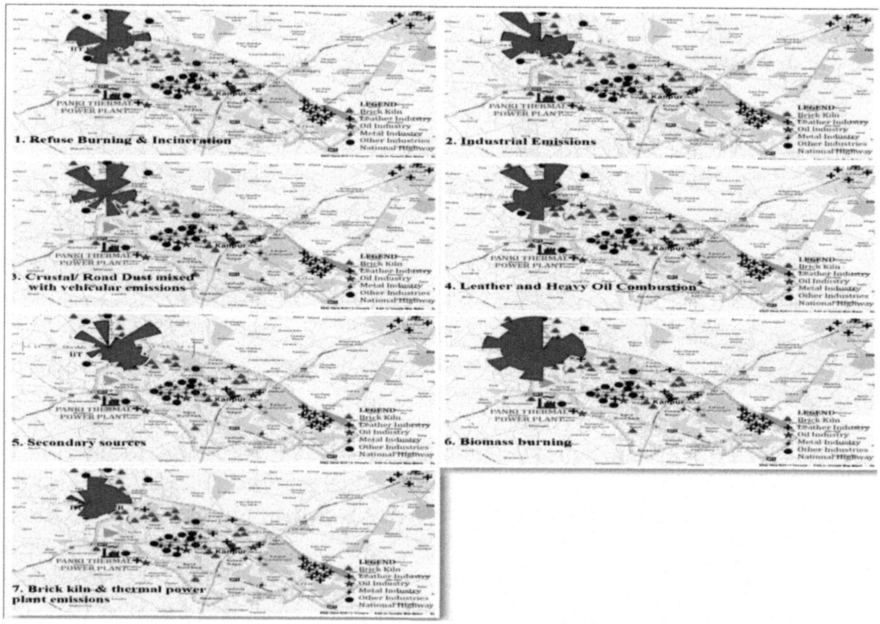


**Fig. 8** CPF plots for the winter season 2009–10, showing the direction of maximum probability of location of individual sources

***CPF Plots*** The CPF plots were superimposed on the map of Kanpur depicting major pollutant sources in and around Kanpur to identify the location of sources obtained from PMF. These plots are shown in Fig. 8.

The CPF plots for brick kiln and thermal power plant emissions pointed towards the N-E-S sector where a lot of brick kilns are located. Also, there is a large coal fired thermal power plant (210 MW generating capacity) situated to the South of sampling site. CPF plot had a peak in this direction. From the opposite direction, however, nearly negligible contributions of emissions were recorded. Hence, our results matched quite well with the actual location of known sources.

Similarly, it was found that the biomass burning emissions were prominent from N-W-S direction which is a rural area, and negligible in the opposite direction (predominantly urban). People in villages burn lots of wood and organic matter to keep themselves warm during winters, which is rather rare in urban areas. The vehicular and crustal dust emissions were prominent from all directions which is obvious for an urban area. The highest CPF values for this factor pointed towards 2 major highways that run through nearby. Refuse burning and secondary sources were also spread out evenly.

Most of CPF plots had shown a remarkable agreement with the location of known sources. Industrial emissions showed slight deviations which could not be justified. Leather and heavy oil combustion also pointed away from the leather industries, which are quite large in no here. It pointed towards west direction. May be heavy oil combustion from the Mathura Oil Refinery, situated about 200 km towards the west might be influencing the results too much, and CPF can't work well with far-off sources.

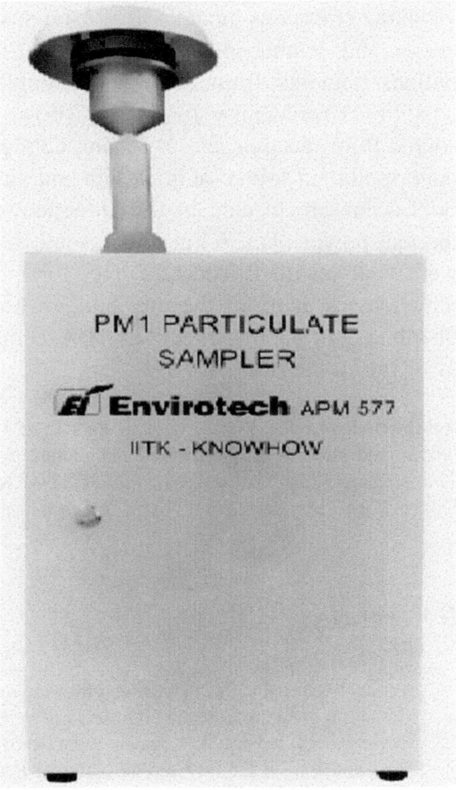


**Fig. 9** Commercialized $PM_1$ sampler

The developed low cost $PM_1$ sampler (Fig. 9) is now commercially available as Envirotech APM 577 (Envirotech Instruments Pvt Ltd., Delhi) in the Indian market. It can be wisely used for various applications including in-cabin air quality measurements in locomotives and other different microenvironments.

## 5 Conclusion

This study describes the lab development of a novel PM1 sampler followed by field evaluation of this novel sampler and successful demonstration of its usage as a tool to identify the key air pollution sources. Submicron particles ($PM_1$) were collected during the winter season in Kanpur, India. The average $PM_1$ mass concentration for the winter season 2009–10 was found to be 192.58 ± 122.29 μg/m$^3$ with a maximum value of 347.59 μg/m$^3$ and a minimum value of 18.44 μg/m$^3$. These samples were chemically characterized for various species like As, Ca, Cd, Cr, Cu, Fe, K, Mg, Mn, Na, Ni, Pb, Se, Ti, V, Zn, $F^-$, $Cl^-$, $NO_3^-$, $SO_4^{2-}$ and $NH_4^+$. From the data analysis, a 7 Factor solution using PMF was found to be optimum. The sources identified along with their respective contributions were brick kiln and thermal power plant emissions (37%), secondary sources (33%), industrial emissions (8%), vehicular emissions mixed with crustal dust (6%), refuse (1%) and biomass burning (13%), and, leather industries and heavy oil combustion (2%). The source contributions obtained from PMF were coupled with the wind direction data using conditional probability function (CPF) to determine the likely direction of each source from receptor site. By using CPF plots, we identified N-E-S direction as a major pollutant source of brick kiln and thermal power plant emissions. Most of the brick kilns are located in this direction, as shown in the map. Also, there is a thermal power plant 6 km to the south of sampling site, at Panki. So our results were well justified. The CPF plot for biomass burning pointed towards N-W-S direction, away from the city and towards rural area. Other sources were well distributed in all the directions. The commercial version of this sampler is now available in the market.

**Acknowledgements** Meteorological data used in this study were produced by the Giovanni online data system, developed and maintained by the NASA GES DISC (GIOVANNI-NASA). The work presented here was part of IITK M.Tech. EEM theses of my students Kamal Kant Ujinwal, Anil Mandariya and Nikhil Rastogi.

## References

1. Ahmad MN et al (2012) Hydrogen fluoride damage to vegetation from peri-urban brick kilns in Asia: a growing but unrecognised problem? Environ Pollut 162:319–324
2. Ashbaugh LL, Malm WC, Sadeh WZ (1985) A residence time probability analysis of sulfur concentrations at Grand Canyon National Park. Atmos Environ 19(0):1263–1270

3. Atkinson RW, Anderson HR, Sunyer J, Ayres J, Baccini M, Vonk JM (2001) Acute effects of particulate air pollution on respiratory admissions. Am J Respir Crit Care Med 164:1860–1866
4. Bachmann JD, Damberg RJ, Caldwell JC, Edwards C, Koman PD (1996) Review of the national ambient air quality standards for particulate matter: policy assessment of scientific and technical information. OAQPS staff paper. Final report. Other Information: PBD: July 1996 (1996), p. Medium: P; Size: 421 p
5. Balachandran S, Meena BR, Khillare PS (2000) Particle size distribution and its elemental composition in the ambient air of Delhi. Environ Int 26:49–54
6. Banerjee T, Murari V, Kumar M, Raju MP (2015) Source apportionment of airborne particulates through receptor modelling: Indian scenario. Atmos Res. doi:10.1016/j.atmosres.2015.04.017
7. Bapna M, Sunder Raman R, Ramachandran S, Rajesh TA (2013) Airborne black carbon concentrations over an urban region in western India—temporal variability, effects of meteorology, and source regions. Environ Sci Pollut Res 20:1617–1631
8. Behera SN, Sharma M (2010) Reconstructing primary and secondary components of PM2.5 composition for an urban atmosphere. Aerosol Sci Technol 44:983–992
9. Brimblecombe P (1999) Air pollution and health history, pp 5–20
10. Buseck PR, Pósfai M (1999) Airborne minerals and related aerosol particles: effects on climate and the environment. Proc Natl Acad Sci 96:3372–3379
11. Buzcu-Guven B, Brown SG, Frankel A, Hafner HR, Roberts PT (2007) Analysis and apportionment of organic carbon and fine particulate matter sources at multiple sites in the midwestern United States. J Air Waste Manage Assoc 57:606–619
12. Chakraborty A, Gupta T (2010) Chemical characterization and source apportionment of submicron (PM1) aerosol in Kanpur region, India. Aerosol Air Qual Res 10:433–445
13. Chandra Mouli P, Venkata Mohan S, Balaram V, Praveen Kumar M, Jayarama Reddy S (2006) A study on trace elemental composition of atmospheric aerosols at a semi-arid urban site using ICP-MS technique. Atmos Environ 40:136–146
14. Echalar F, Gaudichet A, Cachier H, Artaxo P (1995) Aerosol emissions by tropical forest and savanna biomass burning: characteristic trace elements and fluxes. Geophys Res Lett 22:3039–3042
15. GIOVANNI-NASA, MERRA 3-hourly history data collections (3D). http://disc.sci.gsfc.nasa.gov/giovanni
16. Gupta T, Chakraborty A, Ujinwal K (2010) Development and performance evaluation of an indigenously developed air sampler designed to collect submicron aerosol. Ann Indian Natl Acad Eng (INAE) 3:189–193
17. Gupta T, Dubey S (2011) Field performance evaluation of a newly developed PM 2.5 sampler at IIT Kanpur. Sci Total Environ 409:3500–3507
18. Gupta T, Mandariya A (2013) Sources of submicron aerosol during fog-dominated wintertime at Kanpur. Environ Sci Pollut Res 20:5615–5629
19. Hopke PK (2000) A guide to positive matrix factorization. Workshop on UNMIX and PMF as Applied to PM2, p 600
20. Huang S, Rahn KA, Arimoto R (1999) Testing and optimizing two factor-analysis techniques on aerosol at Narragansett, Rhode Island. Atmos Environ 33:2169–2185
21. Hutton M (1983) Sources of cadmium in the environment. Ecotoxicol Environ Saf 7:9–24
22. Jones F, Bankiewicz D, Hupa M (2014) Occurrence and sources of zinc in fuels, Fuel 117 (Part A):763–775
23. Kang CM, Gupta T, Ruiz PA, Wolfson JM, Ferguson ST, Lawrence J, Rohr AC, Goldleski J, Koutrakis P (2011) Aged particles derived from emissions of coal-fired power plants: the TERESA field results. Inhalation Toxicol 23:11–30
24. Karar K, Gupta AK, Kumar A, Biswas A (2006) Characterization and identification of the sources of chromium, zinc, lead, cadmium, nickel, manganese and iron in PM10 particulates at the two sites of Kolkata, India. Environ Monit Assess 120:347–360

25. Kaul D, Gupta T, Tripathi SN, Tare V, Collett J (2011) Secondary organic aerosol: a comparison between foggy and non-foggy days. Environ Sci Technol 45:7307–7313
26. Khandekar RN, Kelkar DN, Vohra KG (1980) Lead, cadmium, zinc, copper and iron in the atmosphere of Greater Bombay. Atmos Environ 14(0):457–461
27. Kim E, Hopke PK (2004) Comparison between conditional probability function and nonparametric regression for fine particle source directions. Atmos Environ 38:4667–4673
28. Kim E, Hopke PK, Edgerton ES (2003) Source identification of Atlanta aerosol by positive matrix factorization. J Air Waste Manage Assoc 53:731–739
29. Kim ND, Fergusson JE (1994) The concentrations, distribution and sources of cadmium, copper, lead and zinc in the atmosphere of an urban environment. Sci Total Environ 144:179–189
30. Krishna RR (2012) Current atmospheric aerosol research in India. Curr Sci 102:440–457
31. Kulshrestha A, Satsangi PG, Masih J, Taneja A (2009) Metal concentration of PM2.5 and PM10 particles and seasonal variations in urban and rural environment of Agra, India. Sci Total Environ 407:6196–6204
32. Lee JH, Hopke PK (2006) Apportioning sources of PM 2.5 in St. Louis, MO using speciation trends network data. Atmos Environ 40:360–377
33. Lestari P, Mauliadi YD (2009) Source apportionment of particulate matter at urban mixed site in Indonesia using PMF. Atmos Environ 43:1760–1770
34. Li J, Pósfai M, Hobbs PV, Buseck PR (2003) Individual aerosol particles from biomass burning in southern Africa: 2. Compositions and aging of inorganic particles. J Geophys Res: Atmos 108:8484
35. Lin JJ, Lee L-C (2004) Characterization of the concentration and distribution of urban submicron (PM 1) aerosol particles. Atmos Environ 38:469–475
36. Mamane Y, Pirrone N (1998) Vanadium in the atmosphere. Adv Environ Sci Technol-New York 30:37–72
37. Mehta B, Venkataraman C, Bhushan M, Tripathi SN (2009) Identification of sources affecting fog formation using receptor modeling approaches and inventory estimates of sectoral emissions. Atmos Environ 43:1288–1295
38. Misra A, Gaur A, Bhattu D, Ghosh S, Dwivedi AK, Dalai R, Paul D, Gupta T, Tare V, Mishra SK (2014) An overview of the physico-chemical characteristics of dust at Kanpur in the central Indo-Gangetic basin. Atmos Environ 97:386–396
39. Niranjan K, Sreekanth V, Madhavan B, Krishna Moorthy K (2007) Aerosol physical properties and Radiative forcing at the outflow region from the Indo-Gangetic plains during typical clear and hazy periods of wintertime. Geophys Res Lett 34
40. Norris G, Vedantham R, Wade K, Brown S, Prouty J, Foley C (2008) EPA positive matrix factorization (PMF) 3.0 fundamentals and user guide. US Environmental Protection Agency, Office of Research and Development, Washington, DC
41. Nriagu JO, Pacyna JM (1988) Quantitative assessment of worldwide contamination of air, water and soils by trace metals. Nature 333:134–139
42. Paatero P (1997) Least squares formulation of robust non-negative factor analysis. Chemometr Intell Lab Syst 37:23–35
43. Paatero P, Hopke PK (2003) Discarding or down weighting high-noise variables in factor analytic models. Anal Chim Acta 490:277–289
44. Paatero P, Tapper U (1994) Positive matrix factorization: a non-negative factor model with optimal utilization of error estimates of data values. Environmetrics 5:111–126
45. Pant P, Harrison RM (2012) Critical review of receptor modelling for particulate matter: a case study of India. Atmos Environ 49:1–12
46. Patil RS, Kumar R, Menon R, Shah MK, Sethi V (2013) Development of particulate matter speciation profiles for major sources in six cities in India. Atmos Res 132–133:1–11
47. Polissar AV, Hopke PK, Paatero P, Malm WC, Sisler JF (1998) Atmospheric aerosol over Alaska: 2. Elemental composition and sources. J Geophys Res: Atmos (1984–2012) 103:19045–19057
48. Pope III C, Dockery D (1999) Epidemiology of particle effects. Air Pollut Health 31:673–705

49. Raman RS, Ramachandran S, Kedia S (2011) A methodology to estimate source-specific aerosol radiative forcing. J Aerosol Sci 42:305–320
50. Reff A, Eberly SI, Bhave PV (2007) Receptor modeling of ambient particulate matter data using positive matrix factorization: review of existing methods. J Air Waste Manage Assoc 57:146–154
51. Sahu LK, Saxena P (2015) High time and mass resolved PTR-TOF-MS measurements of VOCs at an urban site of India during winter: role of anthropogenic, biomass burning, biogenic and photochemical sources. Atmos Res 164–165:84–94
52. Samet J, Dominici F, Zeger S, Schwartz J, Dockery D (2000) The national morbidity, mortality, and air pollution study. Part I: Methods and methodologic issues. Res Rep Health Eff Inst 5
53. Samet JM et al (2000) The national morbidity, mortality, and air pollution study. Part II: Morbidity and mortality from air pollution in the United States. Res Rep Health Eff Inst 94:5–79
54. Sanghi R, Sasi KS (2001) Pesticides and heavy metals in agricultural soil of Kanpur, India. Bull Environ Contam Toxicol 67:0446–0454
55. Shen Z-X et al (2009) Chemical characteristics of submicron particles in winter in Xi'an. Aerosol Air Qual Res 9:80–93
56. Srivastava AK, Pant P, Hegde P, Singh S, Dumka UC, Naja M, Singh N, Bhavanikumar Y (2011) The influence of a south Asian dust storm on aerosol radiative forcing at a high-altitude station in central Himalayas. Int J Remote Sens 32:7827–7845
57. Srivastava M, Tripathi SN, Bhattu D, Jaidevi JJ, Gupta T (2013) CCN closure results from Indian Continental Tropical Convergence Zone (CTCZ) aircraft experiment. Atmos Res 132–133:322–331
58. Uno U, Ekpo B, Etuk V, Etuk H, Ibok U (2012) Comparative study of levels of trace metals in airborne particulates in some cities of the Niger delta region of Nigeria. Environ Pollut 2:p110
59. Wen H, Carignan J (2007) Reviews on atmospheric selenium: emissions, speciation and fate. Atmos Environ 41:7151–7165
60. Xie R, Seip HM, Wibetoe G, Nori S, McLeod CW (2006) Heavy coal combustion as the dominant source of particulate pollution in Taiyuan, China, corroborated by high concentrations of arsenic and selenium in PM10. Sci Total Environ 370:409–415
61. Xie ZM, Wu WH, Xu JM (2003) Study on fluoride emission from soils at high temperature related to brick-making process. Chemosphere 50:763–769

The manufacturer's authorised representative in the EU is Springer Nature Customer Service Centre GmbH, Europaplatz 3, 69115 Heidelberg, Germany. If you have any concerns regarding our products, please contact ProductSafety@springernature.com

Printed and bound by CPI Group (UK) Ltd, Croydon, CR0 4YY
25/09/2026
02222885-0001